T0309575

HUMAN PARASITOLOGY

FIFTH EDITION

HUMAN PARASITOLOGY

FIFTH EDITION

BURTON J. BOGITSH
Vanderbilt University, Nashville, TN, United States

CLINT E. CARTER
Vanderbilt University, Nashville, TN, United States

THOMAS N. OELTMANN
340 Light Hall, Nashville, TN, United States

Academic Press is an imprint of Elsevier
125 London Wall, London EC2Y 5AS, United Kingdom
525 B Street, Suite 1650, San Diego, CA 92101, United States
50 Hampshire Street, 5th Floor, Cambridge, MA 02139, United States
The Boulevard, Langford Lane, Kidlington, Oxford OX5 1GB, United Kingdom

Library of Congress Cataloging-in-Publication Data
A catalog record for this book is available from the Library of Congress

British Library Cataloguing-in-Publication Data
A catalogue record for this book is available from the British Library

ISBN: 978-0-12-813712-3

For information on all Academic Press publications visit our website at https://www.elsevier.com/books-and-journals

Publisher: John Fedor
Acquisition Editor: Linda Versteeg-buschman
Editorial Project Manager: Timothy Bennett
Production Project Manager: Sreejith Viswanathan
Designer: Christian Bilbow

Typeset by Thomson Digital

Contents

5. Visceral Protistans II Flagellates

6. Blood and Tissue Protistans I Hemoflagellates

7. Blood and Tissue Protistans II Human Malaria

8. Blood and Tissue Protistans III Other Protists

9. General Characteristics of the Trematoda

10. Visceral Flukes

11. Blood Flukes

12. General Characteristics of the Cestoidea

13. Intestinal Tapeworms

14. Extraintestinal Tapeworms

Preface

Human Parasitology is a textbook designed specifically for premedical, medical technology, and biology students who require basic knowledge of the biology of parasitism. Several years ago my colleagues and I began to consider collaborating on such a textbook. As the idea germinated, we decided that, while emphasizing the medical aspects of the topic, the book should incorporate sufficient functional morphology, physiology, biochemistry, and immunology to enhance appreciation of the diverse implications of parasitism. It would also explore the potential of certain parasites for producing morbidity and mortality and would present available data regarding the modus operandi of certain modern chemotherapeutic agents. Through considerable discussion and several revisions of the manuscript, the first edition of *Human Parasitology* evolved. One of the goals for *Human Parasitology* was that it would serve as a bridge between classical clinical parasitology texts and the more traditional encyclopedic, advanced treatises that include in-depth consideration of biochemistry and immunology as well as the inclusion of those parasites that infect nonhuman hosts (e.g., strigeids of fishes, amphibians, and birds).

As each new edition of *Human Parasitology* evolved, we have attempted to introduce at least one major advance in the field of parasitology. The subject of opportunistic parasites was expanded in the third edition and with this expansion the section on immunology was accordingly expanded and updated. In the fourth edition, we expanded and simplified the topics of innate and acquired immunity emphasizing the role that toll-like receptors play in the innate immune response. This information was subsequently included in the sections in which host immune responses are discussed relative to the parasite(s) in question.

In the current edition, we have emphasized the influence of genomics in understanding the molecular basis for drug resistance as well as being instrumental in establishing new targets for drug and vaccine development. To illustrate how this technology is starting to produce results, we have included examples from the protists, flatworms, and roundworms.

We still feel that it is important that the biological aspects of parasites be considered. To this end, we have preserved the opening chapter to each major group of parasites and kept the designation as "General Characteristics." In these introductory chapters we deal with those characteristics that are universal within the group as a whole. We deal with the more specialized aspects of individual parasites in their proper context. Ample light and electron micrographs serve to illustrate the various points of the text. An abridged classification of parasites dealing with the general characteristics of each major group has been placed at the end of each chapter. We recognize that the classification of organisms is an ever-changing subject. This is particularly evident as more modern techniques are recognizing new homologies. We have, however, chosen to retain the more classical, morphological-based system that has been used in previous editions. We do present an opening reference at the beginning of each group for those readers who might be interested in the more modern, molecular-based, system.

At the end of most chapters is a section entitled "Suggested Reading." We have attempted to list a few significant publications that present some in-depth information on selected topics pertinent to the topics discussed in the chapter.

The field of chemotherapy is advancing at such a rapid pace that new drugs and regimens become available almost continually. Consequently, the sections dealing with information about current drugs of choice as well as regimens will possibly become obsolete more rapidly than other sections of the book. Nevertheless, in the individual sections and at the end of the book, not only have we included specific, currently prescribed chemotherapeutic drugs and regimens but also we have updated, as of the time of this writing, the newest changes in these protocols.

The current edition, as well as previous editions, is designed for a one-semester course and while the material is somewhat condensed, every effort has been made to include the most recent advances in the field of parasitology. In conclusion, we believe this edition will appeal to those students interested not only in the medical aspects of parasitology but also to those who require a solid foundation in the biology of parasites in order to further their studies in a graduate or a professional school or career of their choice.

Acknowledgments

In addition to those individuals who were acknowledged in the First Edition of Human Parasitology (page xxii), we would like to thank particularly Teresa A. Ward for her invaluable support in the area of information technology. Additionally, we thank Priscilla B. Rosenfeld, BSN, RN, CNRN, for her support in updating the chemotherapeutic protocols.

CHAPTER

1

Symbiosis and Parasitism

Chapter 1 opens with a section in which the four categories of heterospecific relationships (commensalism, phoresis, parasitism, and mutualism) discussed in subsequent portions of the text are defined. Within this context, a comparison between various types of hosts is explained. Included also is a discussion of the roles of reservoir hosts as integral parts of the life cycle of some parasites. The narrative moves into an overview of the broader aspects of parasitism such as ecology, medical implications, and evolution. Topics such as the correlation of diseases caused by parasites relative to the density of a parasitic burden and the effect of influences such as AIDS and cancer treatments on parasitism are discussed. Other areas considered in this chapter include environmental influences that impede the control and/or elimination of harmful parasites as well as those factors that contribute to the prevalence of parasitic diseases. The chapter closes with a discussion of the possible ways that different groups of organisms may have adapted to a parasitic lifestyle.

Parasitology, the study of parasites and their relationships to their hosts, is one of the most fascinating areas of biology. The study of parasitism is interdisciplinary, encompassing aspects of systematics and phylogeny, ecology, morphology, embryology, physiology, biochemistry, immunology, pharmacology, and nutrition, among others. Newly developed techniques in biochemistry and cellular and molecular biology have also opened significant new avenues for research on parasites.

Not only does parasitology touch upon many disciplines, but the varied nature of parasites renders their study multifaceted. While it is entirely proper to classify many bacteria and fungi and all viruses as parasites, parasitology has traditionally been limited to parasitic protozoa, helminths, arthropods, and those species of arthropods that serve as vectors for parasites. It follows, then, that parasitology encompasses elements of protozoology, helminthology, entomology, and acarology. The World Health Organization has proclaimed that, of the six major unconquered human tropical diseases, five—schistosomiasis, malaria, filariasis, African trypanosomiasis, and leishmaniasis—are parasitic in the traditional sense. Leprosy, the sixth major disease, is caused by a bacterium.

DEFINITIONS

The complexity of the host–parasite relationship has often led to misunderstandings of the precise nature of parasitism. In order to avoid such misperceptions, researchers have devised

Human Parasitology. http://dx.doi.org/10.1016/B978-0-12-813712-3.00001-1

the following concepts to distinguish among the several types of associations involving heterospecific organisms.

Any organism that spends a portion or all of its life intimately associated with another living organism of a different species is known as a symbiont (or symbiote), and the relationship is designated as symbiosis. The term *symbiosis*, as used here, does not imply mutual or unilateral physiologic dependency; rather, it is used in its original sense (living together) without any reference to "benefit" or "damage" to the symbionts.

Although the lines of demarcation between them are indistinct, at least four categories of symbiosis are commonly recognized: commensalism, phoresis, parasitism, and mutualism. The scope of this text is limited to relationships of medical importance, and, since parasitism is the major type of symbiosis meeting this criterion, definitions of the other forms are included for clarification only.

Commensalism

Commensalism does not involve physiologic interaction or dependency between the two partners, the host and the commensal. Literally, the term means "eating at the same table." In other words, commensalism is a type of symbiosis in which spatial proximity allows the commensal to feed on substances captured or ingested by the host. The two partners can survive independently. Although at times certain nonpathogenic organisms (e.g., protozoa) are referred to as commensals, this interpretation is incorrect since they are physiologically dependent on the host and are, therefore, parasites. An example of commensalism is the association of hermit crabs and the sea anemones they carry on their borrowed shells.

Phoresis

The term *phoresis* is derived from the Greek word meaning "to carry." In this type of symbiotic relationship, the phoront, usually the smaller organism, is mechanically carried by the other, usually larger, organism, the host. Unlike commensalism, there is no dependency in the procurement of food by either partner. Phoresis is a form of symbiosis in which no physiologic interaction or dependency is involved. Both commensalism and phoresis can be considered spatial, rather than physiologic, relationships. Examples of phoresis are the numerous sedentary protozoans, algae, and fungi that attach to the bodies of aquatic arthropods, turtles, etc.

Parasitism

Parasitism is another type of symbiotic relationship between two organisms: a parasite, usually the smaller of the two, and a host, upon which the parasite is physiologically dependent. The relationship may be permanent, as in the case of tapeworms found in the vertebrate intestine, or temporary, as with female mosquitoes, some leeches, and ticks, which feed intermittently on host blood. Such parasites are considered obligatory parasites because they are physiologically dependent upon their hosts and usually cannot survive if kept isolated from them. Facultative parasites, on the other hand, are essentially free-living

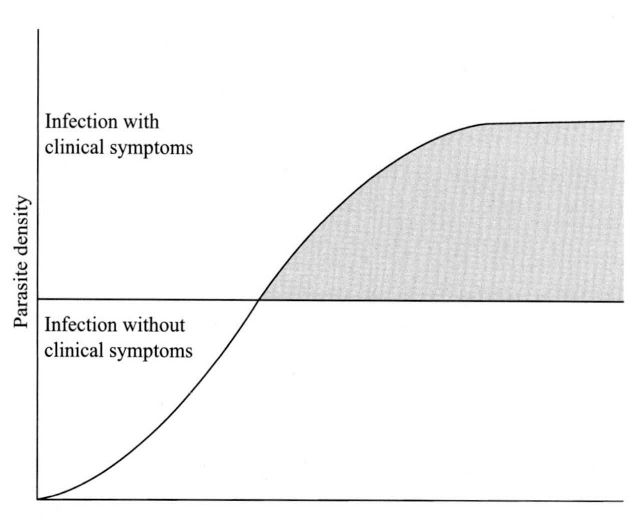

FIGURE 1.1 **Correlation between diseases with clinical symptoms and parasite density.**

organisms that are capable of becoming parasitic if placed in a situation conducive to such a mode. An example of a facultative parasite is the amoeba *Naegleria.*

The physiologic requirements of most parasites are only partially known and understood, but there is sufficient information to indicate certain categories of dependence, such as nutritional. Unlike commensals, parasites derive essential nutrients directly from the host, usually from such nutritive substances as blood, lymph, cytoplasm, tissue fluids, and host-digested food.

The intimate relationship between parasite and host generally exposes the host to antigenic substances of parasite origin. Sometimes these antigens consist of the molecules that make up the surface of the parasite (somatic antigens), or they may be molecules secreted or excreted by the parasite (metabolic antigens). In either case, the host typically responds to the presence of such antigens by synthesizing antibodies. Thus, unlike phoresis and commensalism, parasitism usually involves, in addition to the physiologic dependency of the parasite, immunological responses by the host. The effect upon the host is usually the result of host reaction to the presence of the parasite. One of the more important consequences of such reaction—which may be localized at the site of attachment or deposition or may be more generalized, perhaps throughout the entire host body—is the limitation of the populations of the parasite. It is axiomatic in helminthology that "all species of worms are harmful when present in massive numbers" (Fig. 1.1); therefore, internal defense responses by the host help to reduce pathological effects of the parasite.

While there are numerous systems for classifying host–parasite relationships, the one used here distinguishes between two major types of parasites, endoparasites and ectoparasites, according to location. Endoparasites live within the body of the host at sites such as the alimentary tract, liver, lungs, and urinary bladder; ectoparasites are attached to the outer surface of the host or are superficially embedded in the body surface.

According to its role, the host may be classified as (1) a definitive host, if the parasite attains sexual maturity therein; (2) an intermediate host, if it serves as a temporary, but essential, environment for the development of the parasite and/or its metamorphosis short of sexual maturity; and (3) a transfer or paratenic host, if it is not necessary for the completion of the parasite's life cycle but is utilized as a temporary refuge and a vehicle for reaching an obligatory, usually the definitive, host in the cycle.

Generally, an arthropod or some other invertebrate that serves as a host as well as a carrier for a parasite is referred to as a vector. Unlike the transfer host, the vector is essential for completion of the life cycle. In this text, the term is used to designate an organism, usually an arthropod, that transmits a parasite to the human or vertebrate host; for example, various species of anopheline mosquitoes serve as vectors for the malaria-producing protozoan parasites, *Plasmodium* spp., and transmit the organisms to humans, who serve as vertebrate hosts. From an evolutionary perspective, some intermediate hosts, or vectors as in the case of the *Plasmodium*–mosquito relationship, may once have been definitive hosts, while others may have been transfer or paratenic hosts.

Infected animals that serve as sources of infective organisms for humans are known as reservoir hosts. A wild animal in this role is called a sylvatic reservoir host; a domestic animal, a domestic reservoir host. For example, one type of human filariasis is caused by a filarial worm, *Brugia malayi*, which is transmitted to humans by a mosquito. Although infections are usually transmitted from one person to another via the mosquito, *Brugia malayi* can also be transmitted to humans from cats (domestic reservoirs) or monkeys (sylvatic reservoirs). Often, the reservoir host tolerates the parasitic infection better than the human host does. Thus, the reservoir host, by definition, shares the same stage of the parasite with humans.

The term zoonosis can be used in various contexts; it is used here to denote a disease of humans that is caused by a pathogenic parasite normally found in wild and domestic vertebrate animals. Person-to-person transmission does not normally occur in zoonosis. An example of a zoonotic disease of significant medical importance is trichinellosis, caused by the nematode *Trichinella spiralis*. The worm is found in a variety of sylvatic and domestic reservoirs, notably pigs, bears, and rodents. Humans become infected by consuming raw or undercooked meat from infected animals.

Mutualism

The fourth category of symbiosis, mutualism, is an association in which the mutualist and the host depend on each other physiologically. A classic example of this type of relationship occurs between certain species of flagellated protozoans and the termites in whose gut they live. The flagellate, which depends almost entirely on a carbohydrate diet, acquires nutrients from wood chips ingested by the host termite. In return, the flagellate synthesizes and secretes cellulases, cellulose-digesting enzymes, the end products of which the termite utilizes. Incapable of synthesizing its own cellulases, the termite is, thus, dependent on the mutualist and, in turn, provides developmental stimuli to the mutualist and a hospitable environment for reproduction. If the termite is defaunated, it will die; and, conversely, the flagellate cannot survive outside the termite.

Following the definition of symbiosis and the subcategories of heterospecific relationships, it is important to note that definitions are often arbitrary. They are useful in many cases for categorizing natural symbiotic associations; however, in certain instances, there is considerable overlap. Figure 1.2 illustrates, for example, that some associations qualify as

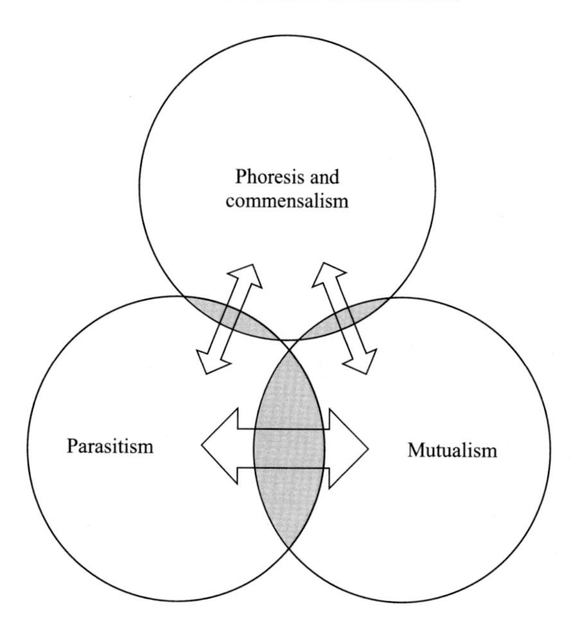

FIGURE 1.2 **Overlap between the major categories of symbiosis.** Note that there is less overlap between phoresis/commensalism and parasitism as well as between phoresis/commensalism and mutualism than between parasitism and mutualism.

both phoretic and commensalistic or as both commensalistic and parasitic. Such overlapping relationships may be regarded as transitional stages that may reflect evolutionary shifts from one category to another. It has been suggested that, theoretically at least, a complete and progressive gradation exists among the various types of symbiosis. This may come about when, for example, in the shift from parasitism to mutualism, the parasite initially gives off some nonessential metabolic by-product that can be utilized by the host; eventually, the host becomes physiologically dependent not only on this by-product of parasite origin but on other factors as well, and the relationship evolves into a mutualistic one.

ECOLOGICAL ASPECTS OF PARASITISM

All species interaction must occur within an ecological and evolutionary context. For example, the body of a host constitutes the environment on or in which the parasite spends some or all of its life. In addition, the physical environment in which both host and parasite exist may affect dramatically the nature and intensity of the host–parasite interaction and the local environment will influence the likelihood and rate of parasite transmission either directly or indirectly through some parasite vector. Finally, the complicated life cycles observed for many parasites that utilize multiple hosts make for exceedingly complicated integrated ecological relationships.

In nature, the range of host and parasite interactions can vary widely. In addition to the common examples of ectoparasites, endoparasites, macroparasites, and microparasites, parasitism may occur in less well-known ways. For example, among birds, females of certain brood or social parasites lay their eggs in the nests of host species, thereby parasitizing the parental care investment of the hosts. Brood parasitism has been implicated as a major factor leading to the declines in songbird populations in the United States. Among the dipteran and hymenopteran insects, flies and wasps in particular, some females deposit their eggs in or on living hosts. The eggs hatch and larvae feed on the host, pupate, and then emerge as adults. These are termed parasitoids, and they can be strong agents of population regulation in natural host populations. While often not evident due to their small size and cryptic behavior, parasitoids are very abundant in nature, and by some estimates they may account for 25% of the world's animal species.

The population biology of parasites is clearly important and some ecologists have proposed the use of techniques employed in quantitative life-history studies to describe the population biology of parasites. By definition, *r*-selection occurs when selective forces upon organisms (termed *r*-strategists or opportunistic species) are unstable and environmental conditions are variable. *K*-selection, on the other hand, prevails when forces influencing organisms (termed *K*-strategists or equilibrial species) remain relatively stable over a period of time. Most *r*-strategists are characterized by high fecundity, high density-independent mortality, short life span, effective dispersal mechanisms, and population sizes that vary over time and which are usually below the carrying capacity of the environment. *K*-strategists are generally characterized by low fecundity, density-dependent mortality that may be low, longer life spans, and more stable population sizes. As an example, digenetic trematodes are considered *r*-strategists, since both their biotic potential for population increase and their mortality rates are high as a result of selective pressures in their environments, which are considered to be unstable because of environmental differences at practically every phase of the life cycle. It is important to understand that the *r*- and *K*-strategy designations are relative. Species B, for example, may be an *r*-strategist when compared to species C, but a *K*-strategist when compared to species A.

Parasites may exert strong control over the population density and population cycling over time in their hosts. For instance, many examples exist in nature of strong regulation of insect populations by parasitic pathogens in which parasitism may be density-dependent. Often parasites may alter the interactions of their hosts with other organisms, such as predators. There is strong evidence that nematode parasitism in snowshoe hares makes the hares more susceptible to their predators, which in turn may destabilize hare–predator–parasite population dynamics. In another complex example, plants being fed upon by herbivorous insects may produce chemical cues that attract parasitoid wasps, which will attack and control the herbivorous insect population. These examples illustrate how organisms on three different trophic levels may be functionally integrated through host–parasite relationships in so-called tritrophic interactions.

A number of influences, such as the presence or absence of certain biological, chemical, and physical factors, dictate the geographic distribution of a parasite. For instance, the ability of a particular species of parasite to survive depends upon the availability of all hosts needed to complete its life cycle; therefore, factors governing survival of hosts indirectly govern the presence of parasites. Another feature governing distribution of a parasite is host specificity, or the adaptability of a species of parasite to a certain host or group of hosts. The degree of specificity varies from species to species. Host specificity is determined by genetic, immunological, physiological, and/or ecological factors.

Many of these ecological aspects play important roles in defining the epidemiology of a disease-producing parasite. Epidemiology is the study of factors responsible for the transmission and distribution of disease. The distribution and characteristics of various hosts (including vectors), host specificity, cultural patterns of human hosts (such as diet, often influenced by religion, economic status, etc.), and density of host and parasite populations all influence the epidemiology of pathogenic parasites.

Ecological modeling has also been very important in understanding the epidemiology of microparasite population dynamics and their control via immunization programs. Many microparasites cause diseases of humans and other animals, as well as in plants, and often the transmission from an infected individual to a susceptible individual occurs through direct contact.

Since pathogens rely on their hosts for survival and reproduction, the population dynamics of the pathogen are influenced by the population dynamics of hosts. The host population will contain three types of individuals, those who are susceptible, those who are infected, and those who have been infected but have recovered. An infection model, such as that of Bulmer (1994), can be devised that predicts the behavior of the pathogen and the various host individuals. Important parameters of such a model include the transmission coefficient, which predicts the likelihood of new infections and the threshold value needed for an infection to persist in the population. Other models may be used to devise control strategies via treatment and immunization procedures in host populations. Models such as those devised by Anderson and May (1991) can predict the level of population immunization necessary to control diseases and parasites with varying reproductive rates of infection, a measure of how many uninfected individuals will be infected over the average lifetime of an infected individual.

This brief discussion shows clearly that a full understanding of host–parasite relationships requires a careful consideration of the ecological context of the relationship. This allows us to appreciate the significance of host–parasite interactions in nature and to understand better the establishment and control of parasite infection in human populations.

MEDICAL IMPLICATIONS

The area of human pathology attributable to parasitic diseases is known as tropical or geographic medicine because so many of these diseases occur most commonly, often exclusively, in the tropical regions of the world. The conditions that make these regions more vulnerable to these diseases will be discussed later in this section.

The causative agents of parasitic diseases of humans include organisms commonly known as protozoans (one-celled life forms, pp. 35–47), flatworms (trematodes and tapeworms, pp. 149, 229), roundworms (nematodes, p. 257), and certain arthropods (insects, ticks, and mites, p. 349). Parasitic diseases differ from viral, bacterial, and fungal diseases in several ways. For example, viral, bacterial, and fungal pathogens generally multiply rapidly, producing large numbers of progeny within the human host, whereas parasites generally reproduce more slowly and produce fewer offspring. Also, viral and bacterial infections usually are more acute, often highly virulent and potentially lethal, while parasitic diseases are usually chronic and if death does result, it commonly comes after a lengthy period of debilitation.

Finally, except for certain viral infections, parasitic diseases are generally more difficult to control than other infectious diseases. Elements in the epidemiology of parasitic diseases are described below, which, singly or in various combinations, help to explain this phenomenon.

Control Impediments

Intermediate Hosts

The life cycles of certain parasites depend upon the presence of at least one intermediate host. Trematodes that parasitize humans, for example, require specific snail intermediate hosts and conditions that ensure host–parasite contact. Specific chemical qualities of the water, physical characteristics of the aquatic environment, type and quantity of aquatic vegetation, etc., are also essential to the successful completion of the life cycle. Many of the intermediate hosts are so specialized that they can protect themselves from a number of control measures. For example, the snail intermediate host of *Schistosoma mansoni*, one of the causative agents of human schistosomiasis, can burrow into the mud, shielding itself from such adverse natural conditions as drought as well as protecting itself from the various molluscicides that are used in an attempt to eliminate it. Also, while alteration of one or more characteristics of the aquatic environment for the purpose of disease control could disrupt the cycle and subvert the transmission of schistosomiasis, such deliberately induced changes in the habitat of the snail intermediate host might also disrupt the biology of cohabitant plants and animals and upset the entire ecological balance of the immediate environment.

Vectors

Plasmodium falciparum is the causative agent of one type of malaria known as malignant tertian malaria (p. 125) in which the vector is a female mosquito of the genus *Anopheles*. These mosquitoes breed by the millions in small, isolated bodies of fresh water. Because of the inaccessibility of many of these breeding sites, the use of insecticides for eradication of the mosquito population is of limited value; therefore, combating the spread of malaria by this method is never completely effective. In addition, strains of these mosquitoes often emerge that are resistant to many insecticides.

Leishmania braziliensis, the protozoan parasite that causes mucocutaneous leishmaniasis (p. 100), is transmitted by sand flies of the genus *Lutzomyia*. Essential to the development of these sand flies is an environment of low light intensity, high humidity, and organic debris upon which the larvae feed. Breeding sites are commonly located under logs and decaying leaves, inside hollow trees, and in animal burrows. As with the mosquito vector of *P. falciparum*, the inaccessibility of breeding sites limits the effectiveness of insecticides in eliminating the vectors. Hence, in this case also, only partial eradication of the disease has been achieved by targeting the vector.

Several genera of mosquitoes serve as vectors for a form of human Bancroftian filariasis caused by the nematode *Wuchereria bancrofti*. The incidence of periodic filariasis, common in areas of dense population and poor sanitation, parallels the distribution of the principal vector, *Culex fatigans*, which breeds in sewage-contaminated water. In economically depressed areas of the world, contaminated water is common and, as previously noted, the widespread use of insecticides to control mosquitoes is not totally reliable. In recent years, the combined use of screens, insect repellents, and insecticides, augmented by mass treatment with chemotherapeutic drugs, have proven effective in combating this type of filariasis in the United States Virgin Islands, Puerto Rico, and Tahiti. Nevertheless, failure to find an effective means of eliminating the vector essential to its transmission has complicated the control of Bancroftian filariasis.

In all of these instances, an inherent danger in the widespread use of insecticides is their indiscriminate effect, destroying or contaminating beneficial life forms along with the vectors.

Resistance and Resurgence

During the past decade, there have been spectacular resurgences of several parasitic diseases that had been considered under control or eradicated, the most dramatic of which has been malaria. This upsurge in malaria is due primarily to the emergence of strains of the malaria-causing parasite resistant to available drugs as well as insecticide-resistant mosquito vectors. In addition, the crowded conditions resulting from resettlement of refugees from war and famine expedite transmission.

Impact of Genomics

In 2002, *P. falciparum* became the first parasitic protist to have its entire genome sequenced. The filarial worm *B. malayi* in 2007 represented the first human-infective multicellular organism so studied. Subsequently, the genomes of a number of parasites have become the subject of intense sequencing. As data accumulates from such studies, the basis for the comprehensive knowledge of the underlying genetic controls required for successful host–parasite relationships is becoming available. Such data also have the potential to provide targets for drug and vaccine development. Expansion of our understanding of the mechanisms that allow for immune evasion and drug tolerance may also come to light. On a population basis, additional information can provide insight relative to parasite evolution and epidemiologic dynamics. With this plethora of possibilities, it is not surprising that much of current research is concentrated on the expansion of DNA sequencing over a large range of parasitic organisms. One of the major ways to control the effects and the spread of diseases caused by parasites is the use of reliable and sustainable intervention programs. The problem that arises with such programs is either that many drugs are not active against all portions of the parasite's life cycle or that drug-resistance occurs after prolonged use of current protocols. These shortcomings require a continual search for new targets for drugs and vaccines and/or means to obviate the rise and continuance of resistant strains. To illustrate how current genomic studies can impact these aforementioned problems, we will discuss two approaches as examples. One approach is to seek those genes within the parasite's genome whose mutations can be linked to resistance. The mutated gene(s) then serves as a probe or warning sign that the parasite population is at a high risk of developing resistance. Researchers can target this population and destroy resistant strains before they can take hold and spread any further. Another approach is to seek proteins that are involved in a metabolic pathway that is essential for the parasite's existence and use them as potential targets for drugs or vaccines.

As models for examining the effectiveness of these approaches, we will examine in subsequent chapters three parasites responsible for medically important diseases. *P. falciparum* representing malaria (see page 111), *S. mansoni* representing schistosomiasis (see page 193), and *B. malayi* representing filariasis (see page 313).

Diagnosis

Diagnosis of human parasitic diseases is often complicated. Emphasis upon medical parasitology in North American medical curricula has traditionally been woefully lacking and continues to be increasingly neglected. Consequently, the actual and potential impact of these diseases is not being addressed, and younger clinicians have little appreciation for the hazard they represent. This lack of awareness results in few, if any, diagnostic tests being ordered. This, in turn, produces a cyclical effect wherein there is a corresponding reduction in the medical technology curricula dealing with diagnostic laboratory tests designed to identify parasitic diseases. The cycle continues in continued failure to recognize the need for diagnostic tests, resulting in misdiagnosis of parasitic diseases. This indictment applies to almost all North American medical schools and hospitals. In contrast, instruction in medical schools in Central and South America in the diagnosis of such diseases is commendable. The greatest deficiency, however, remains in the field clinics in tropical and subtropical Africa and parts of Southeast Asia. Inadequate funding, supplies, and trained staff thwart courageous efforts to achieve significant progress in the struggle against these insidious diseases.

Factors Influencing Prevalence

A variety of conditions contribute to the prevalence of parasitic diseases in the tropics and subtropics, among them unsanitary living conditions, inadequate funding for disease control and treatment, poor nutrition, lack of health education (although this is improving), regional and ethnic customs conducive to infection by parasites, climatic conditions, and compromised immune systems.

Unsanitary Living Conditions

In most countries of the tropical and subtropical belts, construction of modern sewage systems is still in the planning or preliminary stages. Consequently, raw sewage contaminating open trenches and streams remains very common. In rural Southeast Asia, for example, shacks built on stilts overhang streams polluted with human and animal excreta, and vegetation growing in these streams is often gathered for human consumption. Such scenarios create an ideal environment for the transmission of parasitic and other diseases.

Disease Control and Treatment

Third-world nations, including most tropical countries, invariably have limited funds in the national budgets for public health, and the research and other programs essential to improving conditions are costly. The control of snails that transmit schistosomiasis, for example, is an expensive undertaking involving vehicles, pumps, and other machinery, and chemicals. Consequently, in spite of aid from such international agencies as the World Health Organization, funding for disease control is vastly inadequate.

Parasitic diseases most commonly afflict the poor, and, unfortunately, pharmaceutical companies are reluctant to invest large sums in research and development of new drugs, which victims would be unlikely to be able to afford. Where significant progress has been made over the past two decades in developing new drugs, limited production has kept the price high, beyond the means of most of the afflicted population.

Poor Nutrition

Immunological defense mechanisms in all animals, including humans, are influenced by several physiological processes, including nutrition. In most parts of the world where parasitic diseases abound, malnutrition plays an important role in susceptibility and manifestation of clinical symptoms, often severe. Undernourished persons, especially children suffering from protein deficiency, are particularly vulnerable to infection, and physical and physiological deviations from the norm are also markedly more pronounced, especially among the young. For example, hookworm infection (p. 306) exacerbated by malnutrition commonly results in anemia, significant loss of body weight, abdominal distension, and mental malaise.

Health Education

Education of the population in endemic areas concerning methods of reducing or eliminating parasitic infections is probably the most economical approach to disease control. Educational programs usually involve teams that present illustrated lectures to school children in rural areas. Longstanding practices and attitudes often produce stubborn resistance to these endeavors, but, while such efforts alone have limited effect, they can be useful when incorporated into more comprehensive programs involving the media and other advertising ploys such as road signs. Indeed, recent growth in such programs has been helpful in reducing diseases such as schistosomiasis in rural Egypt.

Regional and Ethnic Customs

Epidemiologists have long recognized that certain regional and ethnic customs practiced by inhabitants of third-world countries in the tropics and subtropics contribute significantly to the spread of parasitic diseases. For example, in Muslim countries, ablution is a common practice. The use of communal pools for this ritual bathing of previously unwashed body parts leads to contamination of the water and facilitates the spread of diseases such as schistosomiasis (p. 197).

In many parts of the Orient, raw and lightly pickled crabs and other crustaceans are considered delicacies. These hosts harbor one of the larval stages of the Oriental lung fluke, *Paragonimus westermani* (p. 192). This encysted larva, known as a metacercaria, is ingested in crustacean meat and, upon excystation, penetrates the human gut and enters the peritoneal cavity, eventually reaching the lungs.

In Egypt, especially in parts of the lower Nile valley, there is a high incidence of *Heterophyes heterophyes*, an intestinal fluke of humans. This parasite also occurs in Greece, Israel, Korea, Taiwan, and has been reported in Hawaii among persons of Philippine origin. Human infections are contracted through consumption of raw or poorly cooked fish. The most common, the mullet, a favored delicacy when served raw, is largely responsible for heterophyidiasis, as well as other parasitic diseases.

Climatic Conditions

The climate in tropical and subtropical lands favors the transmission of several parasitic diseases, especially those transmitted by arthropods. Exposed bare skin and perspiration invite insect bites, and, since various insects serve as vectors of protozoan and nematode pathogens, there is a high incidence of insect-transmitted parasitic diseases in warm and hot climates. Going barefoot facilitates invasion of the skin by such parasites as hookworm (p. 306) whose infective stage lives in warm soil and penetrates the bare skin of the victim. In addition, the abundance of bodies of stagnant water, in conjunction with the prolonged elevated temperatures of the tropical and subtropical regions, provides ideal, essentially year-round breeding environments, which enhance survival of intermediate hosts and arthropod vectors. The favorable habitat for intermediate hosts and uninterrupted life cycles of vectors help to establish and perpetuate parasitic diseases in these areas further impeding eradication efforts.

Opportunistic Parasitism

The human immunodeficiency virus and any organism or drug that weakens the immune system of humans increases its vulnerability to "opportunistic" parasites (p. 28) and other disease-causing organisms. The virus responsible for the current worldwide AIDS epidemic so compromises the immune system of its victims that they are left virtually defenseless. Even relatively benign parasites that cause only mild symptoms, if any, in a healthy person, can be quite devastating to a patient suffering from AIDS.

EVOLUTION OF PARASITISM

When and how did parasites arise? While there is no definitive answer, it is agreed that parasites have evolved among very diverse groups of free-living progenitors. One of these earlier organisms probably formed an initially casual association with another organism, and one member of the pair, perhaps due to preadaptation, developed a gradually increasing dependency on the other. The term *preadaptation,* in the context of parasitism, denotes the potential in a free-living organism for adaptation to a parasitic (symbiotic) lifestyle. The predisposition may never become operational as long as the organism remains free-living; however, if for some reason it becomes associated with a potential host, this latent ability becomes critically important for survival should hostile environmental conditions develop. Preadaptations can be structural, developmental, and/or physiological.

Parasites of the alimentary tract probably became such after having been swallowed, either accidentally or intentionally, by the host. If they were preadapted to withstand the environment or were capable of subsequent adaptation to it, they might have become progressively more dependent upon the new environment or might even have migrated to other more hospitable areas, such as the lungs or the liver. It is evident that parasites requiring two or more hosts developed their multihost life cycles sequentially. For instance, blood-inhabiting flagellates first parasitized the alimentary tracts of insects and, when introduced into vertebrate blood while the insect was feeding, adapted to that environment secondarily. Thus, present intermediate hosts once may well have been definitive hosts. Recent studies suggest, however, that the advancement to parasitism is best represented by a bell-shaped curve (Fig. 1.3), in which the increasing number of intermediate hosts signifies higher adaptability only up to a point. Thereafter, elimination of certain hosts is considered a more successful state because simplification of the life cycle may enhance the parasite's chance of reaching the definitive host.

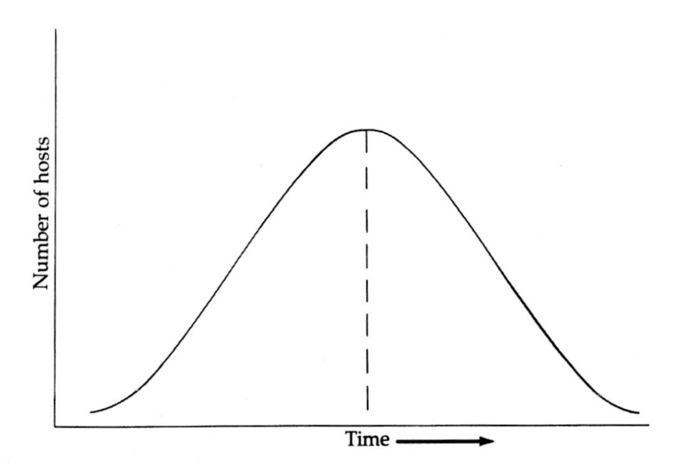

FIGURE 1.3 **Hypothetical scheme showing an increase in the number of obligatory hosts in the evolution of parasites.**

While parasites seem to have evolved independently in numerous animal groups, studies related to the free-living nematode *Caenorhabditis elegans* and its parasitic relative *Pristionchus pacificus* have provided insight into the evolution from a free-living mode to one that is parasitic. An interesting aspect of the life cycle of free-living nematodes is the formation of dauer larvae. Dauer larvae can be defined as an arrested developmental stage of a free-living nematode (i.e., dauer diapause) that is triggered by adverse environmental conditions. Pheromones signal these changes in the environment and cause the worm to develop into the dauer stage. It is intriguing to consider that the existence of dauer larvae and infective parasitic larvae may be part of the means by which parasitic nematode life cycles have evolved from free-living ones. The dauer stage usually attaches itself to other animals and is thus dispersed. In the life cycle of directly transmitted parasitic nematodes, the infective stage is almost always the third or L_3 larval stage (see pp. 283–284). There are striking similarities between dauer and infective L_3 larvae. They are both in an arrested state of development, both are the dispersive stages of their life cycles, and both will emerge from their arrested states under the proper environmental cues. *Strongyloides stercoralis* (see p. 301), by displaying both free-living and parasitic life cycles, may be considered an example of such a mechanism.

References

Anderson, R. M., & May, R. M. (1991). *Infectious diseases of humans: dynamics and control.* Oxford, UK: Oxford University Press.
Bulmer, M. G. (1994). *Theoretical evolutionary ecology*. Sutherland, MA: Sinauer Associates.

Suggested Readings

Blair, D., Campos, A., Cummings, M. P., & Laclette, J. P. (1996). Evolutionary biology of parasitic helminths: the role of molecular phylogenetics. *Parasitology Today*, *12*, 66–71.
Blaxter, M., & Koutsovoulos, G. (2016). The evolution of parasitism in *Nematoda*. *Parasitology*, *142*(Suppl. 1), 826–839.

Brindley, P. J., Mitreva, M., Ghedin, S., & Lustigman, S. (2009). Human genomics: the implication for human health. *PLoS Neglected Tropical Diseases, 3*(10), 22–36.

Ebert, D., & Herre, E. A. (1996). The evolution of parasitic diseases. *Parasitology Today, 12*, 96–101.

Fenwick, A. (2006). Waterborne infectious diseases—could they be consigned to history? *Science, 313*, 1077–1081.

Force, D. C. (1974). Succession of r and k strategists in parasitoids. In P. W. Price (Ed.), *Evolutionary strategies of parasitic insects and mites* (pp. 112–129). New York: Plenum Press.

Hupalo, D. N., Bradic, M., & Carlton, J. M. (2015). The impact of genomes on population genetics of parasitic diseases. *Current Opinion in Microbiology, 2*, 49–54.

Kutz, S. J., Hoberg, E. P., Polley, L., & Jenkins, E. J. (2005). Global warming is changing the dynamics of Arctic host–parasite systems. *Proceedings of the Royal Society B: Biological Sciences, 272*, 2571–2576.

Macpherson, C. N. L. (2005). Human behavior and the epidemiology of parasitic zoonoses. *International Journal of Parasitology, 35*, 1319–1331.

Pietrock, M., & Marcogliese, D. J. (2003). Free-living endohelminth stages: at the mercy of environmental conditions. *Trends in Parasitology, 19*, 293–299.

Poulin, R. (1996). The evolution of life history strategies in parasitic animals. *Advances in Parasitology, 37*, 107–134.

Viney, M. E., Thompson, F. J., & Crook, M. (2005). TGF-β and the evolution of nematode parasitism. *International Journal for Parasitology, 35*, 1473–1475.

CHAPTER

2

Parasite–Host Interactions

Chapter 2 introduces selected aspects of the parasite–host interaction with a range of topics involving their physiology, biochemistry, and immunology. Emphasis in this chapter is given to the rapidly developing field of immunology with an expanded discussion of the host's innate immune system interactions elicited by parasites. In view of its observed importance in parasite–host interactions, this chapter includes a basic overview of the mechanisms of lymphocyte development and differentiation in response to parasite invasion. Both cellular and humoral reactions of B and T cells in response to parasite invasion are discussed with regard to the presentation of endogenous and exogenous antigens. Basic immunoglobulin (Ig) structure and function are discussed along with the role of some of the more important cytokines involved in the immune response. The impact of selected opportunistic parasitic infections on immunocompromised individuals is also discussed.

The study of host–parasite relationships is an integral part of modern parasitology. Knowledge of how the parasite survives in its host often serves as the basis for the eventual control of the parasite. For instance, new chemotherapeutic agents may be synthesized on the basis of biochemical differences between the parasite and its host, or the elaboration of vaccines may be possible if the immunological responses of the host to the parasite and the parasite's evasion of these responses are known.

EFFECTS OF PARASITES ON HOSTS

Parasites trigger varying degrees of change within their hosts. Although not inevitably, disease often results. As previously noted, parasitic diseases, especially when caused by metazoan parasites, are usually functions of parasite density. Several factors commonly influence the onset of recognizable disease symptoms: among these are the number of parasites, the species (or strain) of the parasite, and the physiological condition of the host. Small numbers of parasites often elicit no clinical symptoms. Several of the conditions that most often develop in parasitic diseases are discussed below.

Tissue Damage

Beyond the erosion caused by ingestion or by mechanical disruption of cells by the parasite, three major types of histopathological cell damage occur in parasite-injured tissues.

Human Parasitology. http://dx.doi.org/10.1016/B978-0-12-813712-3.00002-3

Parenchymatous or Albuminous Degeneration

This type of damage is characterized by swollen cells packed with albuminous or fatty granules, indistinct nuclei, and pale cytoplasm. It is often seen in infected liver, cardiac muscle, and kidney cells.

Fatty Degeneration

This type of degeneration results in the deposition of abnormal amounts of fat in cells. This type of degeneration imparts a yellowish color to the cells and is common among parasite-laden liver cells.

Necrosis

Persistent cell degeneration of any type causes death of cells or tissues. The dead cells give tissues an opaque appearance. Encystment and calcification of *Trichinella spiralis* larvae in mammalian skeletal muscle cells cause necrosis of surrounding tissue.

Tissue Changes

Cell and tissue parasites sometimes evoke changes in the growth pattern of the affected tissue. Some of these changes are very deleterious; others are merely structural with no serious systemic consequences to the host organism. Tissue changes of parasitic origin are of four major types.

Hyperplasia

Elevation of the metabolic rate of cells causes cell proliferation by accelerating cell division. This condition, when associated with parasitism, is precipitated by the marked increase in host body repair activity that commonly follows inflammation. For example, inflammation caused by the liver fluke *Fasciola hepatica* stimulates excessive division of epithelial cells lining the bile duct, resulting in thickening of the duct wall.

Hypertrophy

An increase in cell or organ size in victims of parasitic diseases is usually due to engorgement by intracellular parasites. For example, during the erythrocytic phase in the life cycle of the malaria-producing organism, *Plasmodium vivax*, the parasitized red blood cells and the spleen commonly become enlarged.

Metaplasia

One type of tissue may be converted into another without the intervention of embryonic tissue. In patients infected with the lung fluke *Paragonimus westermani*, the parasite is surrounded by a host capsule consisting of cells, in this case, fibrocytes transformed from other types of cells.

Neoplasia

Abnormal cell growth may occur in a tissue, producing an entirely new entity, such as a tumor. The neoplastic tumor is not inflammatory, is not required for the repair of organs, and does not conform to the normal growth pattern. Neoplasms may be **benign** (i.e., may remain localized with no invasion of adjacent tissues) or **malignant** (i.e., may invade adjacent tissues or spread [metastasize] to other parts of the body through the blood or lymph). Cancers are malignant neoplasms. The human blood fluke *Schistosoma haematobium* is sometimes associated with malignant neoplasia of the urinary bladder.

BIOLOGIAL ADAPTATIONS OF PARASITISM

The intimacy of parasite–host associations invariably involves physiological, biochemical, morphological, and immunological adaptations. In this section, aspects of physiology, biochemistry, and immunology will be addressed briefly; morphological adaptations will be considered in subsequent chapters.

Physiology and Biochemistry of Parasitism

The study of parasitism deals with such basic questions as "How are parasites physiologically dependent upon their hosts?", "How do parasites affect their hosts?", and "How do hosts affect the parasites?". Since the antigenicity of parasites has somatic as well as physiological origins, increasing attention is being focused on the chemical composition of parasites as well as their metabolites. It is important, therefore, for anyone interested in parasitism and parasites to understand the enzymatic activities, pathways associated with energy production, protective mechanisms, secretions and excretions, respiration, and general metabolism of parasites. From a practical viewpoint, the prime requisite in modern chemotherapeutic research as it relates to parasitic diseases is the identification of an essential metabolic process in the pathogenic parasite that does not occur in the host or that can be chemically impaired or halted in the parasite with no adverse effect upon the host. After such identification, potential therapeutic agents can be developed to inhibit the process, killing the pathogen without harming the host. The discussion of chemotherapy in subsequent sections includes the information available about targets within the parasite toward which such agents are directed. Chemotherapy must be upgraded continuously as parasites evolve resistance to a particular drug or family of drugs and as research suggests new, more effective targets for drug action.

Immunology

The following discussion is aimed at providing a basic overview of some of the principles of immunology as they relate to the parasite–host relationship. For more detailed information, the reader is directed to the **Suggested Reading** section at the end of this chapter.

The guiding principle in immunology is the recognition by animals of *self and nonself*. Immunoparasitological implications of this principle are manifested in several ways. For instance, either the host recognizes the parasite or portions of the parasite as foreign (nonself) and reacts against it, or the parasite manages to evade immune recognition by the host. In one highly successful type of evasion, known as **molecular mimicry**, the parasite produces host-like molecules on its body surface or its surface becomes covered with host molecules and, as a result, the host is duped into accepting them as self. Immunity, therefore, may be defined as the ability of an animal to protect itself against infectious nonself agents. Immunity may be either **innate** or **adaptive** (**acquired**) (Fig. 2.1). The innate system is an evolutionary older system, highly conserved and present in all vertebrates. It provides protection through a diverse set of cells [e.g., macrophages, dendritic cells, natural killer (NK) cells, etc.]. The adaptive system provides a second level of defense. In evolutionary terms, it is younger than the innate immune system and is found in higher vertebrates. It depends on B and T lymphocytes. The two systems can act in a complementary fashion in that the innate system proceeds as a rapid, incomplete antipathogenic host defense until the slower (4–7 days), more definitive adaptive immune response develops. The innate system may also play a role in determining which antigens the adaptive system responds to and the nature of the response.

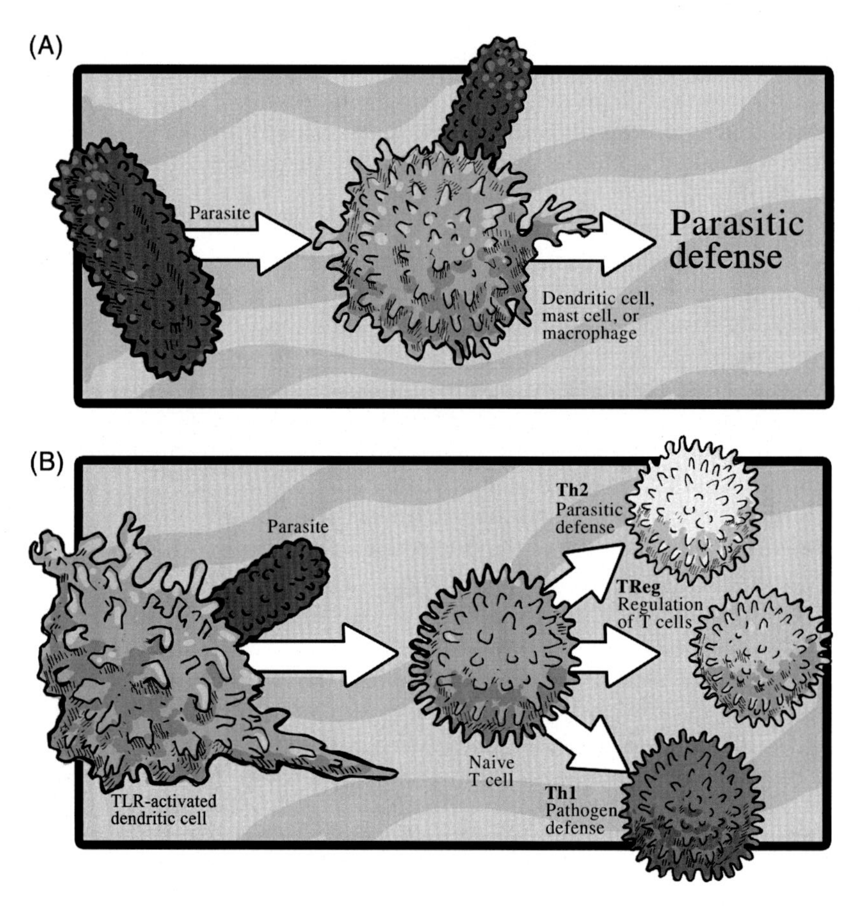

FIGURE 2.1 Roles of APCs in the (A) innate and (B) adaptive immune responses. *Abbreviations: APCs,* antigen presenting cells. *Credit: Image courtesy of Gino Barzizza.*

Toll-like Receptors

The innate immune system is relatively simple. It does not elicit memory as in the adaptive system; instead, it reacts quickly through a class of as many as 13 transmembrane proteins termed **toll-like receptors (TLRs)**. Some TLRs are associated with outer cell membranes, while others are associated with the inner surface of food vacuole membranes. On the TLR surface are specific proteins termed **pattern recognition receptors (PRRs)** which recognize and bind to complementary molecules termed **pathogen associated molecular patterns (PAMPS)** found on the surfaces of invading pathogens (Fig. 2.2). It is the interaction between the PAMP and the PRR that signals the cells of the innate immune system to synthesize and secrete **cytokines** such as interleukin-1 (IL-1) and tumor necrosis factor-alpha (TNF-α). Cytokines subsequently function to stimulate both the innate and adaptive systems. Released cytokines can also produce the classic symptoms of an inflammatory response, namely, fever and flu-like symptoms.

Different combinations of TLRs can be present in the various cell types, and different TLRs can bind to different PAMPs. For example, TLR-2 recognizes bacterial lipoproteins, while TLR-3 binds to double-stranded RNA, etc. TLRs can also act in pairs such as both TLR-1 and TLR-2 binding to the glycosylphosphatidylinositol-anchored proteins found in many parasites.

A parasite, whether protozoan or metazoan, is a mosaic of various molecules that, when recognized by the host as nonself, are known as **immunogens (=antigens)**. For the purpose of this discussion, the terms immunogen and antigen will be considered as synonyms. An antigen is any substance that is capable, under appropriate conditions, of inducing synthesis

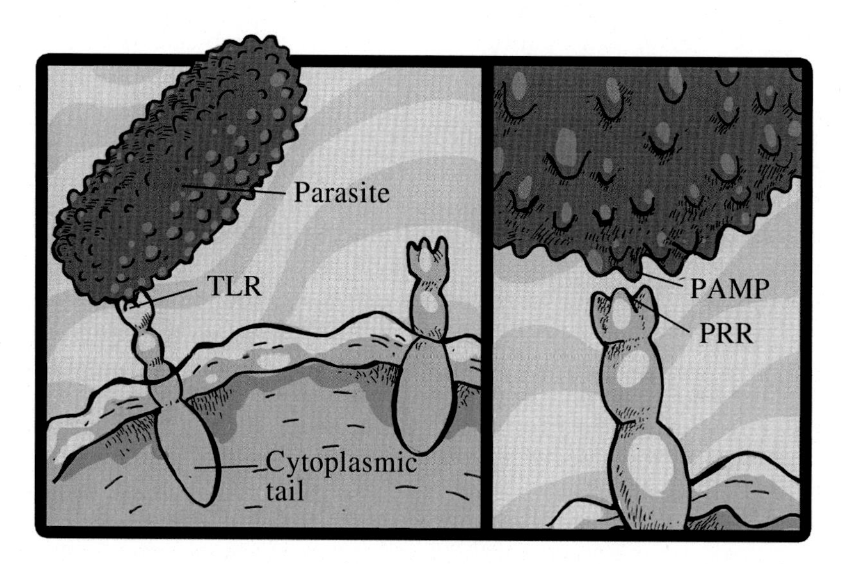

FIGURE 2.2 **In an innate immune response, the parasite binds, via its PAMP to a TLR's PRR.** *Abbreviations: PAMP*, pathogen associated molecular pattern; *TLR*, Toll-like receptor; *PRR*, pattern recognition receptor. *Credit: Image courtesy of Gino Barzizza.*

of **antibodies**. The two substances interact to form an antigen–antibody complex. Antibodies are glycoproteins synthesized in response to an antigen and may react, in varying degrees, to molecules of similar structure to the antigen. A parasite or any portion thereof recognized by the host as antigenic generally elicits the adaptive immune system, which involves **cellular** (or **cell-mediated**) and **humoral reactions**. In a cell-mediated reaction, specialized cells are mobilized to arrest and, eventually, attempt to destroy the parasite; in a humoral reaction, specialized molecules in the circulatory system interact with the parasite, seeking to immobilize or destroy it. Humoral antibodies may also react with excretory–secretory products of the parasite, forming immune complexes, some of which may precipitate causing host organ dysfunction (e.g., kidney disease or nephritis).

When a host is immunologically challenged, it signals functionally specialized types of lymphocytes, **B cells** (**B lymphocytes**) and **T cells** (**T lymphocytes**), to interact and/or produce specific antibodies (Fig. 2.3). B and T cells are generated from a common progenitor stem cell in the bone marrow, although they mature in separate tissues. B cells are so designated because they were originally recognized by their dependence upon maturation in the bursa of Fabricius, an avian lymphoid tissue attached to the intestine near the cloaca. In mammals, however, maturation of B cells occurs in the bone marrow with subsequent activation usually occurring in primary lymph tissue such as spleen, lymph nodes, and gut-associated lymphoid tissue. T cells, on the other hand, are so designated because they must mature in the thymus. B cells produce humoral, or circulating, antibodies, while T cells may elicit cell-mediated reactions usually independent of circulating antibodies. In certain situations, however, humoral antibodies can participate in parasite destruction, as in the case of antibody-dependent, cell-mediated cytotoxicity, described below.

As many as 10^9 B cells, each with different receptor (antibody) specificities, exist in the body prior to contact with a recognizable antigen. This pool of naïve B cells is able to respond to most any potential antigen. Under appropriate conditions, the binding of the antigen to its cognate receptor stimulates the B cell to undergo clonal proliferation. Each cloned cell displays surface receptors with the same specificity as the originally stimulated or naïve cell. This **clonal selection** phenomenon is well-supported experimentally and is widely accepted.

Under appropriate conditions, the direct binding of B-cell receptors to their specific antigens leads to activation. Because antigen-specific T-cell receptors do not recognize free or soluble antigens, T-cell activation requires presentation of antigen via a membrane-bound receptor. These membrane-bound receptors are glycoproteins, products of a complex of genes termed the **major histocompatibility complex** (**MHC**). MHC molecules serve to alert the immune system against foreign substances. Cells of the MHC present protein fragments (=peptides) on their cell surfaces. These fragments may be self or nonself and, therefore, may act as antigens. If cytotoxic T lymphocyte or NK-cell receptors recognize these as antigens, the immune system is activated. MHC molecules can bind only protein fragments and cannot bind carbohydrate or lipid fragments. There are two classes of MHC molecules. Class I molecules are located on the surfaces of all nucleated cells and present peptide antigenic fragments to cytotoxic T lymphocytes and NK cells via the CD8 receptor. Class II molecules are restricted to the surfaces of **antigen presenting cells** (**APCs**) such as macrophages, B cells, and dendritic cells. Antigenic fragments utilize the lysosomal network and are presented to T helper (Th) cells via CD4 binding. It is noteworthy to mention that T lymphocytes can only recognize an intracellular antigen if it is presented via an MHC molecule. The complex of

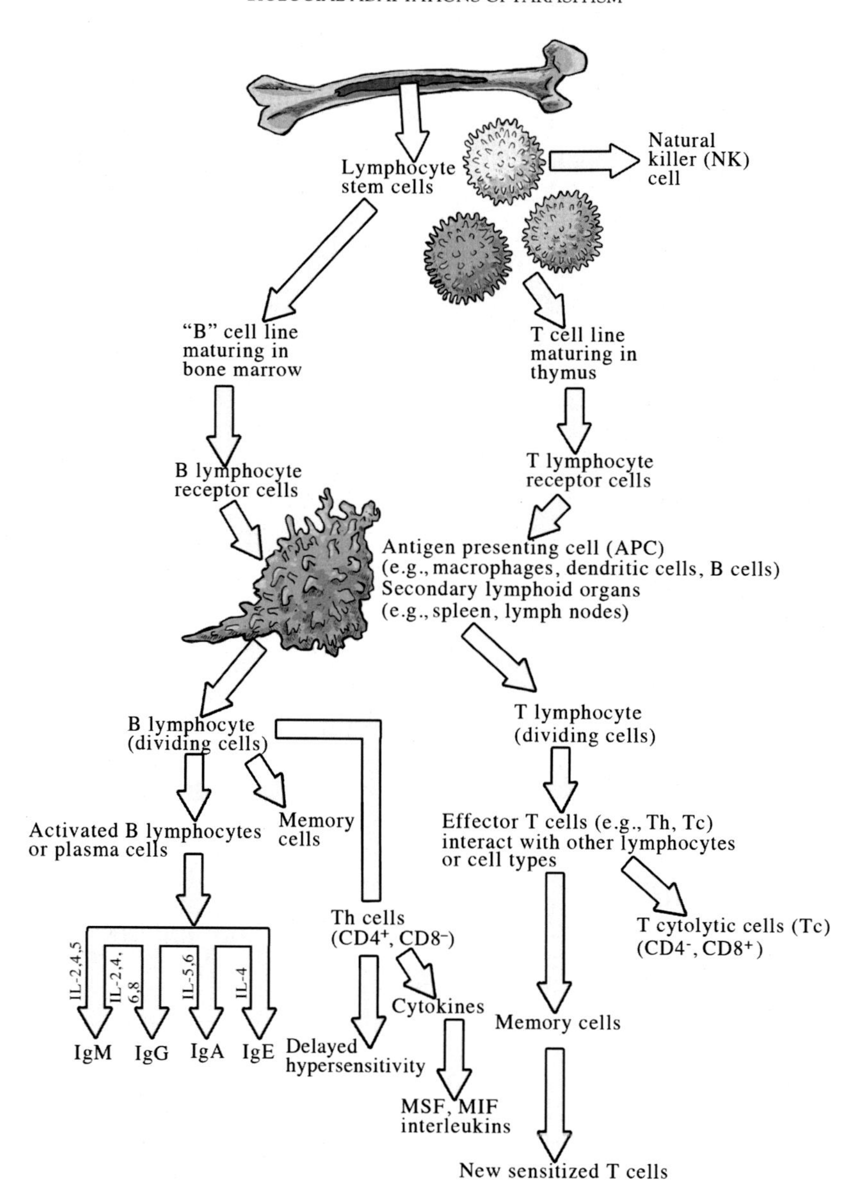

FIGURE 2.3 **Generalized scheme showing the basic patterns of lymphocyte development and differentiation to become immunocompetent.** *Credit: Image courtesy of Gino Barzizza.*

MHC molecule and antigenic peptide is presented at the cell surface and recognized by the appropriate cells via CD8 or CD4 receptors.

The pathways involved in antigen processing and presentation differ depending upon whether the proteins (e.g., antigens) are endogenous or exogenous to cells (Fig. 2.4). Endogenous proteins (e.g., produced by intracellular parasites) associate with Class I MHC

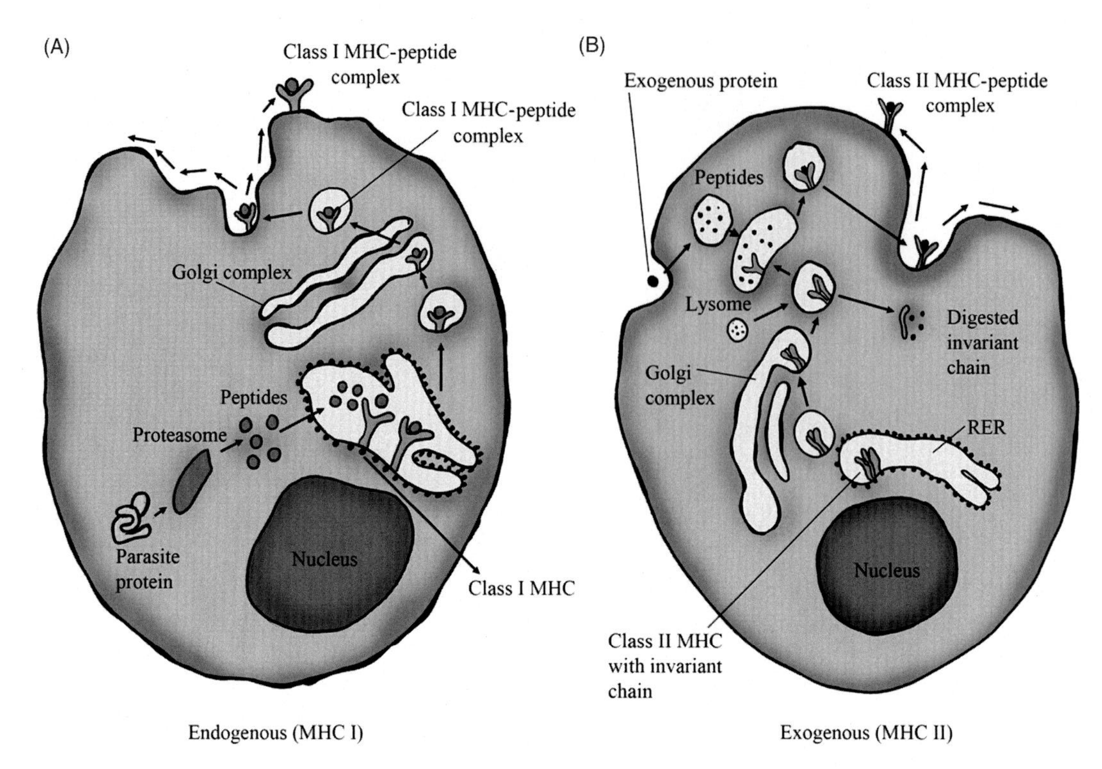

FIGURE 2.4 Diagram of antigen-presenting pathways for (A) endogenous and (B) exogenous antigens.

molecules. Class I MHC molecules are found on the surface of almost all nucleated cells. Exogenous proteins (e.g., produced by extracellular parasites), on the other hand, enter the cell by endocytosis and/or phagocytosis and are presented in association with Class II MHC molecules. Class II MHC molecules are restricted to the surfaces of **APCs** (e.g., macrophages, B cells, dendritic cells) that are involved in the immune system.

Class I Endogenous Pathway

Endogenous proteins generated by intracellular parasites are degraded in the cytosol of all nucleated cells by proteases located in organelles termed proteasomes. Resulting peptides are translocated across the RER membrane into the cisterna of the RER by an energy-dependent transport system. Once in the RER cisterna, the peptides associate with membrane-bound Class I MHC molecules. Vesicles containing the Class I MHC molecule-peptide complex "pinch off" from the RER and are then translocated to the plasma membrane via the Golgi complex.

Class II Exogenous Pathway

Exogenous proteins generated by extracellular parasites can be internalized by APCs via endocytosis and/or phagocytosis. Macrophages can internalize antigens by either process. B cells, on the other hand, utilize receptor-mediated endocytosis very effectively. Once the antigen is internalized, it is degraded via the lysosomal system of the cell. Food vacuoles (endosomes) fuse with primary lysosomes forming digestive vacuoles (secondary lysosomes) in which the protein is degraded to peptides.

Class II MHC molecules are synthesized within the RER. Associated with the Class II MHC molecule within the RER is another protein, the **invariant chain**. This protein stabilizes the Class II MHC molecule in its translocation from the RER through the Golgi complex to the late food vacuole. Once inside the lysosomal pathway, the invariant chain is degraded by lysosomal proteases. The freed Class II MHC molecule associates with the peptides derived from the endocytosed protein, and the resulting Class II MHC-peptide complex moves to the plasma membrane.

T-Cell Activity

Mature T cells equipped with complementary surface receptors enter primary lymph tissues such as spleen and lymph nodes where they most often make contact with antigens. Upon recognition of the antigen, the T cells are induced to proliferate and differentiate into **T memory** and **T effector cells**. On entering the general circulation, these specialized T cells function in several ways. For example, T memory cells revert to a "resting state" and may serve as an increased source of new, antigen-specific T cells whenever the same antigen reenters the body. T effector cells can be divided, functionally, into **Th cells**, **T cytotoxic (Tc) cells**, and **regulatory T (Treg) cells** (see Fig. 2.2). Activation of Th cells via Class II MHC molecules facilitates B and T cell proliferation, differentiation, and macrophage activation. Tc cells can destroy parasite-infected cells by recognition of antigens presented by Class I MHC molecules. Treg cells, formerly considered suppressor T cells, are now considered Treg cells. Their function is to suppress activation of the immune system and shut down T cell mediated immune reactions. Most T cell activity, however, is involved with the synthesis and release of various chemical mediators called **cytokines** (Table 2.1). Cytokines react with a wide variety of cells essential to a number of immunological processes. For instance, cytokines, such as **gamma interferon (IFN-γ)** and **macrophage inhibitory factor** regulate the activity of macrophages, attract other types of cells to an inflamed site (**chemotactic factor**), stimulate proliferation of T cells and B cells, or delay or totally inhibit other types of cell proliferation (**cytostatic factor**).

Naïve Th cells can differentiate into subgroups of cells, which are distinguished by the cytokines they characteristically secrete. Two of the better defined subgroups of Th cells are **Th1** and **Th2** cells.

Th1 cells characteristically secrete **IL-2, IFN-γ,** and **TNF**. These cytokines support the inflammatory process, activate macrophages, and induce proliferation of **NK** cells. Th1 cells stimulate type 1 immunity which is symptomatic of bacteria and intracellular parasitic infections. Th2 cells typically secrete several cytokines, among which are IL-4, IL-5, IL-9, IL-10, and IL-13. Th2 cells stimulate type 2 immunity which is symptomatic of helminthic and other extracellular parasitic infections. It is characterized by activation of eosinophils, mast cells and B cell secretions of antibodies such as IgE.

B-Cell Activity

The B cell follows basically the same pattern of stimulation described for T cells (i.e., interaction with antigen in lymph under tissues). Under appropriate conditions (i.e., the presence of specific cytokines and direct contact with Th cells), individual naïve B cells are induced to proliferate into either **plasma cells** or **memory cells** (Fig. 2.5). The plasma

TABLE 2.1 Classes of Cytokines

Interleukins	Role	Source
IL-1	Primary role—a mediator of the host inflammatory response in innate immunity	Primarily activated mononuclear phagocytes but actually many different cell types (responds to TNF-γ)
IL-2	1. Growth of T cells 2. Stimulates growth of NK cells 3. Growth factor for B cells	Primary—$CD4^+$ T cells Secondary—$CD8^+$ T cells
IL-3	Growth and differentiation of all cell lines	T cells, mast cells, eosinophils
IL-4	1. Production of IgE and IgG by B cells 2. Growth factor for Th2 T cells	Th2 T cells, mast cells, eosinophils
IL-5	1. Activates eosinophils in such a way to kill helminths 2. Ig class switch in B cells	Th2 T cells, mast cells, eosinophils
IL-6	1. Causes liver cells to synthesize various proteins (e.g., fibrinogen) that contribute to acute phase response 2. Growth factor for activating T and B cells	Mononuclear phagocytes, Th2 T cells, bone marrow stromal cells. Responds to IL-1 and TNF
IL-7	Growth and differentiation of lymphocytes of pre-B and pre-T cell lineages	Bone marrow, thymic, spleen cells
IL-8	Attracts and activates neutrophils and promotes angiogenesis	Monocytes, macrophages, fibroblasts, endothelial cells, etc.
IL-9	Supports growth of some T cell lines and bone marrow-derived mast cell progenitors	T cell proliferation
IL-10	Suppresses Th1 T cells and macrophage functions	Monocytes, macrophages, Th2 T cells
IL-11	Similar to IL-6	Bone marrow stromal cells
IL-12	1. Mediates innate immunity 2. Activation and proliferation of T- and NK cells 3. Antiviral cytokine	Mononuclear phagocytes, B cells, macrophages
IL-13	Mimics IL-4 activity but cannot replace IL-4	Th2 T cells
IFN-γ	1. Activator of mononuclear phagocytes 2. Promotes macrophage-rich inflammatory reactions 3. Activates granulocytes, macrophages, NK cells 4. Induces antiviral state in uninfected cells	Th1 T cells and NK cells

cell, which has a half-life of only a few days, secretes large numbers of antibodies of the same antigen-recognition specificity as its cell surface receptors into the circulation. Five classes of structurally similar antibody molecules have been described. Due to their common globular structure these antibodies are called **Igs** (Table 2.2). The first Ig secreted in response to an antigen is the membrane-bound **IgM** (mu). With the aid of Th cells, B cells can shift production to any of four other Ig classes: **IgG** (gamma), **IgE** (epsilon), **IgA**

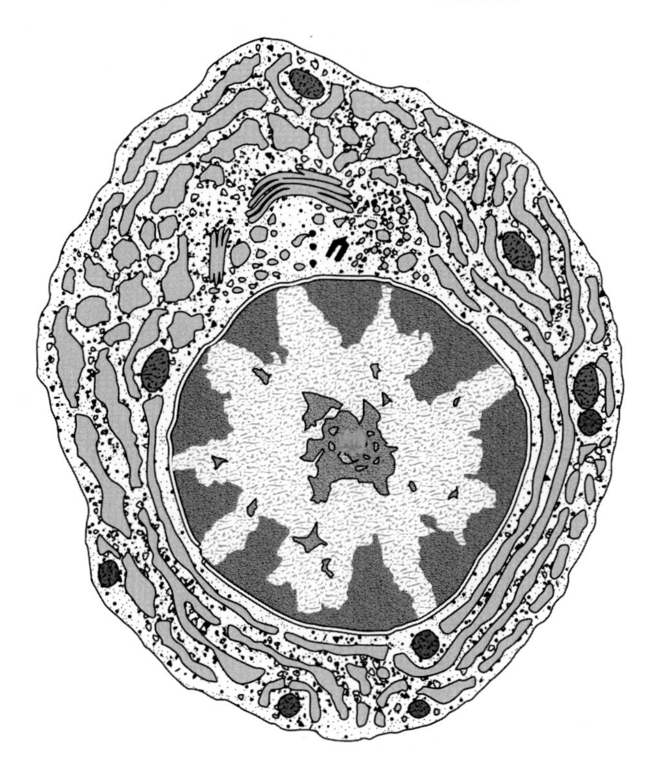

FIGURE 2.5 **Plasma cell.** Note the extensive rough endoplasmic reticulum.

(alpha), or **IgD** (delta). While all classes of Ig can be found in the blood and tissue fluids, they are present in varying amounts. For instance, IgG represents approximately 70% of the serum Ig, IgM represents 10%–15%, IgA represents 10%–15%, while IgE represents less than .002% and IgD less than .02%. IgA is also termed **secretory IgA** because it is the primary Ig secreted across the intestinal wall into the lumen. IgE levels are frequently elevated in helminthic infections.

TABLE 2.2 Classes of Immunoglobulins

Class	Molecular weight	Biological function
IgG	150,000	Fix complement Bind to amine-containing cells (e.g., mast cells and basophils) Bind to macrophages and granulocytes Cross placenta
IgA	170,000	Secreted across mucus surface
IgM	890,000	Fix complement Secreted across mucus surface
IgD	150,000	?
IgE	196,000	Bind to amine-containing cells (e.g., mast cells and basophils)

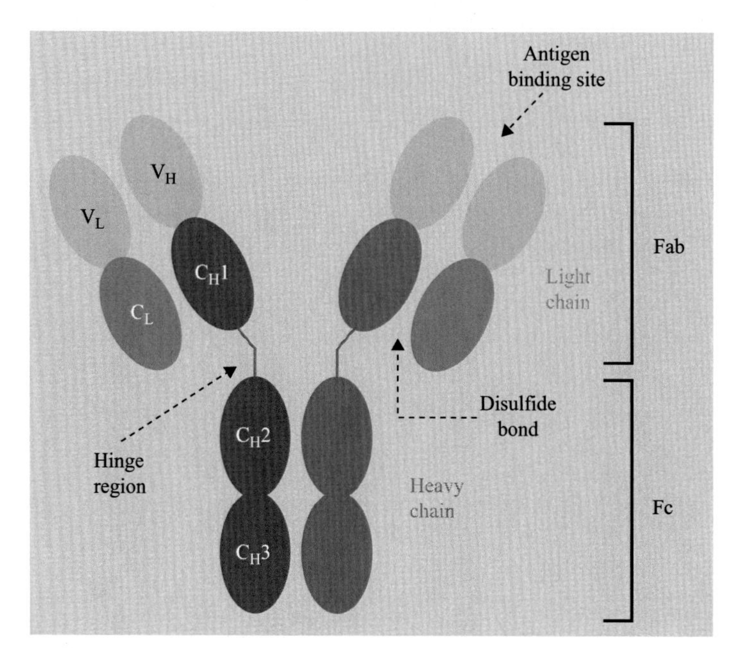

FIGURE 2.6 **IgG molecule.**

The IgG molecule illustrates the basic molecular structure of an Ig (Fig. 2.6). The molecule is composed of two identical light chains and two identical heavy chains linked by disulfide bonds. The light chain contains 220 amino acids and is separated into two 110 amino acid domains. The heavy chain contains either 440 amino acids or 550 amino acids and is separated into either three or four segments. Both heavy and light chains contain amino terminal **variable (V)** regions as well as a **constant (C) region**. IgG, IgD, and IgA have three constant region segments, while IgM and IgE have four. Antigenic specificity lies in the variable regions of both the light and heavy chains (**Fab regions**) and is due to variability in its primary amino acid sequences. This variability appears greatest in three specific regions, termed **hypervariable regions** (Fig. 2.6). Each Fab has six such regions, three on the light chain and three on the heavy chain. These ultimately form the sites for antigen binding. The hypervariable regions are not contiguous within the variable regions. However, three-dimensional folding of the protein brings these regions in close proximity, forming regions complementary in structure to the antigenic determinants (or epitopes), which are termed **complementarity determining regions (CDR1, CDR2, CDR3)**. Regions where primary amino acid sequences remain constant (**Fc regions**) determine the biological properties of the particular class of Ig. Each Ig molecule is **divalent** and **bifunctional**. It is divalent because the Fab portion can bind two identical antigen molecules. It is bifunctional because the Fc portion can bind to specialized receptors (Fc receptors) on the surface of a variety of cells such as macrophages, mast cells, NK cells, etc.

Igs display various mechanisms to react with the surfaces of parasites and parasite-infected cells. For instance, IgG is also known as an **opsonizing antibody** (from the Greek

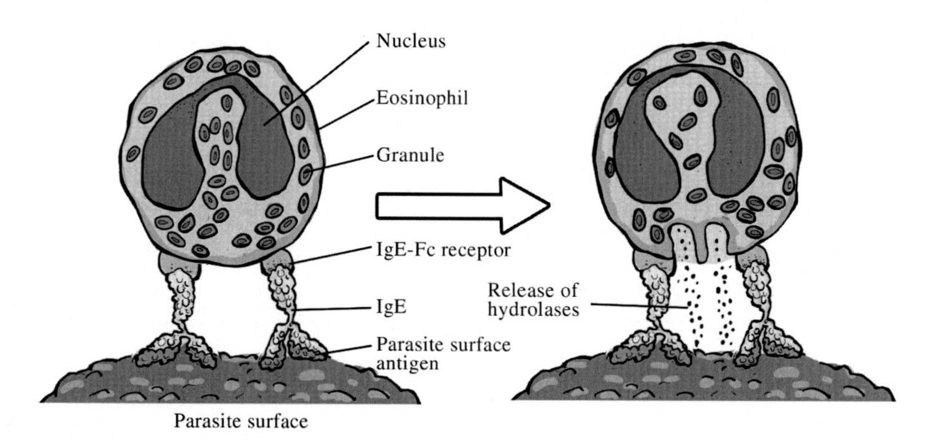

FIGURE 2.7 **Antibody-dependent cell-mediated cytotoxicity.** *Credit: Image courtesy of Gino Barzizza.*

opsonin meaning to prepare for eating). IgGs react with antigenic determinants on various microorganisms via their Fab regions leaving Fc portions to bind with Fc receptors on a variety of phagocytic cells, inducing phagocytosis. Another example of Ig activity is illustrated in the **antibody-dependent cell-mediated cytotoxicity** reaction. The Fab portion of IgG molecules binds to a target cell (e.g., a parasite-infected cell) and the Fc portions bind to Fc receptors on NK cells. As a result of this interaction, the NK cells are stimulated to deliver their cytotoxic components to the infected cell, causing cell and parasite death.

Infections with parasitic helminths commonly result in host eosinophilia (increased numbers of eosinophils). The Fab portions of IgA, IgG, and IgE can react with membrane-bound antigens associated with parasite surfaces. As shown in Fig. 2.7, the Fc portions of these Igs react with their respective Fc receptor sites on the eosinophil surface. This interaction stimulates the eosinophil to release lyosomal hydrolases or other cytotoxic factors causing destruction of the parasite. By a similar mechanism, mast cells and basophils can react with IgE via Fc receptors on their surfaces. The Fab portions of the IgE molecule react with various parasite surface antigens. The resulting interaction stimulates the mast cells and basophils to release histamines and other vasoactive substances increasing host capillary permeability and dilation.

Complement refers to a group of approximately 20 different plasma glycoproteins that interact through a cascade of reactions. The last reaction in the cascade usually results in lysis of the cell membranes of many pathogens. Antigen-bound IgG and IgM can initiate the first step in this cascade.

Genetics of Immunoglobulin Variability

As stated previously, as many as 10^9 different B cells possessing membrane-bond Igs exist in the human body. Each cell's Ig molecule differs in its molecular conformation and, therefore, antigen specificity. Interestingly, the number of genes that contribute to this Ig variability is relatively low. In the early stages of B cell development, DNA coding for the heavy chain of an Ig molecule exists in four different types of gene segments (V, D, J, C) for each of the five classes if Igs. Humans display about 65 different V gene segments, about 27 D gene segments, and about 6J gene segments along with a single C gene segment.

During B-cell maturation, one of each type of gene segment is selected at random and recombined into functional genes. This recombination can result in functional genes encoding for more than 10^9 different variations in the Ig molecules. This random recombination accounts, at least in part, for the diversity of Ig molecules.

OPPORTUNISTIC PARASITES

Immunosuppression in humans is sometimes brought on by normal therapy following successful organ transplants or grafts or used in the treatment of certain types of cancer. It can also be produced by several infectious diseases, such as AIDS and measles. A number of parasites, some of which were formerly considered inconsequential, have become life-threatening to individuals who are immunosuppressed.

One of the major effector mechanisms in the human immune system affected by immunosuppression is the **$CD4^+$ T cell** (Fig. 2.8). The CD4 antigen is found on 20% of the macrophages, 10% of the monocytes, and approximately 5% of the B lymphocytes. There are two well-defined subsets of $CD4^+$ T cells. (1) **Th1 cells** are responsible for the primary inflammatory response through the cytokines (see p. 23) IL-2, IFN-γ, and TNF-β. IL-2 has several functions among which are targeting the proliferation and differentiation of B cells and the proliferation of other T-cell subpopulations, including cytotoxic lymphocytes and NK cells. IFN-γ activates NK cells and macrophages and inhibits the activity of Th2 $CD4^+$ T cells. TNF-β activates neutrophils to stimulate the inflammatory process. (2) **Th2 cells**, which constitute another helper cell subset, secrete a number of cytokines including IL-3, IL-4, IL-5, IL-10, and IL-13. IL-3 promotes the proliferation of mast cells. IL-4, IL-5, and IL-13 stimulate B-cell differentiation into antibody-producing cells, stimulate eosinophil growth and development, foster Th2 cell growth, and promote IgE and IgG synthesis. IL-10 inhibits Th1 cell-mediated inflammation by down-regulating the production of cytokines (e.g., IFNγ).

Macrophage activation is a major mechanism by which the body eliminates infectious organisms. IFN-γ activates macrophages, which, in turn, produce effectors such as toxic oxygen radical intermediates, reactive nitrogen intermediates, TNF, and lysosomal enzymes. Such macrophage activity enables the host to either kill intracellular parasites or, in many instances, keep their numbers in check.

Worldwide, parasitic infections account for a higher incidence of morbidity and mortality than diseases produced by any other group of organisms. In developing countries, the effects of parasitic infections are often exacerbated by poor nutrition. Perhaps, the most insidious disease that compromises the immune system in the greatest number of people throughout the world is AIDS. AIDS victims concomitantly infected with certain parasites are the most severely impacted by immunosuppression. It has been theorized that any parasite infection that activates a chronic immune response is important in the progression of the human immunodeficiency virus (**HIV**) disease. This is especially noticeable in third-world countries where HIV infection is accelerated.

The HIV attacks $CD4^+$ T cells and macrophages. A brief review of the mechanisms used by the virus to infect these cells is as follows. The virus possesses two important **chemokines**

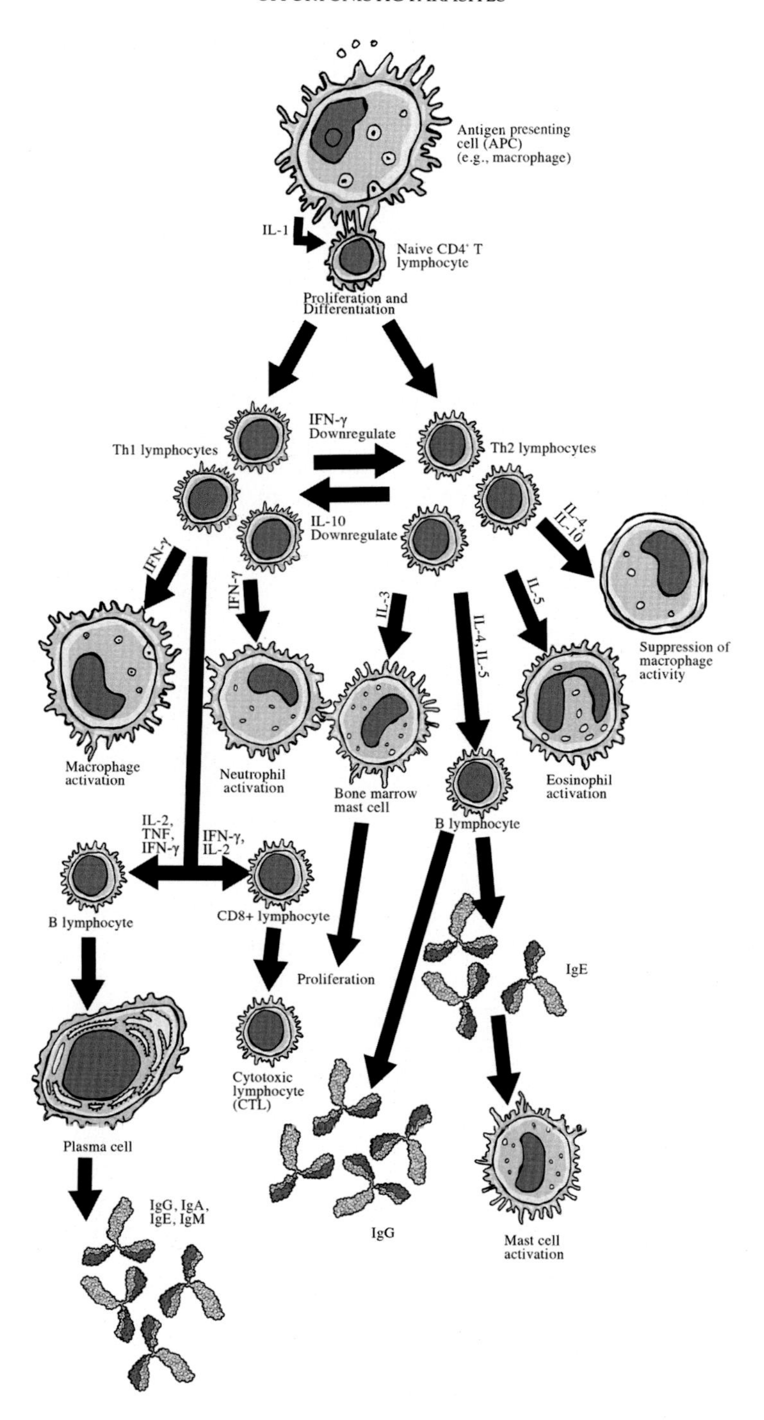

FIGURE 2.8 **Role of CD4+ T cell in the immune response.** *Credit: Image courtesy of Gino Barzizza.*

associated with its outer surface, GP120 and GP41. Viral GP120 binds to the **chemokine receptor CD4** presented on the surfaces of $CD4^+$ T cells and mononuclear phagocytes (e.g., macrophages) permitting one end of the GP41 molecule to penetrate the plasma membrane of the host cell, initiating fusion of the virus with the membrane. Alternatively, GP120 may be expressed on the plasma membranes of infected cells before the virus is released. It can bind to CD4 on another cell, initiating a membrane fusion event between the two cells. The fusion allows HIV genomes to be passed between the cells directly, ultimately causing death of the infected cell subsequent to replication and release of the virus. HIV infection results in a 10-fold reduction in the number of $CD4^+$ and helper T cells in the patient. Of the many effects attributed to a depletion of $CD4^+$ T cells, three are of particular significance in patients with concomitant parasitic infections: (1) killer function of macrophages is impaired, (2) Th1 cell cytokine synthesis decreases and Th2 cell cytokine synthesis increases, and (3) differentiation of B cells and cytolytic T lymphocytes (CTLs or $CD8^+$ cells is inhibited. Any parasitic infection in which a compromised cellular immunity is a major factor is called **opportunistic**. Likewise, such a parasite can be considered an opportunistic parasite. The following discussion addresses those parasites whose effects in human hosts are impacted, to some extent, by immunosuppressive factors such as HIV.

Toxoplasma Gondii (see p. 137)

Toxoplasma gondii is an intracellular parasite that infects a wide variety of cells. About 20% of AIDS patients will display symptoms of toxoplasmosis. As is the case with *Pneumocystis carinii*, large segments of the world's population are likely to have been exposed to *T. gondii*. This observation is based on serological surveys showing that prevalence of antibodies against *T. gondii* in humans increases with age. One response of a normal immune system to *T. gondii* infection is the formation of pseudocysts (or zoitocysts) containing bradyzoites (see pp. 139–140). Bradyzoites, released from ruptured pseudocysts, invade neighboring cells of various tissues including those of the central nervous system. Normally, cysts in the central nervous system are prevented from rupturing either by the protection afforded by immune mechanisms or by the intraneural environment. In either case, if pseudocyst-infected nerve cells bearing CD4 receptors (e.g., some glial cells) are invaded by HIV, bradyzoites are freed to infect new nerve cells, this causes **toxoplasmic encephalitis**.

In an immunocompetent host, NK-cell proliferation is accelerated by production of IL-12 and IFN-γ. IFN-γ probably acts in concert with TNF-α to further stimulate the release of oxidants by macrophages. After several days postinfection, parasite-specific T-cell response occurs. The latest evidence shows that $CD4^+$ T cells and $CD8^+$ T cells act in harmony for long-term host protection. During the latter stages of AIDS, symptomatic *T. gondii* infection is often observed. Concomitantly, $CD8^+$ T cells and IFN-γ levels decline, bradyzoites spread unimpeded, and TE symptoms become evident.

Cryptosporidium Parvum (see p. 142)

Cryptosporidium parvum infects the microvilli of the small intestine of humans and other animals. The infection disrupts the ionic balance in the intestinal tract resulting in an overall increase in ion loss. In immunocompetent hosts, the infection may be severe but is usually self-limiting, leaving the host immune to reinfection. HIV patients infected with *C. parvum* develop cholera-like symptoms. Experimental evidence indicates that, in individuals with normal immune systems, parasite development is controlled by the production of IFN-γ by $CD4^+$ T cells. The protective role of $CD8^+$ T cells, on the other hand, is more uncertain.

Enterocytozoon Bieneusi (see p. 66)

Enterocytozoon bieneusi is a protist of the phylum Microsporidia, which consists of obligate intracellular parasites that infect a wide variety of invertebrates and vertebrates, including mammals. Human infection by microsporidia is rare. However, with the advent of AIDS, there has been a dramatic rise in the number of human cases. *E. bieneusi* is one of the primary causes of chronic diarrhea in such individuals. In infected individuals, there is a decrease in the number of $CD4^+$ T cells, notably Th1 cells. The microbiostatic activity of IFN-γ-activated macrophages has been shown experimentally in mice to be essential for the control of microsporidia. In HIV-infected patients, depletion of Th1 cells reduces the ability of macrophages to control the organisms, allowing the disease to flourish.

Leishmania spp. (see p. 90)

According to the World Health Organization, adult visceral leishmaniasis is considered an AIDS-related opportunistic disease due to "reactivation of latent infections by immunosuppression." The amastigotes of *Leishmania* spp., the cyst-like stages of this polymorphic protozoan, reside exclusively in monocytes/macrophages. Some macrophages, on the other hand, are more resistant to the lytic effects of HIV than other $CD4^+$ T cells, and they can serve as important reservoirs for the virus. Under normal circumstances, *Leishmania* infection is regulated by the infected macrophages themselves, which induce cell-mediated immunity by stimulating Th1 T cells to produce macrophage-activating factors such as IFN-γ. *Leishmania* exposure also results in an increase in Th2 T-cell cytokines such as IL-4, IL-5, and IL-10. *Leishmania* and HIV coinfections, however, alter the Th1/Th2 T-cell balance. For instance, a coinfection depresses Th1 T-cell activity and enhances Th2 T cells. Inhibition of Th1 T-cell activity would tend to depress cure, while stimulation of Th2 T-cell activity (e.g., secretion of IL-10) would tend to inhibit macrophage activation. Therefore, HIV-infected individuals will experience macrophage dysfunction allowing the *Leishmania* infection to further exacerbate. *Leishmania* infection, on the other hand, causes activation of HIV in latently infected macrophages by stimulating the release of TNF-α, which induces HIV replication. Thus, macrophages infected with both *Leishmania* and HIV display an intensification of both infections. Infection with *Leishmania* induces HIV replication. HIV infection suppresses cell-mediated immunity due to T-cell inhibition and lessens the regulatory effects of the macrophages, promoting *Leishmania* multiplication. Infection by *Leishmania* and HIV of macrophages results in the "worst of both worlds."

Trypanosoma Cruzi (see p. 108)

Trypanosoma cruzi amastigotes also infect macrophages of humans (see pp. xxx). As in leishmaniasis, activated macrophages are important in regulating the disease, especially the acute phase, by controlling the number of intracellular parasites. Any breakdown in T cell function serves to upgrade the infection, often resulting in acute central nervous system involvement in immunocompromised individuals. However, *T. cruzi* does not appear to exacerbate the effects of dual infections with HIV as *Leishmania* does.

Plasmodium spp. (see p. 112)

The causative agent of human malaria displays two life cycle phases in its human host, an extraerythrocytic phase initiated when sporozoites infect hepatocytes and an erythocytic phase involving the infection of red blood cells by merozoites. Suppression of $CD4^+$ Th cell functions inhibits $CD8^+$ cytotoxic cell differentiation and release of IFN-γ. Both of these processes are important in the immunology of human malaria, especially the extraerythrocytic stages. Inhibition of cytotoxic cell functions prevents lysing of sporozoite-infected hepatocytes, while suppression of IFN-γ prevents activation of macrophages and results in the release of uncontrolled numbers of merozoites during the extraerythrocytic phase. One of the consequences of these changes is that AIDS patients display an increased incidence of cerebral malaria.

Giardia Lamblia (see p. 71)

The flagellate *Giardia lamblia* is an extracellular parasite of the small intestine. While the infection ostensibly is controlled by IgA and IgM secretions of the host, there is evidence that depletion of $CD4^+$ T cells results in a marked decrease in the clearance of the motile trophozoites. In an immunocompromised patient, prolongation and recrudescence of giardiasis is evident and levels of IgG, IgM, and IgA antibodies are depressed.

Strongyloides Stercoralis (see p. 286)

The nematode *Strongyloides stercoralis* is normally a chronic pathogen infecting the mucosa of the upper intestinal tract. *S. stercoralis* can display an autoinfectious (hyperinfectious) cycle allowing the infection to persist for many years in an immunocompetent host. In such a host the number of parasites is usually kept under control by high levels of IL-5 and high levels of specific IgE antibodies against *S. stercoralis*. In immunocompromised patients, a tremendous proliferation of adult worms and filariform larvae occurs throughout the body and is termed disseminated strongloidiasis. This proliferation ultimately causes injury to almost every organ and tissue, making *S. stercoralis* a significant opportunistic parasite. In such individuals, there is an increase in the levels of IFN-γ and IL-10, a decrease in the levels of IL-4, IL-5, and IgE, and a switch from a predominantly Th2 response to a Th1 response. Normally, Th2 cells secrete IL-10. Recent evidence suggests that the source of IL-10 in immunocompromised patients may not be $CD4^+$ Th2 cells; rather, it may represent a host's attempt to modulate the high levels of IFN-γ production through other cell types.

Schistosoma spp. (see p. 193)

In most helminth infections Th2 cells, the primary helper cell subset, predominate. Infection with *Schistosoma* spp. may also impair $CD8^+$ T cell and Th1 T-cell responses, suggesting that immune protective mechanisms against HIV and other viral infection may be reduced. A schistosomiasis patient coinfected with HIV progresses to AIDS more rapidly than one not infected with schistosomiasis. These patients respond to egg antigens by making less IL-4 and IL-10, indicating that there is a swing in the overall balance of response from Th2 to Th1. Coinfected patients also display an altered pattern of fibrosis of the liver and an increased risk of liver damage due to faulty sequestration of toxins and increased production of pro-inflammatory cytokines. This shift in Th cell activity to a Th1 dominated immune response is conducive to a rapid progression of HIV.

Since the primary pathology of human schistosomiasis is the granulomatous response to eggs in the urinary tract, liver, spleen, and gastrointestinal tract (see pp. 197–198), depletion of the factors responsible for this inflammatory reaction might be considered beneficial to persons infected with schistosomes. The granulomatous response is normally triggered when $CD4^+$ T cells are stimulated by trapped eggs, activating macrophages, which then induces delayed hypersensitivity reactions. The granuloma serves to isolate the eggs in the tissue. Concurrently, the fibrous granulomas may disrupt blood and lymph flow and destroy neighboring cells. Accompanying side effects may include enlargement of organs, cirrhosis, etc. (see pp. 206–207). In immunocompromised patients, there is a reduced response to the eggs, which leads to both a reduction in the cell-mediated response and the granulomatous response. Concomitantly, fewer eggs reach the environment, suggesting that a certain level of inflammatory response is requisite for the movement of eggs through the intestinal and bladder walls. In addition to a reduction in the chronic inflammatory response, reduced Th2 cell activity diminishes the level of concomitant immunity (see pp. 206–207).

Resistance

Defined as the ability of a host to withstand infection by a parasite, **resistance** may develop in several ways:

1. It may result from the presence of some physical or chemical barrier that prevents the parasite from penetrating or migrating in the host. Such barriers are thought to prevent the larval form of avian blood flukes from entering the circulatory system of abnormal hosts, such as humans.
2. Once an invader (e.g., parasite) breaches these physical or chemical barriers, it may be exposed to a **natural** or **innate immunity** or both. Natural immunity can result from a host's chemical, physiological, or nutritional condition that is incompatible with a parasite. An example of a host's incompatible chemical condition is the failure of *P. vivax* to infect erythrocytes lacking the Duffy blood group antigens (see p. 120). The Duffy antigen is a chemokine receptor that the parasite uses to gain entry into the erythrocyte. Physiological conditions such as pH and temperature can also have impact on parasitic infections. Other factors important in resistance are serum complement proteins, natural antibodies, and phagocytic cells. Natural antibodies are developed in response to antigens commonly encountered in nature (i.e., by previous exposure to bacterial antigens or other antigens possessing antigenic epitopes similar to those expressed on parasite surfaces). Complement proteins can react either directly with a parasite surface antigen or with bound natural antibodies to elicit their lytic action (see p. 27). As an example, *Leishmania enreitti* is killed by normal guinea pig serum due to natural antibodies against β-D-galactosyl determinants

on the parasite surface. In addition, many protozoan parasites and some intermediate stages of helminth parasites can be phagocytosed and destroyed by tissue macrophages.

3. **Acquired immunity** is conferred by a host's specific immune response developed as a result of a previous parasitic infection. This type of resistance, like natural immunity, is attracting a great deal of attention in the development of protective vaccines. **Premunition**, a form of acquired immunity, is resistance to reinfection dependent upon retention of the infectious agent. This type of resistance is seen in regions where malaria is endemic, and low parasitemia is maintained in the victim when the disease is left untreated.

Suggested Readings

Bloom, B. R. (1979). Games parasites play: how parasites evade immune surveillance. *Nature, 279*, 21–26.
Campbell, W. C. (1986). The chemotherapy of parasitic infections. *Journal of Parasitology, 72*, 45–61.
Cox, F. E. G., & Liew, E. Y. (1992). T-cell subsets and cytokines in parasitic infections. *Parasitology Today, 8*, 371–374.
Damian, R. (1964). Molecular mimicry: antigen sharing by parasite and host and its consequences. *American Naturalist, 98*, 129–149.
Elliot, D. E., Summers, R. W., & Weinstock, J. V. (2007). Helminths as governors of immune-mediated inflammation. *International Journal of Parasitology, 37*, 457–464.
Hall, R. (1994). Molecular mimicry. *Advances in Parasitology, 34*, 81–132.
Kasper, L. H., & Buzoni-Gatel, D. (1998). Some opportunistic parasitic infections in AIDS: candidiasis, pneumocystosis, cryptosporidiosis, toxoplasmosis. *Parasitology Today, 14*, 150–156.
Kawai, T., & Akira, S. (2010). The role of pattern-recognition receptors in innate immunity: update on Toll-like receptors. *Nature Immunology, 11*, 373–384.
Leder, P. (1982). The genetics of antibody diversity. *Scientific American, 246*, 102–115.
Pham, T. S., Mansfield, L. S., & Turiansky, G. W. (1997). Zoonoses in HIV-infected patients: risk factors and prevention. *The AIDS Reader, 7*, 41–52.
Trager, W. (1986). *Living together*. New York: Plenum Press.
Zambrano-Villa, S., Rosales-Borjas, D., Carrero, J. C., & Ortiz-Ortiz, L. (2002). How protozoan parasites evade the immune response. *Trends in Parasitology, 18*, 272–278.

CHAPTER

3

General Characteristics of the Euprotista (Protozoa)

Chapter 3 consists of an overview of the Protista. After a brief introduction to the group, the chapter moves into a discussion of the various means of locomotion displayed by these single-celled organisms. These include pseudopodia, cilia, and flagella. The information is augmented by electron micrographs and line drawings. The morphology of the locomotory organelles is supplemented with a clear, concise explanation of their functions. A similar format is followed with an explanation of those organelles involved with the physiological activities of the organisms. These include a discussion of the roles of mitochondria and Golgi complexes in energy metabolism as well as the role of lysosomes in the acquisition and utilization of food. Encystation and various types of reproduction such as the several types of asexual and sexual reproduction strategies are considered as they pertain to this diverse group of organisms. Again, the information is supplemented with ample graphic material. The chapter concludes with a presentation of a reduced classification scheme. Within this scheme, the relationship between the various protozoans that will be discussed in subsequent chapters is presented.

Despite immense diversity, organisms of the kingdom Euprotista share a number of characteristics. Perhaps the most distinctive of these characteristics are: (1) the organisms are all single celled or colonial and (2) they are eukaryotic. Beyond these broad criteria, however, sufficient diversity exists to cause frequent confusion in taxonomy. In this text, the designation Protista will be used for all members of the kingdom Euprotista. Protistans parasitic to humans are assigned to seven phyla, Retortamonada, Percolozoa, Parabasalia, Euglenozoa, Ciliophora, Apicomplexa, and Rhizopoda with the mode of locomotion being one criterion for classification. In the phylum Rhizopoda, for example, the amoebae move by means of pseudopodia, the flagellates (Euglenozoa, Retortamonada) use flagella as their primary locomotor apparatus, and the ciliates (Ciliophora) propel themselves by cilia. Members of the phyla Percolozoa and Parabasalia may display multiple locomotory organelles during different stages of their life cycles. Apicomplexans, primarily intracellular parasites, require no locomotor organelles except at some stages of their life cycles (e.g., flagellated gametes in some species). In some species, limited movement is accomplished by contraction of intracellular microfilaments.

Human Parasitology. http://dx.doi.org/10.1016/B978-0-12-813712-3.00003-5

Light microscopy studies have given rise to a terminology unique to the protistans in many ways. Increased use of electron microscopy, however, has shown that many of the structures typical of protistans are ubiquitous among eukaryotic cells. To avoid confusion from the use of two sets of terms, those used in cell biology will be used in this text with attempts to correlate them to their older counterparts.

Structurally, protistans are unicellular organisms, with each cell a self-sufficient unit capable of carrying out all the metabolic functions of which multicellular organisms are capable. Each protistan is surrounded by a unit membrane chemically similar to the plasma membrane common to all eukaryotic cells (i.e., a lipid bilayer associated with a variety of proteins). Among the Rhizopods, this tends to be a very thin, flexible layer often called the **plasmalemma**. On the other hand, a more rigid body wall, usually supported by microtubules and characteristic of some flagellates and most ciliates, termed a **pellicle**, results in a more constant and uniform shape than that of the more amorphic amoebae.

The cytoplasm is usually divided into two areas: the cortical ectoplasm and the medullary endoplasm. The consistency, extent, and appearance of these two zones differ among species. Typically, the ectoplasm is a **gel** containing the basal bodies of cilia or flagella, microfilaments, and, in some protistans, microtubules for rigidity and/or contractility. The endoplasm, or **sol**, is more fluid than ectoplasm and contains such organelles as nuclei, mitochondria, and vacuoles and vesicles of various types.

LOCOMOTOR ORGANELLES

Flagella

The Euglenozoa, commonly known as flagellates, include all protistans usually exhibiting in their **trophozoite** (motile) stage one or more flagella (Fig. 3.1). The ability to swim has facilitated the flagellates' adaptation to a variety of habitats in their hosts. Unlike amoebae, which require a substrate on which to move, flagellates thrive in a liquid medium and, thus, are well suited for survival in the blood, lymph, and cerebrospinal fluid of the host. Their elongate, torpedo-shaped form enables them to swim in the host's body fluids with little resistance, a further adaptation for life in a liquid medium.

While flagella occur in the developmental stages of some amoebae and in the microgametes of some members of the phylum Apicomplexa, they are always present in the trophozoite stage of flagellates. The number of flagella per organism varies widely according to species.

The single flagellum is a filamentous cytoplasmic projection. Electron microscopy reveals this projection to be a sheath consisting of a cytoplasmic matrix enclosed by a **plasma membrane** within which is embedded an axial filament or **axoneme** (Fig. 3.2). The axoneme, extending the length of the flagellum, is comprised of a series of regularly oriented microtubules arranged in a specific pattern consisting of two central microtubules surrounded by an outer circle of nine pairs of microtubules or doublets (Fig. 3.3). Each flagellum is anchored in the cytoplasm by a **basal body** (also called a **blepharoplast** or **kinetosome**) (Fig. 3.4A). Ultrastructural studies demonstrate that the basal body is morphologically identical to the centriole of the cell, and it is from the former organelle that the flagellum (or cilium) arises. The centriole and basal body consist of microtubules arranged in a circle of nine triplets, with two members of each triplet probably giving rise to and extending distally as one of the

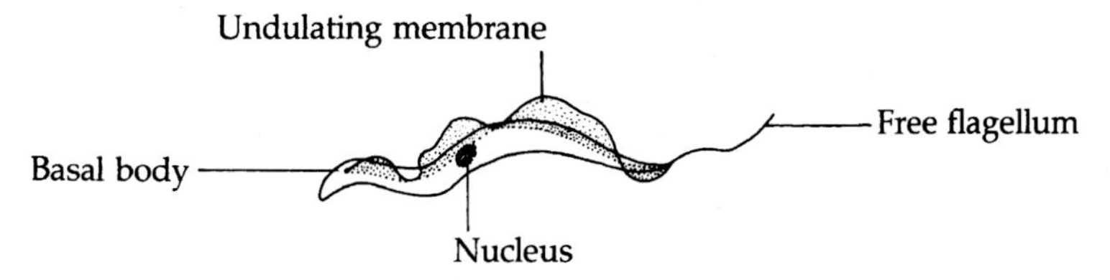

FIGURE 3.1 **A typical flagellate as viewed with a light microscope, showing the flagellum, nucleus, basal body, and undulating membrane.**

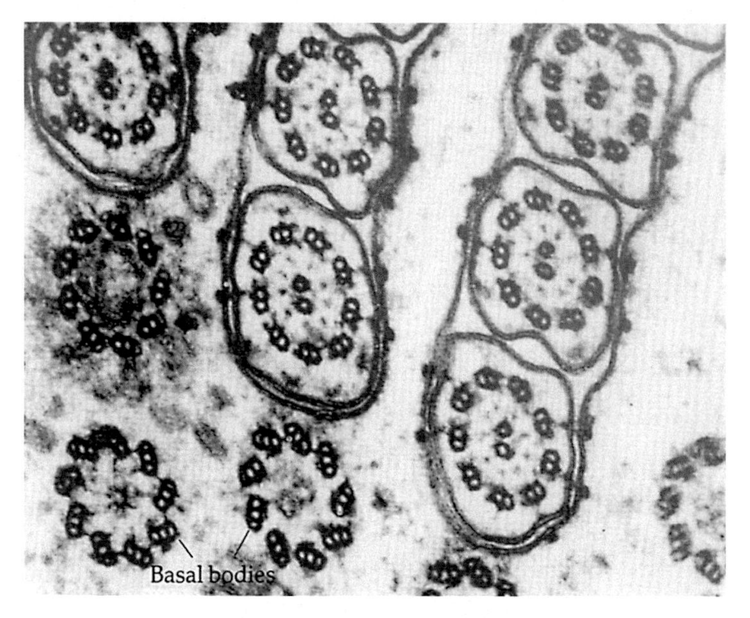

FIGURE 3.2 **Transmission electron micrograph of cross sections of flagella of *Pseudotrichonympha from the gut of a termite.*** Note that the two median microtubule ones are double. Sections through basal bodies are seen below the flagella.

peripheral doublets of the flagellum (Fig. 3.4B). The basal bodies may be found at the base of a depression of the plasma membrane termed the **flagella pocket**.

In most flagellated cells, the flagellum extends from the basal body to the exterior; however, one or more flagella may loop back in a complete reversal of their original direction. A flagellum of this type is known as a **recurrent flagellum** (Fig. 3.1). A recurrent flagellum may extend into a cytostome, where it aids in the procurement of food. Or, it may be attached to the plasma membrane by a series of desmosomes, in which case, during the beating process, it pulls the plasma membrane and a portion of the cytoplasm away from the body of the cell, producing an **undulating membrane** (Figs. 3.1 and 3.5).

Flagellar movement propels and directs the organism and at times assists in procuring food. The movement may also promote tactility and secretion of mating substances.

The so-called "beat" of a flagellum (or cilium) is actually the propagation of a series of wave-like bends along the length of the organelle. These surges are associated with the connections between outer microtubules and the inner sheath containing the two central microtubules.

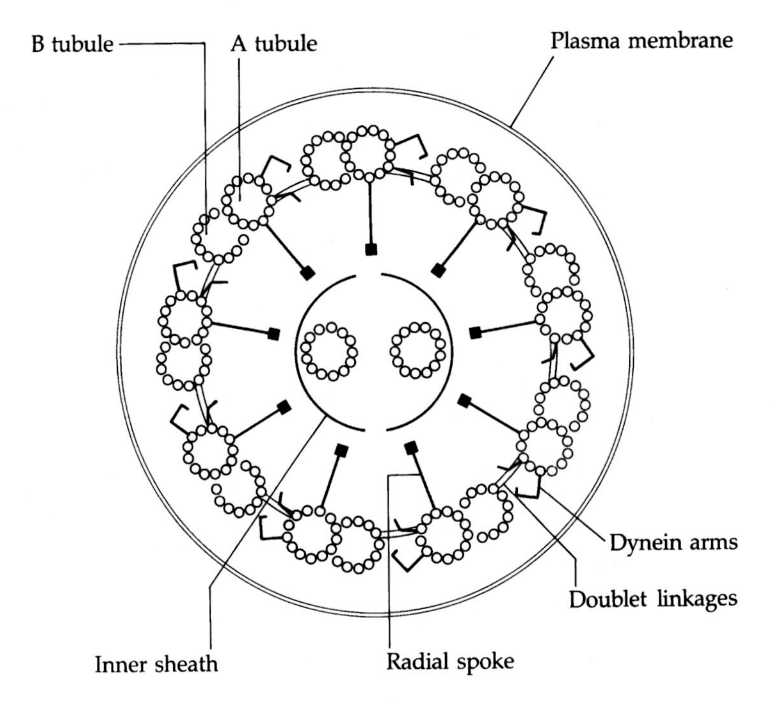

FIGURE 3.3 **Schematic representation of flagellar organization.**

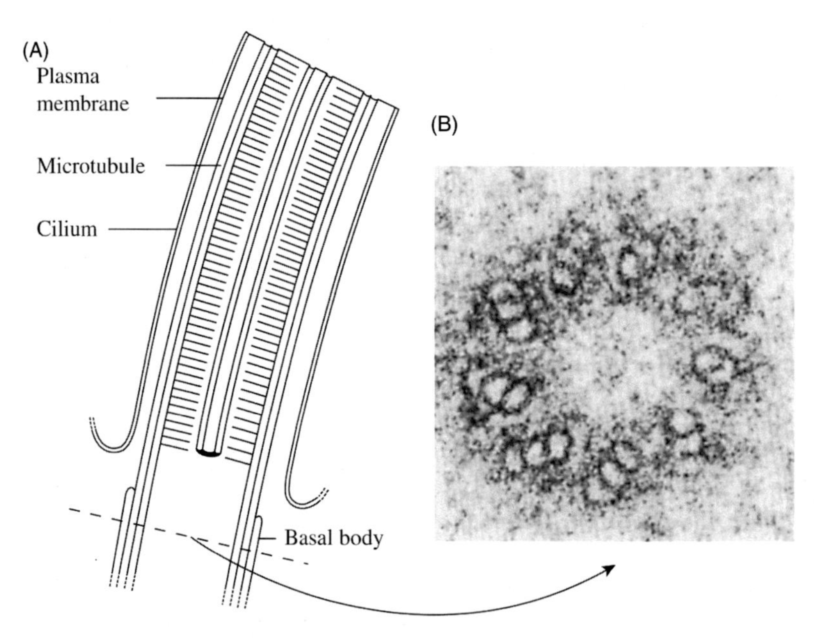

FIGURE 3.4 **General structure of a cilium or flagellum.** (A) Longitudinal aspect of the cilium and basal body. (B) Transmission electron micrograph of a cross section through the basal body.

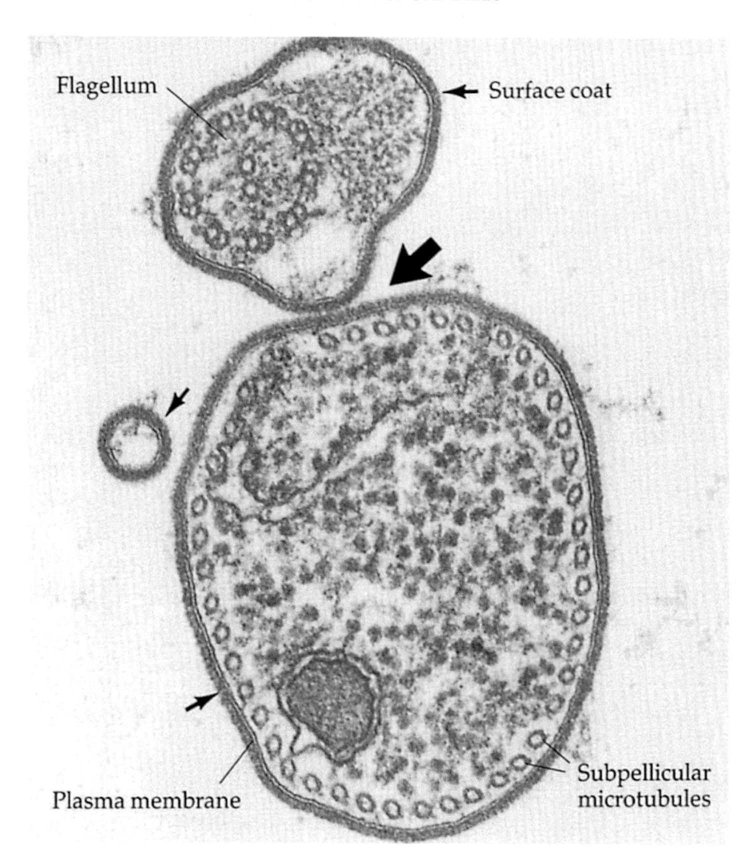

FIGURE 3.5 **Transmission electron micrograph of *Trypanosoma brucei rhodesiense* (slender form) in transverse section.** The surface coat (*small arrows*) is seen as a dense layer covering the plasma membrane of both the body (below) and the flagellum (above). Large arrowhead points to the desmosome-like attachment of the flagellar membrane to the surface membrane. On the left, also in transverse section, is one of the streamer-like extensions that may represent a mechanism for shedding the variable antigen coat.

Energy to fuel this movement comes from the ATPase activity of the **dynein** arms associated with one member of each of the outer doublets (Fig. 3.3). This energy allows the doublet microtubules to slide past each other; however, since this action is restricted by chemical cross-links between the two microtubules, a bending occurs, which becomes a wave as the cross-links are broken and reformed along the axoneme (Fig. 3.6). When this phenomenon occurs sequentially, a regular beat pattern emerges, resulting in directional movement of the cell.

Cilia

The fine structure of the axoneme of flagella and cilia is identical. Cilia may, therefore, be considered miniature flagella. In addition to noticeable differences in length, however, there are other fundamental differences between these two types of organelles, the most obvious of which is their number. Flagella usually number no more than 10 on a given cell surface, while there may be literally thousands of cilia on a surface. There are a few exceptions within the

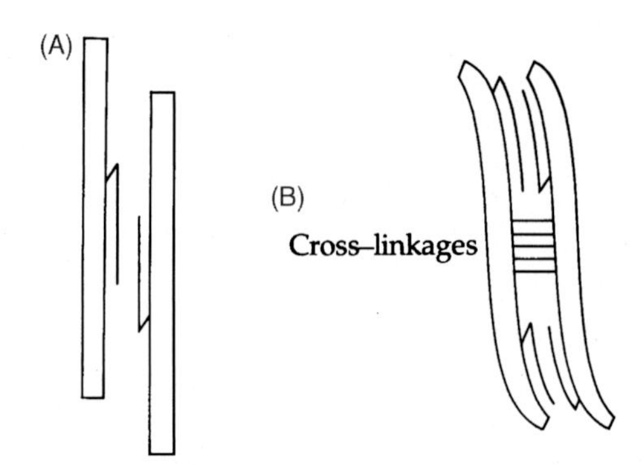

FIGURE 3.6 **Sliding filament mechanism of ciliary and flagellar bending.** (A) With no resistance to sliding the doublets slide past one another. (B) Cross-links cause a resistance to sliding at one region. Displacement of doublets in the nonlinked regions is accommodated by bending.

Euglenozoa group whose members possess large numbers of flagella. In ciliates, the numerous basal bodies are interconnected by a series of subpellicular **microfilaments** or **neurofibrils**, forming an **infraciliature** believed to be responsible for either coordinating the ciliary beat of the cell or providing support for the ciliary beat; however, the precise mechanics of this coordination is not presently understood.

Pseudopodia

Amoebae are usually capable of producing **pseudopodia**, which are used as locomotor and food-acquiring organelles. These transitory body extensions depend for their function on the association of actin and myosin. These two molecules function in the amoebae in a manner similar to their roles in the contraction of vertebrate muscle. Activated by ATP-derived energy and certain cations, such as calcium and magnesium, actin and myosin become intimately associated at the tip of the forming pseudopodium. This association produces a localized contractile response in the cytoplasm, whereupon the cytoplasm everts at the plasma membrane and moves posteriad in the cell, forming an outer zone of cytoplasm known as the **ectoplasm**. At the rear of the cell, actin and myosin become dissociated; the ectoplasm reverts to the relaxed state, becoming more fluid, and moves inward to form **endoplasm**. When the endoplasm streams forward under the pressure of the contractile ectoplasm, actin again becomes associated with myosin, producing anew the contractile state. The overall effect, then, is a recurrent outward and posteriad flow of ectoplasm away from the direction of movement and a concomitant movement of the endoplasm from the rear of the cell in the direction of the forming pseudopod (Fig. 3.7).

Morphologically, pseudopodia can be assigned to one of four types: **filopodia**, **lobopodia**, **rhizopodia**, and **axopodia**. Lobopodia (Fig. 3.8), the most common form among parasitic amoebae, are blunt and may be composed of both ectoplasm and endoplasm or of ectoplasm only. In most species, lobopodia form slowly. Observation of living specimens clearly shows the gradual flow of granular endoplasm, when present, into the broad projection. *Entamoeba histolytica*, an important parasite of the human intestine, is exceptional in that the lobopodia are produced abruptly and withdrawn almost as quickly. Another exception is observed among certain amoebae that appear to move on a substratum with no obvious cytoplasmic

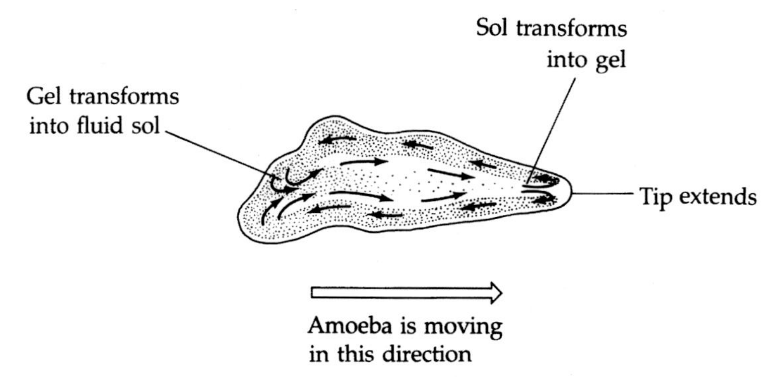

FIGURE 3.7 **One type of amoeboid motion.** The extension of a pseudopodium by an amoeba is accompanied by the constant flowing of cytoplasm in the direction of the extension. There appears to be a continual transformation of gelated ectoplasm to fluid endoplasm at the trailing end and a reverse transformation at the leading end.

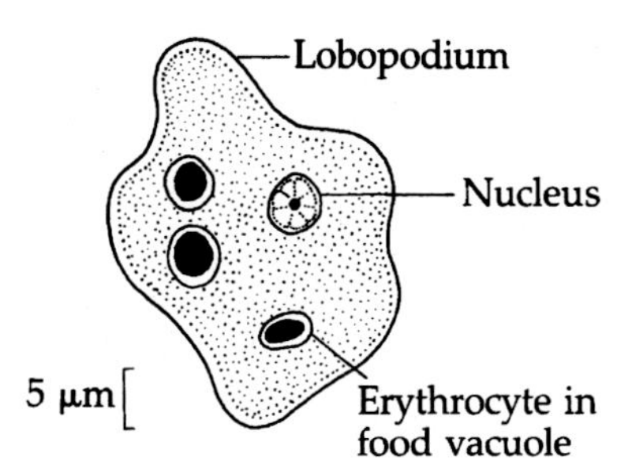

FIGURE 3.8 **Lobopodial type of pseudopodium in an Entamoeba histolytica trophozoite.**

protrusions. Such amoebae are termed **limax forms** after the slug, *Limax* spp., whose movement they appear to mimic.

Although the formation of pseudopodia by trophozoites usually is considered a distinguishing characteristic of amoebae, some flagellates also are capable of pseudopodial movement at some stage during their life and, conversely, some amoebic forms possess flagella during their developmental stages (e.g., *Naegleria*). As a general rule, however, among flagellates the principal means of locomotion is flagellar, while among amoebae it is pseudopodial.

Most amoebae cannot swim; pseudopodial locomotion requires a substrate on which these organisms can glide. Parasitic species, therefore, are commonly found in the alimentary tracts of their hosts, intimately associated with the epithelial lining.

OTHER ORGANELLES

Nucleus

Well-defined nuclei bounded by nuclear envelopes are characteristic of all protozoa. Some protozoa have a single nucleus; others have two or more, essentially identical, nuclei; still others, such as the ciliophorans, have two different types of nuclei: a **macronucleus** and one or more **micronuclei**. As the name indicates, the macronucleus of ciliophorans is larger than the micronuclei, and it is involved with trophic activities of the cell. Micronuclei, whether one or more, are usually involved in reproductive functions, both asexual and sexual.

In addition to size classification as macro- or micronuclei, nuclei may be defined morphologically as either **vesicular** or **compact**. This distinction is often used in identification of species infecting humans, since nuclei of such species are commonly vesicular.

Vesicular Nucleus

The nuclear envelope of a vesicular nucleus, although delicate in appearance, is visible by light microscopy. The term *vesicular* denotes numerous clear areas resulting from the irregular distribution of chromatin, creating the impression of many small sacs or vesicles. The chromatin areas may be concentrated peripherally or internally. The nucleoplasm contains one or more **endosomes** or **karyosomes**, which are DNA negative and probably analogous to metazoan nucleoli; unlike nucleoli, however, they do not disappear during mitosis.

Compact Nucleus

The compact nucleus contains a larger amount of more densely packed chromatin than does the vesicular nucleus. A nucleus of this type is generally larger than a vesicular nucleus and may vary in shape from round to ovate. Compact nuclei are found in the ciliophorans, where they are involved in the sexual process called **conjugation**. During this process, two cells join (or "conjugate"), one micronucleus of each cell undergoes meiosis, and some of the resulting haploid micronuclei are exchanged between the two individuals. Following meiosis and exchange, the micronuclei fuse, resulting in a genetically new set of micronuclei for each partner cell. During the meiotic division, exchange, and fusion of micronuclei, the macronuclei disappear and subsequently reform. They seem to serve as directors of the phenotypic expressions of the cells. Following conjugation, the two cells separate and then usually divide mitotically.

Mitochondria

Double membrane-bound mitochondria provide sites for aerobic metabolism and are similar in ultrastructure to those of most eukaryotes. A feature peculiar to euprotistan mitochondria is the tubular shape of the cristae (Fig. 3.9). Similar cristae are observed in mitochondria of multicellular eukaryotes but less consistently than in those of euprotists. The significance of this structural variation is presently unexplained. Mitochondria may also be present in some euprotistans as a single, branched organelle or may be numerous in others. A number of parasitic protistans (such as *E. histolytica*) possess no mitochondria, a deficiency associated with anaerobic metabolism.

Golgi Complex

The Golgi complex is a cytoplasmic organelle whose specific function in protozoans is essentially identical to that in other eukaryotes. The Golgi is the seat of glycosylation of a number of secretory products of the cell. It is in the cisternae of the Golgi complex, for

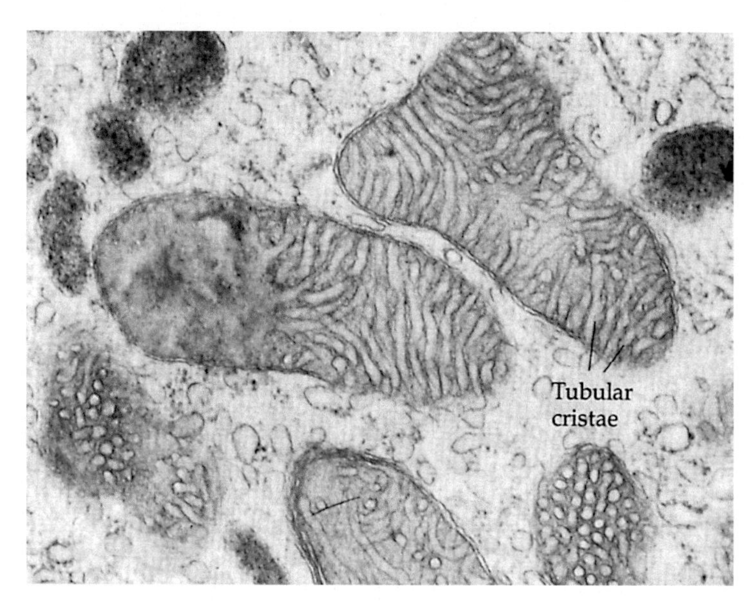

FIGURE 3.9 **Several mitochondria showing profiles of tubular cristae.**

instance, that the final carbohydrate moieties are added to the glycocalyx associated with the plasma membrane. The arrangement and number of Golgi complexes vary during the life cycle of many protozoans. Thus, cyst-forming protistans may lose their Golgi complexes during encystation, only to resynthesize them when they excyst. The so-called **parabasal body** of protozoans is homologous to the Golgi complex of other eukaryotic cells but with several morphological differences, the most notable of which is the frequent presence of a fibril, the **parabasal filament**, running from the cisternae of the Golgi complex to one or more basal bodies.

Lysosomes

The lysosome, an organelle ubiquitous among eukaryotes, is bounded by a single membrane enclosing various hydrolytic enzymes whose optimum activities occur in the acid pH range (Fig. 3.10). Such enzymes, therefore, are designated **acid hydrolases**. Lysosomes, with their battery of acid hydrolases, function in autophagy as well as in intracellular digestion of exogenous foodstuffs. In certain parasitic amoebae, for instance, **food vacuoles** are formed by the engulfment of exogenous food, including host cells. The food vacuoles then fuse with lysosomes, forming a **digestive vacuole**, and the lysosomal enzymes mix with and degrade the ingested material. Among a number of protistans, ingestion occurs at a specialized site on the plasma membrane called the **cytostome** (Fig. 3.11). After digestion, undigested residues are egested through the plasma membrane. Among amoebae such egestion may occur anywhere on the plasmalemma, while in ciliates there is a permanent pellicular site, the **cytopyge**, through which undigested residues are emitted.

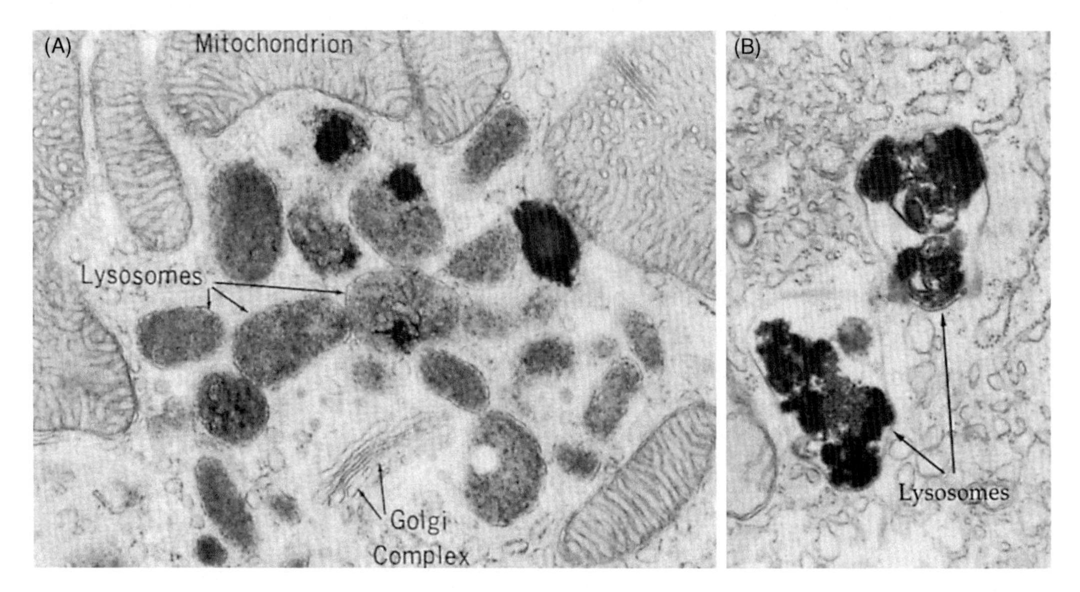

FIGURE 3.10 **Transmission electron micrographs of lysosomes.** (A) A cluster of lysosomes. (B) Two secondary lysosomes enclosing dense inclusions.

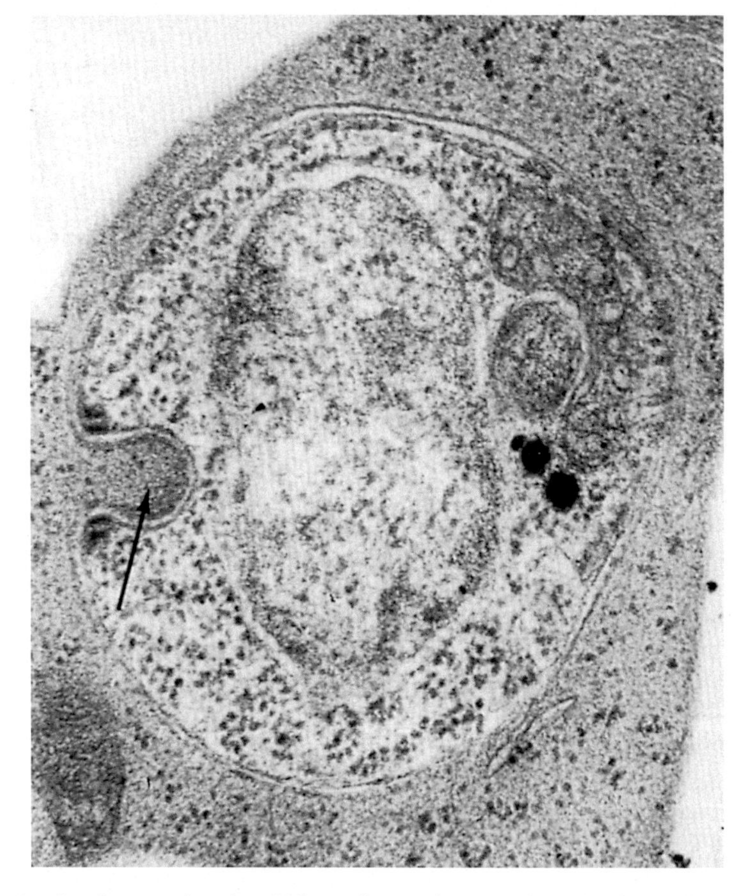

FIGURE 3.11 **A uninucleate trophozoite of *Plasmodium cathemerium* ingesting host cell, cytostome (arrow).**

Cytoplasmic Food Storage

Reserve food inclusions are seen in various species of parasitic protistans. The nature and amount of stored nutrients vary with the environment and the species involved. For example, glycogen and/or amylopectin are sometimes found in cysts of amoebae that inhabit the human intestine. This stored food, accumulated shortly before encystation and used during the nonfeeding stage, is usually completely exhausted by the time of excystation. In addition to these polysaccharides, lipid droplets and nucleic acid reserves may be present in both trophozoites and cysts.

Ribosomes

These organelles, the sites of cellular protein synthesis, are part of the organelle population of all eukaryotes and are abundant in those cells that actively synthesize protein for either secretion or internal use. They occur in association with the endoplasmic reticulum or free, either singly or in clusters known as **polyribosomes**, in the cytoplasm. During encystation, much of the ribosomal constituency of the cell is exhausted.

Costa, Axostyle, and Vacuoles

Associated with the basal bodies of many flagellates is a prominent, striated rod, the **costa**. This structure usually courses from one of the basal bodies along the base of the undulating membrane. It can best be described as a modified, **striated rootlet**. Rootlets are found at the bases of many cilia and flagella in other eukaryotes, penetrating deep into the cytoplasm where they are believed to serve as anchors. In addition to the costa, a sheath of microtubules in the shape of a tube, the **axostyle**, is observed in many flagellates. This structure extends posteriorly from the basal body and may appear to protrude through the plasma membrane. The function of this organelle is unknown, and it has no counterpart in other eukaryotic cells.

Some parasitic protozoans, such as the ciliate *Balantidium coli*, possess fluid-filled vesicles called **contractile vacuoles**. In protistans generally, these are considered osmoregulatory organelles, ridding the cell of excess water as well as some dissolved metabolic wastes. However, since most parasitic protistans, like their marine counterparts, are isoosmotic to their environment, contractile vacuoles are not commonly present.

ENCYSTATION

Many parasitic protistans are capable of encystation, during which the rounded cytoplasmic mass is surrounded by a rigid or semirigid cyst wall secreted by the organism. The cyst wall may be single- or multilayered. Cysts of parasitic protistans serve three primary functions: (1) protection against unfavorable external environmental conditions, (2) the site of morphogenesis and nuclear division, and (3) the means of transmission from one host to another.

Examples of the first function are seen in the human pathogens *E. histolytica*, an amoeba, and *Giardia lamblia*, a flagellate, which form cysts in the intestinal tract that are expelled in fecal material. Such cysts may remain viable for many weeks under normal conditions and for days at higher and lower temperatures or during periods of desiccation. Further, the cyst wall protects the ingested organism as it passes through the host's hostile gastric fluids.

Cysts of many parasitic protistans also serve as sites for nuclear and cytoplasmic reorganization and division. Shortly after the trophozoite encysts, cytoplasmic reorganization occurs, sometimes followed by nuclear divisions, after which the mature cyst may enclose from one (in the absence of nuclear division) to eight vesicular nuclei. In *E. histolytica*, for instance, two consecutive mitotic divisions result in four vesicular nuclei (see Fig. 4.2, p. 54). If mature cysts are reintroduced into a suitable host, excystation occurs, and the escaping motile trophozoite usually divides once more, resulting in eight small trophozoites produced from the single, tetranucleated mass of cytoplasm encased within the cyst wall. Following excystation, the newly excysted trophozoites begin a period of active feeding, followed by rapid growth and binary fission.

Finally, intestinal protistans are transmitted to a new host (or become reestablished in the same host) when that host swallows the cysts. Thus, the cysts serve as a vehicle for transmission.

The precise environmental conditions that trigger encystation are not totally defined. In many species, the process often occurs in response to a deficiency in the host of nutrients essential to the parasite. In addition, increased osmotic pressure, temperature changes, low pH, accumulation of waste products in the medium, and crowding all appear to stimulate encystment.

REPRODUCTION

Parasitic protistans most commonly reproduce by means of an asexual process called **fission**, a type of mitosis whereby each parent gives rise to two progeny. The plane of division is random among amoebae, usually longitudinal in flagellates, and transverse in ciliates. The sequence of division in a typical protistan is as follows: organelles, nucleus, and, finally, cytoplasm.

In apicomplexans, two types of **multiple fission** occur, **schizogony** (=**merogony**) and **endopolyogeny**. Both are characterized by rapid organelle and nuclear divisions, followed by cytokinesis. In schizogony, the multinucleated cell is called the **schizont** or **segmenter**. After cytoplasmic division, the nuclei, with their attendant cytoplasm, form separate organisms, **merozoites**, at the periphery of the mother cell, which usually break away from the aggregate to infect new host cells. Once a merozoite enters a new host cell, it may either enter another schizogonic cycle or undergo **gametogony**, becoming either a macro- or microgametocyte. **Syngamy**, the union of gametes derived from the gametocytes, initiates the sexual cycle. The resulting zygote undergoes **sporogony**, which produces **sporozoites**. The organisms that cause malaria are apicomplexans capable of both schizogonic (asexual) and sporogonic (sexual) reproduction. In fact, many apicomplexans are considered unique among protistans in exhibiting alternation of generations, a characteristic more commonly associated with plants and some invertebrate animals (e.g., cnidarians). The other type of asexual reproduction, **endopolyogeny**, is sometimes considered a form of internal budding. It differs from schizogony only in the location of the daughter cells relative to the mother cell. In endopolyogeny, the daughter cells form in the center of the mother cell rather than at the periphery. The form of endopolyogeny in which the mother cell produces only two daughter cells is termed **endodyogeny**.

Conjugation, the specialized sexual mechanism in the ciliates, has already been discussed; it is distinguishable from syngamy in that conjugation involves nuclear exchange and union, whereas syngamy involves the union of entire cells (e.g., gametes).

CLASSIFICATION OF THE PROTOZOA*

The following system of classification follows the more traditional morphological pattern and is intended primarily as a means to identify those parasitic members discussed in this text while noting some shared characteristics. For those readers interested in a more recent treatment of the classification of the protista, reference is made to the following publication.

Adl, S.M., Simpson, A.G.B., Lane, C.E., Lukes, J., Bass, D., Bowser, S.S., et al. The revised classification of eukaryotes. *Journal of Eukaryotic Microbiology*, 59, 2012, 429–514.

Kingdom Euprotista

Unicellular, plasmodial or colonial protists. Golgi and peroxisomes present. Free living and symbiotic. Mitochondria usually with tubular cristae.

Phylum Euglenozoa

Chloroplasts absent; one to many flagella; amoeboid forms, with or without flagella, in some groups; sexuality known in few groups; a polyphylectic group. One or two flagella arising from depression; flagella typically equipped with paraxial rod in addition to axoneme; single mitochondrion (nonfunctional in some forms) extending length of body as a single tube, hoop, or network of branching tubes, usually containing conspicuous Feulgen-positive (DNA containing) kinetoplast located near flagellar kinetosomes; Golgi complex typically in region of flagellar depression, not connected to kinetosomes and flagella; parasitic (majority of species) and free living. Single flagellum either free or attached to body by undulating membrane; kinetoplast relatively small and compact; parasitic (genera mentioned in text: *Leishmania*, *Trypanosoma*).

Phylum Retortamonada

Two to four flagella, one directed posteriorly and associated with ventrally located cytostomal area bordered by fibril; mitochondria and Golgi apparatus absent; cysts present; parasitic. One or two karyomastigonts; genera with two karyomastigonts exhibit two-fold rotational symmetry or, in one genus, primarily mirror symmetry; individual mastigonts with one to four flagella, typically one recurrent and associated with cytostome, or, in more advanced genera, with organelles forming cell axis; mitochondria and Golgi apparatus absent; intra-

*Only those taxa that include parasitic species discussed in this text are defined.

nuclear division spindle; cysts present; free living or parasitic. Two karyomastigonts; body with two-fold rotational symmetry, or bilateral symmetry in one genus; with four flagella, one recurrent; with variety of microtubular bands; cysts present; free living or parasitic (genera mentioned in text: *Chilomastix, Retortamonas, Giardia*).

Phylum Parabasalia

Typically, karyomastigonts with four to six flagella, but one genus exhibiting only a single flagellum and another none; karyomastigonts and akaryomastigonts in one family with permanent polymonad organization; in mastigont(s) of typical genera, one flagellum recurrent, free, or with proximal or entire length adherent to body surface; undulating membrane, if present, associated with adherent segment of recurrent flagellum; pelta and noncontractile axostyle in each mastigont, except in one genus; hydrogenosomes present; true cysts infrequent, known in very few species; all or nearly all parasitic (genera mentioned in text: *Dientamoeba, Trichomonas, Pentatrichomonas*).

Phylum Rhizopoda

Locomotion by lobopodia, filopodia, or reticulopodia, or by protoplasmic flow without production of discrete pseudopodia. More or less finely tipped, sometimes filiform, often furcate hyaline pseudopodia produced from a broad hyaline lobe; not regularly discoid; cysts usually formed; nuclear division mesomitotic or metamitotic (genera mentioned in text: *Hartmannella, Entamoeba, Endolimax, Iodamoeba*).

Phylum Percolozoa

Body with shape of monopodial cylinder, usually movement accomplished by means of more or less eruptive, hyaline, hemispheric bulges; typically uninucleate, nuclear division promitotic; temporary flagellated stages in most species (genus mentioned in text: *Acanthamoeba, Balamuthia, Naegleria*).

Phylum Apicomplexa

Apical complex (visible with electron microscope), generally consisting of polar ring(s), rhoptries, micronemes, conoid, and subpellicular microtubules present at some stage; micropore(s) generally present at some stage; cilia absent; sexuality by syngamy; all species parasitic. Conoid, if present, forming complete cone; reproduction generally both sexual and asexual; oocysts generally containing infective sporozoites resulting from sporogony; locomotion of mature organisms by body flexion, gliding, or undulation of longitudinal ridges; flagella present only in microgametes of some groups; pseudopods ordinarily absent, but if present used for feeding, not locomotion (genera mentioned in text: *Toxoplasma, Isospora, Plasmodium, Babesia, Enterocytozoon*).

Phylum Ciliophora

Simple cilia or compound ciliary organelles typical in at least one stage of life cycle; with subpellicular infraciliature present even when cilia absent; two types of nuclei, with rare exception; binary fission transverse, but budding and multiple fission also occur; sexuality involving conjugation, autogamy, and cytogamy; contractile vacuole typically present; most species free living, but many commensal, some parasitic, and a large number found as phoronts on a variety of hosts (genus mentioned in text: *Balantidium*).

Suggested Readings

Bailey, G. B., Day, D. B., & McCoomer, N. E. (1992). *Entamoeba* motility: dynamics of cytoplasmic streaming, locomotion and translocation of surface-bound particles, and organization of the actin cytoskeleton in *Entamoeba invadens*. *Journal of Protozoology, 39*, 267–272.

Brenier-Pinchart, M. -P., Pelloux, H., Derouich-Guergour, D., & Ambroise-Thomas, P. (2001). Chemokines in host-protozoan-parasite interactions. *Trends in Parasitology, 17*, 292–296.

Corliss, J. O. (1991). Introduction to the protozoa. In F. W. Harrison, & J. O. Corliss (Eds.), *Microscopic anatomy of invertebrates* (pp. 1–12). (1). New York: Wiley-Liss.

Kirsch, J., & Schmidt, H. J. (2001). Genetically controlled expression of surface variant antigens in free-living protozoa. *Journal of Membrane Biology, 180*, 101–109.

Sacks, D., & Sher, A. (2002). Evasion of innate immunity by parasitic protozoa. *Nature Immunology, 3*, 1041–1047.

Satir, P. (1974). How cilia move. *Scientific American, 231*, 45–52.

Stossel, T. P. (1994). The machinery of cell crawling. *Scientific American, 71*, 54–63.

CHAPTER

4

Visceral Protista I

Rhizopods (Amoebae) and Ciliophorans

Chapter 4 focuses mostly on *Entamoeba histolytica*, an anaerobic parasitic protistan part of the genus *Entamoeba*. Predominantly infecting humans and other primates, *E. histolytica* is estimated to infect about 50 million people worldwide. Previously, it was thought that 10% of the world population was infected, but these figures predate the recognition that at least 90% of these infections were due to a second species, the nonpathogenic *Entamoeba dispar*. Mammals such as dogs and cats can become infected transiently but are not thought to contribute significantly to transmission. Only one ciliophoran infects humans, *Balantidium coli*. This parasitic species of ciliate protozoan causes the disease balantidiasis. Balantidiasis is a zoonotic disease and is acquired by humans via the fecal-oral route from the normal host, the pig, where it is asymptomatic. Contaminated water is the most common mechanism of transmission. Several pathogenic free-living amoebae are also discussed including *Acanthamoeba* and *Balamuthia*.

AMOEBAE

This category of parasites consists of members of the phylum Rhizopoda. All species in this phylum that are parasitic in humans are, for the most part, nonpathogenic or produce only minor diseases with the exception of *E. histolytica*; however, they require special emphasis in order to distinguish them from the potentially highly pathogenic *E. histolytica* (Table 4.1). Such attention is amply justified since *E. histolytica* can produce extreme illness and even death. Furthermore, since side effects from chemotherapy may be pronounced, it is of great importance that diagnosis of the condition be precise and accurate in order to assure treatment only when absolutely necessary, not merely to eliminate a protistan that resembles *E. histolytica*.

At least eight species of amoebae belonging to three genera are known to parasitize humans. These are *E. histolytica*, *Entamoeba hartmanni*, *E. dispar*, *Entamoeba coli*, *Entamoeba polecki*, *Entamoeba gingivalis*, *Endolimax nana*, and *Iodamoeba bütschlii*. All inhabit the large intestine except *E. gingivalis*, which is found in the mouth. In addition, amoebae belonging to at least three genera of the phylum Percolozoa, *Naegleria*, *Acanthamoeba*, and *Balamuthia*, normally free-living, have been shown on occasion to parasitize humans by accident.

Human Parasitology. http://dx.doi.org/10.1016/B978-0-12-813712-3.00004-7

TABLE 4.1 Some Important Enteric Amoebae of Humans

	Entamoeba histolytica	***Entamoeba coli***	***Endolimax nana***	***Iodamoeba bütschlii***	***Entamoeba gingivalis***	***Dientamoeba fragilis***[a]
TROPHOZOITE						
Size (range)	25 μm (15–60)	25 μm (15–40)	9 μm (5–14)	10 μm (6–25)	15 μm (5–35)	10 μm (6–25)
Motility	Active, directional, progressive	Sluggish, nondirectional, nonprogressive	Similar to *E. coli*	Similar to *E. coli*	Moderately active, progressive	Active, progressive
Pseudopodia	Fingerlike, explosive	Short, blunt, broad, slow	Similar to *E. coli*	Similar to *E. coli*	Blunt, rapidly formed	Thin, leaflike, multiple, rapidly formed
Nucleus (stained)	Delicate envelope and chromatin; central endosome	Coarse envelope and chromatin, eccentric endosome	Large endosome; no peripheral chromatin	Large endosome surrounded by granules	Similar to *E. histolytica*	Similar to *E. histolytica*; endosome divided into 4–6 granules
CYST						
Size (range)	12 μm (10–20)	17 μm (10–33)	9 μm (5–14)	10 μm (5–18)	None	None
INCLUSIONS						
Glycogen	Diffuse	Ill defined	Absent	Large mass		
Chromatoidal bars	In young cysts; rounded ends	In young cysts; splintered ends	Occasionally as granules	Usually absent		
Number of nuclei	1–4	1–8	1–4	1		

[a]*Mastigophora.*

Entamoeba histolytica

Undoubtedly, the best known species of amoebae parasitizing humans is *E. histolytica* (Fig. 4.1), the causative agent of amoebic dysentery or amoebiasis. First discovered in Russia by Losch in 1875, it is global in distribution although its prevalence varies markedly from one area to another. For example, it has been reported to infect 85% of the population of Merida, Yucatan, Mexico, but no more than an average 13.6% (range, 0.8%–38%) of several populations surveyed in the United States. It is, nevertheless, important to remember that amoebiasis is not restricted to the tropics and subtropics; it is found also in temperate and even in Arctic and Antarctic zones. The usual mode of infection—ingestion of cysts from contaminated hands, food, or water—causes the incidence to increase considerably in densely populated areas where contact with infected individuals is more likely. Children's homes and mental health institutions often produce conditions favorable to transmission of these organisms.

Life Cycle

The uninucleate trophozoite of *E. histolytica* inhabits the colon and rectum and, at times, the lower end of the small intestine of humans and other primates (Fig. 4.2). The motile trophozoite

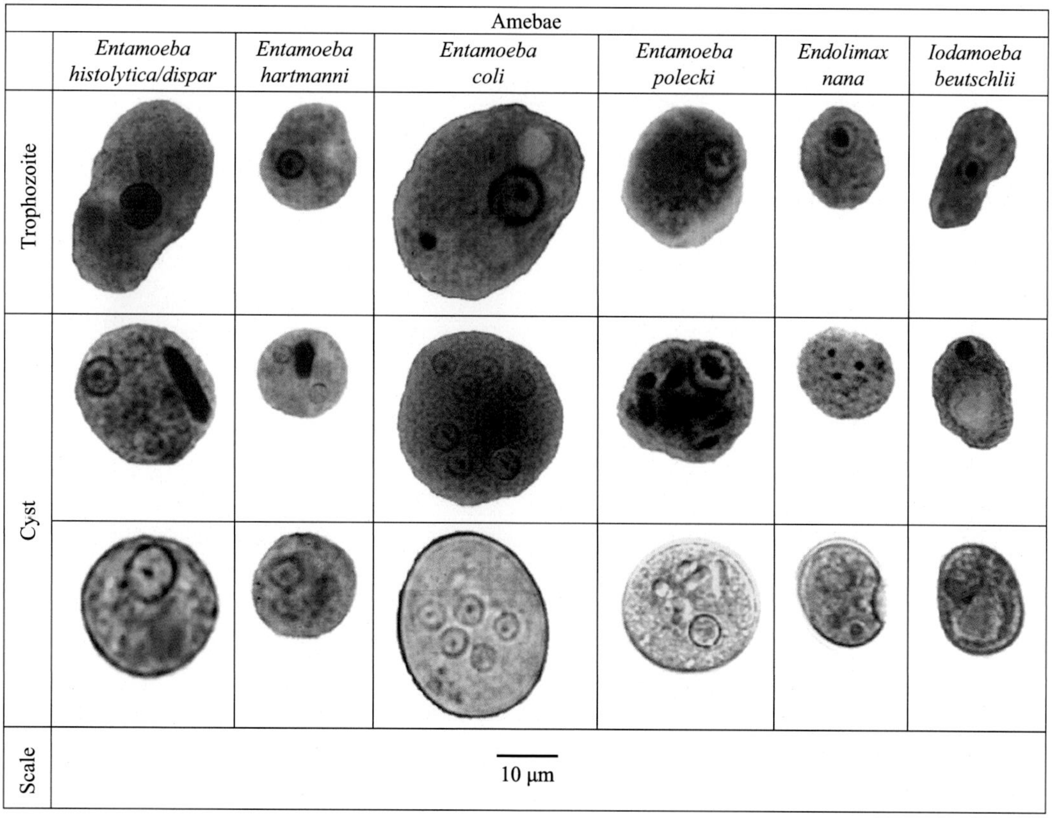

	Amebae					
	Entamoeba histolytica/dispar	*Entamoeba hartmanni*	*Entamoeba coli*	*Entamoeba polecki*	*Endolimax nana*	*Iodamoeba beutschlii*
Trophozoite						
Cyst						
Scale	10 μm					

FIGURE 4.1 **Photomicrographs of life cycle stages of various parasitic amoebae.**

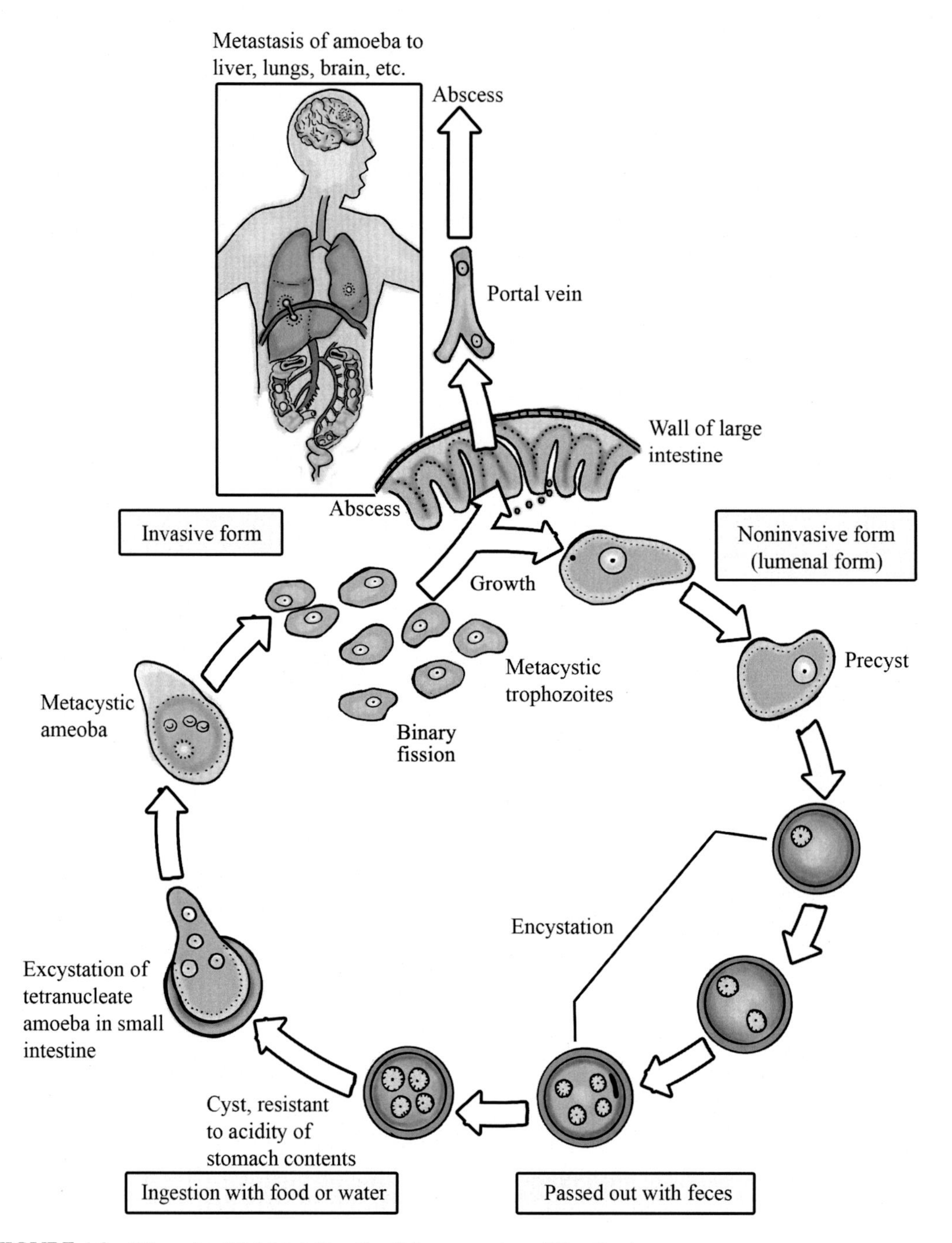

FIGURE 4.2 **Life cycle of *E. histolytica*.** *Credit: Image courtesy of Gino Barzizza.*

measures an average 25 μm in diameter (range, 15–60 μm) and is typically monopodial, producing one large, fingerlike pseudopodium at a time. The single pseudopodium erupts and is withdrawn so rapidly that, in prepared slides, trophozoites with pseudopodia extended are rarely seen. The cytoplasm is differentiated into two zones: a clear, refractile ectoplasm and a finely granular endoplasm in which food vacuoles occur. Such vacuoles may contain host erythrocytes, leukocytes, and epithelial cells, as well as bacteria and other intestinal material. Trophozoites proliferate mitotically (binary fission) within the host's gut.

The nucleus is of special importance in differentiating *E. histolytica* from most of the other intestinal amoebae. In saline preparations, the nucleus has a barely discernible nuclear envelope. However, in stained preparations, the vesicular nucleus is clearly visible. Ideally, it has a well-defined envelope, lined on the inner surface with fine peripheral chromatin granules and a minute, centrally located endosome. Unfortunately, this "ideal" morphology is not confined to *E. histolytica*. Often, other species of *Entamoeba*, notably *E. dispar*, show similar nuclear morphologies.

Under certain adverse environmental and/or physiological circumstances, trophozoites assume precystic characteristics by becoming more spherical and, as food vacuoles are extruded, shrinking in size. Pseudopodia, if formed, are sluggishly extended, and there appears to be no progressive movement. Encystation begins with the secretion by the precyst trophozoite of a thin, surrounding hyaline membrane to form a cyst wall. At this stage, the cyst is usually spherical, an average 12 μm in diameter (range, 10–20 μm), with a single nucleus. At times, glycogen masses and chromatoidal bars may be observed. The latter structures are considered to be deposits of nucleoproteins such as RNA and may vary in shape but always have smoothly rounded ends in *E. histolytica* (Fig. 4.1). This characteristic distinguishes *E. histolytica* cysts from those of *E. coli*, in which the chromatoidal bars have jagged or splintered ends. The nucleus undergoes two mitotic divisions to produce four vesicular nuclei in the mature cyst of *E. histolytica* (Fig. 4.1). Such cysts represent the infective form and pass out of the host in feces, after which the glycogen and chromatoidal substance are slowly metabolized and disappear.

Cysts of *E. histolytica* are highly resistant to desiccation and even to certain chemicals (e.g., chlorinated compounds, fluorides). Cysts in water can survive for a month, while those in feces on dry land can survive for more than 12 days; they tolerate temperatures up to a thermal death point of 50°C.

When food or water contaminated with *E. histolytica* cysts is ingested by a host, the cysts pass through the stomach (protected from the harsh environment by the cyst wall) to the ileum, where excystation occurs. The neutral or slightly alkaline environment afforded by the small intestine is apparently requisite for this phenomenon. However, in vitro studies suggest that excystation does not occur immediately; cysts placed in fresh culture medium at body temperature require 5–6 h for excystation. Upon excystation, a single tetranucleate organism immediately undergoes mitosis, giving rise to eight small, metacystic trophozoites, which pass downward to the large intestine where they feed, grow, and reproduce. Reduction in intestinal peristalsis often allows the trophozoites to become established in the cecal area of the colon. The greater the number of organisms, the greater the likelihood that they will attain a foothold in the intestinal epithelium. Conversely, greater intestinal motility and/or large volumes of ingested food reduce the potential for establishment of the amoebae.

Multiplication of this species is thus seen to occur at two stages during the life cycle: by binary fission in the intestine-dwelling mature trophozoite stage and by nuclear division followed by binary fission in the metacystic stage.

Epidemiology and Pathology

E. histolytica is cosmopolitan, with an estimated incidence of human infection exceeding 500 million cases. However, this figure is possibly misleading; up to 90% of reported human infections may be due to intestinal colonization with the morphologically identical *E. dispar*. *E. histolytica*, therefore, is probably responsible for only 50 million cases worldwide, with *E. dispar* accounting for the remainder. Although the prevalence of human infection varies widely, certain groups appear more susceptible than others (e.g., patients in mental institutions). Transmission depends upon ingestion of contaminated food and/or drinking water. Areas with low standards of sanitation and those where night soil is used as fertilizer display the highest prevalence of human infections. The main source of infection is the cyst-passing, asymptomatic carrier, or chronic patient. The infection in these individuals is called **luminal amoebiasis**. Acutely ill patients, those with **invasive amoebiasis**, are not significant transmitters since they pass the noninfective trophozoite (noninfective because, unlike the cyst, it is unable to survive outside the intestinal environment) in their diarrheic feces. In addition, flies and cockroaches have been implicated as mechanical vectors in the spread of *E. histolytica* and other amoebae since cysts can survive for lengthy periods in their digestive tracts, later to be regurgitated or passed out in feces upon human food.

The explanation for the apparent nonpathogenicity in certain human hosts remains elusive. In humans living in temperate zones, the organism often produces the nonpathogenic, luminal form of the disease, while in the tropics and subtropics the invasive form is more common.

Under pathogenic conditions, the food vacuoles of *E. histolytica* trophozoites characteristically contain host erythrocytes along with leukocytes and epithelial cells. At the height of its pathogenicity, the trophozoite attaches to the mucosal surface of the intestine. This mucosal surface, whose major component is MUC2, presents a physical barrier between the intestinal contents and the underlying epithelium. The attachment process is fostered by a cell-to-cell recognition via lectins between trophozoite surface proteins and complementary sugar residues on the mucosal epithelial surface. Within the trophozoite is a complement of granules containing a battery of cysteine polypeptides termed **amoebapores** of which three have been isolated (amoebapores A, B, and C). Amoebapore A is involved with liver abcesses (see below), while amoebapores B and C are the most efficient in lysing erythrocyte and bacterial membranes. This lytic process enables the organism to invade the submucosal tissue causing amoebic colitis. In the infected individual who develops dysentery, the mucosal ulceration may penetrate deeper into the intestinal tissue, causing vast areas of tissue to be destroyed. The overlying mucosal epithelium may be sloughed off, exposing these necrotic areas (Fig. 4.3). This destructive process is usually followed by a regenerative period, resulting in a thickening of the intestinal wall as a result of the deposition of fibrous connective tissue. Trophozoites also may be carried to the liver, chiefly by the hepatic portal system, producing amoebic liver abscess and even amoebic hepatitis. Genomic studies have shown that the expression of amoebapore A by *E. histolytica* trophozoites is required for amoebic liver abscess. Trophozoites that do not express amoebapore A, however, can still cause colitis.

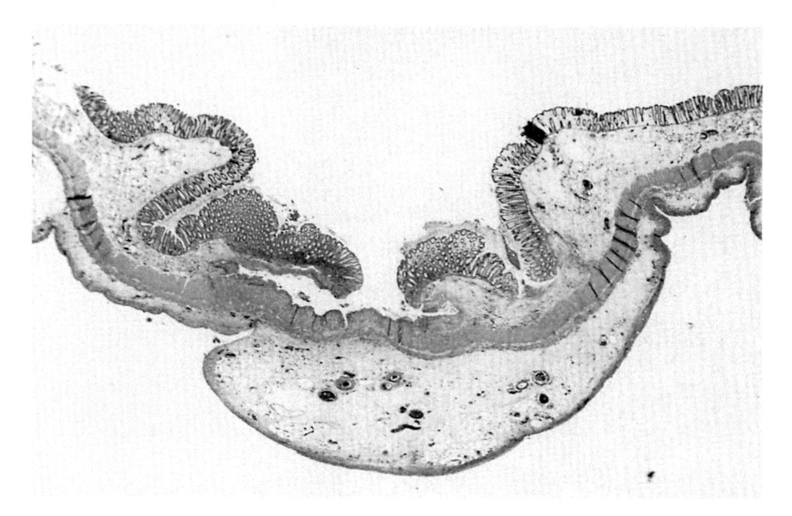

FIGURE 4.3 **Section of human colon showing chronic amoebic ulcer.**

The first sign of hepatic involvement is the formation of an early hepatic abscess containing a matrix of necrotic hepatic cells, which eventually become liquefied. Hepatic abscesses may be single or multiple (Fig. 4.4). While the liver appears to be the visceral organ most often affected (about 5% of all cases), other organs such as the lungs, heart, brain, spleen, gonads, and skin may also be invaded, resulting in secondary amoebiasis. The reaction in the liver is due not only to the trophozoite and its secretions but also to emanation of toxic material due to the ulcerative changes in the intestine.

Second only to the liver in frequency as an extraintestinal site are the lungs. Pulmonary amoebiasis is relatively rare, however, and, when seen, is probably a direct result of hepatic infection. Unlike most amoebic abscesses, which are commonly bacteriologically sterile, the pulmonary abscess is often vulnerable to secondary bacterial infections.

Symptomatology and Diagnosis

Among victims of amoebiasis, symptoms vary widely; in some individuals even the more highly pathogenic, tropical forms can be symptomless. Pathogenic responses that do occur are highly variable, the severity depending upon the location and intensity of the infection.

Invasive amoebiasis may be manifested in two ways; **amoebic dysentery** (=**acute intestinal amoebiasis**) and **chronic amoebiasis**. In the former type, severe diarrhea (i.e., blood and mucus in liquid feces) usually develops after an incubation period of 1–4 weeks and is commonly accompanied by a fever of 100–102 °F. Diagnosis requires differentiation of amoebic dysentery from other types of dysentery and, ultimately, identification of the parasite; one diagnostic criterion is the presence of characteristic trophozoites and/or cysts in the stools. In the chronic form, on the other hand, there may be continuous attacks of diarrhea or recurrent attacks with intervening periods of milder intestinal problems. While hepatic amoebiasis is the most serious consequence of either form, intestinal abscesses may perforate the abdominal wall or extend through the diaphragm into the lungs. Any of these manifestations may be fatal. *E. histolytica* has been incriminated in a few cases of cerebral,

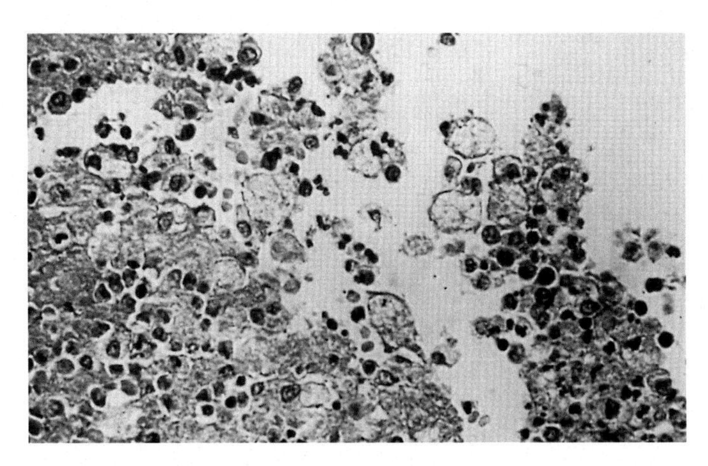

FIGURE 4.4 **Abscess in human liver due to *Entamoeba histolytica*.**

optic, and facial infections exhibiting often severely damaging or even fatal consequences. For instance, although cerebral amoebiasis occurs in less than 0.1% of patients, its onset is abrupt and usually fatal. In such cases, transmission of the organism is accomplished through direct, fecal contamination of the skin, eye socket, etc., rather than metastasis from the intestine or liver.

Laboratory diagnosis of amoebiasis depends upon identification of the cysts and/or trophozoites of *E. histolytica*. Examination of stools for intestinal forms requires both direct smears and concentration procedures such as zinc sulfate flotation or formalin-ether. Such examinations should be performed for 3 consecutive days unless positive results are obtained in a shorter period. The chance of finding cysts in infected persons almost triples after 3 days. Combining direct smears with concentration methods of detection more than doubles the diagnostic effectiveness of a single examination. At the present time, serious consideration is being given to the development of cost-effective means to differentiate *E. histolytica* infections from *E. dispar* infections in clinical laboratories. *E. histolytica* and other parasitic amoebae (e.g., *E. dispar*) can be effectively differentiated using the polymerase chain reaction method. This method is highly effective using species- and strain-specific primers. Monoclonal antibody-based enzyme-linked immunosorbent assay (ELISA) is currently a popular alternative method. Different diagnostic procedures are required for patients with extraintestinal amoebiasis since stool specimens may not disclose the presence of the parasite. Trophozoites may be detected in sputum samples and tissue biopsies. Serological tests and X-ray scans may prove useful in revealing abscesses of the liver. It should be reemphasized at this point that positive identification of the parasite is requisite for an accurate diagnosis before chemotherapy is undertaken.

Chemotherapy

Appropriate chemotherapy should be employed to destroy trophozoites, relieve symptoms, and control secondary bacterial infections. The drug of choice for the entire spectrum of symptoms is either metronidazole or tinidazole. Some shortcomings are associated with metronidazole treatment. The drug is reported to be mutagenic in bacteria and a possible carcinogen when administered to mice in high doses. As an alternative treatment, a combination of tetracycline and diiodohydroxyquin has proven beneficial.

Common sense dictates complete bed rest in cases of severe diarrhea accompanied by fever, regardless of the cause. In addition, a bland diet, low in carbohydrates such as sugar and high in liquids and proteins, is recommended. In symptomless carriers, it is essential that the trophozoites be destroyed since they are the precursors of cysts that pass out of the host. For such patients, either iodoquinol, diloxanide, or paromomycin are the drugs of choice. In cases of iodine sensitivity, iodoquinol is contraindicated. Metronidazole is contraindicated for pregnant women, especially in their first trimester, since it is a known carcinogen and mutagen in rodents and bacteria, respectively.

To combat secondary bacterial infections, antibiotics such as tetracycline are used in combination with either metronidazole or tinidazole. Hepatic amoebiasis also responds well to metronidazole, although the treatment is not totally effective. Emetine or dehydroemetine is used in instances when metronidazole treatment has been unsuccessful. Chloroquine can be used where there are contraindications for either of these two drugs if it is kept in mind that chloroquine has no effect upon trophozoites in the intestine. Ornidazole, a drug closely akin to metronidazole, is reported to have cured hepatic amoebiasis with a single dose.

Physiology

Knowledge of the physiology of *E. histolytica* is fragmentary. When the metabolism of the organism is better understood, perhaps more effective chemotherapeutic methods will be developed. Since it grows best in an oxygen-free atmosphere, *E. histolytica* was once considered an obligate anaerobe. However, it has been shown that the organism can use oxygen in low concentrations even though it does not possess the usual organelles (i.e., mitochondria) or metabolic pathways, such as the cytochrome system or a functional tricarboxylic acid cycle, normally associated with oxygen utilization. In vitro, an oxygen concentration of 10% or higher is lethal to the organism, while carbon dioxide is required for growth, a characteristic *E. histolytica* shares with most intestine-dwelling organisms. Glucose and galactose are the major carbohydrates used by the organisms, from which they produce ethanol, acetate, and carbon dioxide. In the presence of sublethal concentrations of oxygen, the same end products are produced but in different proportions.

Host Immune Response

E. histolytica infection in the colon can initiate an intense post inflammatory response, both acute and chronic. In acute amoebiasis, an increase in the Th2 response is indicated by an elevation of IL-4. In chronic amoebiasis, patients exhibit little or no change in their $CD4^+$/ $CD8^+$ ratio. The Th1 and Th2 responses in these patients remain unchanged as well. In the few asymptomatic patients investigated, high levels of IFN-γ were observed, indicating a bias toward a type 1 immune reaction.

There are reports that some of *E. histolytica*'s excretory and secretory products (e.g., proteins) may alter macrophage metabolism and thus reduce the efficacy of the patient's immune response, especially in the case of invasive amoebiasis. For instance, it has been shown that *E. histolytica* trophozoites produce a small peptide, **monocyte locomotion inhibitory factor**, which inhibits the motility of host monocytes and macrophages and also suppresses monocyte and neutrophil nitric oxide production. It is probable that these factors contribute to the ability of trophozoites to survive within the host and even establish prolonged infections.

A degree of naturally acquired immunity to *E. histolytica* has been reported in humans. This immunity has been linked to a mucosal antiadherence lectin IgA response.

Prevention

Food and water contaminated with feces containing the cysts of *E. histolytica* are the most common vehicles for transmission. Prevention, therefore, depends upon interruption of the contamination-ingestion cycle. One such measure is the boiling or iodination (1.25 g iodine/liter drinking water, allow to stand 2–3 h) of drinking water in endemic areas. In many areas, fruits and vegetables become contaminated when human excrement (night soil) is used as fertilizer. The rule of thumb for Westerners traveling in third-world countries is to drink only bottled water and avoid ice cubes, salads, and those fruits not peeled by the person consuming them. Broad education to improve sanitation coupled with a ban on the use of untreated human excrement as fertilizer are perhaps the most effective means for curbing transmission of pathogenic protozoa such as *E. histolytica*.

Transmission of *E. histolytica* by infected food handlers can be controlled by local ordinances requiring periodic physical examinations, including stool examinations, for all food handlers.

Entamoeba dispar

It is now generally accepted that *E. histolytica* is in reality a species complex with the pathogenic form retaining the species name of *E. histolytica*. On the basis of isoenzyme electrophoresis as well as clinical epidemiological evidence, *E. dispar*, although morphologically identical to *E. histolytica*, is considered its nonpathogenic equivalent (Fig. 4.1). For instance, *E. histolytica*, whether symptomatic or asymptomatic, can elicit a serological response in humans, while *E. dispar* cannot. Therefore, the presence of quadrinucleate cysts in feces combined with either a negative serological test or PCR results would indicate the presence of the nonpathogenic *E. dispar*.

Entamoeba hartmanni

Since verification of the fact that, within this complex of intestinal amoebae parasitizing humans, only one, *E. histolytica*, causes disease, it has been reported that this amoeba occurs in two sizes, one ("small race") with trophozoites measuring 12–15 μm in diameter and cysts 5–9 μm in diameter, the other ("large race") with trophozoites measuring 20–30 μm in diameter and cysts 10–20 μm in diameter. Some investigators regard the small race of *E. histolytica* as a separate species, *E. hartmanni*, whose life cycle and overall morphology are very similar to those of *E. histolytica*. *E. hartmanni* (Fig. 4.1) is considerably smaller and nonpathogenic; therefore, no erythrocytes, etc. are to be found in vacuoles. As with *E. dispar*, it is important that diagnosticians differentiate between these organisms and *E. histolytica* in order to preclude unnecessary chemotherapy.

Entamoeba coli

This widely distributed intestinal amoeba (Fig. 4.1) is generally considered nonpathogenic in humans. The trophozoite averages 25 μm in diameter (range, 15–40 μm) and forms a cyst averaging 17 μm diameter (range, 10–33 μm). The prevalence of infection, like that for all parasites, varies in different localities, seasons, etc. For example, one report shows 37.1% of a sampling of Tennesseans harboring *E. coli*; another survey reveals that 26.1% of the inhabitants of Wise Count, Virginia harbor this amoeba. Still another, involving an extensive study of all sections of the United States, found infection rates exceeding 19%, making *E. coli* possibly the most common intestinal amoeba in the United States.

The trophozoite of *E. coli* does not ingest or invade host tissues. The food vacuoles observed in its heavily granulated cytoplasm usually contain bacteria, yeast, and fragments of intestinal debris. The nucleus may be visible in saline preparations. When stained, the nuclear envelope appears coarse with irregularly dispersed, peripheral chromatin on its inner surface, and there is a large, eccentric endosome. However, there is wide variation in this morphology. As in *E. histolytica*, cysts constitute the usual means of identification. The mature cyst characteristically contains eight vesicular nuclei each with an eccentrically situated endosome. Younger cysts may contain one, two, or four nuclei. Chromatoidal bodies, when present in cysts, may be needlelike or irregular with distinctly splintered terminals. Cysts are found frequently in diarrheic stools, but there is no evidence that this amoeba is the cause of the diarrhea.

The life cycle of *E. coli* parallels that of *E. histolytica* (Fig. 4.2), including the precystic, metacystic, and trophozoite stages, with infection of the host initiated by ingestion of cysts.

Entamoeba polecki

An intestinal amoeba, rarely found in humans but usually reported in pigs, goats, monkeys, and dogs, this parasite can be confused with *E. histolytica*. In size, its trophozoite and cyst are intermediate between *E. histolytica* and *E. coli* (Fig. 4.1), with the cyst stage being almost always uninucleate. Some investigators consider the parasite identical to *E. coli*, while others place it in a separate species; hence, its inclusion here. It is generally considered to be nonpathogenic.

Entamoeba gingivalis

E. gingivalis (Fig. 4.5) is cosmopolitan in distribution, commonly found in the tartar and debris associated with the gingival tissues of the mouth. It was the first parasitic amoeba reported in humans. There is little indication that it is pathogenic, and, while it abounds in people with unhealthy oral conditions (i.e., gingivitis or periodontitis), a cause and effect relationship has not been established. Food vacuoles may contain oral epithelial cells, leukocytes, occasionally erythrocytes, and various microbial organisms. No cyst is formed by *E. gingivalis*, and it is transmitted either directly (kissing) or indirectly via trophozoite-contaminated food, chewing gum, toothpicks, etc.

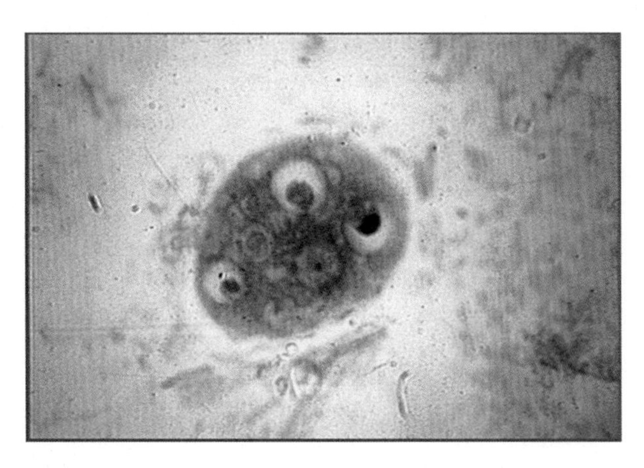

FIGURE 4.5 ***Entamoeba gingivalis.***

Iodamoeba bütschlii

I. bütschlii (Fig. 4.1) is transmitted by a cyst that is very distinctive, facilitating identification. It varies from a rounded to a somewhat angular shape, usually 10 μm in greatest diameter (range, 5–18 μm). The nucleus is large with a large, ovoid, usually eccentric endosome. Within the cyst is a large glycogen body, which stains deeply with iodine. The cyst and the emerged trophozoite are uninucleate. The amoeba escapes through a pore in the cyst wall and moves rapidly. Moisture and warmth are the only known requirements for excystation. Mature trophozoites average 10 μm in diameter (range, 6–25 μm) and reside in the large intestine, feeding on bacteria and yeast, as is evident from the contents of their food vacuoles. This species is not considered a pathogen. In the trophozoite, as in the cyst, the large, vesicular nucleus is a prominent feature.

Endolimax nana

E. nana (Fig. 4.1) is the smallest of the intestine-dwelling amoebae infecting humans, its trophozoite averaging only 8 μm in diameter (range, 6–15 μm). The trophozoite lives in the host's colon and is generally considered to be nonpathogenic. According to some surveys, prevalence may be as high as 30% in some populations. The life cycle is identical to that of other cyst-forming amoebae, with the cyst being the infective stage. *E. nana* cysts can be identified and distinguished from other cysts by their smaller size (9 μm in greatest diameter; range, 5–14 μm, ovoid shape, and 1–4 vesicular nuclei, each usually containing a large, eccentric endosome). The nuclear envelope is very thin and is difficult to see even in stained preparations. A tetranucleate metacystic amoeba escapes through a pore in the cyst wall and undergoes a series of cytoplasmic divisions in which a portion of cytoplasm is passed on to each uninucleate product. Trophozoites actively feed upon bacteria and multiply rapidly by binary fission.

PATHOGENIC FREE-LIVING AMOEBAE

In recent years, there has been a great deal of interest in a group of small, free-living amoebae belonging to the genera *Naegleria*, *Balamuthia*, and *Acanthamoeba*. These amoebae, normally free-living in fresh water and soil, are capable of facultative parasitism in humans and are highly pathogenic. Their pathogenicity in humans was first noted in 1965 when fatal cases of **primary amoebic meningoencephalitis (PAM)** due to *Naegleria fowleri* were diagnosed simultaneously in Australia and Florida. Since then, more than 150 cases have been reported worldwide, including cases from many parts of the United States. Most victims have a history of recent exposure to stagnant, warm, fresh, or brackish water, such as in swimming pools, ponds, and lakes.

Naegleria Fowleri

N. fowleri (Fig. 4.6) appears to be the principal causative agent for PAM. Its life cycle includes flagellated and amoeboid trophozoites and cysts, with rapid transformation from one form to the other. The flagellated trophozoite possesses two flagella while the amoeboid trophozoite displays a single, blunt pseudopodium with minutely pointed extensions on its end. In the free-living state, the organisms display contractile vacuoles. Binary fission occurs only

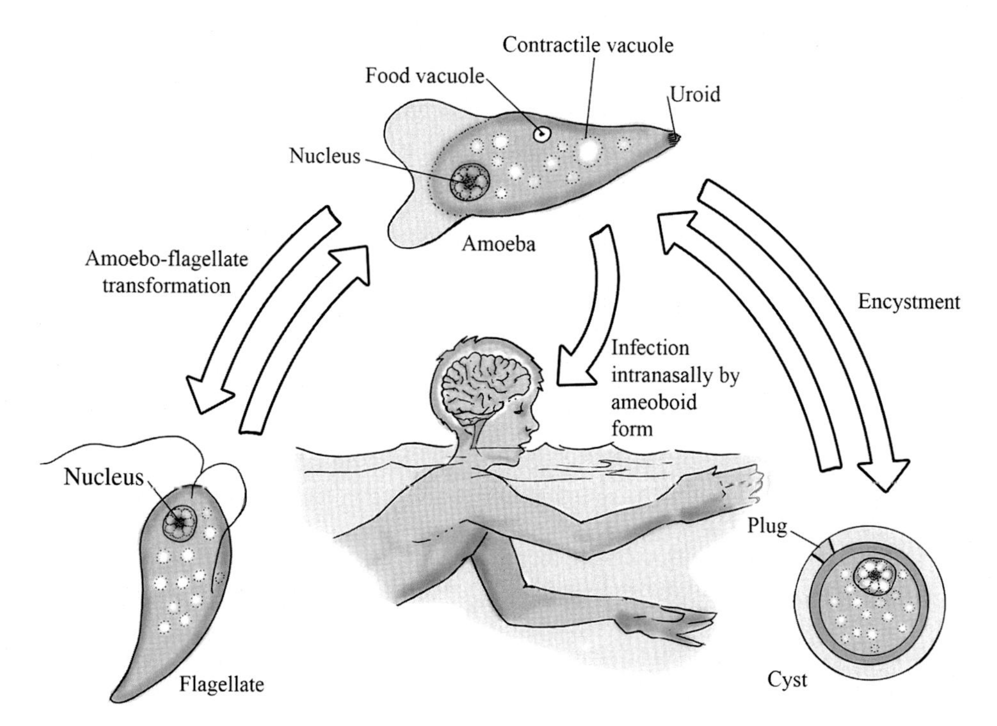

FIGURE 4.6 **Life cycle of *N. fowleri*.** *Credit: Image courtesy of Gino Barzizza.*

in the amoeboid trophozoite stage. The flagellated trophozoites are capable of rapid movement through the water, and transmission to humans most likely occurs when the amoeboid trophozoites invade the nasopharyngeal mucosa. The trophozoite secretes cytolytic naegleriapores A and B that exert membrane permeabilizing activity in a role similar to the amoebopores of *E. histolytica*. The amoeboid trophozoite migrates through the nervous system, via the cribriform plate, to the brain, where inflammation occurs and death usually ensues. No cyst stage occurs in the human host. Amoeboid trophozoites may be observed in the cerebrospinal fluid and in the tissues of the brain. Flagellated trophozoites may be observed in the cerebrospinal fluid. Diagnostic confirmation requires identification of the trophozoites in cerebrospinal fluid, but, since death may occur in 5–7 days, most cases are diagnosed at autopsy. If diagnosed in time, the treatment of choice is an intravenous and intrathecal use of both amphotericin B and miconazole, plus rifampin.

Acanthamoeba and Balamuthia

Four species of *Acanthamoeba* (Fig. 4.7) (*Acanthamoeba culbertsoni, Acanthamoeba polyphaga, Acanthamoeba castellanii*, and *Acanthamoeba rhysodes*) and *Balamuthia mandrillaris* are capable of causing **granulomatous amoebic encephalitis (GAE)** in individuals with compromised immune systems. Acanthamoeba is also the causative agent of ocular keratitis. Therefore, they can be considered opportunistic free-living amoebae. It is possible that some of the various species of *Acanthamoeba* may represent strains of a single species, *A. castellanii*. Additionally, recent genomic studies have identified several genotypes of *Acanthamoeba*. Of those, a subunit of a single molecular genotype, T4, has been reported responsible for Acanthamoeba keratitis.

Except for the absence of a flagellated trophozoite stage, the life cycles of *Acanthamoeba* and *Balamuthia* are very similar to that of *Naegleria*. The amoeboid trophozoites of the three genera can be identified by their distinctive pseudopodia. Those of *Naegleria* form a single, lobose pseudopodium and move rapidly; *Acanthamoeba* amoeboid trophozoites, on the other hand, form small, pointed pseudopodia, acanthopodia, and move sluggishly. The trophozoites of *Balamuthia* can display two forms of pseudopodia: either broad lobose or fingerlike. *Acanthamoeba* infections can be distinguished from those of *Naegleria* and *Balamuthia* by the characteristic cysts of *Acanthamoeba* found in affected tissues. The cysts of *Naegleria* and *Balamuthia* are round while those of *Acanthamoeba* are square. Less common than *Naegleria, Acanthamoeba* is a facultative parasite of humans responsible for symptoms similar to but less severe than those of *Naegleria* infections. Central nervous system involvement, GAE, is relatively rare, but, in immunologically compromised patients, usually fatal. *Acanthamoeba* and *Balamuthia* cysts and trophozoites are found in tissue. *B. mandrillaris* has only been isolated from autopsy specimens of humans and animals. While there is no uniformly effective treatment for *Naegleria* infections, in a few cases there has been favorable response to intravenous administration of amphotericin B. New infections of *Acanthamoeba* respond favorably to sulfonamides, and established infections appear to respond to amphotericin B. Ocular keratitis, attributed to *Acanthamoeba* infecting soft contact lens wearers, is refractory to these drugs. Patients with ocular keratitis have been treated successfully with intensive topical propamidine isethionate plus neomycin-polymyxin B-gramicidin ophthalmic solution.

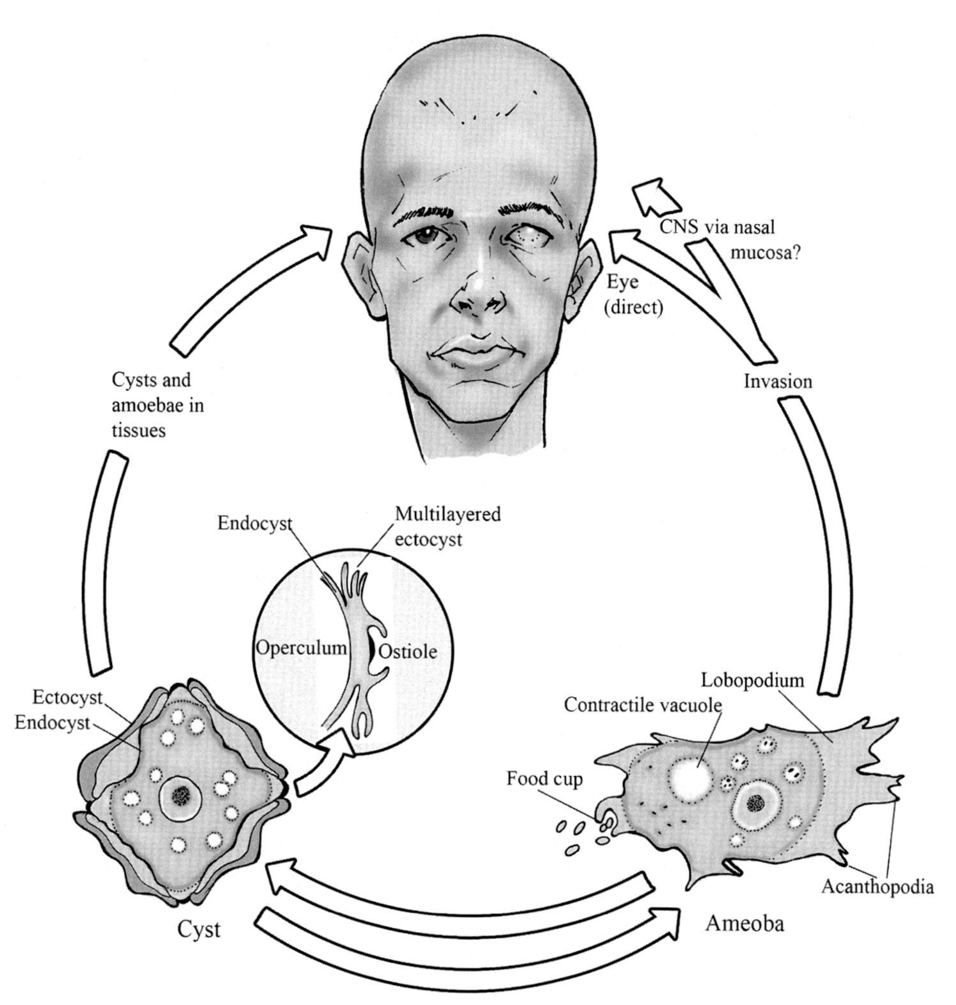

FIGURE 4.7 **Life cycle of *Acanthamoeba*.** *Credit: Image courtesy of Gino Barzizza.*

MICROSPORIDIANS

Several protistan species of Microsporidia, a taxon in a state of flux, parasitize a wide range of invertebrates and some vertebrates (e.g., fish, birds, and mammals). Human infection was rare until the advent of AIDS. Since 1985, a number of microsporidia species have been reported in AIDS patients. Phylogenetically, microsporidians are similar to prokaryotic cells in that they lack mitochondria, Golgi complexes and have similar ribosomal RNA sequences. They resemble eukaryotes in their nuclear organization. The life cycle consists of an intracellular, divisional phase followed by a spore-producing phase. When the infected cell dies, mature spores are released into the immediate environment. Morphologically unique, the spore is equipped with an extrusive polar tube by which sporoplasm can be inoculated into new host cells. *Enterocytozoon bieneusi* infects intestinal cells, causing chronic diarrhea, cramps, and nausea. In AIDS patients, it is associated with severe weight loss, malabsorption, and zinc deficiency. Several other species belonging to the genus *Encephalitozoon* are capable of infecting a wide variety of cells, including macrophages and neurons, causing diarrhea, bronchitis, nephritis, hepatitis, inflammation of the cornea, and peritonitis among AIDS patients. *Enterocytozoon bieneusi*, however, with a prevalence of 40%, is the most common microsporidium infecting AIDS patients.

Immunodiagnosis, using either indirect immunofluorescent antibody tests or Western-blot techniques, is the most sensitive diagnostic technique available. Albendazole eradicates all *Encephalitozoon* species by disrupting microtubules, thereby inhibiting cell division. Fumagillin, 5-fluorouracil is also effective against *Encephalitozoon* intestinal infections. No effective therapy against *E. bieneusi* has yet been developed, although a 2-week oral regimen of fumagillin provides symptomatic relief.

CILIATES

Members of the phylum Ciliophora are protistans possessing cilia in at least one stage of their life cycle and having two different types of nuclei: one macronucleus and one or more micronuclei. Only one ciliophoran, *B. coli*, infects humans.

Balantidium Coli

A distinctive feature of *B. coli* (Fig. 4.8) is the presence of a depression, or **peristome**, leading into the cytosome. *B. coli* is commonly considered a pathogen of humans that also parasitizes pigs and monkeys. Some investigators, however, classify the organism parasitic to pigs as a distinct species, *Balantidium suis*.

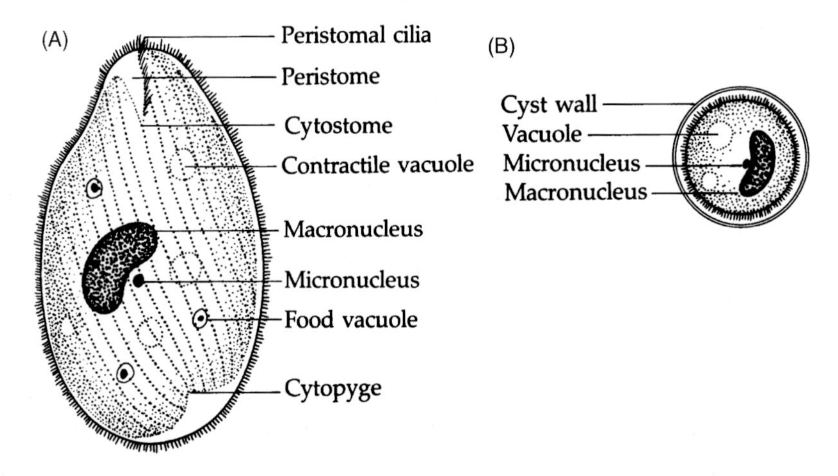

FIGURE 4.8 ***Balantidium coli*, an intestinal parasite of pigs, monkeys, and humans.** (A) Trophozoite and (B) cyst.

Life Cycle

Both a motile trophozoite stage and a cyst stage occur in the life cycle of *B. coli* (Fig. 4.9). The trophozoite inhabits the cecum and colon of humans and is the largest known protozoan parasite of humans, measuring 50–130 μm by 20–70 μm. The conspicuous vestibulum leads into a large cytosome at the anterior end of the cell, opposite to which lies a cytopyge. Coarse cilia line the peristomal area. The macronucleus is typically elongate and kidney shaped, while the vesicular micronucleus is spherical. There are two prominent contractile vacuoles, one in the middle of the cell and the other near the posterior end. The presence of contractile vacuoles, unique among parasitic protozoa, indicates a degree of osmoregulatory capability. Food vacuoles in the cytoplasm contain debris, bacteria, starch granules, erythrocytes, and fragments of host epithelium. While the organism typically reproduces asexually by transverse fission with the posterior daughter cell forming a new cytosome after division, conjugation also occurs in this species.

Transmission of *B. coli* from one host to another is accomplished via the cyst. The cysts are round, measuring 40–60 μm in diameter, with a heavy cyst wall consisting possibly of two layers. Cilia, the large macronucleus, and contractile vacuoles are readily visible within the cyst. Encystation usually occurs in the large intestine but may also occur outside the body of the host. Cysts are common in the feces of infected hosts and are generally not considered sites of reproduction although cysts containing two individuals sometimes occur. Infection occurs when cysts are ingested by the host. Excystation occurs in the small intestine.

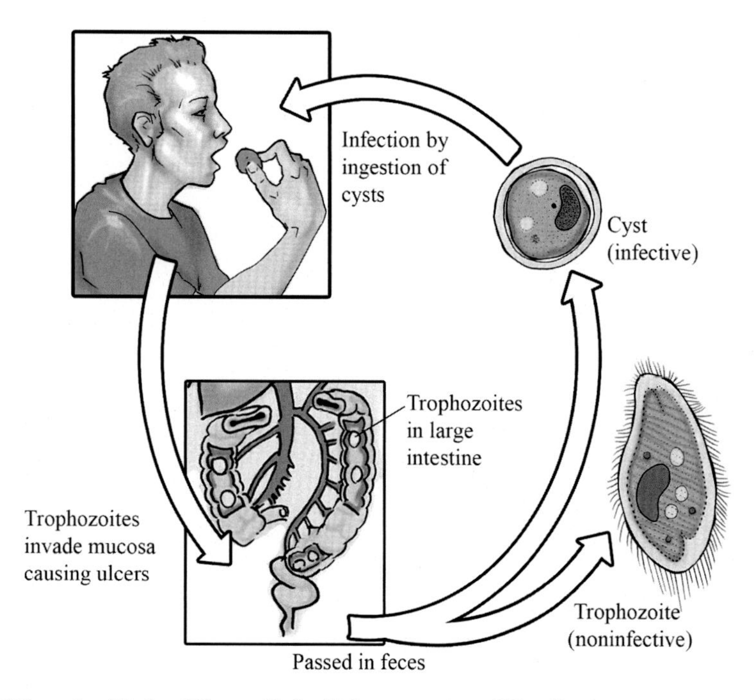

FIGURE 4.9 **Life cycle of *Balantidium coli*.** *Credit: Image courtesy of Gino Barzizza.*

Epidemiology

Balantidiosis is most often found in tropical regions throughout the world; however, with an infection rate of less than 1%, it is not a common human disease. The parasite, while nonpathogenic in pigs, is far more common among these animals than among humans in these regions, with a prevalence among pigs ranging from 20% to 100%. Human infection is most common where malnutrition is widespread, where pigs share habitation with human families, and where fecal contamination of food and water occurs.

Symptomatology and Diagnosis

The trophozoite resides in the cecal area and throughout the large intestine. It thrives in an environment rich in starch, such as the small intestine; however, in such an environment, the trophozoite does not invade the intestinal mucosa. This proclivity for carbohydrate may be the reason for the trophozoite's invasive character once it becomes established in the human cecal region, a region low in carbohydrate content; in the pig's intestine, where carbohydrate is more abundant, the organism remains in the lumen. It is believed that the trophozoite secretes proteolytic enzymes that act upon the mucosal epithelium, facilitating invasion.

Results of infection range from asymptomatic to severe. Parasitic invasion of the mucosal epithelium is followed by hemorrhage and ulceration, hence the name **balantidial dysentery** often given to this condition. While symptoms such as colitis and diarrhea may resemble amoebiasis in many respects, extraintestinal disease is rare. Occasionally, *B. coli* is transported by the blood into the spinal fluid. Fatalities are rare, although one case of fatal myocarditis in Russia has been attributed to *B. coli*. A few deaths have been reported from Mexico and Central America.

The usual diagnostic procedure consists of stool examination for the presence of trophozoites and cysts. The trophozoites are readily identified by their large size and the fact that *B. coli* is the only ciliophoran parasitic in humans. Cysts can be identified by their large size, heavy cyst wall, large macronucleus, and the presence of cilia within the cyst.

Chemotherapy

The infection may disappear spontaneously, or the host may become asymptomatic, remaining as a carrier. Drug treatment usually consists of oral administration of tetracycline or, as alternatives, metronidazole or iodoquinol. The use of tetracycline is contraindicated in pregnant patients and in children less than 8 years old.

Suggested Readings

Begum, S., Queech, J., & Chadee, K. (2015). Immune evasion mechanisms of *Entamoeba histolytica*: progression to disease. *Frontiers in Microbiology, 6*, 1394–1423.

Canning, E. U., & Hollister, W. S. (1987). Microsporidia of mammals—widespread pathogens or opportunistic curiosities? *Parasitology Today, 3* 276-273.

Diamond, L. S., & Clark, C. G. (1993). A redescription of *Entamoeba histolytica* Schaudinn, 1903 (emended Walker, 1911) separating it from *Entamoeba dispar* Brumpt, 1925. *Journal of Eukaryotic Microbiology, 40*, 340–344.

Elsdon-Dew, R. (1968). The epidemiology of amoebiasis. *Advances in Parasitology, 6*, 1–62.

Gill, E. E., & Fast, N. M. (2006). Assessing the microsporidia-fungi relationship: combined phylogenetic analysis of eight genes. *Gene, 375*, 103–109.

Gitler, C., & Mirelman, D. (1986). Factors contributing to the pathogenic behavior of *Entamoeba histolytica*. *Annual Reviews in Microbiology*, *40*, 237–262.

Marciano-Cabral, F., MacLean, R., Mensahg, A., & LaPat-Polasko, L. (2003). Identification of *Naegleria fowleri* in domestic water sources by nested PCR. *Applied Environmental Microbiology*, *69*, 5864–5869.

Ravdin, J. I. (1990). Cell biology of *Entamoeba histolytica* and immunology of amoebiasis. In D. G. Wyler (Ed.), *Modern parasite biology: cellular, immunological and molecular aspects* (pp. 126–150). New York: W.H. Freeman & Company.

Reeves, R. E. (1984). Metabolism of *Entamoeba histolytica* Schaudinn, 1903. *Advances in Parasitology*, *23*, 106–142.

Sanchez-Guillen, M. C., Perez-Fuentes, R., Salgado-Rosas, H., Ruiz-Arguelles, A., Ackers, J., Shire, A., & Talamas-Rohana, P. (2002). Differentiation of *Entamoeba histolytica/Entamoeba dispar* by PCR and their correlation with humoral and cellular immunity in individuals with clinical variants of amoebiasis. *American Journal of Tropical Medicine and Hygiene*, *66*, 731–737.

Schuster, F. L., & Visvesvara, G. S. (2002). Free-living amebas present in the environment can cause meningoencephalitis in humans and other animals. *Encyclopedia of environmental microbiology*. New York: John Wiley & Sons pp. 1343-1350.

Sharma, M., Vohra, H., & Bhasin, D. (2005). Enhanced pro-inflammatory chemokine/cytokine response triggered by pathogenic *Entamoeba histolytica*: basis of invasive disease. *Parasitology*, *131*, 783–796.

Stanley, S. L., Jr. (2001). Pathophysiology of amoebiasis. *Trends in Parasitology*, *17*, 280–285.

Warhurst, D. C. (1985). Pathogenic free-living amoebae. *Parasitology Today*, *1*, 24–28.

CHAPTER

5

Visceral Protistans II

Flagellates

Chapter 5 introduces eight species of flagellated protistans that infect the human digestive and reproductive systems. In the ensuing discussion, this group of organisms is divided into two groups, the nontrichomonad flagellates and the trichomonad flagellates. Within these two groups, eight species of flagellates inhabit the aforementioned sites, but only two organisms are usually considered pathogenic, *Giardia lamblia* and *Trichomonas vaginalis*. The former infects the small intestine of humans, while the latter infects the reproductive system. The epidemiology of *G. lamblia* is discussed emphasizing the wide variety of wild animals that can serve as reservoir hosts for the organism. Backpacker's disease is presented as an example of the role that wild animals can play in transmitting the organism to humans. The host's immune response is considered as is the current chemotherapeutic regimen. Of the trichomonads, only *T. vaginalis* is considered pathogenic to humans. Following a brief overview of the trichomonad morphology, the physiology and epidemiology of *T. vaginalis* is discussed in detail. In considering the latter, the roles of males and females in the spread of the infection is considered. A brief discussion of the host's immune response is followed by a paragraph outlining the current chemotherapeutic regimen for *T. vaginalis*.

Members of the phyla Retortamonada and Parabasalia, the flagellates infecting the digestive and reproductive systems of humans, belong to seven genera. As in the case of amoebae, only a few are pathogenic, but it is important to distinguish the nonpathogenic from the pathogenic forms. The nonpathogenic organisms are *Chilomastix mesnili*, *Retortamonas intestinalis*, *Enteromonas hominis*, *Trichomonas tenax*, *Pentatrichomonas hominis*, and *Dientamoeba fragilis*; forms pathogenic in humans are *G. lamblia* and *T. vaginalis*. Of the eight organisms listed above, all but two species, *T. tenax* (mouth) and *T. vaginalis* (reproductive tract), are intestinal parasites.

NONTRICHOMONAD FLAGELLATES

Giardia Lamblia (=G. Duodenalis)

The bilateral symmetry of members of this genus (Figs. 5.1 and 5.2) is distinctive among the protistans. The trophozoite is rounded at the posterior end, tapered posteriorly, and flattened dorso-ventrally. It is 14-μm long (range, 8–16 μm) by 10-μm wide (range, 5–12 μm)

Human Parasitology. http://dx.doi.org/10.1016/B978-0-12-813712-3.00005-9

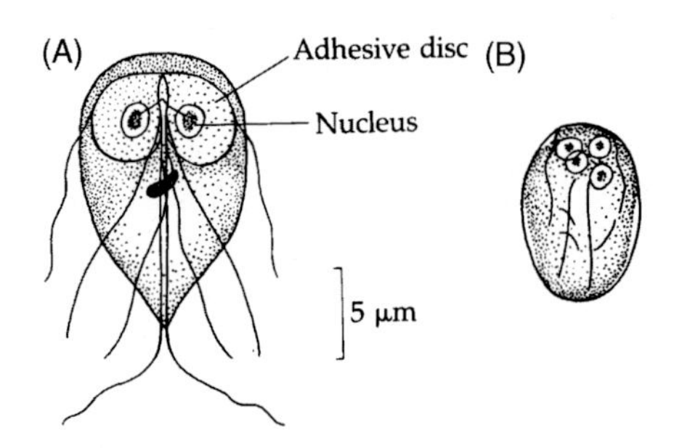

FIGURE 5.1 ***Giardia lamblia.*** (A) Trophozoite. (B) Cyst.

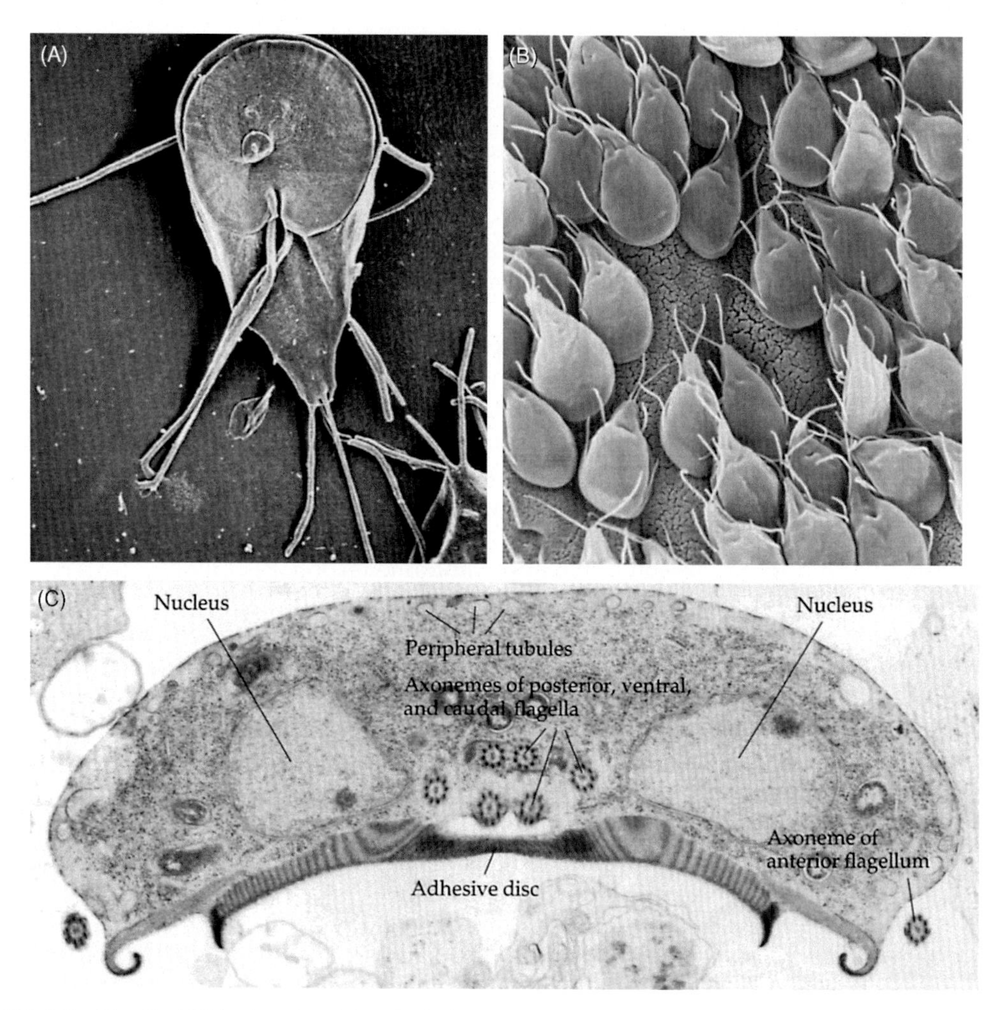

FIGURE 5.2 ***Giardia.*** (A) Trophozoite of the *Giardia intestinalis* type. Scanning electron micrograph of the ventral surface showing the attachment organelle. Bar = 1 mm. (B) Transmission electron micrograph of a cross section of a *Giardia muris* trophozoite in the small intestine of an infected mouse. The marginal groove is the space between the striated rim of cytoplasm and the lateral ridge of the adhesive disc. This specimen bears endosymbionts, which apparently are bacteria. (X 15,350) (C) Scanning electron micrograph of an intestinal villus. The microvillus border of the epithelial cells is almost obscured by attached trophozoites.

(Fig. 5.1). The dorsal surface is convex; the ventral surface is usually concave but occasionally flat and is dominated by a large, binucleate adhesive disc with a nucleus in the center of each half (Fig. 5.2A). The rim of the adhesive discs is supported by microtubules and clusters of microfilaments (Fig. 5.2B), and four pairs of flagella arise from basal bodies clustered between the two nuclei. One pair extends down the midline of the cell, emerging posteriorly as trailing flagella; the ventral pair emerges at the posterior edge of the adhesive disc. Of the remaining two pairs, one emerges anteriolaterally and one laterally. Two prominent, slightly curved **median bodies** are distinctive to the genus *Giardia*. Their function is unknown, although it has been suggested that they may act as supportive structures.

Life Cycle (Fig. 5.3)

The trophozoite of *G. lamblia* reproduces by longitudinal fission. Its organelles undergo division in the following order: nuclei, adhesive disc, and cytoplasm. In the duodenum and bile duct of its host, the trophozoite can either maintain position by attaching its large adhesive disc to the epithelial cells (Fig. 5.2C) or use its flagella to swim rapidly in the lumen. Attachment is facilitated by the two ventral flagella working with the flexible rim of the disc. As trophozoites pass through the digestive tract, they usually encyst in the colon. The cystic transmission stage is typically ovoid and averages 11-μm long (range, 9–12 μm) (Fig. 5.1). In saline smears, refractile granules can be seen in the cysts, and, at times, the cytoplasm appears to be detached from the cyst wall in several places. In cysts stained with iodine or hematoxylin, two to four nuclei are visible in addition to numerous fibrils (probably flagellar remnants) and median bodies.

In victims of giardiasis, massive infection is common. The presence of up to several billion trophozoites in a single diarrheic stool sample is not unusual. Cysts are rarely encountered in such stools, being found instead in either formed or partially formed stools. Infection results

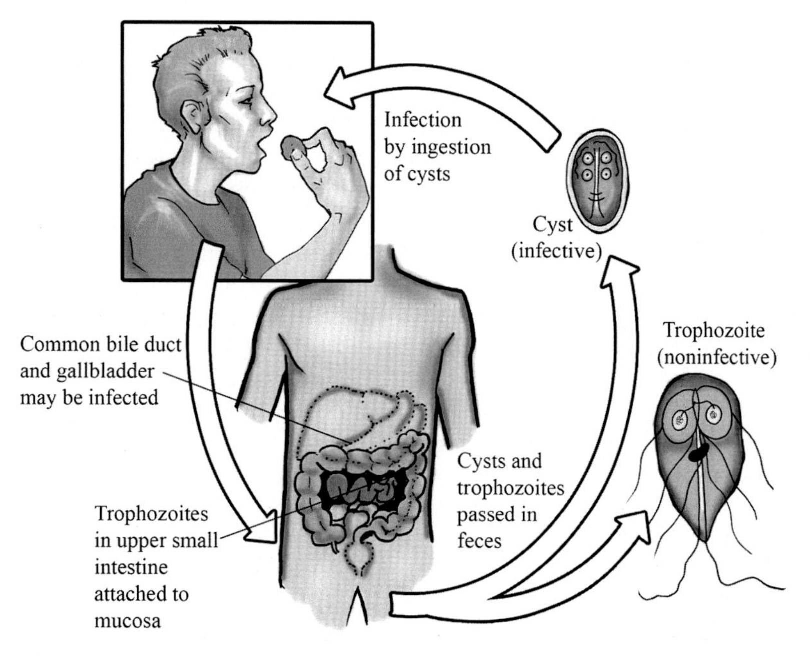

FIGURE 5.3 **Life cycle of *Giardia lamblia*.** *Credit: Image courtesy of Gino Barzizza.*

from ingestion of cyst-contaminated food or water or from direct hand-to-mouth contact. Ingestion of 100 or more cysts is considered infective. Following ingestion, cysts pass through the stomach to the small intestine where they excyst and begin the cycle anew.

Epidemiology

Giardia is the most prevalent intestinal parasite in humans. It is cosmopolitan and is common in children 6–10 years of age but is also seen often in older children and adults, with a high incidence in homosexual males. Outbreaks are frequent in daycare nurseries and other institutions where sanitation may be inadequate. An outbreak of giardiasis occurred in the ski resort town of Aspen, Colorado, when a water supply line was inadvertently crossed with a sewage line, and 11% of the skiers present that season became infected. Among 59 persons whose stools were positive, 56 experienced clinical symptoms of the disease. Giardiasis is common among tourists (an infection rate of approximately 23%) returning from Russia. An increase in *Giardia* infection has been noted among wilderness campers in the United States, probably due to drinking water polluted beyond the line above which human contamination is unlikely. The term "Backpackers Disease" has been coined in reference to the disease contracted by this group. Such outbreaks have led epidemiologists to suspect that wild animals may harbor species of *Giardia* capable of infecting humans. Surveys have implicated beavers, dogs, and sheep as potential reservoirs for human infections. Significant differences in size and structure among species of the genus *Giardia* have led to the assumption that each different host species has a different parasite species. It now appears more likely that the variable morphology of *Giardia* is due to host diet rather than genetic variation, so that many of the described "species" are invalid distinctions.

Symptomatology and Diagnosis

G. lamblia infection causes severe intestinal disorders, most commonly diarrhea and related symptoms due to malabsorption. Attachment of the trophozoite to the mucosal surface by means of its adhesive disc (Fig. 5.2C) causes shortening of the villi of the small intestine, inflammation of the crypts and lamina propria, and lesions on mucosal cells. Occasionally, trophozoites penetrate the mucosa, but this is rare. Since *Giardia* has not been known to produce toxins, it appears that symptoms result from combined mechanical and chemical factors. Severe *Giardia* infections produce a malabsorption syndrome characterized by the inability of the small intestine to absorb such essential, fat-soluble substances as carotene, vitamin B_{12}, and folate. These absorptive abnormalities may be accompanied by reduced secretion of a number of small-intestinal digestive enzymes, such as disaccharidase. Additional symptoms of infection are diarrheic stools, steatorhea, abdominal distension, nausea, flatulence, and eventual weight loss. Occasionally, bile duct and gall bladder involvement may produce jaundice and colic. These symptoms may become evident as early as 3–25 days (average 10 days) after ingestion of cysts.

Identification of characteristic cysts in the stool is used in diagnosis of this parasite. Either saline or iodine smears can be employed for initial diagnosis, but a concentration method is commonly used to enhance detection. Examination for trophozoites is rare since their detection depends upon almost immediate inspection or fixation of diarrheic stool samples. Duodenal aspiration, either by intubation or by the enteric capsule method, is an alternate, more satisfactory technique for trophozoite detection, especially in early stages of infection. An enzyme-linked immunosorbent assay to detect salivary IgA antibodies to *G. lamblia* has been used successfully to test school children.

Chemotherapy

Treatment with either metronidazole, tinidazole, or quinacrine is recommended. Complete cure is usually effected within a week after treatment begins. The limitations regarding metronidazole usage are discussed on page 58. If the bile duct or gall bladder is infected, "relapses" may occur for years. Because of the ease with which the cyst is transmitted, all members of the household should be treated simultaneously. Untreated patients can pass cysts from a few weeks to months postinfection.

Physiology

Little is known of the physiology of *Giardia*, due in large measure to the inadequancy of in vitro culture methods. An in vivo study to determine the method of uptake of macromolecular markers such as ferritin by *Giardia* indicates rapid transfer of the marker from the host's intestinal lumen into vacuoles close to the surface of the organism, suggesting a means by which *Giardia* obtains nutrients. Other studies using radiolabeled sugars show *Giardia* capable of incorporating certain monosaccharides into glycogen. *Giardia* has no mitochondria but can use oxygen when available. There is no evidence of either an electron transport system or a Krebs cycle. The organisms rely on a flavin-dependent substrate-level phosphorylation as their major means of obtaining adenosine triphosphate (ATP), metabolizing carbohydrates to ethanol, CO_2, and acetate as principal end products. This pathway is blocked by quinacrine and chloroquine. In the presence of oxygen, excretion of acetate is favored, while ethanol is favored in the absence of oxygen.

Host Immune Response

In the majority of human giardia infections, the disease appears to be self-limiting, although reoccurrences do occur. While specifics of the immune defense are not well understood, both humoral and cellular mechanisms appear to be important for parasite clearance. Human *G. lamblia* infections result in the production of antigiardial antibodies from mucosal secretions and serum, and IgA-dependent host defenses are central for eradicating *Giardia* trophozoites. IgA is believed to bind to the surface of the trophozoite, inhibiting attachment to the intestinal epithelium as opposed to direct killing. Specific IgA has been detected bound to trophozoite surfaces in human jejunal biopsies and jejunal fluid.

G. lamblia trophozoites display surface antigenic variation mediated by a unique family of cysteine-rich proteins, **variant-specific surface proteins (VSPs)**. Unlike the variant surface glycoproteins of trypanosomes (see pp. 106–107), the expression of *G. lamblia* trophozoite VSPs do not involve gene movement but may be influenced by epigenetic mechanisms. Evidently, variations can occur spontaneously and selections for and against different variations may be determined by physiological and immunological factors of the host.

Prevention

Generally, preventive measures recommended for *Entamoeba histolytica* are applicable to *G. lamblia* as well. The prescribed amount of iodine added to drinking water should be doubled to ensure killing of *G. lamblia* cysts.

Chilomastix Mesnili

Cosmopolitan in distribution, this organism (Fig. 5.4) infects about 6% of the world's human population. Usually considered nonpathogenic, *C. mesnili*, like most protistan parasites, when

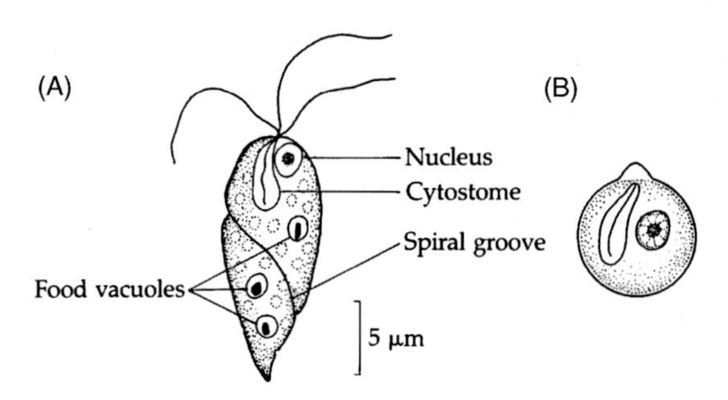

FIGURE 5.4 ***Chilomastix mesnili.*** (A) Trophozoite. (B) Cyst.

present in sufficient numbers may cause intestinal disorders, most commonly diarrhea. The stage found in the human colon, the motile, pyriform trophozoite, averaging 12-μm long (range, 5–20 μm), has a blunt anterior end from which extend three free flagella. A spiral groove extends the length of the cell, terminating at the pointed posterior end. A prominent cytostome enclosing a fourth, recurrent flagellum is located in the anterior portion of the cell, as is the large nucleus. A prominent, curved, supporting fibrillar structure, the so-called Shepherd's Crook, lies just under the cytostome wall.

The parasite utilizes a resistant cyst stage for transmission. The lemon-shaped, relatively thick-walled cyst, approximately 8 μm in diameter, is identifiable by its single nucleus and the cytostome containing the remnant of the recurrent flagellum and Shepherd's Crook. When the cyst is stained, basal bodies, one for each of the four flagella, may be seen.

Retortamonis Intestinalis

Although only about one-third its size, this nonpathogen (Fig. 5.5) closely resembles *C. mesnili.* The trophozoite has one free, anterior flagellum and a recurrent, cytostomal flagellum that emerges as a free, posteriorly trailing flagellum. As in *C. mesnili*, the trophozoite resides in the colon, and a cyst, approximately 6-μm long by 3-μm wide, serves as the transmission stage.

Enteromonas Hominis

This rare human intestinal parasite (Fig. 5.6) also has trophozoite and cyst stages, but human hosts experience no clinical symptoms with the infection. The pyriform trophozoite, 4–10-μm long by 3–6-μm wide, has three anterior flagella and one recurrent flagellum, the latter extending posteriorly along one side and trailing free. The mature cyst, approximately 7 by 4 μm, is ovoid with two to four nuclei, which are usually situated at the ends of the cyst. Most cysts are binucleate.

Dientamoeba Fragilis

The current system of classification of the Protista places *D. fragilis* (Fig. 5.7) with certain amoeboid forms that may or may not possess flagella. Although *D. fragilis* exists only in the

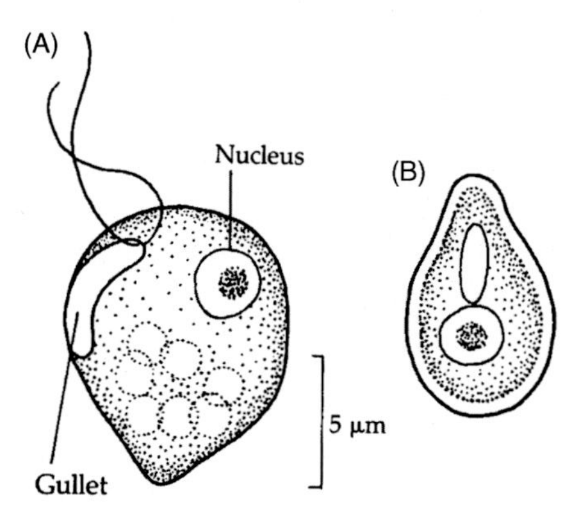

FIGURE 5.5 ***Retortamonas intestinalis***. (A) Trophozoite. (B) Cyst.

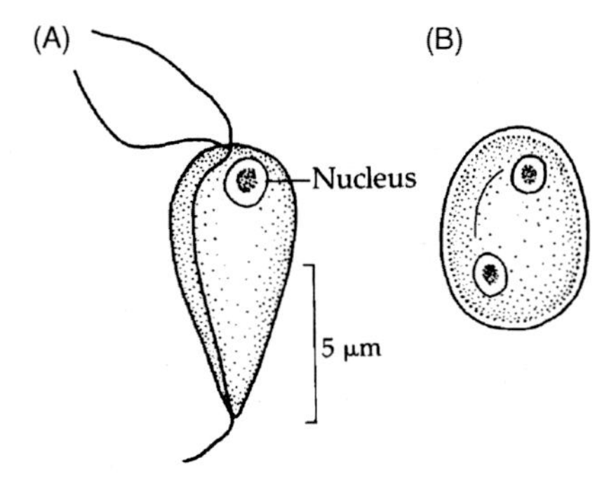

FIGURE 5.6 ***Enteromonas hominis.*** (A) Trophozoite. (B) Cyst.

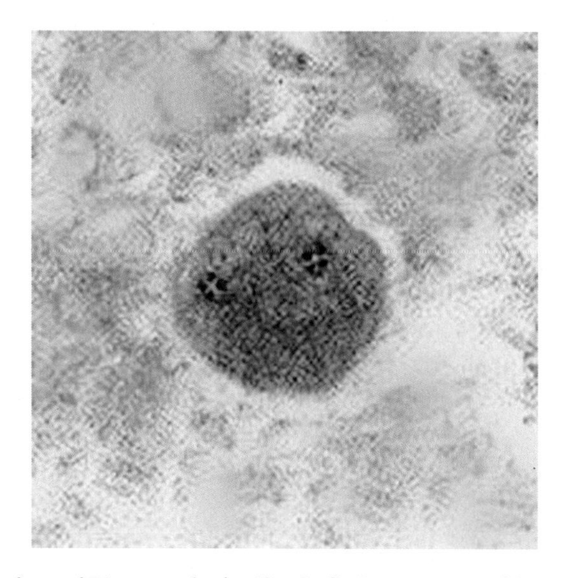

FIGURE 5.7 **Binucleate form of *Dientamoeba fragilis.*** *Credit: Image courtesy of the Center for Disease Control.*

amoeboid form, it is assigned to this group on the basis of ultrastructural and immunological affinities.

D. fragilis occurs worldwide infecting about 4% of the world population. While the organism is often considered nonpathogenic and intestinal lesions attributed to the organism have never been demonstrated, patients with gastrointestinal disturbances experience relief from discomfort when the organism is destroyed by chemotherapy. Fibrosis of the appendiceal wall in all *D. fragilis* infections of the appendix constitutes further strong evidence of the pathogenic potential of *D. fragilis*. Also, *D. fragilis* shows a decided preference for erythrocytes when they are available.

The trophozoite, 6–12 μm in diameter, lives in the cecal area of the large intestine and moves sluggishly by means of thin, leaflike pseudopodia. It is frequently binucleate, with a thin nuclear envelope visible only after staining. The prominent endosome is surrounded by minute clumps of chromatin, giving it a beaded appearance. Since no cyst form has been reported, the mechanism of transmission is unknown, although the eggs of the intestinal nematode *Enterobius* have been suggested as possible carriers. While the trophozoite is highly viable and is capable of motility up to 48 h after leaving the host in feces, it cannot survive the digestive juices in the upper regions of the digestive tract.

From 20% to 80% of the trophozoites recovered from human feces are binucleate, a condition that may represent merely an arrested telophase stage of mitosis. Identification of the trophozoite in the feces serves as diagnosis of infection. When placed in water, the trophozoite swells and then returns to normal size. In the swollen state, numerous cytoplasmic granules exhibit Brownian movement. This feature, called the "Hakansson Phenomenon," is peculiar to *D. fragilis* and occasionally is used in its identification.

THE GENUS TRICHOMONAS AND RELATED FORMS

Of the trichomonads that infect humans, two species, *T. tenax* and *T. vaginalis*, possess four free, anterior flagella. A third species, formerly called *T. hominis*, has five free, anterior flagella and is accordingly placed in the genus *Pentatrichomonas*. All trichomonads possess certain common features (Fig. 5.8), among which are three to five anterior flagella and a recurrent flagellum in the form of an undulating membrane. All flagella in these forms originate from anteriorly situated basal bodies. The costa also originates from the region of the basal bodies and extends along the base of the undulating membrane. In all three species, associated with the costa and/or axostyle is a row of granules, the **hydrogenosomes** (paracostal or paraxostylar granules). An axostyle extending the length of the trichomonad appears to protrude from its posterior end although it is covered by the plasma membrane. A prominent Golgi complex (parabasal body) lies anteriorly near the single nucleus. There are no known cyst stages in the life cycles of these organisms; while venereal or oral contact are obvious methods of transmission for *T. vaginalis* and *T. tenax*, that of *P. hominis* remains obscure.

Trichomonas Tenax

This flagellate (Fig. 5.9A) is commonly found in the tartar and gums of the mouth, as well as in the nasopharyngeal region. Trophozoites are very small (5–16 by 2–15 μm), with four

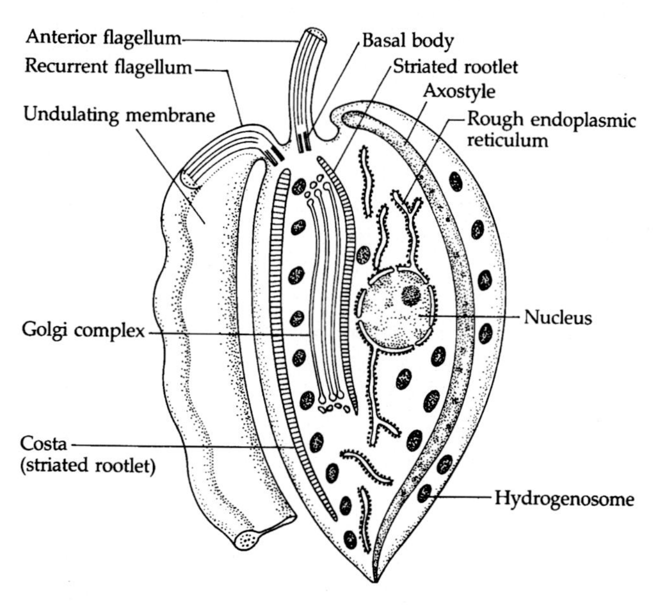

FIGURE 5.8 **Ultrastructural morphology of a generalized trichomonad.**

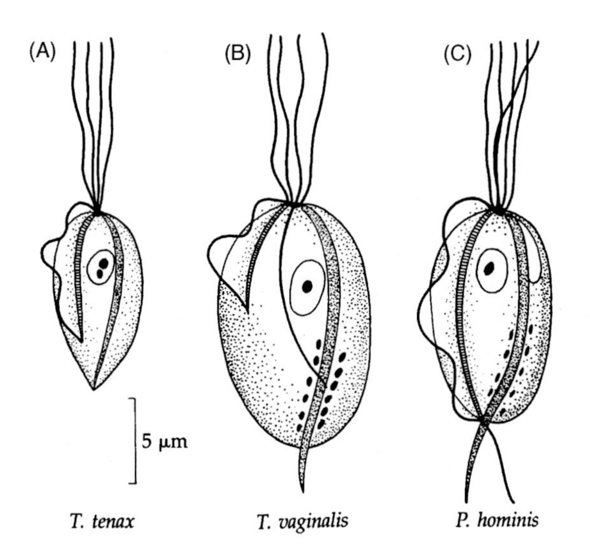

FIGURE 5.9 (A) *Trichomonas tenax*. (B) *Trichomonas vaginalis*. (C) *Pentatrichomonas hominis*.

free flagella and a fifth flagellum recurved as an undulating membrane that extends about two-thirds of the length of the cell. The costa runs parallel to the undulating membrane. Transmission is necessarily by direct contact, usually kissing or using contaminated eating utensils. Drinking contaminated water from a community source may be another means of transmission since some investigators have shown that this flagellate can live in drinking water for several hours. The organism is not considered pathogenic and can be avoided through proper oral hygiene. Like *Entamoeba gingivalis*, it tends to flourish in unhealthy environments fostered by poor oral hygiene and is most easily found in patients who practice poor hygiene.

Trichomonas Vaginalis

Of the three human-infecting trichomonads, *T. vaginalis* (Fig. 5.9B) is the only pathogen, although a heavy infection of *P. hominis* may cause diarrhea. *T. vaginalis* inhabits the vagina in the female and the urethra, epididymis, and prostate gland in the male. Morphologically, it is distinguishable from the other two trichomomads by its larger size (7–32 by 5–12 μm) and its shorter undulating membrane, which extends only one-third the length of the cell. The trophozoite occasionally produces pseudopodia. Clusters of hydrogenosomes extend along both the costa and the axostyle.

Life Cycle

Typical of flagellates, *T. vaginalis* reproduces by longitudinal binary fission. The optimum pH range for the organism to reproduce is approximately 5 or 6. While the normal pH of the vagina is 4–4.5, when the level of acidity is disturbed, an environment is created in which *T. vaginalis* thrives. Normally, the pH of the vagina is maintained by the activity of a group of lactic acid-producing bacteria, but *T. vaginalis* can disrupt such bacteria, causing the pH to rise above 4.9.

Epidemiology

The prevalence among women is approximately 10%–25%, varying inversely with the level of hygiene practiced. While about 15% of women with trichomoniasis complain of symptoms, altered vaginal secretions are evident in many more. In infected households, the recorded incidence of infection among men is much lower than among women from the same household. This statistic is misleading, however, since the flagellate is much more difficult to detect in men; in fact, positive identification sometimes requires the examination of prostate exudate. Transmission is by direct contact, usually through sexual intercourse. Damp washcloths and similar items also are sources of infection among children and adults, viable trophozoites having been recovered from wet washcloths 24 h after contamination. Trichomoniasis among newborns indicates that the fetus can acquire the organism while passing through the birth canal.

Symptomatology and Diagnosis

T. vaginalis produces deterioration of the cells of the vaginal mucosa, resulting in low-grade inflammation and persistent vaginitis. The condition is characterized by a yellowish discharge accompanied by persistent itching and burning. In males, symptoms are much less noticeable, although there may be urethritis and swelling of the prostate gland. These symptoms are sometimes confused with gonorrhea.

Diagnosis in females is confirmed by microscopical identification of motile trophozoites in vaginal discharge smears. Examination of the urine of both sexes and examination of prostate secretions of the male following prostate massage are also helpful diagnostic procedures.

Chemotherapy

Metronidazole or tinidazole are the most effective drugs, although they are contraindicated in pregnant patients. Restoration of the normal pH of the vagina by periodic douches with a dilute solution of vinegar is an effective preventive method and can control mild infections. It is recommended that sexual partners be treated simultaneously.

Physiology

While trichomonads are anaerobic organisms, deriving much of their energy from the incomplete degradation of simple sugars accompanied by the production of short chain organic acids such as lactic and acetic acids, the presence of oxygen has little effect on this process. Glucose and maltose are the most effective growth stimuli in vitro. One of the products of carbohydrate metabolism is acetic acid, which is anaerobically produced in the hydrogenosome from part of the pyruvic acid pool. Pyruvic acid is produced in the cytoplasm via glycolysis, and a portion enters the hydrogenosome while the remainder is reduced in the cytoplasm to lactic acid and excreted. ATP is formed in the cytoplasm and in the hydrogenosome by substrate level phosphorylation. Trichomonads lack mitochondria; however, it has been suggested that the hydrogenosome may be a modified mitochondrion since it shows morphological and functional similarities to such organelles, such as a double membrane and regulation of cell calcium. This organelle is also considered by some to be a specialized microbody.

In culture, *T. vaginalis* feeds on bacteria and, occasionally, erythrocytes. The predilection for bacteria suggests a mechanism for the breakdown of the normal pH of the infected vagina, since the lactic acid bacilli act to maintain normal pH levels.

Pentatrichomonas (Trichomonas) Hominis

This trichomonad (Fig. 5.9C) is a smaller (5–14 by 7–10 μm), highly motile organism with an anterior cytostome and three to five free flagella. Typically, four flagella beat synchronously, while the fifth beats independently. A sixth, a recurrent flagellum, is associated with the undulating membrane and extends the length of the cell, the flagellum protruding beyond the posterior end as a trailing flagellum. *P. hominis* is generally considered a nonpathogen of the human colon, and while it is often associated with diarrhea, there is no definite evidence that it causes the condition. *P. hominis* has no cyst stage; so, transmission must occur via trophozoites, and flies may be implicated as mechanical vectors. The ability of trophozoites to survive for at least 24 h in feces-contaminated milk suggests that transmission may occur through contaminated food and drink and that trophozoites are able to withstand the acidic environment of the stomach en route to the intestine. Reproduction is by longitudinal fission. *P. hominis* infects dogs, cats, and mice and other rodents, with such hosts serving as reservoirs in nature.

Identification of trophozoites in fresh fecal preparations provides the most accurate means of diagnosis. It is important that only fresh samples be used since old stools may contain atypical or degenerating trophozoites resembling amoebae, which could result in their misidentification.

Suggested Readings

Adam, R. D. (1991). The biology of *Giardia* spp. *Microbiology Reviews*, *55*, 706–732.

Camp, R. R., Mattern, C. F. T., & Honigberg, B. M. (1974). Study of *Dientamoeba fragilis* Jepps and Dobell. I. Electron-microscopic observations of the binucleate stages. II. Taxonomic position and revision of the genus. *Journal of Protozoolgy*, *21*, 69–82.

Flanagan, P. A. (1992). *Giardia* diagnosis, clinical course and epidemiology: a review. *Epidemiology and Infection*, *109*, 1–22.

Honigberg, B. M. (1978). Trichomonads of importance in human medicine. In J. P. Kreier (Ed.), *Parasitic protozoa* (Vol 3). New York: Academic Press.

Kabnick, K. S., & Peattie, D. A. (1991). *Giardia*: a missing link between prokaryotes and eukaryotes. *American Scientist*, *79*, 34–43.

Mirhaghani, A., & Warton, A. (1996). An electron microscope study of the interaction between *Trichomonas vaginalis* and epithelial cells of the human amnion membrane. *Parasitology Research*, *82*, 43–47.

CHAPTER

6

Blood and Tissue Protistans I

Hemoflagellates

Chapter 6 discusses the family of blood feeding hemoflagellates that includes two genera, *Leishmania* and *Trypanosoma*, both of which require a blood feeding insect vector and infect humans. The chapter discusses the defining characteristics of the hemoflagellates including the subpellicular microtubular network, the kinetoplast, and the metamorphic transitions of the four morphologic forms observed in this family that include the amastigote, the promastigote, the epimastigote, and the trypomastigote. The chapter provides electron micrographic images and illustrative drawings of the various stages of development and key organelles like the flagellum and the mitochondrion with the associated kinetoplast DNA. The morphologic changes during the life cycle are well illustrated. This chapter further provides a more lengthy discussion of the physiology, pathology, epidemiology, symptomatology, and the host immune response of both genera of hemoflagellates.

Hemoflagellates belonging to two genera, *Leishmania* and *Trypanosoma*, infect humans. Both require blood-feeding insect vectors in their life cycles. The term **hemoflagellate** denotes the protozoan's site of residence in the human host: the blood and/or closely related tissues such as spleen and liver. During their life cycle, hemoflagellates may assume as many as four distinct morphologic forms. While these forms appear to be successive stages, there is no specific sequential pattern of progression from one form to the next. Indeed, it appears that any of the forms is capable of developing into any other. Each of the forms is discussed in detail below.

Certain organelles are common to all forms. Underlying the plasma membrane is a system of microtubules, the **subpellicular microtubular network**, which forms a spiral framework just beneath the surface of the pellicle and provides limited structural support (Fig. 3.5, p. 39). However, this subpellicular network is not associated with the lining of the flagellar pocket, the site of uptake of exogenous macromolecules (Fig. 6.1). A flagellum arises from a basal body in close association with a prominent kinetoplast forming a kinetoplast-basal body complex (Figs. 6.1 and 6.7). The organelle is rich in DNA that resembles mitochondrial DNA of other organisms, being composed of a limited number of nucleotides arranged in linked circlets. Kinetoplast DNA (kDNA) appears to be responsible for the elaboration of mitochondria in many of the forms as well as for the metamorphosis of one stage to another.

Human Parasitology. http://dx.doi.org/10.1016/B978-0-12-813712-3.00006-0

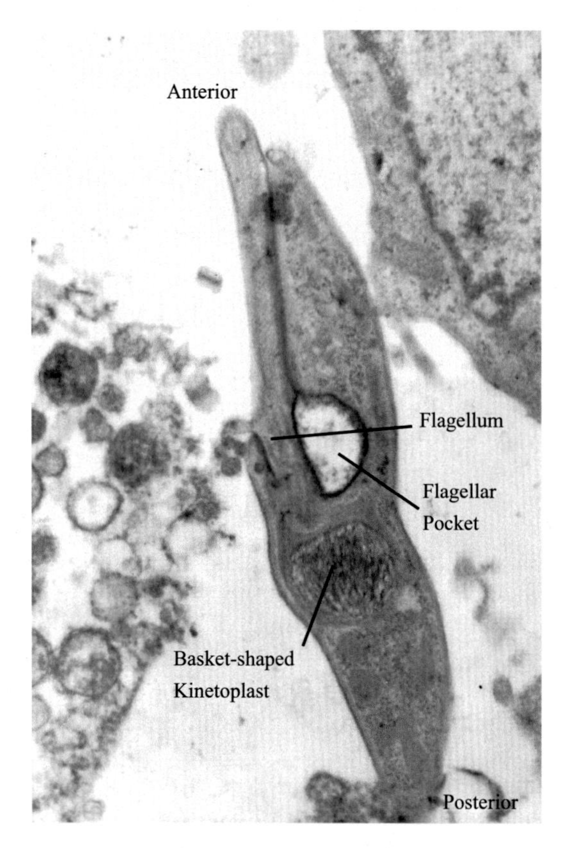

FIGURE 6.1 **Transmission electron micrograph of a developing trypomastigote exposed to exogenous protein.** Note the accumulation of protein in the flagellar pocket.

MORPHOLOGIC FORMS

Amastigote

The amastigote (Fig. 6.2) is ovoid in form and usually develops in vertebrate host cells. It is characterized by a single prominent nucleus and a very short flagellum projecting barely (if at all) beyond the cell surface. The organelles are as described in all other forms, although somewhat reduced in size and possibly in function (see above).

Promastigote

The promastigote (Fig. 6.3), which occurs only in the insect vector, differs morphologically from the amastigote in two significant aspects: (1) it is more elongated and (2) its long flagellum is free anteriorly and serves the function of both locomotion through the medium and attachment to the insect gut wall. In addition, protruding from the kinetoplast of promastigotes are two mitochondrial branches; a prominent posterior one, often extending the length of the cell, and a shorter, anterior one.

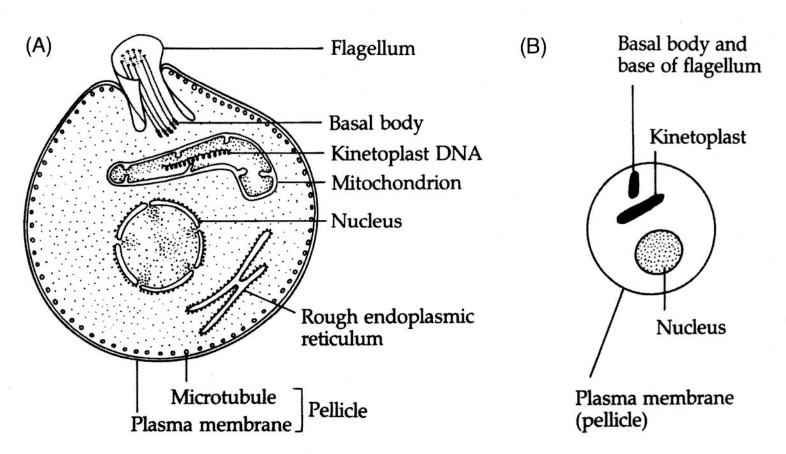

FIGURE 6.2 (A) Ultrastructure of a hemoflagellate amastigote. (B) Amastigote as it would appear by light microscope.

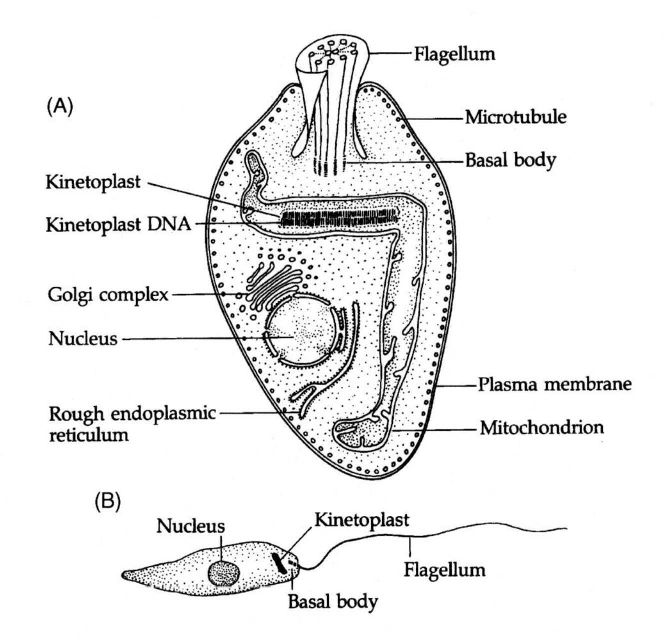

FIGURE 6.3 (A) Ultrastructure of a hemoflagellate promastigote. (B) Promastigote as it would appear by light microscopy.

Epimastigote

In this form (Fig. 6.4), the kinetoplast-basal body complex is situated more posteriorly but remains anterior to the nucleus. From its point of origin near the kinetoplast-basal body complex to its emergence at the anterior tip of the cell, the flagellum is attached to the pellicle, producing an undulating membrane. The distal, free portion of the flagellum projects anteriorly, and anterior and posterior mitochondrial branches remain well developed.

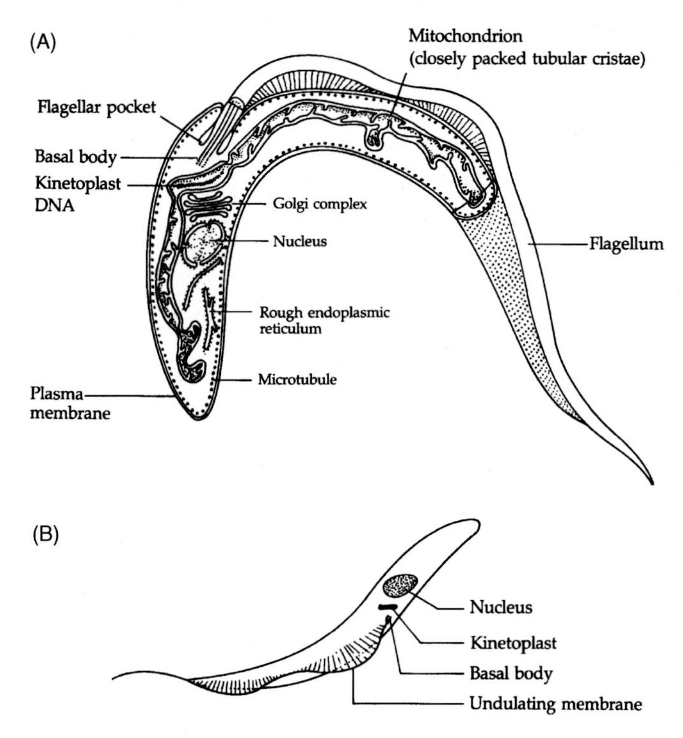

FIGURE 6.4 (A) Ultrastructure of a hemoflagellate epimastigote. (B) Epimastigote as it would appear by light microscopy.

Trypomastigote

The fourth morphological form (Fig. 6.5) discernible among hemoflagellates, the trypomastigote, exhibits varying degrees of polymorphism. One type, the **long, slender trypomastigote** (Figs. 6.6 and 6.7) is characterized by (1) lengthening of the body, (2) elongation of the undulating membrane and flagellum, and (3) migration of the kinetoplast-basal body complex to a site posterior to the nucleus. In this form, mitochondria are greatly diminished in function, and glycosomes, membrane-bound, microbody-like organelles containing at times crystalline cores are numerous in the cytoplasm. Another type, the **stumpy trypomastigote** (Fig. 6.6), is relatively shorter and thicker in form and has either a shorter or no free flagellum. The significance of polymorphism in the trypomastigote will be treated later.

Genus *Leishmania*

Through the use of molecular and immunological techniques, a number of species and subspecies of *Leishmania* have been partially characterized (Table 6.1). Of those that infect humans, three clinical manifestations are evident: **visceral**, **cutaneous**, and **mucocutaneous leishmaniasis**. While their life cycles are identical and they are morphologically indistinguishable, they differ in the type and location of primary lesions they produce in the human host. Leishmaniases are now endemic in 88 countries on five continents with a total of 350 million people at risk.

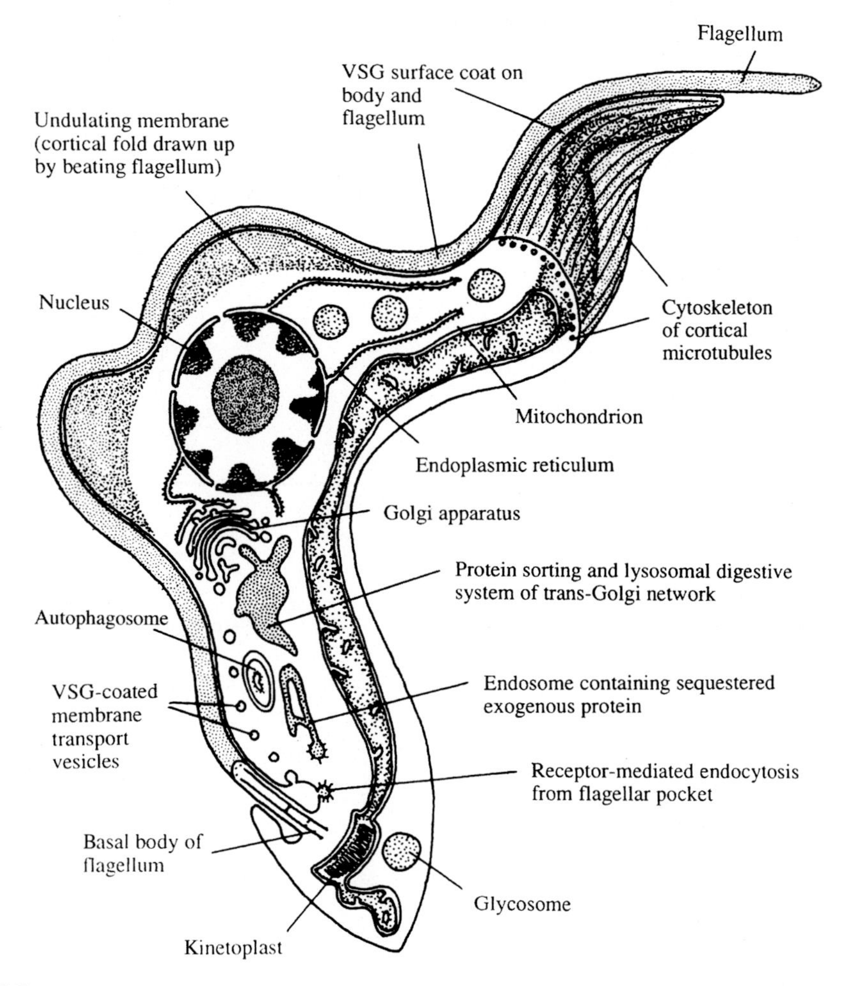

FIGURE 6.5 **Ultrastructure of a *hemoflagellate trypomastigote*.**

Life Cycle

For all species of *Leishmania* (Fig. 6.8), the portion of the life cycle spent in mammalian hosts is paradoxical in that the amastigote infects macrophages (Fig. 6.9), the very cells of the mammalian host that constitute its primary defense against invasion by foreign organisms. The parasite, upon entering the macrophage, establishes itself in an endocytotic vacuole called a **parasitophorous vacuole**. Lysosomes fuse with this vacuole, producing a variation of a secondary lysosome (=digestive vacuole). The amastigote, impervious to the lytic action of the lysosomal enzymes, lives and reproduces within the parasitophorous vacuole. A number of mammals act as natural reservoir hosts for the parasite, the most common being canines, both wild and domestic, and rodents. Leishmaniasis in humans is therefore considered a **zoonosis**.

In the course of obtaining a blood meal from a mammalian host, any of a wide variety of species of sand flies belonging to the genera *Phlebotomus* and *Lutzomyia* ingests infected cells

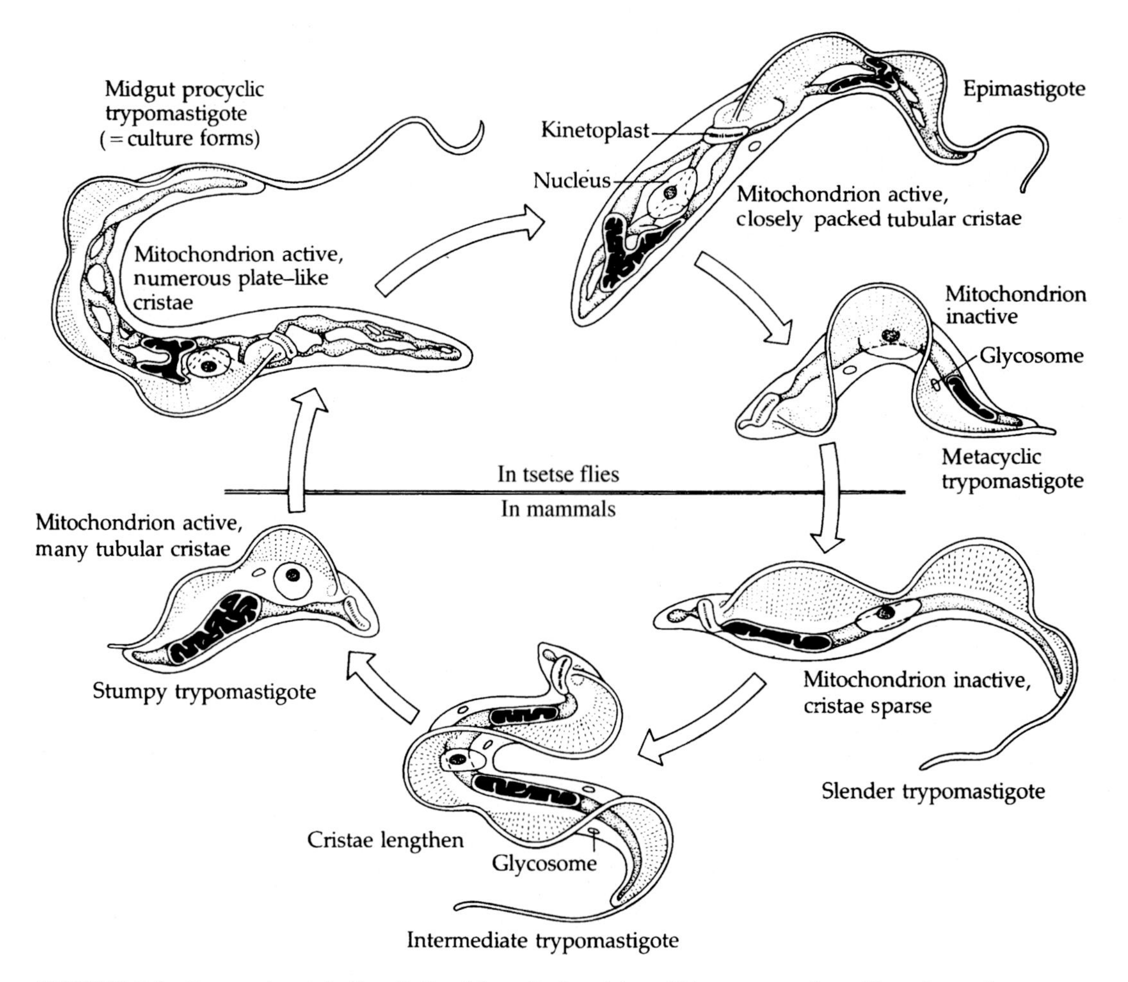

FIGURE 6.6 **Form and metabolic activity of the mitochondrion of *Trypanosoma brucei brucei* at various stages of its life cycle.**

containing the amastigotes. Following ingestion by the vector, the amastigote transforms into a promastigote in the gut of the insect. If the arthropod is a suitable vector, the promastigote attaches to the midgut epithelium; if otherwise, the organism passes out of the arthropod's gut. This attachment, therefore, serves dual purposes. It serves to retain the parasites in the vector's gut during passage of food, and it is essential for the transformation of the promastigote to the mammalian infective stage, the metacyclic promastigote. While attached to the wall of the gut, the promastigotes multiply by longitudinal binary fission. The reproductive rate is so rapid that, after 1–3 weeks, the anterior gut and the pharynx of the insect become clogged with promastigotes. The promastigotes, as they transform to infective metacyclic promastigotes, detach from the gut wall and are subsequently deposited in the skin of the mammal when the sand fly feeds again. Macrophages of the mammalian host quickly engulf the promastigotes, which then revert to the intracellular amastigote form. Reproduction of the amastigotes by longitudinal binary fission, followed by rupture of the infected host cells, produces large numbers of amastigotes, which are engulfed by other phagocytic cells, thus

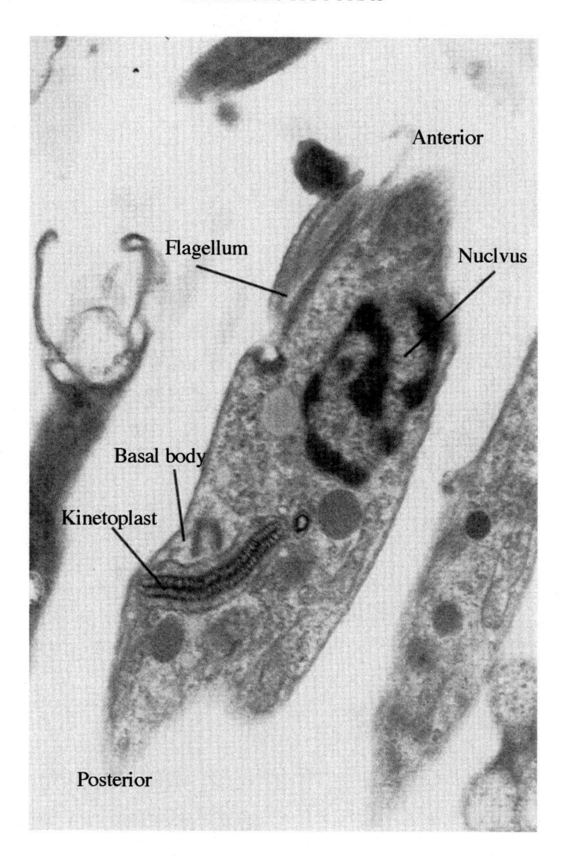

FIGURE 6.7 **Transmission electron micrograph of a developing trypomastigote.** Note the position of the basal body/kinetoplast complex just posterior to the nucleus.

TABLE 6.1 *Leishmania* Species and Forms of Human Leishmaniasis

OLD WORLD FORMS
L. major "wet" cutaneous: widespread in rural areas of Asia and Africa
L. tropica "dry" cutaneous: uncommon; urban areas of Europe, Asia, and North Africa
L. aethiopica "diffuse" cutaneous: Ethiopia and Kenya, associated with rock rabbits
L. donovani donovani visceral (kala-azar): Africa and Asia
L. donovani infantum infantile visceral: Mediterranean region
NEW WORLD FORMS
L. donovani chagasi cutaneous: South America
L. braziliensis braziliensis mucocutaneous: South America, especially Brazil
L. braziliensis guyanensis cutaneous: South America
L. braziliensis panamensis cutaneous: South and Central America
L. mexicana mexicana, *L. mexicana amazonensis*, *L. mexicana pifanoi* cutaneous: South and Central America
L. peruviana cutaneous: South America, mainly Andean region

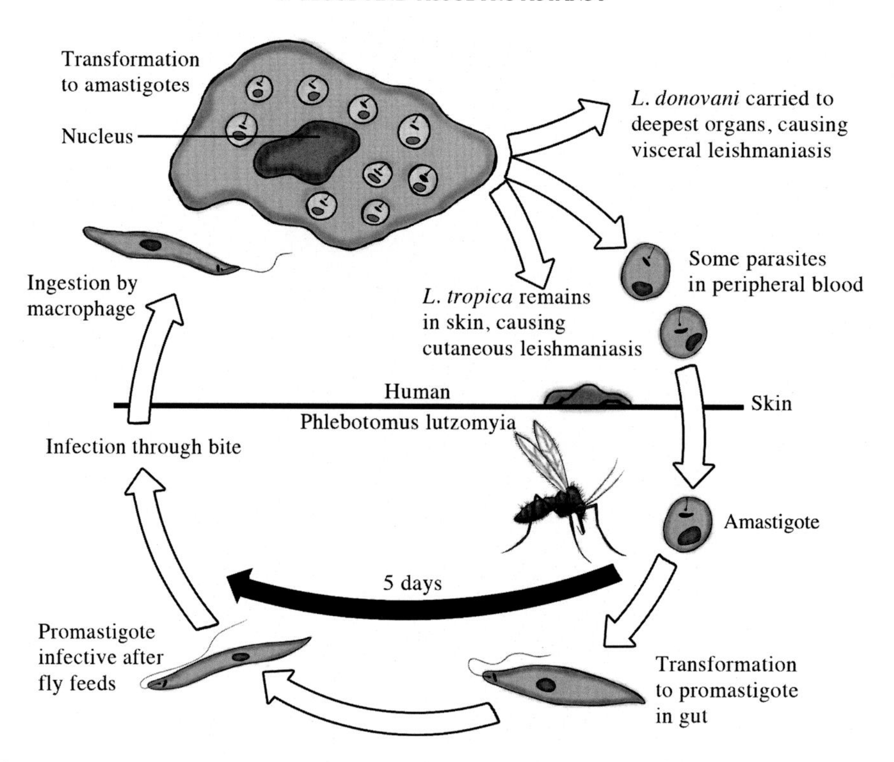

FIGURE 6.8 **Life cycle of *Leishmania*.** *Credit: Image courtesy of Gino Barzizza.*

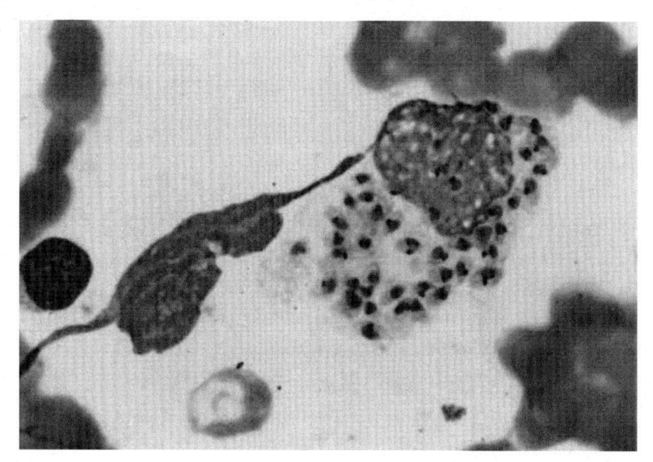

FIGURE 6.9 **Amastigotes of *Leishmania* spp. in macrophage.** *Credit: Image courtesy of Gino Barzizza.*

spreading the infection. Factors such as the species of *Leishmania* involved, temperature, immune status of the host, and even behavioral characteristics of the insect vector may determine the extent and site of infection in the mammalian host.

Reservoir hosts play an important role in the prevalence of leishmaniasis. In many regions of the world, domestic reservoir hosts, such as dogs, serve as a link between the sylvatic, or wild, reservoir hosts and the human population, via the sand fly vector. The reservoir hosts are usually unaffected by the parasites; thus, they serve as a constant source of infection for the human population. Where such reservoir hosts are present, they, rather than other infected humans, serve as primary sources for human infection via the bite of infected sand flies.

Physiology

Carbohydrate metabolism in members of the genus *Leishmania* is inextricably linked to the kinetoplast, the mitochondrion, and glycosomes of the amastigote and promastigote forms. For example, since the poorly developed mitochondrion of the amastigote includes neither a cytochrome system nor a functional Krebs cycle, the amastigote processes carbohydrates incompletely by anaerobic metabolism. This process occurs in glycosomes and cytosol, producing short chain, organic acids as end products and ATP by substrate-level phosphorylation. When the amastigote is ingested by the sand fly or subjected to in vitro culture conditions simulating conditions within the vector, the amastigote transforms to a promastigote. With this transformation, the mitochondrion grows, the number of cristae increases, and it becomes functionally and morphologically well developed with an active cytochrome system and functional TCA cycle. Under such conditions, the cell utilizes aerobic metabolism, producing ATP by oxidative phosphorylation. Such mitochondrial proliferation is influenced by kDNA.

The chemical components of the amastigote pellicle apparently protect the cell from the hydrolytic action of the macrophage lysosomal enzymes. Knowledge of the physiology of these organisms has not, to date, led to the development of effective chemotherapeutic agents or vaccines. However, the pentavalent antimony compound, antimony sodium gluconate (Pentostam), is currently being used effectively against most forms of cutaneous leishmaniasis, although its mode of action is not presently understood. Intravenous application of amphotericin B is considered investigational in the United States. Lipid-encapsulated amphotericins B (AmBisome, Abelcet, Amphotec), however, have been approved by the FDA for treatment of visceral leishmaniasis and some mucocutaneous forms.

Host Immune Response

Leishmaniasis is now considered an opportunistic infection (see p. 31) and is grouped among those diseases that take advantage of immunocompromised individuals.

Visceral Leishmaniasis (*Leishmania donovani*)

Leishmania donovani is the causative agent for visceral leishmaniasis (**VL**), also known as **dumdum fever** or **kala-azar**, the most severe and often fatal form of leishmaniasis. If left untreated, VL has a mortality rate of almost 100%. In the mammalian host, amastigote-infected cells are found at numerous sites (e.g., spleen, liver, bone marrow, lymph glands, and intestinal mucosa).

Epidemiology

Recent epidemiologic and clinical studies reveal the existence of at least three varieties or strains of *L. donovani* (Fig. 6.10D). The Mediterranean-Middle Asian variety occurs throughout

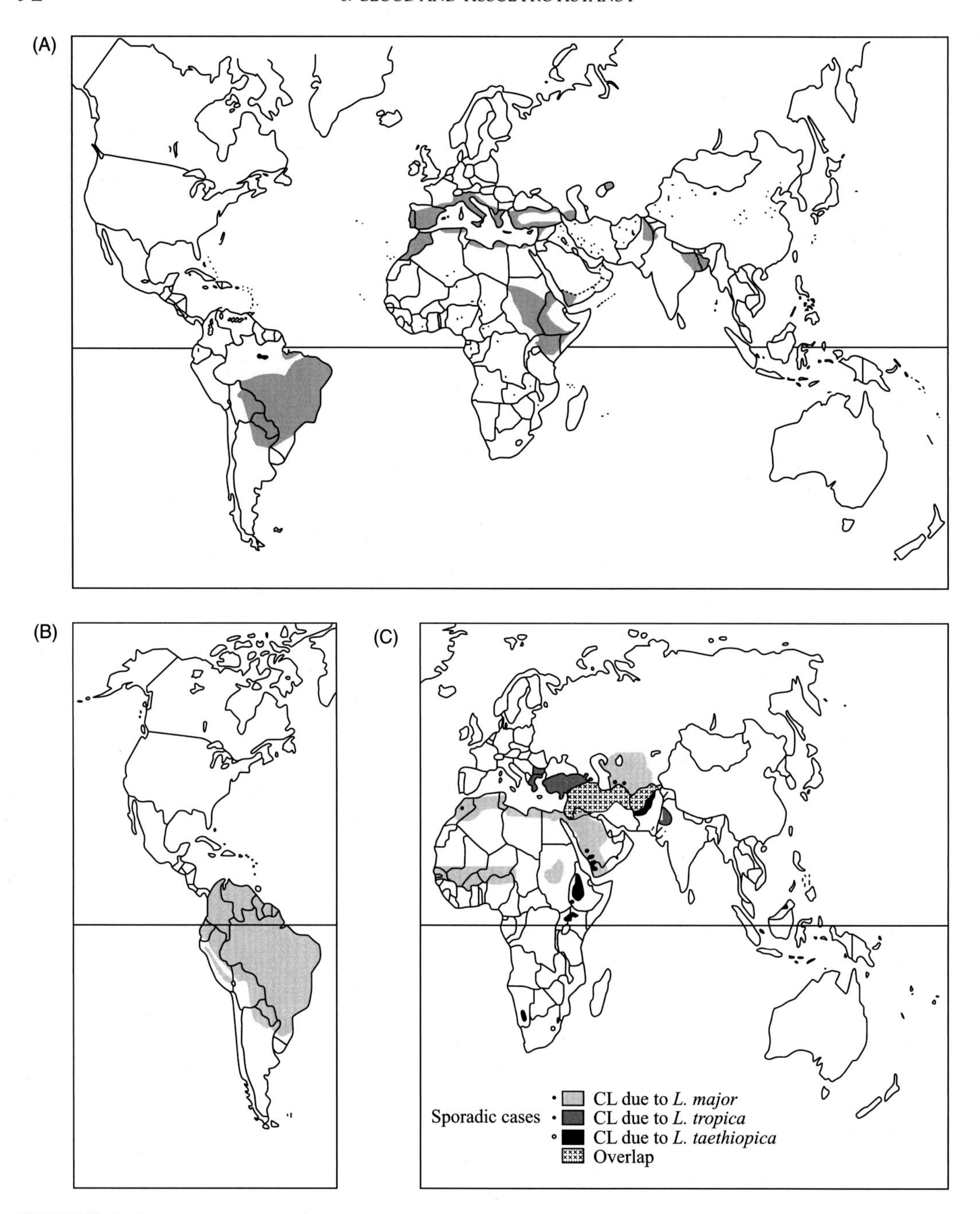

FIGURE 6.10 ***Global distribution of the various forms of leishmaniasis.***

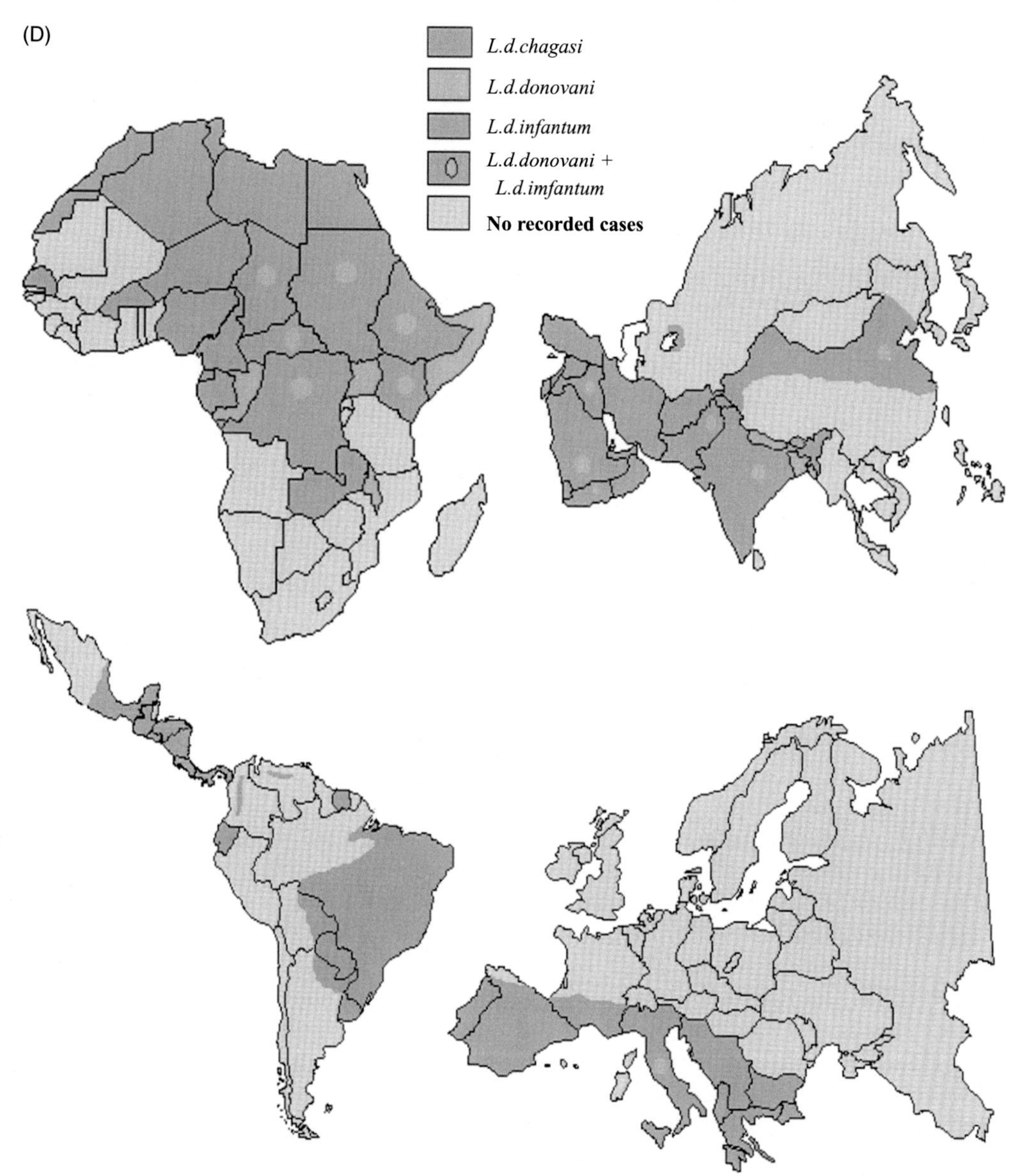

FIGURE 6.10 (*Cont.*).

the Mediterranean Basin and extends through southern Russia to China. The common sand fly vectors are *Phlebotomus major*, *Phlebotomus chinensis*, *Phlebotomus perniciosus*, and *Phlebotomus longicuspis*. A number of canines serve as both sylvatic and domestic reservoir hosts, and, since young children are the most frequent human victims, the disease is known as **infantile kala-azar**.

A second variety, classic kala-azar, occurs in northeast India and Bangladesh. The usual vector for this strain is *Phlebotomus argentipes*. The amastigote mainly infects adult and adolescent humans and involves no reservoir hosts. A more virulent, but clinically similar, variety is transmitted by different *Phlebotomus* species in east Africa and wild rodents serve as reservoirs.

A third variety, widespread in Central and South America, uses *Phlebotomus longipalpis* as the vector and both sylvatic and domestic canines as reservoir hosts.

Symptomatology and Diagnosis

Since leishmaniasis is primarily a disease of the reticuloendothelial system (macrophage system), replacement of infected cells produces hyperplasia and consequent enlargement of visceral organs associated with the system, such as the spleen and liver (splenomegaly and hepatomegaly). A concomitant decrease in red and white blood cell production results in anemia and leukopenia, facilitating secondary bacterial infection. Without medical treatment, the condition is usually fatal. Surviving individuals, however, commonly acquire long-lasting immunity. In India, a **post-kala-azar dermal leishmanoid** may develop in which numerous parasite-laden nodules appear in the skin. Such nodules are found in no more than 10% of fully recovered kala-azar patients.

In endemic areas, classic initial symptoms of kala-azar are fever and chills, which may persist for several weeks. The fever chart typically shows two fever spikes per day (a "dromedary" curve), a pattern that provides a useful diagnostic tool. In more advanced cases, enlargement of the liver and spleen produces abdominal distention (Fig. 6.11). Definitive diagnosis is the positive identification of intracellular amastigotes from blood or tissue smears. When such smears are inconclusive, other diagnostic techniques must be employed. One technique is **xenodiagnosis**, that is, the direct inoculation of laboratory animals, such as hamsters, with tissue homogenates from the patient. Signs of infection in the animal within 1 month constitute positive diagnosis. Biopsy and punctures of such organs as the spleen, liver, or sternum to reveal parasites are also useful diagnostic procedures. Immunological tests are used, but these are difficult to evaluate since postrecovery cases are indistinguishable from active cases. Further, such tests cannot differentiate among the various species of *Leishmania* and *Trypanosoma cruzi*.

Chemotherapy

Proper nursing care and complete bed rest are essential, especially in more acute cases; blood transfusions are frequently required. Chemotherapy consists of closely monitored intramuscular or intravenous injection of pentavalent antimony compounds such as antimony sodium gluconate. Extreme caution is essential in the administration of such treatment, since not only can antimony produce serious side effects but also insufficient treatment may result in relapses or post-kala-azar dermal leishmanoid. Lipid-encapsulated amphotericin B has been shown to be successful for treatment.

Cutaneous Leishmaniasis (*Leishmania tropica and Leishmania mexicana*)

Cutaneous leishmaniasis, a relatively mild skin disease commonly known as **oriental sore**, is caused by *Leishmania tropica* in the Old World and *Leishmania mexicana* in the New World.

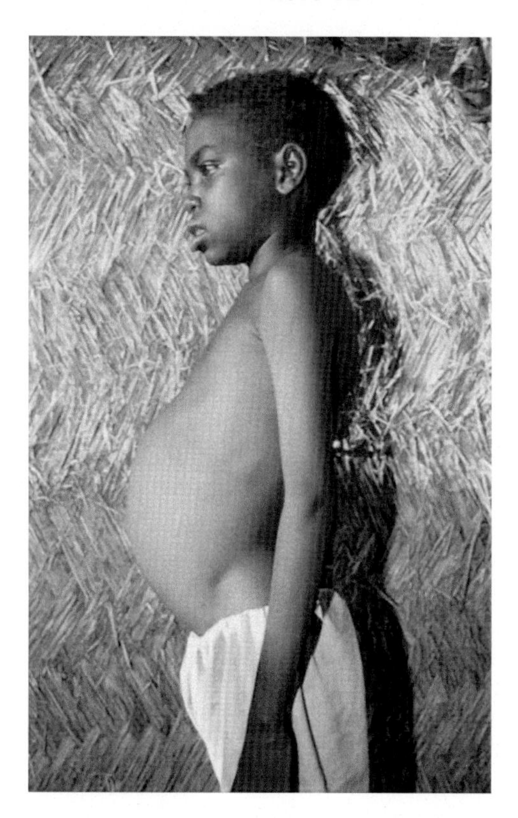

FIGURE 6.11 **Youngster infected with visceral leishmaniasis.** Note distention of abdomen.

It is the most common of the leishmaniases and represents 50%–75% of all new cases. Unlike the amastigote of *L. donovani*, those of *L. tropica* and *L. mexicana* are found primarily in macrophages around cutaneous sores. Sand flies must feed at these sites in order to acquire the infective amastigotes.

Epidemiology

L. tropica is endemic to those countries of Europe and northern Africa bordering the Mediterranean Sea and to the Asian countries of Syria, Israel, southern Russia, China, Vietnam, and India. *L. mexicana* has been reported from Peru, Bolivia, Brazil, the Guianas, and Mexico (Fig. 6.10C). A variant clinical form of cutaneous leishmaniasis occurring in South and Central America and Ethiopia is referred to as **diffuse cutaneous leishmaniasis (DCL)** (Fig. 6.10B). DCL never heals spontaneously and there is a tendency to relapse after treatment.

The vectors for *L. tropica* and *L. mexicana* are members of the sand fly genera *Phlebotomus* and *Lutzomyia*, respectively. The life cycles of *L. tropica* and *L. mexicana* parallel that of *L. donovani*. In addition to humans, *L. tropica* infects dogs and cats in China and a few Mediterranean countries. Natural infections are known to occur in monkeys, bullocks, and brown bears in the Middle East, as well as horses and gerbils. In some areas of the Middle East, the infection is endemic among rodents in whose burrows the sand flies live and breed. Humans, intruding in these areas, are readily infected. A few instances of *L. tropica* infecting the spleen and lymph glands (viscerotropic infection) have been reported in American military personnel in the Middle East following the Persian Gulf War (1990–91). In the Western Hemisphere, dogs serve as the primary domestic reservoir, while armadillos and arboreal rodents may also serve as sylvatic reservoirs.

Symptomatology and Diagnosis

In humans, the initial sign of the infection is the appearance of a vascularized papule or nodule on the skin at the feeding site of the insect. The papule becomes ulcerated after a few weeks, erupts, and spreads, forming cutaneous lesions most commonly on the hands, feet, legs, and face (Fig. 6.12). The usual incubation period varies from 1 to 2 weeks up to several months or even, in rare instances, several years. Two types of oriental sore are produced by different strains of the protozoan: (1) the chronic, dry (or urban) type (*Leishmania minor*, which is often considered a different species), producing delayed ulceration with numerous amastigotes, and (2) the acute, moist (or rural) type (*L. major*), characterized by early ulceration with few amastigotes.

In the absence of secondary bacterial contamination, sores tend to heal within a year, but disfiguring scars often remain. The lesions in diffuse cutaneous leishmaniasis, however, differ from those accompanying the usual infection in that they are disseminated as multiple nodules under the skin and contain numerous parasites in the associated macrophages, in this manner resembling lepromatous leprosy. This form of cutaneous leishmaniasis is found in patients with deficiency in their cell-mediated immune processes.

Characteristic features of the lesions of cutaneous leishmaniasis, such as an elevated and hardened rim of the ulcer, are useful in diagnosis. Positive diagnosis requires identification of amastigotes in infected cells. The most reliable diagnosis is achieved by in vitro culturing of lesion scrapings or aspirates and subsequent identification of promastigotes in the medium. An immunological test is available, but, as in *L. donovani*, its diagnostic value is limited.

Chemotherapy

Healing may eventually occur without chemotherapy, but the process is long and can produce disfiguring scars, especially if proper hygienic practices are not observed. Secondary microbial infections are a constant danger as long as the ulcer is open. Treatment of choice is a daily intramuscular injection of pentavalent antimony compounds for approximately

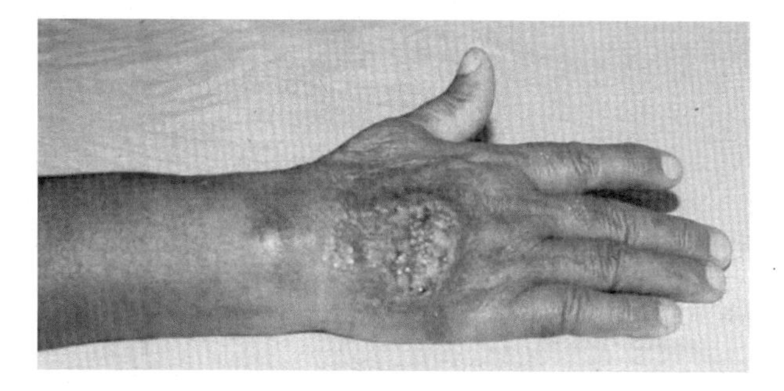

FIGURE 6.12 **Oriental sore or cutaneous leishmaniasis.**

1 week. A second or third course of treatment may be required. Concomitant topical antibiotic treatment is employed in cases of microbial contamination of skin lesions. Topical paromomycin can be used in regions where mucosal spread is low. A topical treatment in a paraffin-based combination of 15% paromomycin and 12% methylbenzethonium chloride (Leshcutan) has been reported to be effective against *L. major* infections.

Mucocutaneous Leishmaniasis (*Leishmania braziliensis*)

Leishmania braziliensis causes mucocutaneous leishmaniasis. Amastigotes are found in macrophages in ulcerations at mucocutaneous junctures of the skin. This disease is also known by various other names, including **American leishmaniasis**, **espundia**, **uta**, **pian bois**, and **chiclero ulcer**.

Epidemiology

The disease is common in humans in an area extending from the Yucatan peninsula in Mexico south to Argentina. While human infections have occurred in the Sudan, Kenya, Italy, China, and India, the disease is far more common in the Western Hemisphere, hence the name **American leishmaniasis**. Interestingly, its vector is not found in the high Andes mountains; therefore, the disease does not occur in that region (Fig. 6.10A).

A primary skin lesion appears following the bite of an infected sand fly of the genus *Lutzomyia*. Geographic location determines the site of secondary lesions. For instance, in Mexico and Central America the secondary lesion usually appears on the ear, causing chiclero ulcer, a condition common among the chicleros, forest-dwelling natives who harvest the gum of chicle trees. Recent investigations suggest that the variety of the parasite that causes chiclero ulcer may be a separate species, *L. mexicana*. This variation of the disease, like many forms of leishmaniasis, is zoonotic, and various forest rodents, dogs, cats, and kinkajous serve as reservoirs. Mucocutaneous involvement is seldom seen in the geographic region where *L. mexicana* is acknowledged as the causative agent.

In its southern range, the disease caused by *L. braziliensis* follows a different course. The secondary lesions erupt at the mucocutaneous junctures of the skin, with nasal and buccal tissues most often affected. In these geographic areas, the disease is commonly called espundia or uta.

Symptomatology and Diagnosis

As indicated earlier, clinical manifestations in the Western Hemisphere vary so widely that considerable confusion exists concerning the identity of the parasite responsible (Table 6.1). Some investigators attribute the entire battery of symptoms occurring at multiple loci to a single species, *L. tropica*; others list four or more species. Regardless, introduction of promastigotes into humans by the sand fly typically results in a small red papule on the skin, the **primary lesion** (Fig. 6.13), which ulcerates in 1–4 weeks and heals within 6–15 months. In Venezuela and Paraguay, primary lesions often appear as flat, ulcerated plaques that remain open and ooze. The disease is termed pian bois in these areas. A secondary lesion invariably appears elsewhere on the body. The sites of secondary lesions are usually distinctive. For instance, chiclero ulcer is associated with degeneration of the pinna of the ear, while degeneration of the cartilaginous and soft tissues of the nasal and buccal areas is characteristic of

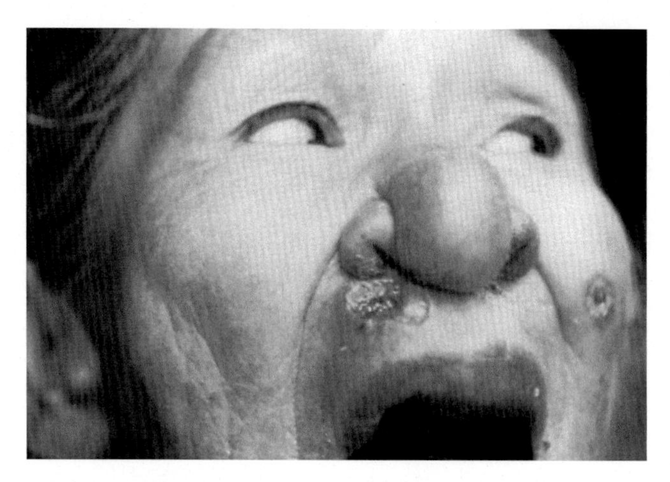

FIGURE 6.13 **Mucocutaneous leishmaniasis.** Lesions caused by *L. braziliensis*.

espundia and uta. Occasionally, infections metastasize to adjacent tissues, forming satellite lesions. Secondary bacterial and fungal infections are common.

Distinguishing the lesions of leishmaniasis from those of other skin diseases, such as yaws, syphilis, and chronic skin diseases, is essential for accurate diagnosis. The surest diagnosis is identification of the protozoan in infected cells and in cultures. Material collected by either aspiration or scrapings from edges of the lesions is suitable for such diagnostic studies. The most efficient method providing the most accurate results is histological examination of biopsy material. When possible, it is preferable to examine early lesions, since they yield more parasites than older lesions. The same problems exist with immunological diagnostic procedures for this form of leishmaniasis as are encountered in *L. donovani* and *L. tropica* infections.

Chemotherapy

Generally, the treatment with pentavalent antimony compounds used for the other forms of leishmaniasis is also used for this form. When there is mucocutaneous involvement, more extensive chemotherapy is indicated, since these lesions are most resistant to the usual regimen. For the most intractable cases, daily intravenous injections of amphotericin B for up to 10 days is recommended. Amphotericin B is toxic to some patients, and its administration should be closely monitored. If treatment is insufficient or is discontinued too soon, the parasite may remain dormant for many years then reemerge to cause relapse.

The author and journalist Douglas Preston, while accompanying an expedition searching for a lost city in the rain forests of Honduras, contracted mucocutaneous leishmaniasis. In his book, "The Lost City of the Monkey God," he graphically describes his infection with and treatment for the disease. His treatment with amphotericin B, or what he called *Amphiterrible B*, consisted of daily (4–5 h) administrations of the drug over a period of 6 days. The treatment was traumatic but successful. Since it was not prophylactic, he is still vulnerable to relapse especially if his immune system becomes compromised.

GENUS *TRYPANOSOMA*

Members of the genus *Trypanosoma* infecting humans can be divided into two major groups according to geographic distribution and characteristic pathogenicity. The African varieties, indigenous to that continent, cause a disease commonly known as **African sleeping sickness**. The other variety, confined to the Western Hemisphere, causes **American trypanosomiasis**, or **Chagas' disease**. The New World disease involves intracellular parasitism; however, the parasite also affects blood and other tissue fluids.

African Trypanosomiasis (*Trypanosoma brucei rhodesiense and Trypanosoma brucei gambiense*)

The two organisms responsible for African sleeping sickness are subspecies of *Trypanosoma brucei*, namely *Trypanosoma brucei rhodesiense* and *Trypanosoma brucei gambiense*. The life cycles of the organisms are essentially identical, the major differences being (1) which of the 21 species of the insect genus *Glossina* serve as vectors, (2) what animal serves as vertebrate host, (3) what time intervals are required for development within host and vector, and (4) what length of time is required for evolution of the disease in the vertebrate host. The following generalized life cycle, therefore, can be used for both organisms (Fig. 6.14).

Life Cycle

Introduction of the infective stage of the protozoan into the human host occurs with the bite of an infected tsetse fly vector belonging to the genus *Glossina*. In preparation for its blood meal, the insect secretes parasite-laden saliva into the dermis of its victim to dilate the blood vessels and prevent coagulation of the blood, simultaneously introducing the **metacyclic trypomastigote**, the infective form of the protozoan. Morphology and physiology of the mitochondrion distinguish the metacyclic trypomastigote from most of the other life cycle forms that occur in the insect vector. The mitochondrion of the metacyclic trypomastigote has few cristae and contains none of the electron transport complexes. The metacyclic trypomastigote is blunt, with a short free flagellum. Once introduced into the mammalian circulatory system, metacyclic trypomastigotes transform to long, slender trypomastigotes (Fig. 6.15), spread rapidly within the host, eventually entering the cerebrospinal fluid. In the mammalian bloodstream, trypomastigotes exhibit three forms: (1) a long, slender form with a free flagellum extending from the undulating membrane; (2) a short, stumpy form lacking a prominent free flagellum; and (3) a form intermediate between the two.

In order for the parasite's life cycle to be completed and for the tsetse fly to transmit sleeping sickness, the insect must ingest in its blood meal the short, stumpy trypomastigote, which is physiologically adapted for existence within the insect vector. The presence of a mitochondrion with prominent cristae and a functional electron transport system enables this form to live in the environment of the insect midgut where food may be at a premium compared to the vertebrate host's bloodstream. Once ingested by the insect, the stumpy trypomastigotes elongate, lose their surface coats and antigenic identity, and become **procyclic trypomastigotes**. These multiply by longitudinal binary fission and invade the extraperitrophic spaces.

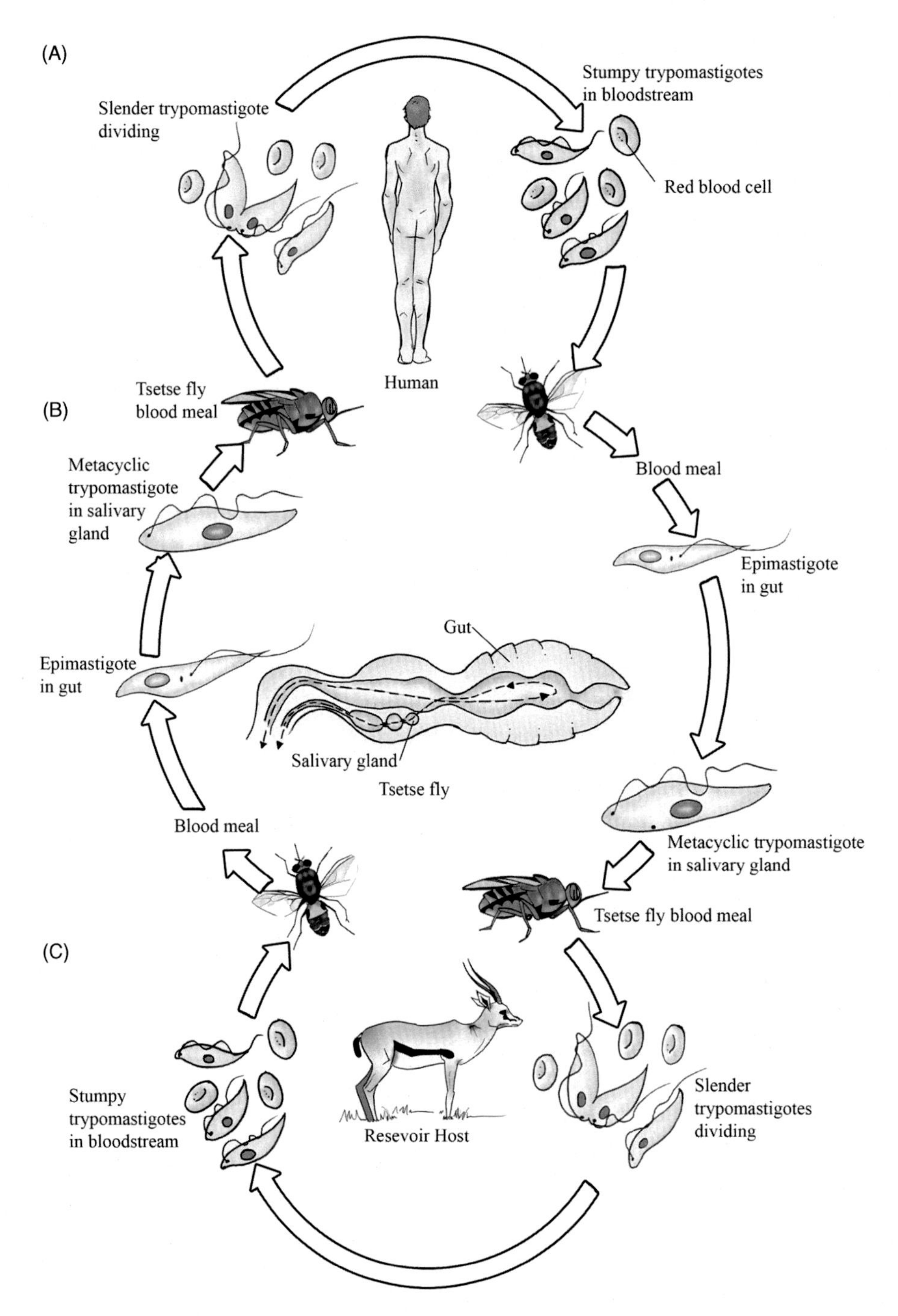

FIGURE 6.14 **Life cycles of *Trypanosoma brucei gambiense* and *T. b. rhodesiense*.** (A) Development of trypanosomes in the peripheral blood of humans: infection of central nervous system. (B) Development in tsetse flies (*Glossina* spp.). (C) Development similar to (A) in the peripheral blood of the reservoir host (e.g., an antelope). ***Credit: Image courtesy of Gino Barzizza.***

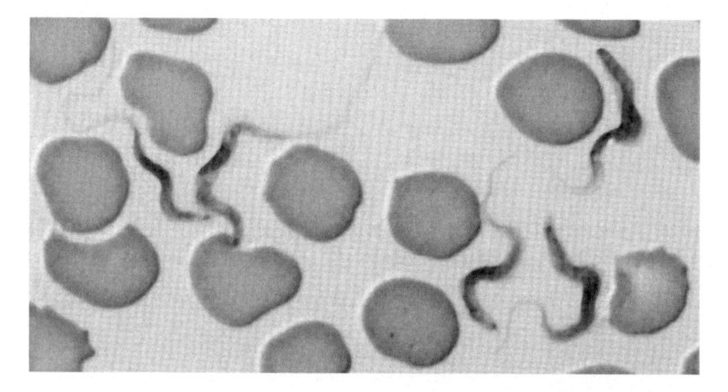

FIGURE 6.15 **Slender form of *Trypanosoma brucei rhodesiense*, found in the bloodstream of its mammalian host.**

As their numbers increase, they migrate anteriorly, and by the 10th day after ingestion, they enter the proventriculus.

Metamorphosis during this migration to the insect's foregut produces the **epimastigote**, the dominant form in the esophagus and buccal cavity of the fly. By the 20th day, the epimastigotes move into the salivary gland ducts, attach to the epithelium by their flagella, multiply, shortly after transform into metacyclic trypomastigotes, and detach into the lumen of the gland. Thus, in trypanosomes that cause African sleeping sickness, two distinct forms, the trypomastigote, with its morphological and physiological variants, and the epimastigote, are essential for completion of the life cycle.

Epidemiology

African sleeping sickness has probably plagued human inhabitants of Africa since humans first encroached upon the domain of the tsetse fly. At the turn of the 20th century, three quarters of a million people died in Central Africa from African trypanosomiasis. Subsequently, the colonial regimes in Africa were very successful in curbing the disease and almost eradicating it. However, since the era of independence of the 1960s, control measures have deteriorated in many African nations. Since 1990, the rate of infection has risen 10-fold and may now surpass 1 million cases each year.

Its pathological effects on humans and domestic animals (primarily cattle) have sometimes brought productive activity in certain areas to a virtual standstill. So devastating is the disease that the area of Africa between the 15th northern parallel and the 15th southern parallel (an area approximately the size of the continental United States), with the potential for supporting 125 million head of livestock, still lies essentially useless (Fig. 6.16). The toll in human victims is staggering: more than 20,000 new cases are diagnosed annually, about 50% fatal, and of the remaining 50% many resulting in permanent brain damage. The disease has reached epidemic proportions in Zaire and in Angola.

Since the disease also affects cattle, sheep, and goats in a region where human diet has historically been acutely deficient in protein, its effects are compounded. A related disease in domestic animals, called **nagana**, is caused by similar organisms, *Trypanosoma brucei brucei*, *Trypanosoma vivax*, and *Trypanosoma congolense*. Authorities disagree as to whether humans are susceptible to *T. b. brucei*. Fortunately for the rest of the world, the tsetse fly appears to be the only insect capable of transmitting African trypanosomiasis that causes human sleeping sickness, and Africa appears to provide its only habitat.

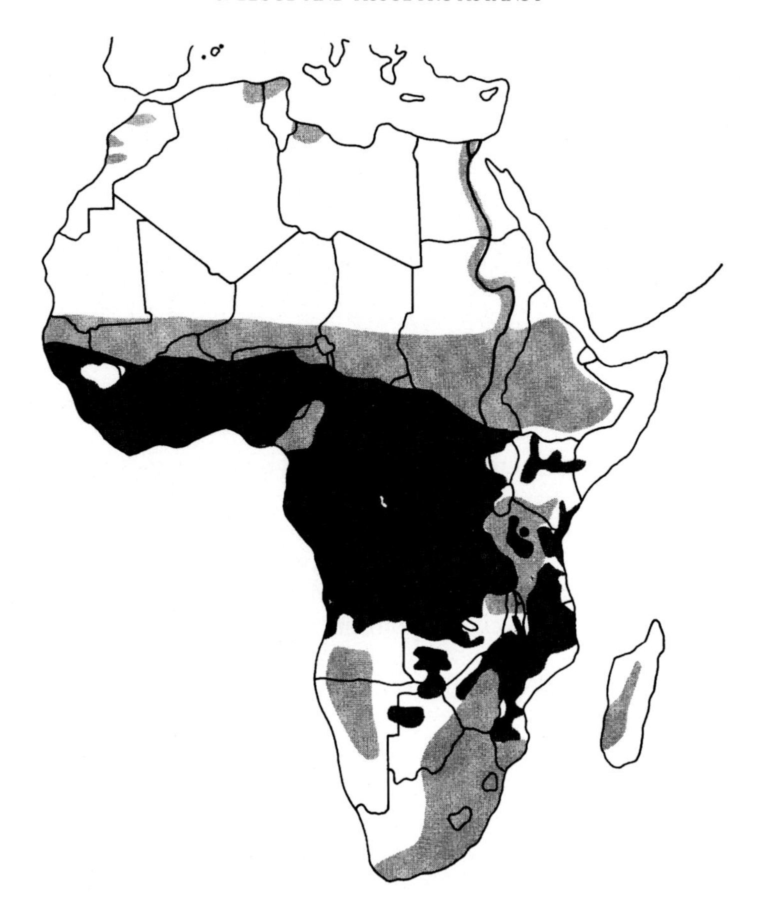

FIGURE 6.16 **Africa's cattle-raising country (*light shading*) and tsetse fly-infested areas (*dark shading*) show virtually no overlap.** Trypanosomiasis spread by tsetse flies has kept 10 million square kilometers of grazing land out of production.

T. b. rhodesiense causes the more virulent, **East African** or **Rhodesian**, form of African sleeping sickness and is usually transmitted by *Glossina morsitans*. In addition to human hosts, wild game and domestic animals in which this parasite is endemic may serve as reservoir hosts and sources of human infection. The disease runs its course so rapidly (2–6 months) in humans that person-to-person transmission via the tsetse fly is uncommon.

T. b. gambiense causes **West African**, **Equatorial African**, or **Gambian** sleeping sickness, a more chronic form of the disease. Domestic and wild animals, such as pigs, antelopes, buffaloes, and reed bucks, may serve as reservoir hosts. The insect vectors are *Glossina palpalis* and *G. tachinoides*. Human-to-human transmission via the bite of the tsetse fly is common since the trypomastigotes can remain in circulating blood for 2–4 years, providing ample opportunity for vectors to transmit them.

Symptomatology and Diagnosis

The diseases caused by the two organisms are very similar except for the interval required for their development in humans. In general, shortly after the introduction of metacyclic trypomastigotes through the bite of the tsetse fly, an inflammatory reaction of 1–2 days' duration occurs at the site of the bite. The characteristic reaction, a **trypanosomal chancre**, includes reddening of the skin, a swelling of 1–5 cm diameter, and enlargement of adjacent lymph nodes. When the blood and lymph are invaded, headache and irregular fever develop. These symptoms are accompanied by further enlargement of lymph glands, especially in the neck and supraclavicular areas. During this period, the patient may show a number of neurological symptoms, such as tremors of the tongue and eyelids, and some mental dullness manifested as progressive apathy. From this point, neurological symptoms dominate the clinical picture along with increased apathy, loss of appetite, extended daytime sleeping, and concurrent involvement of the muscular system, progressing to paralysis. Classic symptoms are rapid weight loss due to anorexia, anemia induced by malnutrition, drowsiness, and, finally, irreversible coma.

Typically, diagnosis of the disease is a multistep procedure. Step 1 is clinical assessment, especially when there are telltale neurological signs and/or mental dullness accompanied by enlarged and sensitive cervical lymph nodes (known as **Winterbottom's sign**). Step 2 is examination of blood smears, marrow, or cerebrospinal fluid for trypomastigotes. Finally, Step 3, if results from the previous two steps are inconclusive, is testing for specific antibodies in the blood. A complete history of the patient is also required for verification of contact with known tsetse fly habitats. It is estimated that this diagnostic regimen produces accurate results in more than 80% of cases.

Chemotherapy

In his search for a successful trypanocidal agent, Paul Ehrlich, a pioneer in therapeutic research and immunology in the early 20th century, developed a series of compounds such as trypan blue and trypan red. These compounds were used as chemotherapeutic agents for some time until it was discovered that their level of toxicity in humans was too great to justify their continued use. Suramin sodium, one of the current drugs of choice (see below), evolved from these earlier agents. The research by Ehrlich, in spite of its failure at the time to produce a satisfactory drug for curing African sleeping sickness, proved that chemotherapeutic agents could be effective against disease-producing organisms; Ehrlich eventually succeeded in developing salvarsan, one of the first drugs to treat syphilis.

At the present time, therapeutic drugs are most effective against African sleeping sickness. Treatment should be initiated early in the course of the disease, prior to central nervous system involvement, since most effective agents do not pass the blood–brain barrier. Chemotherapy begun later becomes increasingly complex with diminished effects. During the hemolymphatic stage, it is recommended that six intravenous injections of suramin sodium be administered over a 3-week interval. For the late, central nervous system stage, the arsenic-containing compound melarsoprol is administered intravenously for several weeks. Berenil is effective against early infections and, in combination with melarsoprol, is used for late human African trypanosomiasis infections. These two drugs, as well as a number of alternatives, can cause various toxic side effects. Severe reactions are rare, however, and the usually mild adverse effects are not sufficient to contraindicate treatment when weighed against the virulence of a disease that is almost invariably fatal when left untreated. As a result of field tests in Africa, a new drug, difluoromethylornithine, has shown promise against *T. b. gambiense* infection during the late, nervous system infective stage.

Host Immune Response

In human African sleeping sickness, especially Gambian sleeping sickness, fluctuation in the number of parasites in the blood is common, producing periods of remission alternating with periods when the parasite census is high. These fluctuations are attributed to the ability of the organism to vary the chemical composition of its **surface glycoprotein (VSG)** (Fig. 6.17). As the number of parasites increases, the VSGs of the parasites elicit an effective immunoglobulin M (IgM) response by the host. The resultant antibody-dependent complement-mediated lysis of the parasites rapidly depletes the parasite population. A small

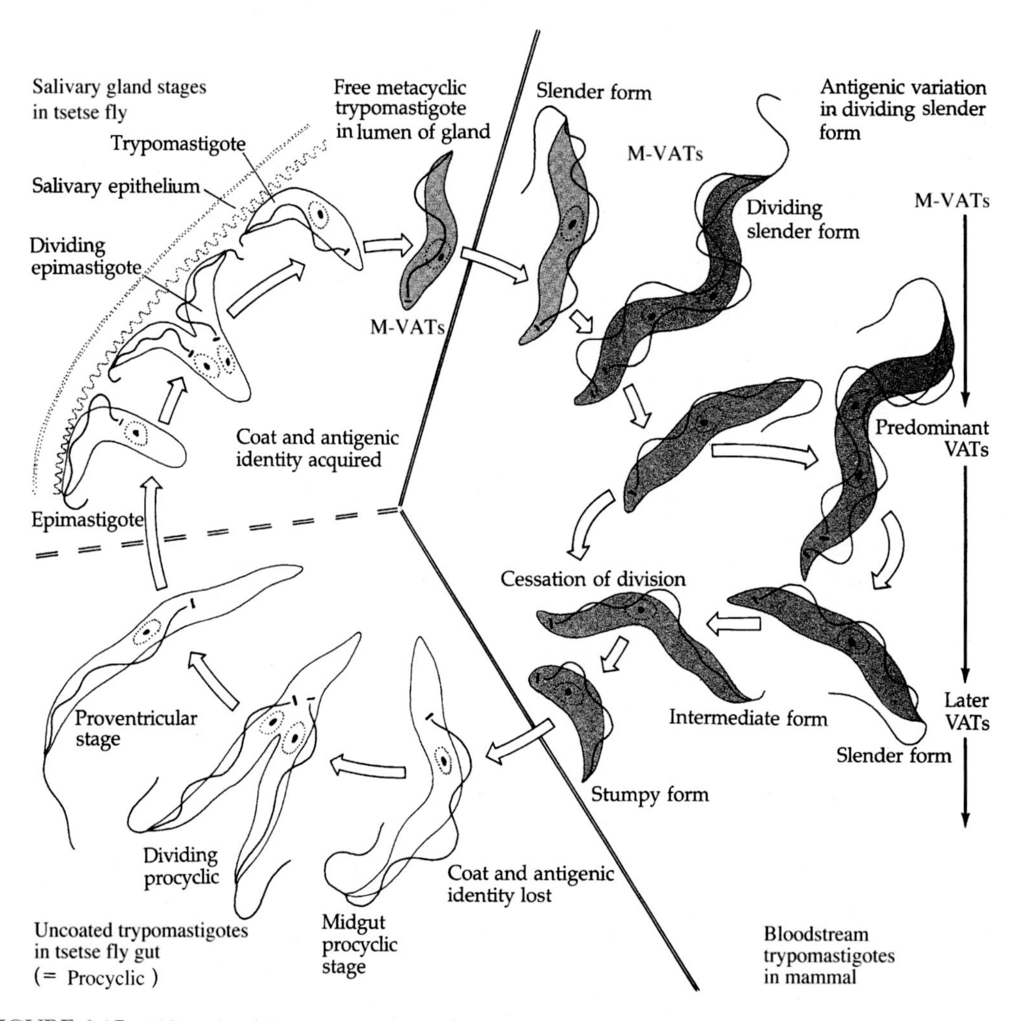

FIGURE 6.17 **Life cycle of *Trypanosoma brucei brucei* showing phases of multiplication.** Shaded stages possess the variable antigen-containing surface coat (VAT). The coat is acquired in the salivary glands of the tsetse fly, when free-swimming metacyclic trypomastigotes arise from vector-attached trypomastigotes. Only the coated metacyclic stage can infect the mammal. After ingestion by the vector, the stumpy trypomastigotes transform into procyclic trypomastigotes, simultaneously losing the variable antigen coat.

number (less than 1%) of the surviving parasites survive due to their possession of a variant VSG. These survivors proliferate until they, in turn, elicit a host response causing the cycle to continue. Successive waves of parasite populations thus can temporarily escape the host's antibody response directed against the preceding VSG. The end result is a veritable parade of parasites of **variant antigenic types (VATs)** in the vertebrate host. It is estimated that more than 1000 variants transcribed by more than 1000 genes are theoretically possible for *T. b. gambiense*. Of these variants, only one is genetically dominant at any one time, and the maximum number of trypanosome VATs that has been shown to develop in an untreated host is 101. The expression site of each gene is accomplished through monoallelic transcription. Each site is adjacent to a chromosomal telomere, but only a single site will be actively transcribed in any given organism. Usually the active VSG gene is duplicated and translocated to one of approximately 20 potential blood stream VSG expression sites. Recent genomic evidence indicates that the active site is located in an extranucleolar body where it is transcribed by RNA polymerase I. With each alteration of the coat, the immunological mechanism of the vertebrate host is activated, gradually depleting the ability of the host immune system to respond. The antigenic variability of these parasites makes the search for an effective vaccine conferring lasting protection an unpromising avenue for the control of this disease.

Physiology

The physiology of the long, slender trypomastigote found in the circulatory system of the vertebrate host differs from that of the epimastigote observed in the insect vector. Morphological changes related to metabolic characteristics occur in the mitochondrion of each form. For example, the epimastigote with its well-developed mitochondrion utilizes oxidative phosphorylation for synthesis of ATP in addition to glycolysis. A dearth of metabolizable substrate in the lumen of the insect gut dictates that the epimastigote evolve an efficient system for synthesis of energy-rich compounds. On the other hand, the long, slender trypomastigote, living in the organic cornucopia of the mammalian blood stream, derives no selective advantage from substrate conservation; so, it can obtain energy-rich compounds by the far less efficient system of substrate-level phosphorylation via glycolysis. Glycolysis, regardless of the morphological stage of the parasite, occurs in specialized organelles called **glycosomes**. Reduced nicotinamide adenosine dinucleotide (NAD) is oxidized in the glycosome indirectly by an α-glycerophosphate oxidase system, part of which is localized in the mitochondrion.

American Trypanosomiasis (*Trypanosoma cruzi*)

Using specific probes for *T. cruzi* kDNA and tissue specimens from human mummies from Chile and Peru, researchers were able to show that the sylvatic cycle of Chagas' disease was probably well established in humans more than 9000 years ago. However, it was not until 1909 that the life cycle was elucidated. The Brazilian physician and scientist Carlos Chagas found that thatched roof huts in a small village in Brazil were infested with large, blood-sucking insects whose digestive tracts were laden with flagellates able to infect laboratory animals. Diseased children inhabiting these infested huts were found later to harbor the same flagellates. Today, this disease is recognized as an American form of trypanosomiasis caused by *T. cruzi*. The disease is named **Chagas' disease** in recognition of its discoverer.

Life Cycle

The stage of *T. cruzi* infective to humans, the metacyclic trypomastigote, develops in the hindgut of the cone-nosed insect, or "kissing bug," *Panstrongylus megistus* and many related hemipteran insects (Fig. 6.18). Since development to infectivity occurs in the hindgut rather than the salivary glands, *T. cruzi* is placed in the section Stercoraria. Metacyclic forms are passed with the feces of the bug, usually as it is taking a blood meal from a vertebrate host, and infection occurs when infected fecal material is rubbed into the bite wound, eyes, or mucous membranes. Mammalian reservoir hosts may become infected by ingestion of infected insects or small mammals.

Upon entering the bloodstream of the vertebrate host, trypomastigotes are phagocytosed by a variety of cells, including macrophages, and rapidly transform into the amastigote form. The transformation is activated, in part, by the low pH of the lysosomal contents during phagocytosis. The resulting amastigote evades the lysosome system by escaping into the cytosol of the infected cell. This is accomplished by a *T. cruzi* pore-forming molecule activated by the low pH of the lysosome thereby causing a rupture of the lysosomal membrane.

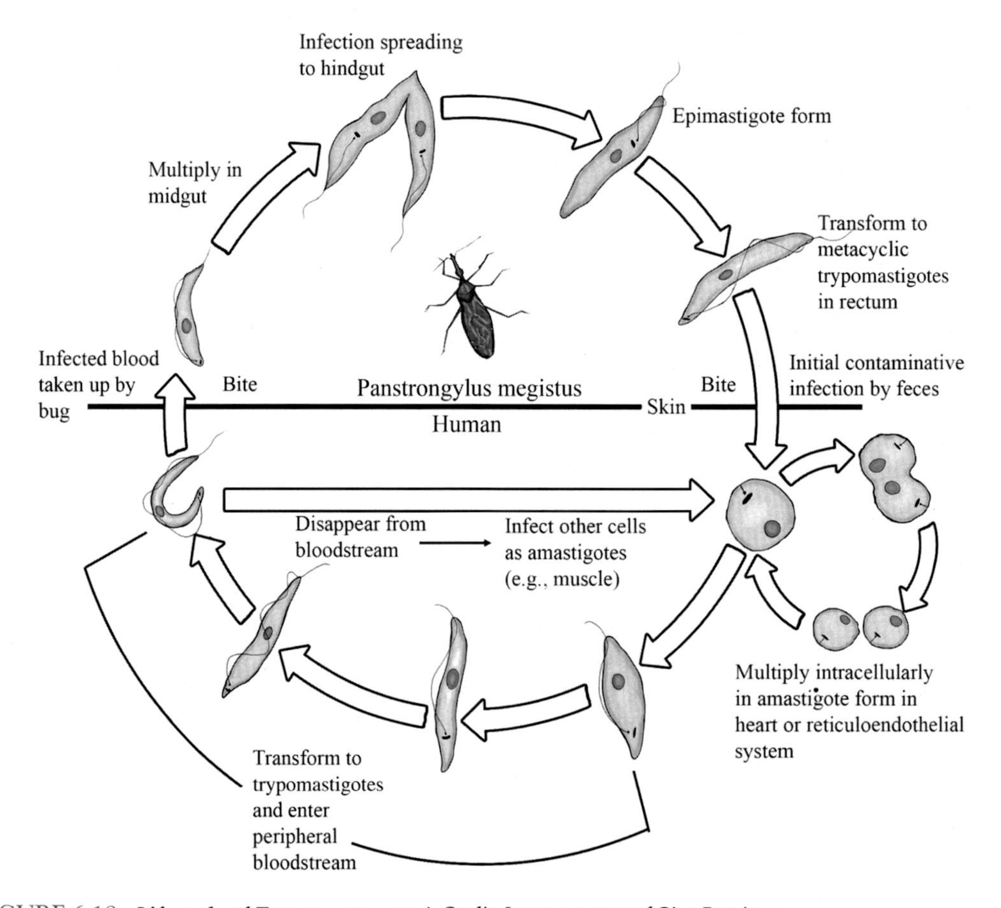

FIGURE 6.18 **Life cycle of *Trypanosoma cruzi*.** *Credit: Image courtesy of Gino Barzizza.*

The organs most vulnerable to infection are the spleen, liver, lymph glands, and all types of muscle cells. Other areas, such as the nervous and reproductive systems, intestine, and bone marrow, are occasionally invaded. Following repeated longitudinal binary fissions by the amastigotes, the infected cell ruptures and the released amastigotes enter other cells. When a cluster of amastigotes occurs in a cardiac muscle fiber, the aggregate is termed a **pseudocyst**. Some of the released amastigotes revert to the trypomastigote form and enter the circulatory system, but, unlike other trypanosomes such as the African trypanosomes, the trypomastigote form of *T. cruzi* never reproduces in mammalian blood plasma. In chronic cases, trypomastigotes are rarely observed in the blood since they can be effectively destroyed by circulating antibodies. Unlike salivarian trypanosomes, *T. cruzi* appears unable to produce variable surface antigens. This, in turn, apparently protects amastigotes in various host cells from antibody reactions. Trypomastigotes of *T. cruzi* also differ morphologically from those of African trypanosomes, being shorter (about 20 μm long) and showing a characteristic "U" or "C" shape in stained blood preparations (Fig. 6.19).

Insects become infected by ingesting blood containing trypomastigotes, which then undergo repeated longitudinal fission during passage through the digestive tract of the insect. By the time they reach the midgut, they have metamorphosed to the epimastigote stage. Still replicating, the epimastigotes pass into the insect hindgut where they attach by their flagella to the epithelium of the rectal gland. As transformation to the metacyclic trypomastigote forms occurs, the flagellar epithelial attachment is lost. By the 10th day after ingestion, the infective metacyclic trypomastigotes appear free in the lumen of the rectum.

Epidemiology

T. cruzi infection is prevalent throughout Latin America, affecting an estimated 24 million people, with approximately 100 million people at risk of contracting the infection. The most common mode of acquiring *T. cruzi* infection is through either the blood-sucking triatomine

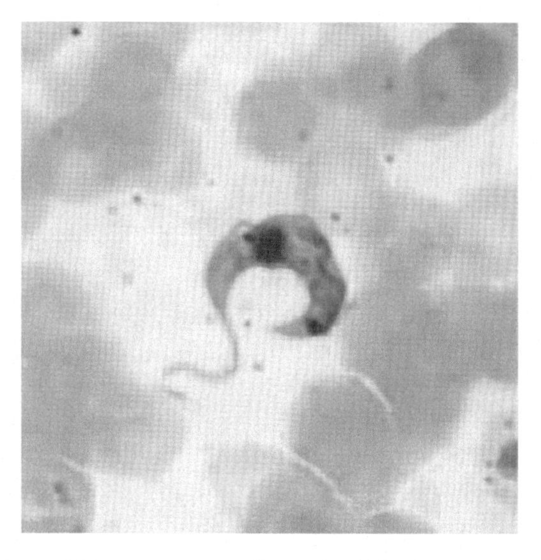

FIGURE 6.19 **Trypomastigote of *Trypanosoma cruzi* in a stained blood smear.** Note the "C" shape of the parasite.

bug or blood transfusion. Incidence is greatest in rural areas, especially among the poor, whose primitive living conditions facilitate bites by the insect vectors. American trypanosomiasis is a typical zoonotic disease that affects dogs, cats, bats, armadillos, rodents, and other mammalian reservoir hosts. Rodent infections also occur in the southwestern United States and in raccoons as far north as Maryland and Illinois. However, only a few cases of naturally acquired human infections have been reported in the United States and only in Texas and California. Several explanations have been advanced to account for the dearth of human infection in the United States. One maintains that the insect vectors are **zoophilic**; that is, they prefer animal hosts to human hosts (converse = **anthropophilic**). Another suggestion is that perhaps defecation by northern insect vectors occurs not during a blood meal but well afterward, thereby reducing the probability of infection.

Symptomatology and Diagnosis

Upon introduction into the human, the parasites invade macrophages of the subcutaneous tissue at the site of infection, causing a local, edematous swelling called a **chagoma**. In endemic areas, as the disease progresses, human patients often exhibit edematous patches over the body, which occur most frequently on one side of the face. This unilateral edema is often periorbital and associated with conjunctivitis, a syndrome known as **Romaña's sign** (Fig. 6.20). In early stages of the disease, parasites abound in infected tissues as well as in the circulating blood. As the infection becomes more chronic, the number of parasites in the circulation diminishes greatly, to the extent that they are almost undetectable. In acute cases, approximately 1–3 weeks after infection, fever, headaches, malaise, and prostration may develop. Enlargement of the liver and spleen as well as myocardial damage may follow, but cardiac involvement and gastrointestinal symptoms (**megasyndrome**) may not become apparent until many years after the primary infection.

Examination of fresh blood within the 1st month or two following infection may reveal *T. cruzi* trypomastigotes, particularly if the blood is drawn during a fever episode. Blood

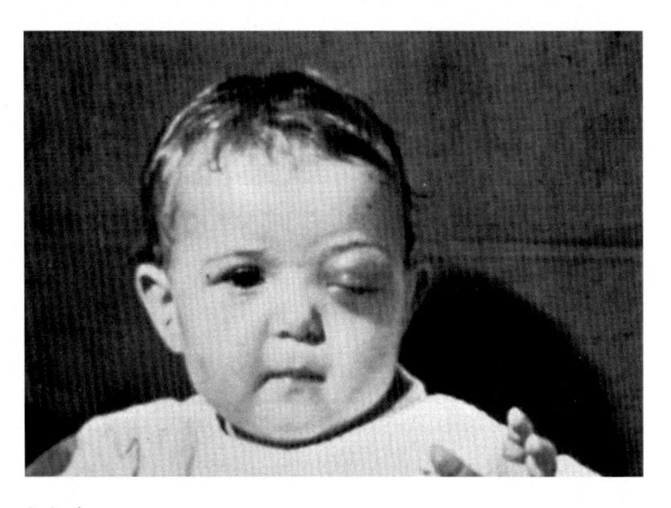

FIGURE 6.20 **Romaña's sign.**

cultures may also yield incriminating organisms, and serological tests and clinical examinations sometimes provide accurate diagnosis. Direct agglutination tests with the IgM serum fraction are sensitive enough for diagnosis in the acute cases when infectious organisms are scarce. Normally, antibodies do not develop until several months following initial infection, rendering other serological tests, such as complement fixation and immunofluorescence, impractical. Another simple, but practical, method employed in public health surveys is called **xenodiagnosis**. This procedure involves allowing a laboratory-raised vector to feed on a suspected patient, dissecting the insect after 2–3 weeks, and examining it for intestinal flagellates. A **polymerase chain reaction** (**PCR**) technique has been developed but, because it is still expensive and difficult to perform in a field situation, it has not come into general use.

Chemotherapy

Current lack of an effective chemotherapeutic agent makes treatment of Chagas' disease extremely difficult. A Bayer product, nifurtimox, has shown promise in treating early chronic and acute cases. Once the protozoan invades the host cell, it apparently is shielded from the action of any drug. Limited success against the bloodstream form has been achieved by treatment with the antimalarial drug primaquine phosphate. It is felt that, even though the intracellular amastigotes are shielded from drug activity, any reduction in the number of circulating infective trypomastigotes is beneficial in that it reduces the number of potential cell invaders.

Physiology

T. cruzi differs physiologically from the African trypanosomes. For instance, the occurrence of well-developed mitochondrial cristae in all stages of the life cycle of *T. cruzi* suggests that there is little difference in oxygen metabolism in the various stages. Indeed, recent data indicate that oxygen consumption is the same in the intracellular amastigote, the bloodstream trypomastigote, and the insect stages. A complete glycolytic pathway has been reported in all stages, with glucose the major carbohydrate. Also, at least some intermediates of a functional TCA cycle have been reported in all stages. Of the glucose consumed by the organisms, some is degraded entirely to carbon dioxide, while some is incompletely degraded to organic acids such as succinic and acetic.

Host Immune Response

In all clinical forms of Chagas' disease, cell-mediated immunity is of major importance. Specific antibodies and the activation of phagocytes by interferon-γ (IFN-γ) are the main elements of the immune response of the infected host. During the very early acute infection, there is an increase in production of interleukin-12 (IL-12) and natural killer (NK) cell activity. During the acute phase of the infection, intracellular amastigotes express glutathione S-transferase into the blood stream of the host. A decreased T-cell response to the infection ensues with a concomitant decreased secretion of IL-2 and IFN-γ enabling the parasites to evade immune surveillance. Recent investigations have shown that the early control of the parasite burden is due to the expression of the Th1 response in spleen cells, while the increased levels of IL-12 and IFN-γ are related to the initiation of tissue damage in the chronic stage of the disease.

Suggested Readings

Alexander, J., & Russell, D. G. (1992). The interaction of *Leishmania* species with macrophages. *Advances in Parasitology, 31*, 176–254.

Al-Rajhi, A. A., Ibrahim, E. A., DeVol, E. B., Faris, R. M., & Maguire, J. H. (2002). Flucanozole for treatment of *Leishmania major* cutaneous leishmaniasis. *New England Journal of Medicine, 346*, 891–895.

Aufderheide, A. C., Salo, W., Madden, M., Streitz, J., Buikstra, J., Guhl, F., Arriaza, B., Renier, C., Wittmers, L. E., Fornaciari, G., & Allison, M. (2004). A 9,000-year record of Chagas' disease. *Proceedings of the National Academy of Sciences, 101*, 2034–2039.

Bloom, B. R. (1979). Games parasites play: how parasites evade immune surveillance. *Nature, 279*, 21–26.

Bringaud, F., Riviere, L., & Coustou, V. (2006). Energy metabolism in tryposomatids: adaptation to available carbon sources. *Molecular Biochemical Parasitology, 149*, 1–9.

Donelson, J. E., & Turner, M. J. (1985). How the trypanosome changes its coat. *Scientific American, 252*, 44–51.

Gruszynski, A. E., van Deursen, F. J., Albareda, M. C., et al. (2006). Regulation of surface coat exchange by differentiating African trypanosomes. *Molecular Biochemical Parasitiology, 147*, 117–223.

Nantulya, V. M. (1986). Immunological approaches to the control of animal trypanosomiasis. *Parasitology Today, 2*, 168–173.

Pays, E. (2005). Regulation of antigen gene expression in *Trypanosoma brucei. Trends in Parasitology, 21*, 517–520.

Peterson, A. T., Sanchez-Cordero, V., Beard, C. B., & Ramsey, J. M. (2002). Ecologic niche modeling and potential reservoirs for Chagas' disease, Mexico. *Emerging Infectious Diseases, 8*, 266–267.

Rittig, M. G., & Bogdan, C. (2000). *Leishmania*-host-cell interaction: complexities and alternative views. *Parasitology Today, 16*, 292–297.

Turner, M. J., & Donelson, J. E. (1990). Cell biology of African trypanosomes. In D. J. Wyler (Ed.), *Modern parasite biology: cellular, immunological and molecular aspects* (pp. 51–63). New York: W.H. Freeman & Company.

Vanhamme, L., Pays, E., McCulloch, R., & Barry, J. D. (2001). An update on antigenic variation in African trypanosomes. *Trends in Parasitology, 17*, 338–343.

Weina, P., Neafie, C., Wortmann, G., et al. (2004). Old world leishmaniasis: an emerging infection among deployed US military and civilian workers. *Clinical Infectious Diseases, 39*, 1674–1680.

Welburn, S. C., Fevre, E. M., Coleman, P. G., Odiit, M., & Maudlin, I. (2001). Sleeping sickness: a tale of two diseases. *Trends in Parasitology, 17*, 19–24.

CHAPTER

7

Blood and Tissue Protistans II

Human Malaria

Chapter 7 deals with the second phase of "Blood and Tissue Protistans, Human Malaria." This section opens with an overview of the group of organisms that possess structures known collectively as the apical complex. Within this group are the protozoans that belong to the genus *Plasmodium*, the causative agents of the disease termed malaria. Four species of *Plasmodium* are the primary causative agents for human malaria, *Plasmodium vivax*, *Plasmodium malariae*, *Plasmodium ovale*, and *Plasmodium falciporum*. A fifth species, *Plasmodium knowlesi*, hitherto considered a parasite limited to monkeys, has been implicated with the human form of the disease. Following a brief study of the effect that human malaria has had in human history, a generalized life cycle of the organism is presented. Ample electron micrographs illustrate the various stages of the protozoan as it passes through the mosquito and human hosts. This generalized depiction of the life cycle is followed by a discussion of the life cycle variations that exist among the four species. Following a review of the epidemiology of the disease, the concepts of relapse and recrudescence are presented. The chapter concludes with a detailed discussion of the history of the chemotherapy that has been used through the ages and the latest regimens in use. Ample maps displaying world incidence of the four species is presented.

The organisms that cause human malaria, babesiosis, toxoplasmosis, cryptosporidiosis, isosporiasis, and cyclosporidiosis belong to the phylum Apicomplexa. This taxonomic group was established to accommodate protozoans possessing structures known collectively as the **apical complex** (Fig. 7.1). This complex of organelles is found in the sporozoite and merozoite stages of the life cycles of these organisms. At the protozoan's anterior end, immediately beneath the plasma membrane, are one or two electron-dense structures called **polar rings** (Fig. 7.1). In certain members of the phylum (including *Toxoplasma gondii*, the causative agent of toxoplasmosis), a truncated cone of spirally arranged fibrillar structures (the **conoid**) lies within the polar rings. The **rhoptries** (singular, rhoptry) are two or more electron-dense bodies located within the polar rings (and the conoid, when present) and extending posteriorly from the plasma membrane.

Except in members of the genus *Babesia*, subpellicular microtubules radiate from the polar rings parallel to the long axis of the cell. These organelles probably serve as support elements and possibly facilitate the limited motility of these parasitic cells. A group of smaller, more

Human Parasitology. http://dx.doi.org/10.1016/B978-0-12-813712-3.00007-2

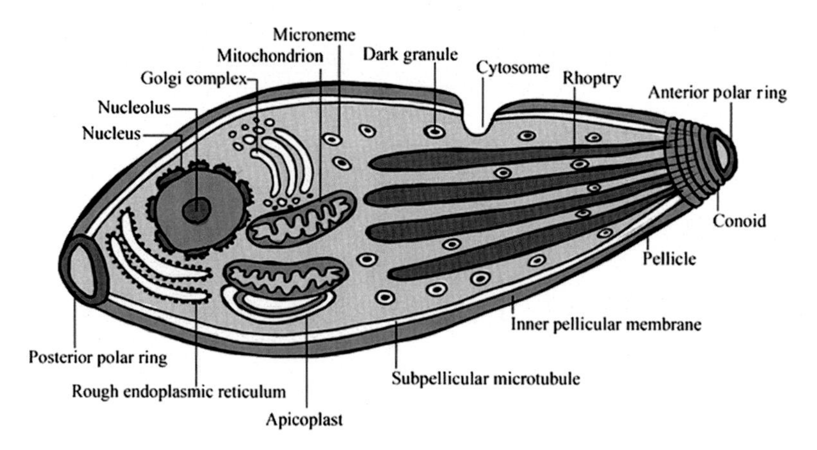

FIGURE 7.1 **Apicomplexan sporozoite or merozoite showing constituents of apical complex.** *Credit: Image courtesy of Gino Barzizza.*

convoluted structures, called **micronemes**, lies parallel to the rhoptries and appears to merge with them at the apex of the cell. The function of the rhoptries and micronemes has not been firmly established, but they appear to secrete proteins that probably alter the host cell's plasma membrane, facilitating the parasite's incorporation into the host cell.

All apicomplexans, except *Cryptosporidium* spp., possess an organelle, the **apicoplast**, that is believed to have evolved from an endosymbiotic bacterium. Its vulnerability to antibacterial compounds such as tetracycline lends credence to this premise. Its function is still unknown although it is thought to supply carbon and energy to the cell.

Located at the lateral edges of the parasite are one or more **micropores**. These organelles are analogous to cytostomes as they seem to be the sites of endocytosis of nutrients during the intracellular life of the organism. At the edges of the micropore are two concentric, electron-dense rings, situated directly beneath the plasma membrane. Host cytoplasm is drawn through the microporal rings into the parasite where a food vacuole forms and is pinched off from the plasma membrane, whereupon the process of intracellular digestion begins. Once transformation from the sporozoite or merozoite to the trophozoite stage has occurred following incorporation into a host cell, all the aforementioned organelles, with one exception, lose their physical integrity and disappear; the micropores persist through all succeeding stages.

PLASMODIUM AND HUMAN MALARIA

Malaria, one of the most prevalent and debilitating diseases afflicting humans, has played a major role in shaping history and civilizations. Its ravages probably contributed to the fall of the ancient Greek and Roman empires. In medieval times, crusaders often fell victim to malaria during their expeditions, and probably more of their casualties can be attributed to the disease than to the infidels they fought. The manner in which malaria was introduced into the Western hemisphere is uncertain. Writings from ancient civilizations, such as the Mayan, make no reference to any malaria-like diseases. It appears most likely that the Spanish conquistadors and their African slaves first brought the parasite to the New World. United States troops in both the American Civil War and the Spanish–American War were severely incapacitated by this disease; more than one-quarter of all persons admitted to hospitals during those conflicts were malaria patients. During World War II, malaria epidemics critically affected both the Japanese and the Allied forces in the Pacific Islands and in Southeast Asia. It has been claimed that during the Vietnam conflict, malaria was second only to battle wounds as the most common cause for hospitalization among American forces.

While malaria is often regarded as a tropical disease, it is by no means confined to the tropics. As recently as 1937, there were at least 1 million cases of malaria annually in the United States. The disease has been reported in more than 90 countries, inhabited by 2.4 billion people. More than 90% of all malaria cases are from countries in mostly sub-Sahara Africa. Two-thirds of the remaining cases are concentrated in six countries, the largest number in India, followed by Brazil, Sri Lanka, Viet Nam, Colombia, and Solomon Islands. It has been estimated that worldwide prevalence of the disease in 2006 was between 300 and 500 million clinical cases each year and an annual death toll of approximately 2 million. Although control programs sponsored by several cooperating nations and the World Health Organization (WHO) of the United Nations have made great inroads in the fight against this disease, it obviously remains a major health problem in many parts of the world.

The more than 50 species of *Plasmodium* infect a wide variety of animals, but only four, *P. vivax*, *P. falciparum*, *P. malariae*, and *P. ovale*, commonly cause malaria in humans. Recently, a fifth species, *P. knowlesi*, commonly a parasite of Old World monkeys, has been reported to infect humans in Southeast Asia. Regardless of the species responsible, certain facets of the disease, such as life cycle of the infective organism, chemotherapy, and epidemiology, are similar enough that the following discussion will make no distinction among the five species except where dissimilarities are medically significant.

Life Cycle

Although Ronald Ross received the Nobel Prize in Medicine in 1902 for studies on the life cycle of the malaria-producing organism, a number of others, such as Manson, Lavern, Bignami, Grassi, and others, are also credited with making significant contributions. The entire life span of the four species of *Plasmodium* that infect humans is spent in two hosts [i.e., the insect vector, a female mosquito belonging to the genus *Anopheles*, and a human host (Fig. 7.2)]. Only female mosquitoes serve as vectors. The mouthparts of males cannot penetrate human skin; hence, they feed solely on plant juices. Females, on the other hand, also feed on blood, which is usually required for oviposition. A significant feature of the life cycle is the alternation of sexual and asexual phases in the two hosts. The asexual phase, termed **schizogony**, occurs in the human, whereas the sexual phase, **gamogony**, occurs in the mosquito; subsequent to the sexual stage, another asexual reproductive phase, termed **sporogony**, occurs in the mosquito. The infective form in humans is the slender, elongated **sporozoite**, about 10–55 μm in length and about 1 μm in diameter.

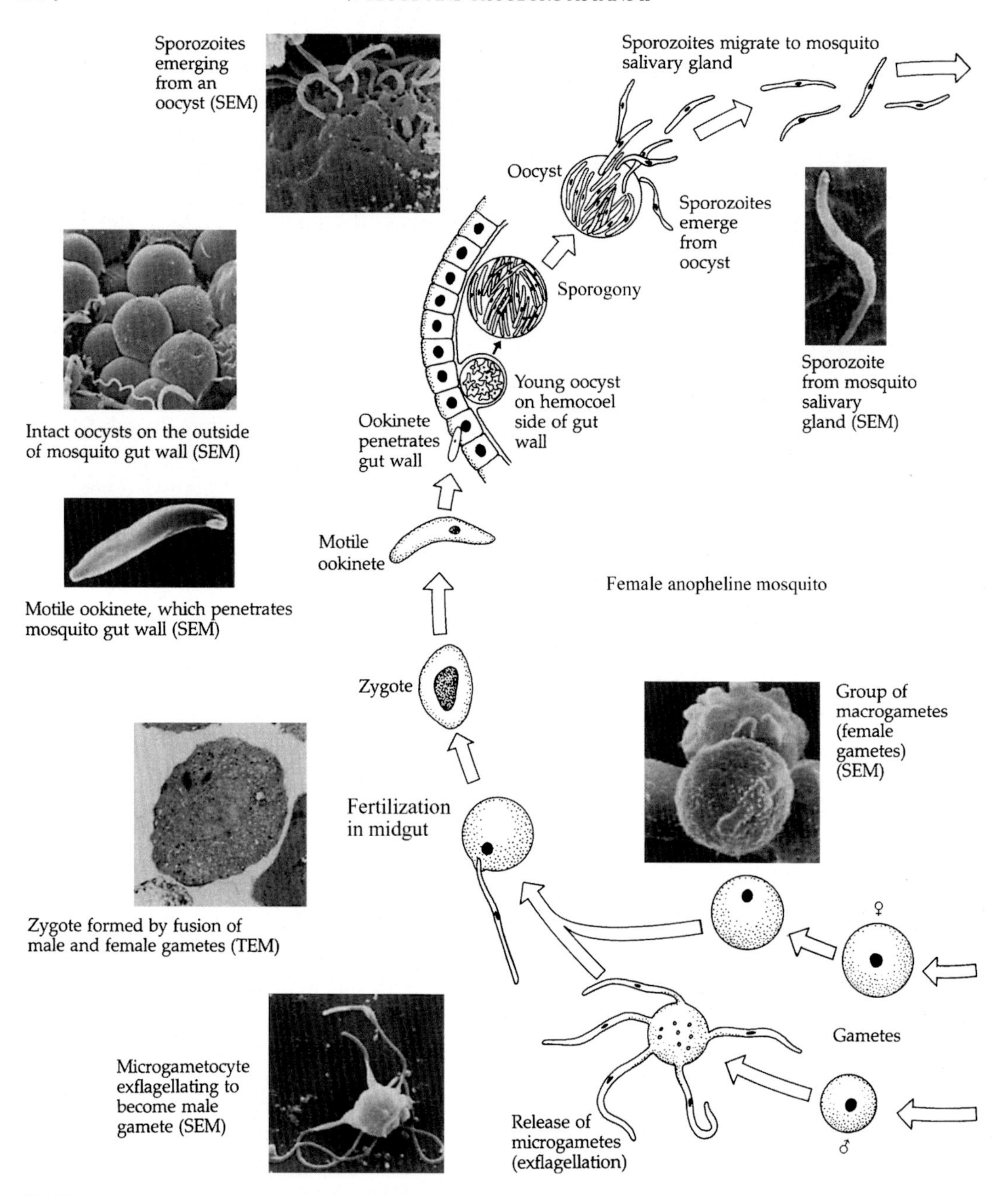

FIGURE 7.2 **Life cycle of Plasmodium spp.** *Credit: Image courtesy of Gino Barzizza.*

During feeding, the mosquito secretes sporozoite-bearing saliva beneath the epidermis of the human victim, thus inoculating the sporozoites into the bloodstream. After approximately 1 h, the sporozoite disappears from the circulation, re-emerging 24–48 h later in the parenchymal cells of the liver where the **exoerythrocytic schizogonic phase** begins.

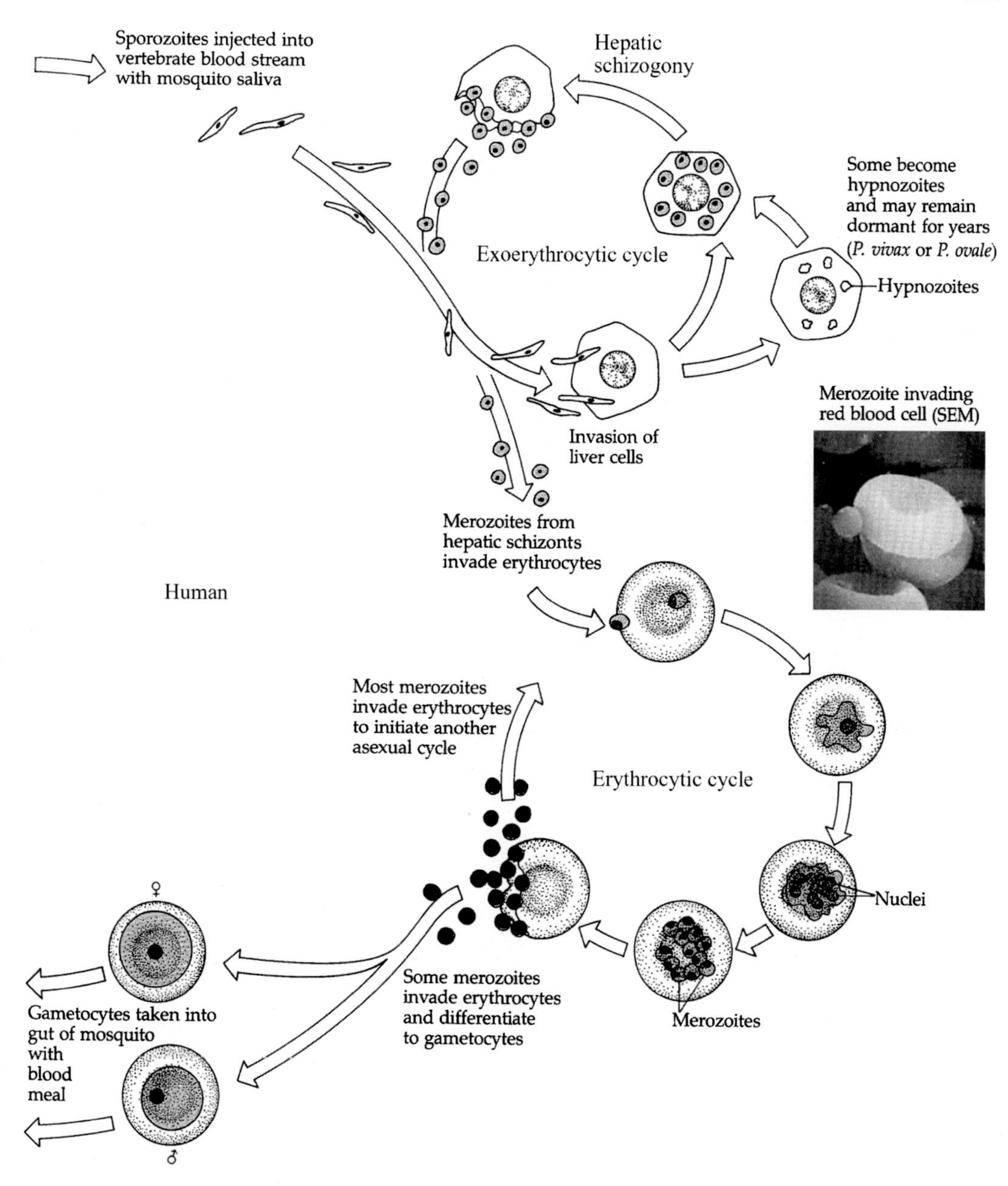

FIGURE 7.2 (*Cont.*).

The specificity of the relationship of the sporozoite with hepatocytes rather than with other cells of the body is due, in part, to the recognition of the surface coat of the sporozoite (**circumsporozoite coat**) by receptors on the surface of the hepatocytes. Electron microscopy has confirmed that sporozoites and merozoites interact with the plasma membrane of the host cell and actively participate in their incorporation (Fig. 7.3). During this process, rhoptries

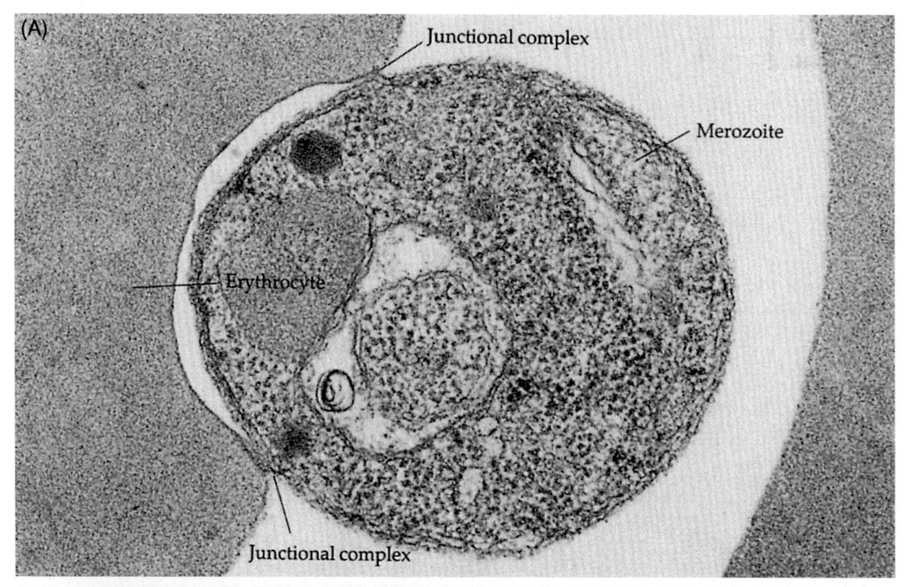

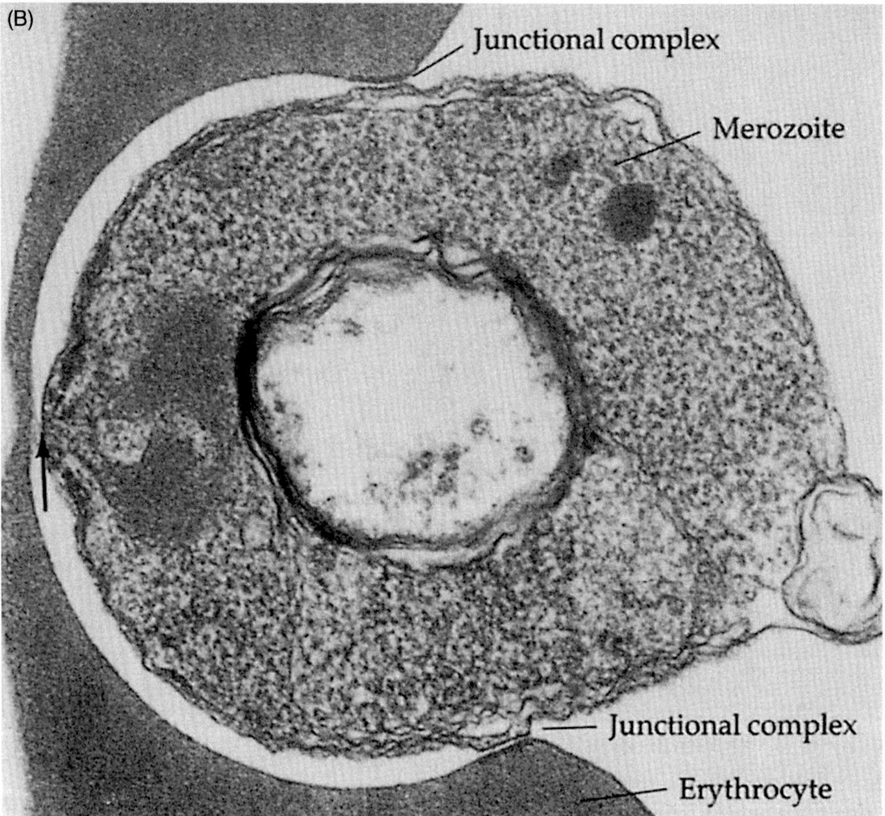

FIGURE 7.3 (A) Merozoite entering erythrocyte. Note the junction complexes at each side of entry. (B) Merozoite further along in penetrating erythrocyte. The *arrow* points to a projection connecting the merozoite's apical end and the erythrocyte membrane. Again, note the junctional complexes.

and micronemes are believed to secrete surface-active molecules that cause the host-cell plasma membrane to expand and then invaginate to form a **parasitophorous vacuole** that envelopes the parasite.

Once inside the hepatocyte, the sporozoite develops into a trophozoite, feeding on host cytoplasm with its now functional micropore. There is evidence that additional nutrients enter the trophozoite by pinocytosis. After 1–2 weeks (depending upon the species of *Plasmodium*), the nucleus of the trophozoite divides several times, followed by division of the cytoplasm. This multiple fission process produces thousands of merozoites, each approximately 2.5 μm in length and 1.5 μm in diameter. The merozoites rupture from the host cell, enter the blood circulation, and invade red blood cells, initiating the **erythrocytic schizogonic phase**. As with the engulfment of the sporozoite by the hepatocyte, the endocytosis of the merozoite into the red blood cell is dependent on surface recognition between the two cells. Studies of *P. vivax* show that the membrane receptor site for the invasion phenomenon is determined by the type of antigen present on the surface of the red blood cell. For instance, merozoite engulfment requires at least one of two Duffy antigens (Fy^{a+} or Fy^{b+}). Humans lacking the Duffy antigens, practically all West Africans and approximately 70% of American blacks, are resistant to *vivax* malaria. *P. falciparum* and *P. ovale* malarias, on the other hand, are not influenced by Duffy antigens, thus accounting for their prevalence in West Africa.

Inside the erythrocyte, the merozoite grows to the early trophozoite stage. Under light microscopy, the early trophozoite appears to consist of a ring of cytoplasm and a dotlike nucleus. Due to its resemblance to a finger ring, this stage is called the **signet ring stage**. In reality, the ring stage trophozoite is cup-shaped with a large vacuole filled with host hemoglobin in varying stages of digestion (Fig. 7.4). This early form develops to the mature trophozoite stage and then undergoes multiple fission into schizonts, producing a characteristic number of a new generation of merozoites in each infected erythrocyte. As in the liver, each of these merozoites is capable of infecting a new erythrocyte. One of two fates await this new penetrant; it may become another signet ring trophozoite and begin schizogony anew, or it may become a male **microgametocyte** or a female **macrogametocyte**. The determinants for the courses these parasites take have not yet been identified.

The sexual phase occurs in the female *Anopheles* and begins when the mosquito takes a blood meal that contains macrogametocytes and microgametocytes. These stages are unaffected by the digestive juices of the insect. Lysis of the surrounding erythrocytic material releases gametocytes into the lumen of the stomach. There, microgametocytes undergo a maturation process known as **exflagellation** during which the nucleus undergoes three mitotic divisions, producing 6–8 nuclei that migrate to the periphery of the gametocyte. Accompanying the nuclear divisions are centriolar divisions, following which one portion joins each nuclear segment to become a basal body, providing the center from which the axoneme subsequently arises. Almost simultaneously, the nucleus with the axoneme and a small amount of adhering cytoplasm form a **microgamete**, which detaches from the mass and swims to the macrogametocyte. During this period, the macrogametocytes develop into female **macrogametes**, each of which forms a membrane-derived fertilization cone to be penetrated by the microgamete.

The fusion of male and female pronuclei (**syngamy**) produces a diploid zygote that, after 12–24 h, elongates into a motile, microscopic wormlike **ookinete**. This ookinete penetrates the gut wall of the mosquito to the area between the epithelium and basal lamina, where it

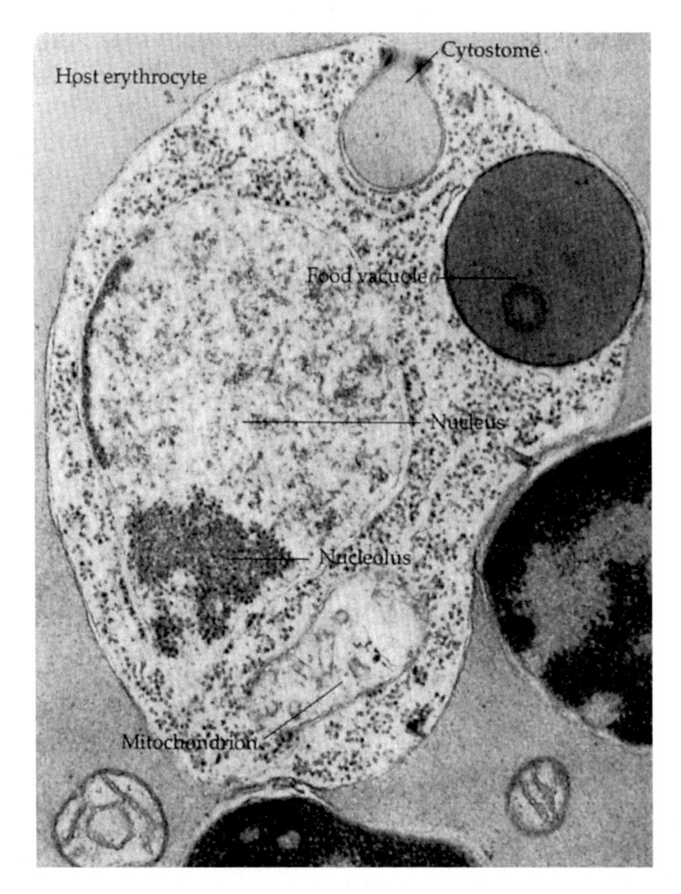

FIGURE 7.4 **An erythrocyctic trophozoite of *Plasmodium gallinaceum* showing host cell hemoglobin being ingested at cytostome site.** The bulge is prominently limited by a double membrane, whereas the food vacuole near the cytostome is limited by a single membrane. The content of the food vacuole is dark compared to the host cell contents, indicating that digestion is taking place. A mitochondrion, endoplasmic reticulum, ribosomes, and a large nucleus with a nucleolus are also present in the parasite (×22,500).

develops into a rounded **oocyst**. Formation of the oocyst occurs approximately 40 h after the mosquito has taken its blood meal. Following a period of growth during which its diameter increases 4–5 times, the oocyst is seen as a bulge on the hemocoel side of the gut. Growth of the oocyst is due, in part, to the proliferation of haploid cells, called **sporoblasts**, within the oocyst. Sporoblast nuclei undergo numerous divisions, producing thousands of sporozoites enclosed within the sporoblast membranes. As the membranes rupture, sporozoites enter the cavity of the oocyst.

Within 10–24 days after the mosquito ingests the gametocytes, the sporozoite-filled oocysts themselves rupture, releasing the sporozoites into the hemocoel. The sporozoites are carried to the salivary gland ducts of the insect and are then ready to be injected into the next person from whom the mosquito draws a blood meal.

TABLE 7.1 Diagnostic Differences Among the Four Species of Human-Infecting *Plasmodium*

	Plasmodium vivax	***Plasmodium malariae***	***Plasmodium ovale***	***Plasmodium falciparum***
Duration of schizogony	48 h	72 h	49–50 h	36–48 h
Motility	Active amoeboid until about half grown	Trophozoite slightly amoeboid	Trophozoite slightly amoeboid	Trophozoite active amoeboid
Pigment (hematin)	Yellowish-brown; fine granules and minute rods	Dark brown to black; coarse granules	Dark brown; coarse granules	Dark brown; coarse granules
Stages found in peripheral blood	Trophozoites, schizonts, gametocytes	Trophozoites, schizonts, gametocytes	Trophozoites, schizonts, gametocytes	Trophozoites, gametocytes
Multiple infection in erythrocyte	Common	Very rare	Rare	Very common
Appearance in infected erythrocyte	Greatly enlarged; pale with red Schuffner's dots	Not enlarged; normal appearance with Ziemann's dots	Slightly enlarged; outline oval to irregular, with Schuffner's dots	Normal size; greenish; basophilic Maurer's clefts and dots
Trophozoites (ring forms)	Amoeboid; small and large rings with vacuole and usually one chromatin dot	Small and large rings with vacuole and usually one chromatin dot; also young band forms	Amoeboid; small and large rings with vacuole	Very small and large rings with vacuole, commonly with two chromatin dots; amoeboid
Segmented schizonts	Fills enlarged RBC; merozoites irregularly arranged around mass of pigment	Almost fills normal-sized RBC; 6–12 merozoites regularly arranged around central pigment mass	Fills approx. 3/4 of RBC; 6–12 merozoites around centric or eccentric pigment mass	Not usually seen in peripheral blood
Gametocytes	Round; fills RBC; chromatin undistributed in cytoplasm	Round; fills RBC; chromatin undistributed in cytoplasm	Round; fills 3/4 of RBC; chromatin undistributed in cytoplasm	Crescentic- or kidney-shaped; chromatin undistributed in cytoplasm

See color plates 1–4 for further reference.

Life Cycle Variations

While the life cycles of the various species of *Plasmodium* that infect humans are basically similar, there are a number of differences, some of which are important in clinical diagnosis. These differences are summarized in Table 7.1.

Plasmodium vivax *and* P. ovale *(Benign Tertian Malaria)*

Plasmodium vivax was first described by Grassi and Feletti in 1890 and is the most common species of the genus in the Americas. *Plasmodium ovale* was first described by Stephens in 1922. Both species have a predilection for immature erythrocytes (reticulocytes). Less than 1% of the total erythrocyte population in each victim is parasitized by *P. vivax* or *P. ovale.* A diagnostically significant characteristic is the larger size of these infected erythrocytes, probably due to the fact that the parasites prefer to invade relatively larger reticulocytes. This enlargement of infected cells is less pronounced in *P. ovale* malaria than in *P. vivax* infections. Cells infected with *P. ovale* also tend to be somewhat ellipsoid in shape. In all *Plasmodium*-infected erythrocytes, two types of granules are found. One type (**Schüffner's dots** in *P. vivax* and *P. ovale*) is distributed throughout the cytoplasm of the erythrocyte and usually stains pink to red when subjected to traditional hematological stains, such as Giemsa's, Wright's, or Romanovsky's. Electron microscope studies indicate that such pigmented granules are small surface invaginations surrounded by small vesicles. The source of these granules in infected cells is uncertain; they may be products of degenerative changes in the infected erythrocyte. The second type is the coarser, dark **hemozoin** granules, the by-products of hemoglobin degradation by the parasite. Hemozoin is usually found more closely associated with the parasite than with erythrocytic cytoplasm.

The cytoplasm of the trophozoite stages is very irregular and displays active amoeboid-like movement, hence the species name (*P. vivax*, from Latin meaning "vigorous"). During schizogony, 12–24 (average, 16) merozoites are produced, each measuring about 1.5 μm in diameter. These rupture from the infected erythrocyte synchronously at 48-h intervals, with accompanying fever. The designation *tertian* is derived from the ancient Roman custom of counting the days of an event in sequence; therefore, the first day of the fever peak is designated "day one," the intervening day is "day two," and the day of the next fever episode is "day three," although the time interval between peaks, in *P. vivax* malaria, is only 48 h; so "day three" then becomes "day one" when counting the next interval.

Gametocytes begin to appear in approximately 4 days. Macrogametocytes, which outnumber their somewhat smaller male counterparts about 2 to 1, measure about 10 μm in diameter, each almost completely filling the infected erythrocyte.

Plasmodium malariae *(Quartan Malaria)*

Plasmodium malariae, the first parasite to be recognized as a cause of malaria, was described in 1880 by a French army physician, Charles Louis Alphonse Laveran. While *P. vivax* and *P. ovale* selectively parasitize young cells, *P. malariae* shows an affinity for older cells, parasitizing about 0.2% of the victim's total erythrocyte population.

Following incorporation into erythrocytes, early trophozoites begin to accumulate hemozoin and the pink-staining **Ziemann's dots**. The cytoplasm of the trophozoite is compact, often appearing as a band across the infected cell. Morphologically, mature trophozoites resemble macrogametocytes and are, therefore, difficult to distinguish. No change in diameter is evident in the infected erythrocyte, probably due to the parasite's affinity for older erythrocytes.

The number of merozoites, following schizogony, varies from 6 to 12 (average, 8). Hemozoin usually accumulates as a dense mass in the center of the schizont. Merozoites rupture from the infected cell synchronously every 72 h with an accompanying fever paroxysm (quartan malaria). Recrudescence (see pp. 125) has been reported as long as 53 years after initial infection.

Plasmodium falciparum *(Malignant Tertian Malaria)*

Plasmodium falciparum is responsible for most cases of human malaria worldwide (80%) and is deeply entrenched in tropical Africa. Examination of blood smears of infected patients, originally described by William Welch in 1897, shows that *P. falciparum* differs significantly from the preceding three species. Typically, only ring trophozoites and gametocytes are seen in the peripheral circulation, the later stages of schizogony being trapped in capillaries of muscle and visceral organs. Plasma membranes of infected erythrocytes undergo alteration that causes them to adhere to the walls of capillaries. Infected erythrocytes are not enlarged and represent about 10% of the total erythrocyte population. *P. falciparum* infects erythrocytes of any age indiscriminately. Multiple infections of single erythrocytes are common, and the presence of more than one ring trophozoite in a cell is not unusual. Double nuclei also occur frequently in the ring stage. The schizonts, rarely seen in peripheral blood, produce 8–32 (average, 20) merozoites. Rupture of merozoites from infected erythrocytes is erratic, with accompanying fever paroxysms occurring at 48-to-72-h intervals. Gametocytes are elongated or crescent-shaped cells that stretch, but remain within, the erythrocyte. Macrogametocytes are slightly longer than microgametocytes, the two measuring 12–14-μm and 9–11-μm long, respectively. Hemozoin, as well as the pigment characteristic of *P. falciparum*-infected erythrocytes, **Maurer's dots** or **clefts**, tend to aggregate around the nuclear region of gametocytes.

Epidemiology

Endemicity of human malaria is usually determined by the geographic distribution of its arthropod vector, an anopheline mosquito. Areas where the vector is not present are free of the disease. For instance, since there are no anopheline mosquitoes in Hawaii, many southeastern Pacific islands, and New Zealand, malaria does not occur in these areas (Fig. 7.5A).

Local environmental factors determine which particular species of mosquito transmits malaria in a given area; therefore, local epidemiological surveys are used to assay the prevalent transmitters. Precipitin tests of ingested blood from infected mosquitoes reveal whether the vectors have zoophilic or anthropophilic feeding preferences.

Statistical computation of the average number of bites per person per night yields the **critical density**. A continuously declining critical density indicates that malaria in a survey area is waning and may eventually disappear. An accurate critical density assessment must include not only the number of mosquitoes and their feeding preferences, but also the frequency of feeding and the life expectancy of the mosquito species.

Critical density is influenced by environmental factors that affect breeding and/or sporogony. These functions require temperatures between 16 and 34°C and a relative humidity in excess of 60%. Water dependency for breeding varies greatly; some species of *Anopheles* favor small bodies of water, others require large bodies of water such as ponds and even lakes, and still others have intermediate requirements. Females' feeding habits vary also, even as to the preferred locale, that is, indoor (**endophilic**) or outdoor (**exophilic**).

Mosquito populations in areas of low critical density produce a pattern of stable malaria, a universal low-grade infection with little, if any, disease symptoms. The immunity level in the human population is unaffected by environmental and climatic changes. Incidence is usually transient, alternately producing periods of high mortality and then declining as optimal conditions subside. Since there is insufficient time for immunity to become established under such circumstances, a pattern of recurring epidemics develops.

Although the areas affected by malaria have diminished over the past 50 years, control is becoming progressively more difficult, and earlier gains are being eroded. Under the auspices of the WHO, malaria was either under control or drastically diminished in many parts of the world by the 1960s due primarily to such factors as availability of antimalarial drugs, use of screens on houses to keep out mosquitoes, proper use of insecticides, elimination of mosquito breeding sites, mosquito eradication, and other environmental measures. However, there has been a marked resurgence in the disease since the 1970s. Several elements have contributed to this, most important of which are the development of widespread resistance by anopheline mosquitoes to insecticides and the evolution of chloroquine-resistant *P. falciparum* (Fig. 7.5B,C). Another factor is the reduction in personnel trained to maintain the WHO-established standards for control. Increased prevalence is also linked to "progress" in many third-world countries (i.e., activities such as road-building, mining, logging, and the introduction of new agricultural and irrigation projects). Wars, which have caused disintegration of health services and mass movements of refugees from malarial zones, have also contributed to the upsurge.

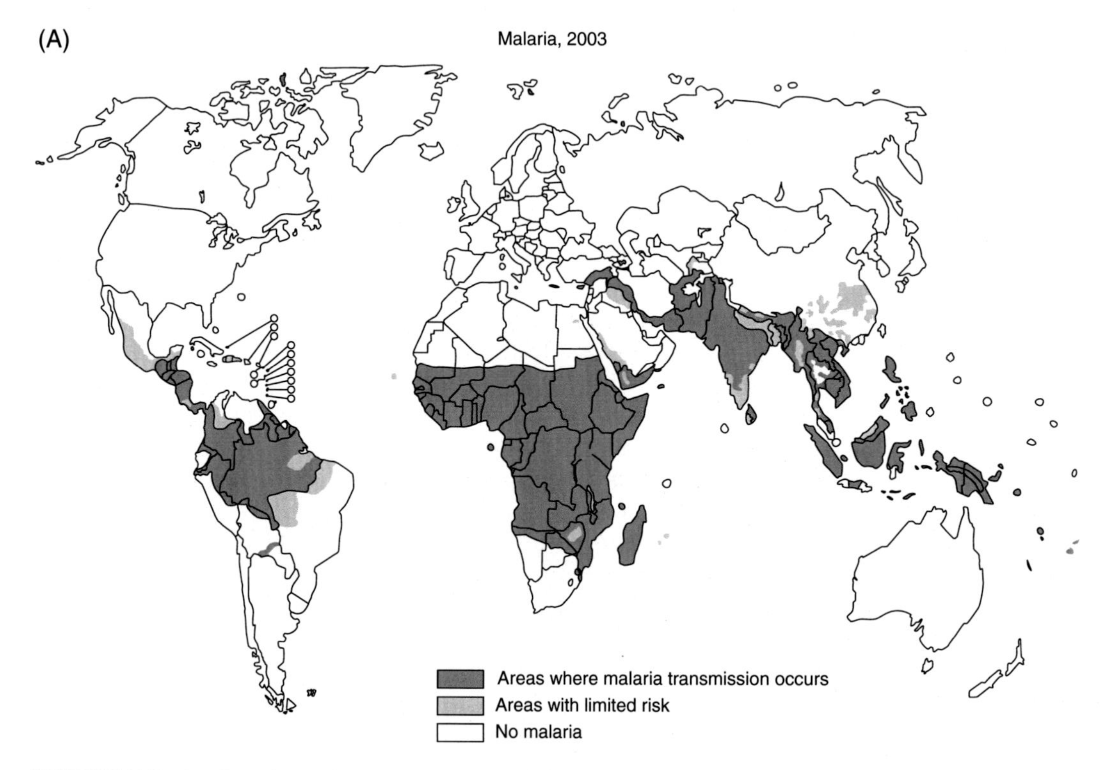

FIGURE 7.5 (A–C) Malaria: global distribution, 2003, and suggested prophylaxis.

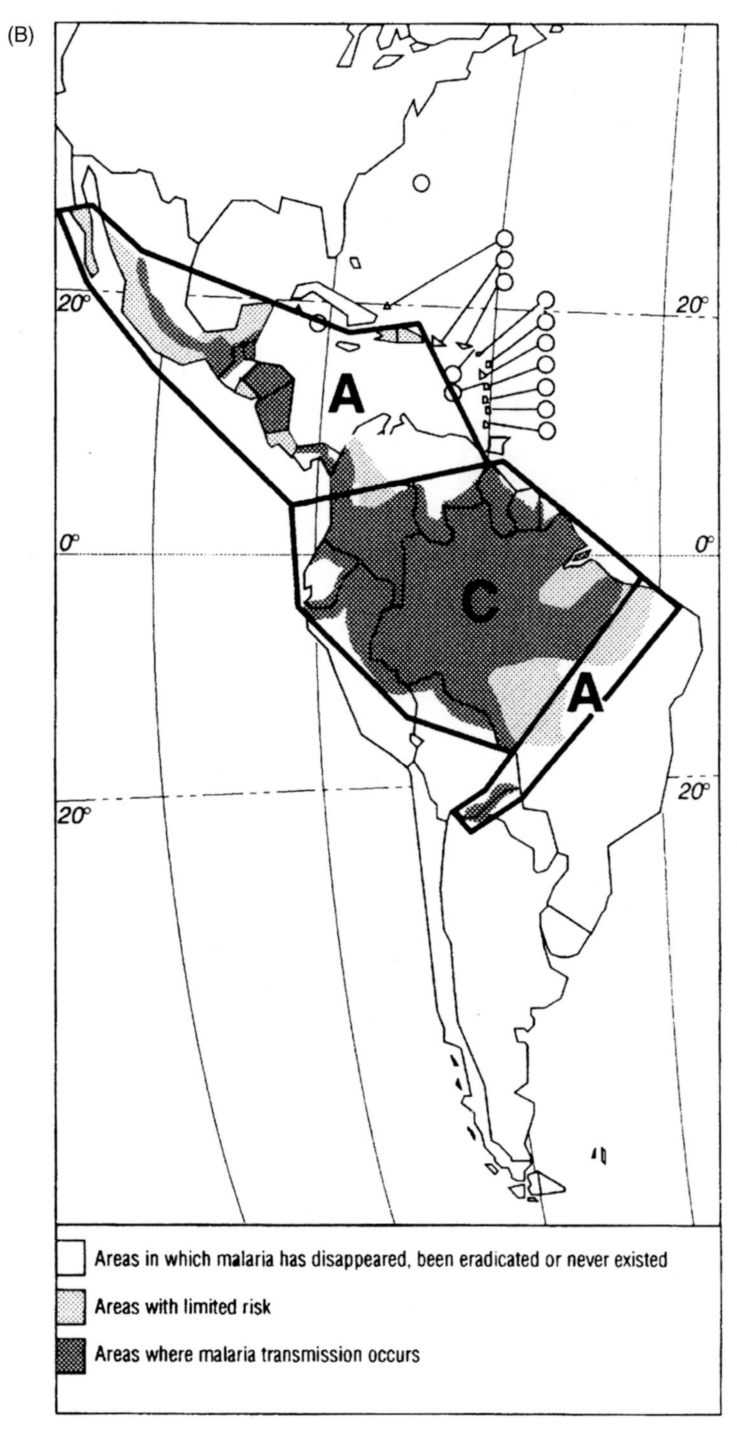

FIGURE 7.5 (*Cont.*).

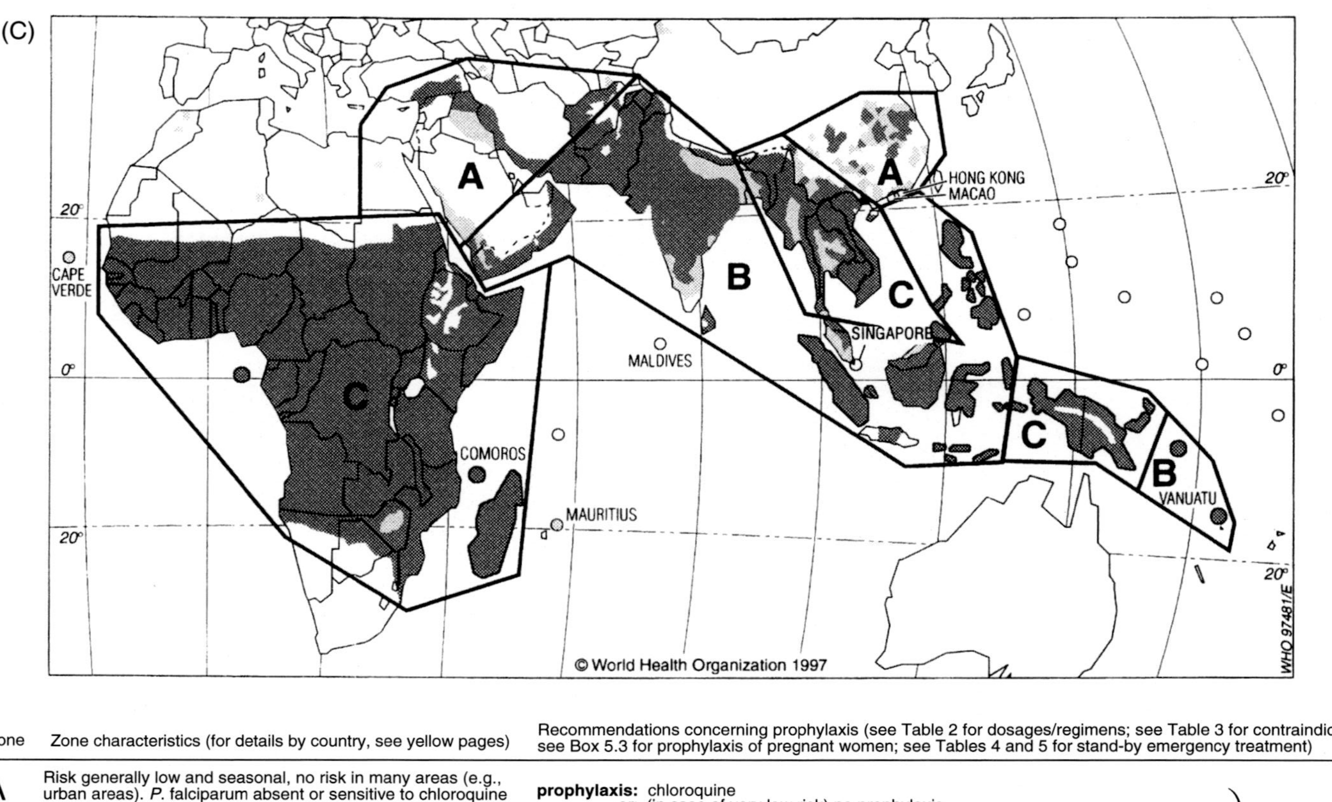

Zone	Zone characteristics (for details by country, see yellow pages)	Recommendations concerning prophylaxis (see Table 2 for dosages/regimens; see Table 3 for contraindications; see Box 5.3 for prophylaxis of pregnant women; see Tables 4 and 5 for stand-by emergency treatment)	
A	Risk generally low and seasonal, no risk in many areas (e.g., urban areas). *P.* falciparum absent or sensitive to chloroquine	**prophylaxis:** chloroquine *or:* (in case of very low risk) no prophylaxis	Protection from mosquito bites should be the rule in all situations, even when prophylaxis is taken
B	Low risk in most areas. Chloroquine alone will protect against *P. vivax* Chloroquine with proguanil will give some protection against *P. falciparum* and may alleviate the disease if it occurs despite prophylaxis	**prophylaxis:** chloroquine *or:* (in case of very low risk) no prophylaxis	
C	Risk high in most areas of this zone in Africa, except in some high-altitude areas. Risk low in most areas of this zone in Asia and America, but high in parts of the Amazon basin (colonization and mining areas). Resistance to sulfadoxine–pyrimethamine common in zone C it. Asia, variable in zone C in Africa and America.	**prophylaxis:** first choice – mefloquine second choice – chloroquine + proguanil border areas Cambodia/Myanmar/Thailand – doxycycline (for details, see yellow pages) *or:* (in case of very low risk) no prophylaxis	

FIGURE 7.5 (*Cont.*).

Relapse and Recrudescence

It has long been known that victims of *P. vivax* or *P. ovale* malarias, after apparent recovery, may suffer **relapse**. Originally, such relapse was thought to be due to populations of cryptozoites entering the exoerythrocytic cycle. While one population progressed to the usual erythrocytic phase, underwent schizogony, and released merozoites into the circulating blood causing malaria, the other population was thought to maintain an ongoing exoerythrocytic cycle known as a para-erythrocytic cycle. It was believed that parasites in the hepatic stages of the cycle remained protected from host antibodies until activated by some physiological change within the host that allowed them to erupt from the hepatocytes, precipitating another bout of malaria.

Now, it is recognized that there are two different populations of sporozoites. **Short-prepatent sporozoites** (**SPPs**), upon entering the human host, undergo the usual exoerythrocytic phases of development and cause malaria. **Long-prepatent sporozoites** (**LPPs**) or **hypnozoites**, remain dormant in the hepatocytes for an indefinite period. When a stimulus, such as the physiological fluctuation cited above, activates hypnozoites into the exoerythrocytic and erythrocytic cycles, relapse occurs. The ratio of LPPs to SPPs in *P. vivax* infections in a given human population appears to vary according to strain. For instance, in a North Korean strain found in temperate zones, LPP sporozoites are far more numerous than SPP sporozoites. On the other hand, in strains common to tropical regions, the relative proportions are equal or sometimes reversed.

Recurrence of malaria among victims infected by *P. malariae* many years after apparent cure fostered an enduring belief that this species produced relapses like those produced by *P. vivax* and *P. ovale*. However, it has been shown that the periodic increase in numbers of parasites results from a residual population persisting at very low levels in the blood after inadequate or incomplete treatment of the initial infection. The number of parasites is usually so small that infected individuals remain symptomless. This situation has been known to persist for as long as 53 years before something, such as splenic dysfunction, triggers a parasite population explosion with accompanying disease manifestations, a phenomenon termed **recrudescence**. The difference, therefore, between relapse and recrudescence is that the former results from exoerythrocytic stages in the liver as a result of the activation of hypnozoites, while the latter is due to a sudden increase in what was a persistent, low-level parasite population in the blood.

Symptomatology and Diagnosis

Pathology in human malaria is generally manifested in two basic forms: host inflammatory reactions and anemia. Of the four species of *Plasmodium* responsible for human malaria, *P. falciparum* is the most virulent and causes, by far, the highest mortality.

The initial symptoms of malaria, such as nausea, fatigue, a slight rise in temperature, mild diarrhea, and muscular pains, are often mistaken for influenza or gastrointestinal infection. Host inflammatory reactions are triggered by the periodic rupture of infected erythrocytes, which releases malarial pigment such as hemozoin, cellular debris, and parasite metabolic wastes into the circulatory system (Fig. 7.6). These ruptures are accompanied by fever paroxysms that are usually synchronous except during the primary attack. As explained

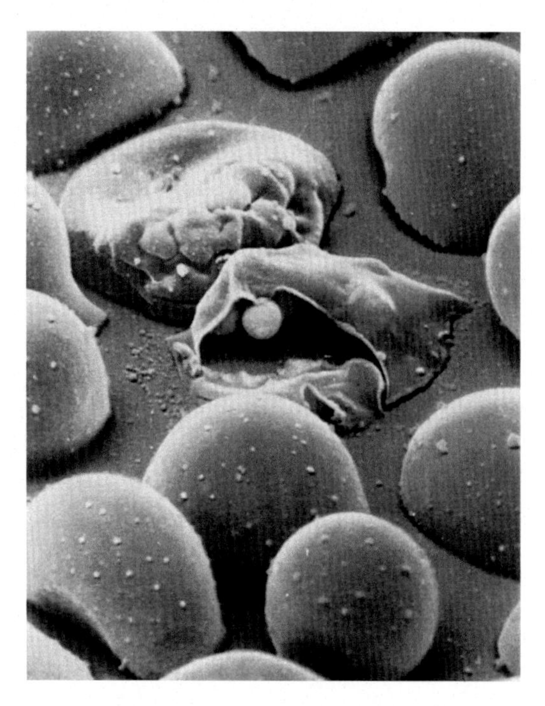

FIGURE 7.6 **Scanning electron micrograph of *Plasmodium*-infected red blood cells.** One cell has burst open, releasing merozoites.

earlier, the interval between paroxysms is species specific. However, during the primary attack, since the infection may arise from several populations of liver merozoites at different stages of development, synchrony may not be evident. The mechanism by which the synchronous pattern gradually emerges remains unexplained, but *P. falciparum* tends to be the most persistently erratic. Macrophages, particularly those in the liver, bone marrow, and spleen, phagocytose released pigment. In extreme cases of falciparum malaria, the amount of pigment is so great that it imparts a dark, reddish-brown hue to visceral organs such as liver, spleen, and brain. With increased erythrocyte destruction, accompanied by the body's inability to recycle iron bound in the insoluble hemozoin, anemia develops.

One pathological element unique to *P. falciparum* is vascular obstruction. Plasma membranes of erythrocytes infected with schizonts, the more mature stages of the organism, develop electron-dense "knobs" by which they adhere to the endothelium of capillaries in visceral organs (Fig. 7.7). Engorged with hordes of infected erythrocytes, the capillaries become obstructed, causing the affected organs to become anoxic. In terminal cases, blocked capillaries (**ischemia**) in the brain cause it to become swollen and congested. Erythrocyte surface antigens are believed to regulate growth of the malarial parasite and to be also involved in establishing chronic infections. The acquisition of the aforementioned adhesive properties in *P. falciparum* might have evolved as a secondary event.

A condition known as **blackwater fever** often accompanies *falciparum* malaria infections. Characterized by massive lysis of erythrocytes, it produces abnormally high levels of hemoglobin in urine and blood. Fever, vomiting with blood, and jaundice also occur, and there

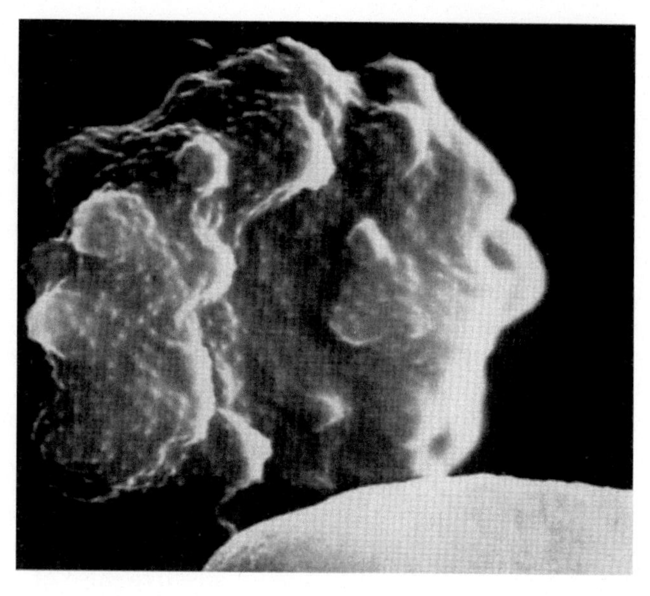

FIGURE 7.7 **Erythrocyte, parasitized by *Plasmodium falciparum*, showing surface knobs.**

is a 20%–50% mortality rate usually due to renal failure. The exact cause of this condition is uncertain; it may be a reaction to quinine, or it may result from an autoimmune phenomenon in which hemolytic antibodies are produced.

Diagnosis of malaria usually consists of microscopical demonstration of the parasites in stained thick and thin blood smears of peripheral blood. Even though synchrony is obvious, diagnostic blood samples can be drawn almost anytime, since a few "stragglers" from the previous episode always remain in the blood. Smears, however, should be made at regular intervals over a period of several days, especially if one smear fails to show parasites. Most recently, rapid diagnostic tests for diagnosis in *P. falciparum* endemic areas are being used. This procedure uses small blood samples to detect *P. falciparum* histidine rich protein II antigen. It is sensitive enough to detect low parasitemia in patients with or without symptoms. Results can be received almost instantly at a low cost making it more efficient in prescribing drugs. Other serological tests in areas other than those where *P. falciparum* is prevalent are of little clinical value since they make no distinction between present and past infections.

Chemotherapy

Malaria control requires effective treatment of the disease in humans and continuous efforts to control mosquito populations. The first known effective antimalarial drug was quinine, an extract from the bark of the cinchona tree of South America and other tropical areas. The tree was named by Linnaeus in the 17th century in honor of the wife of the Peruvian envoy, Countess del Chinchón, who was treated for malaria with the bark. The drug destroys the schizogonic stages of malaria, but it has little or no effect on exoerythrocytic stages or gametocytes. During World War II, when the Japanese occupied the cinchona plantations in Indonesia, it became essential that researchers develop alternative drugs for the Allied forces. A synthetic drug, atabrine dihydrochloride (quinacrine hydrochloride), developed in Germany in 1936, proved effective against the erythrocytic stages of all species of *Plasmodium* and in suppressing clinical symptoms. Like quinine, atabrine is ineffectual against exoerythrocytic stages; consequently, *vivax* and *ovale* malaria patients treated with this drug are susceptible to relapse. Atabrine also produces a number of undesirable side effects, such as jaundice and gastrointestinal disturbances. Since World War II, research has yielded a number of other synthetic drugs, of which the most commonly used are chloroquine, amodiaquin, and primaquine. Each, however, has limited effect upon *Plasmodium* and, for maximum benefit, should be administered in combination with one or more other drugs. For instance, while chloroquine and amodiaquin act to suppress clinical symptoms by destroying the erythrocytic stages, the slow-acting drug primaquine destroys the exoerythrocytic stages. Currently, the optimal chemotherapeutic regimen for treatment of malaria includes taking chloroquine for 3 days followed by a single dose of primaquine after leaving endemic areas.

The folic acid cycle provides a suitable metabolic pathway for chemotherapeutic management of malaria. This cycle is vital for synthesizing bases for nucleic acid formation. Fansidar is a combination of pyrimethamine and sulfadoxine. It inhibits two enzymes of the cycle, dihydrofolate reductase and dihydropteroate synthetase. The dual sequential action of the two components reduces the minimum effective dose of each agent making each less harmful to the host but more lethal to the parasite. In 1984, a new antimalarial drug, mefloquine, was tested and approved for use by health authorities. This drug acts against blood schizonts. Currently, mefloquine is being added to the pyrimethamine–sulfadoxine combination in a one-step treatment for chloroquine-resistant *falciparum* malaria. Mefloquine is prescribed as a prophylactic drug for travelers to areas where chloroquine-resistant *falciparum* malaria has been reported. The combination of quinine and tetracycline is used in Southeast Asia as standard treatment for uncomplicated malaria, while quinine has been re-established as an alternative drug for the treatment of chloroquine-resistant strains. Malarone was approved as a prophylactic against *Plasmodium* spp. in July, 2000 in the United States and in 2002 in Canada. Malarone, a combination of atovaquone and proguanil, interferes with two different malarial pathways for the biosynthesis of pyrimidines required for nucleic acid replication. The drug is active against the erythrocytic and exoerythrocytic stages of all species of *Plasmodium*-infecting humans. However, it may not be effective against recrudescent malaria.

As noted earlier, an alarming phenomenon in the treatment of malaria is the increasing resistance of the parasites to chemotherapy (Fig. 7.5C), most likely the result of mutagenic changes in some strains of *P. falciparum*. Chloroquine-resistant strains of *P. falciparum* are now common throughout Africa, pyrimethamine–sulfadoxine-resistant strains are present in Southeast Asia and South America, and strains resistant to mefloquine have been reported in Thailand, Cambodia, and Myanmar. Thus, the development of antimalarial drugs must be a continuous process. Similarly, the use of insecticides to eradicate mosquitoes has led to the appearance of resistant mosquito strains in areas that have been sprayed extensively. In Greece, for example, only a few years after apparently successful efforts to control *Anopheles* with DDT, it was necessary to alternate DDT application with dieldrin in order to control the malaria-carrying species.

Chinese wormwood (*Artemisia annua*) or *qing hao* has been used in traditional Chinese medical practice to treat fevers and a variety of ailments for more than 2000 years. The antimalarial active ingredient of the plant, termed artemisinin, was isolated in 1971 and the results of human trials were published in the Chinese Medical Journal in 1979. Subsequently, several active ingredients of artemisinin have been synthesized and reported to be fast acting and effective against drug-resistant strains of malaria. Artemisinin combination therapy using companion drugs such as mefloquine, sulfadoxine–pyrimethamine, or lumefantrine have also been reported to reduce the rate of recrudescence. An artemisinin derivative, artesunate, is reported to be an active blood schizonticide, especially against the ring stage trophozoites of *P. falciparum*. However, it is not the recommended treatment for *P. vivax*, *P. malariae*, or *P. ovale* infections.

A recent treatment for uncomplicated *P. falciparum* is a combination of chlorproguanil and dapsone known as "Lapdap." It is currently undergoing clinical trials in Africa. Lapdap is effective against drug-resistant malarial organisms, is cheap to produce, and has a short half-life. The latter property presents a smaller chance for selection of resistance than drugs with a longer half-life. The antimalarial efficacy, therefore, can be retained for a longer period of time.

Ideally, the search for new antimalarial drugs and new insecticides should be directed toward development of compounds that block a critical metabolic pathway within the parasite or mosquito. Further, such new compounds need to be inexpensive, safe, and able to produce long-lasting effects. Such an approach, however, must be predicated upon a thorough understanding of the biochemical processes occurring within these parasites, knowledge requiring years of intensive research.

A promising area of research is in the control of inflammation and its accompanying symptoms such as fever and aches and pains. In vivo chemotherapeutic studies using E6446 have shown that inhibition of specific nucleic-acid-sensing toll-like receptors (e.g., TLR-9) responses (see below) resulted in prevention of the severe symptoms of cerebral malaria.

Genomic Studies

A by-product of the acidic enzymatic digestion of host hemoglobin in the malaria parasite's digestive vacuole is the toxic ferric heme which is normally converted to nontoxic hemozoin. Administration of the drug chloroquine raises the pH within the vacuole to become more alkaline thereby affecting the action of the enzyme resulting in the accumulation of the toxic ferric heme. The ferric heme ultimately causes lysis of the vacuole and eventual death of the parasite.

Exacerbating the virulent effects of malaria is the fact that chloroquine-resistant strains have arisen in endemic areas. Chloroquine had been the major antimalarial drug until the 1950s when *P. falciparum* chloroquine-resistant strains emerged. Genomic studies have identified a gene (*Pfcrt*) on the parasite's chromosome 7 whose protein is associated with the vacuolar membrane and likely functions in part to maintain the normal acidic pH of the vacuolar contents. Mutation of the gene is believed to affect the vacuolar membrane permeability to chloroquine inhibiting its accumulation within the digestive vacuole thereby reducing the effect of the drug. Further research has shown that resistant strains do indeed display less accumulation of the drug. A number of other genes linked to *Pfcrt* have subsequently been identified and they may also be involved with chloroquine-resistance either by themselves or more likely in association with *Pfcrt*.

Artemisinin-resistant strains have now been reported in South East Asia. Fortunately, as of now, artemisinin resistance has not been reported in the other major endemic zone, sub-Saharan Africa. This conundrum has led researchers to use gene screening of the parasite populations in the two zones to determine whether any anomalies in the parasites' genetic populations exist. After extensive investigations comparing resistant strains of *P. falciparum* with normal strains, a mutated gene has been identified on chromosome 13 (*Kelch 13*) in drug resistant strains. It is suggested that the *Kelch 13* mutation may only represent one of a larger number of mutations within the genome. Currently, four other genes appear to be associated with Kelch 13 and all seem to mutate at the same time. With this information from genetic screening, a "cause and effect" phenomenon has become apparent. The "cause" of the mutation(s) is theorized to be due to the intense selective pressure derived from the widespread use of drugs (e.g., artemisinin) and vaccines in the South East Asian population. The African population has not been subjected to such intervention pressure. The mutation is the "effect" caused by the over usage of malarial control interventions. This is an example of how the use of genomic studies in association with population studies makes it possible to track the possible emergence and spread of future drug-resistant strains.

While the findings related above show how a combination of genomic studies and population studies can reveal limits for drug intervention, similar studies can also reveal how such studies along with proteonomics can expose targets for possible drug therapy or vaccine formulation. A *P. falciparum* gene family (*var*) is known to encode a specific erythrocytic membrane adhesive protein (PfEMP). This protein is expressed at the surface of infected red blood cell's knobs (see page 126). The localized protein initiates cytoadhesion of the infected red blood cell to the host's endothelial receptors resulting in capillary occlusion. When this occurs in the capillary walls of the brain, cerebral malaria can ensue. The parasite is also capable of altering its *var* gene generating antigenic variation of infected red blood cells avoiding immune recognition. Regulation of this gene family may serve as a viable target for drug or vaccine development.

Host Immune Response

The immune response of the human host differs somewhat for each of the two stages in the malarial life cycle (i.e., the pre-erythrocytic stage and erythrocytic stage). It is believed that T cells, notably $CD8^+$ T cells, play an important role in pre-erythrocytic immunity. The target for innate immune activity to this stage appears to be parasite–antigen bound molecules expressed on the surface of infected hepatocytes. On the other hand, $CD4^+$ T cell regulation appears to play a critical role in acquired immunity to the erythrocytic stage of malaria infection. The severity of the infection seriously affects the balance between Th1 (IFN-γ) and Th2 (IL-4) mediated immune functions, with the latter increasing significantly as severity of infection increases. Malaria fever (ague) is often associated with high levels of TNF-α, while asymptomatic intervals between relapses may be due to loss of reactivity to TNF-α.

The innate immune system plays an important role in defense against numerous pathogens including *Plasmodium*. TLRs are activated by recognizing the PAMPs of *Plasmodium* glycosylphosphatidylinositols (TLR-2, TLR-4) and hemozoin (TLR-9). The subsequent binding of these ligands to their respective TLRs precipitates the release of proinflammatory cytokines resulting in fever, aches, and pains.

In addition to research in chemotherapy, development of a protective vaccine against malaria is being vigorously pursued. Indeed, the development of vaccines and immunodiagnostic tests are two of the major priorities of WHO. The basis for development of a successful vaccine is identifying those stages that stimulate protective immune responses in the vertebrate host. As many of the developmental stages of malarial parasites in the vertebrate host are intracellular and, therefore, are protected from the host's immune mechanism, the extracellular forms (i.e., sporozoites and merozoites) become the targets for vaccine. The main types of vaccines currently being studied are **antisporozoite vaccines**, directed against the sporozoites when introduced by the mosquito to the vertebrate host, **anti-asexual blood stage vaccines**, directed against various blood-stage antigens such as those attributed to merozoites as well as those introduced to the surface of infected erythrocytes; and **transmission-blocking vaccines**, aimed at arresting development of the parasite in the mosquito. A number of approaches such as gene cloning and genetic engineering are creating a degree of cautious optimism in the quest. Certain singular characteristics of the sporozoite surface coat already have been identified. The coat acts as a renewable "decoy" to the vertebrate host's immune system, stimulating the production of antibodies. When the sporozoite is attacked and its "decoy" coat sloughs off, a replacement coat is synthesized, and the "decoy" effect continues. This system provides ideal protection for the sporozoite, which resides only briefly in the blood before it enters a liver cell where it is protected from circulating antibodies. It also suggests a brief "window of opportunity" for targeting a potential vulnerable developmental stage. Protection afforded the sporozoite by its surface coat in its brief transit through the host circulatory system contrasts with that in African trypanosomes (p. 104), which are exposed to the immune system for a long time throughout a lengthy residence in the blood. The surface antigens of African trypanosomes, it will be recalled, undergo continual change, each dominant population keeping one step ahead of the vertebrate host's immune system.

In endemic areas, premunition (p. 34) is the basis for protective immunity as long as low-level infection persists. With complete cure, however, the victim regains susceptibility. Also, while nursing infants in endemic areas are protected through antibodies in the mother's milk, at the time of weaning the children are at greatest risk, and the highest mortality rate in such regions is among children. Also, *P. falciparum* has been reported to cross the placenta and cause infection of the fetus.

Factors other than immunological ones also may affect susceptibility among humans. Several genetic conditions affect the malarial organism. Susceptibility conferred by the presence of Duffy antigens has already been discussed. Genetic deficiency in glucose-6-phosphate dehydrogenase in erythrocytes (**favism**) creates an inhospitable environment for the parasites. This enzyme is rate-limiting in the pentose pathway which, among other functions, provides reducing potential to protect the erythrocyte's plasma membrane against toxic by-products such as those produced by the parasite. The lack of protection causes excessive leakage of potassium from the infected cell as well as allowing excess sodium to enter. Since the malarial parasite may require a higher level of potassium than is available, it dies. Humans heterozygous for **sickle cell anemia** possess a selective advantage over individuals with normal hemoglobin in regions where *P. falciparum* is endemic. The tendency to "sickle" is accelerated when affected erythrocytes are subjected to low oxygen tensions. One explanation for such a selective advantage is that *P. falciparum*-infected erythrocytes, when trapped in visceral capillaries, will sickle, damaging the erythrocyte membrane, causing excessive leakage of potassium from the infected cell. In a heterozygous host, up to 40% of the cells are of this type and are, therefore, unsuitable for the parasite's development, thus accounting for the resistance to the disease among such persons. Ironically, the severe pathological effects of *falciparum* malaria have resulted in the maintenance of this highly deleterious mutation in the human population. The requirement of malarial organisms for intracellular potassium has been questioned. In vitro experiments have indicated that the parasite may grow normally in erythrocytes demonstrating high sodium and low potassium content. An alternate explanation is that such genetic conditions coupled with the presence of the parasite enhance the possibility of detection and removal by host phagocytic cells.

Physiology

Metabolic characteristics of *Plasmodium* provide targets for drug action and immunological control. Unfortunately, many gaps remain in the understanding of several metabolic aspects of the intracellular stages of the life cycle due, in part, to the fact that, until recently, there was no culture technique for reproducing these stages in vitro. For example, glucose is the chief carbohydrate required by the parasite, and most of its energy appears to derive from glycolysis. However, while some intermediates of the TCA cycle have been demonstrated, there is no evidence of functional mitochondria in the erythrocytic stages, although the mosquito stages do possess these organelles. In the erythrocytic stages, the parasites are facultative anaerobes (i.e., using oxygen when available, primarily during the synthesis of nucleic acids).

The end-products of carbohydrate metabolism are lactic acid, formic acid, and a limited amount of acetic acid. The parasite does fix carbon dioxide, and the enzymes that catalyze this process are believed to be vulnerable to quinine and chloroquine.

Hemoglobin is essential for the parasite's development, although it is uncertain precisely which components are required. The parasite digests hemoglobin intracellularly, producing an insoluble by-product, hemozoin. In addition to its possible disruption of carbon dioxide fixation, chloroquine appears to interfere with the intracellular digestive processes of the parasite. Chloroquine and quinine are both weak bases that raise the pH of the lysosomal compartment, reducing the ability of the parasite to digest host hemoglobin efficiently.

Although the malarial parasite depends on host erythrocytes for many essential molecules, it does have the ability to synthesize folic acid, a key compound in pyrimidine synthesis, from basic molecules. Host cells, on the other hand, require outside sources of folic acid. Drugs that block the synthesis of folic acid (**antifols**) by the parasite, therefore, possess immense chemotherapeutic potential.

Suggested Readings

Aidoo, M., Terlouw, D. J., Kolczak, M. S., McElroy, P. D., terKuile, F. O., Kariuki, S. K., et al. (2002). Protective effects of the sickle cell gene against malaria morbidity and mortality. Lancet, *359*, 1311–1312.

Carter, R. (2003). Speculations on the origins of *Plasmodium vivax* malaria. Trends in Parasitology, *19*, 214–219.

Desowitz, R. S. (1976). How the wise men brought malaria to Africa. Natural History, *85*, 36–44.

Friedman, M. J., & Trager, W. (1981). The biochemistry of resistance to malaria. Scientific American, *244*, 154–165.

Godson, G. N. (1985). Molecular approaches to malaria vaccines. Scientific American, *248*, 52–59.

Good, M. F. (2005). Vaccine-induced immunity to malaria parasites and the need for novel strategies. Trends in Parasitology, *21*, 29–34.
Gowda, D. C. (2007). TLR-mediated cell signaling by malarial GPIs. Trends in Parasitology, *23*, 596–604.
Greenwood, B. (2005). Malaria vaccines: evaluation and implementation. Acta Tropica, *95*, 298–304.
Hermentin, P. (1987). Malaria invasion of human erythrocytes. Parasitology Today, *3*, 52–55.
Jide, J., Ying, H., Wenyue, X., & Fusheng, H. (2009). Toll-like receptors, a double-edged sword in immunity to malaria. Journal of Medical Colleges of PLA, *24*, 118–124.
Leete, T. H., & Rubin, H. (1996). Malaria and the cell cycle. Parasitology Today, *12*, 442–444.
Lingelbach, K., & Joiner, K. A. (1998). The parasitophorous vacuole membrane surrounding *Plasmodium* and *Toxoplasma*: an unusual compartment in infected cells. Journal of Cell Science, *111*, 1467–1475.
Meis, J. F. G. M., & Verhave, J. P. (1988). Exoerythrocytic development of malarial parasites. Advances in Parasitology, *27*, 1–61.

CHAPTER

8

Blood and Tissue Protistans III

Other Protists

Chapter 8 considers blood and tissue protistans not considered in the prior two chapters. *Babesia* is a protozoan parasite of the blood that causes a hemolytic disease known as Babesiosis. People who contract Babesiosis suffer from malaria-like symptoms and as a result, malaria is a common misdiagnosis for the disease. Approximately 100 species of *Babesia* have been identified, only a few have been documented as pathogenic in humans. In the United States, *Babesia microti* is the most common strain associated with humans. *Toxoplasma gondii* can be carried by many warm-blooded animals (birds or mammals, including humans), but the definitive host of *T. gondii* is the cat. Toxoplasmosis is usually minor and self-limiting but can have serious or, in rare cases, fatal effects on a fetus whose mother contracts the disease during pregnancy or on an immunocompromised individual.

BABESIA

Infections of humans by *Babesia* spp. have been known since 1957; however, human babesiosis has, in recent years, become sufficiently common on Nantucket Island and Martha's Vineyard in Massachusetts, at Shelter Island on Long Island, New York, and in Wisconsin to warrant the attention of medical parasitologists. In each of these locales, the causative agent has been identified as *B. microti*, a natural parasite of the meadow vole and other rodents, with the vector in all instances being the tick, *Ixodes dammini*. *I. dammini* also serves as the vector for Lyme disease among humans (see p. 355). Humans acquire the infection when an infected tick accidentally feeds on a human host. Splenectomized persons seem especially vulnerable to babesiosis; indeed, such individuals appear to be susceptible to more than one species of *Babesia*, and most of the recorded fatalities have occurred in splenectomized individuals.

Life Cycle (Fig. 8.1)

Infection in the vertebrate host is initiated when **vermicles** (=sporozoites) are introduced through the bite of an infected tick. The vermicle is approximately 2-μm long and varies in shape from pyriform to spiral. There is neither an exoerythrocytic nor a sexual phase in the

Human Parasitology. http://dx.doi.org/10.1016/B978-0-12-813712-3.00008-4

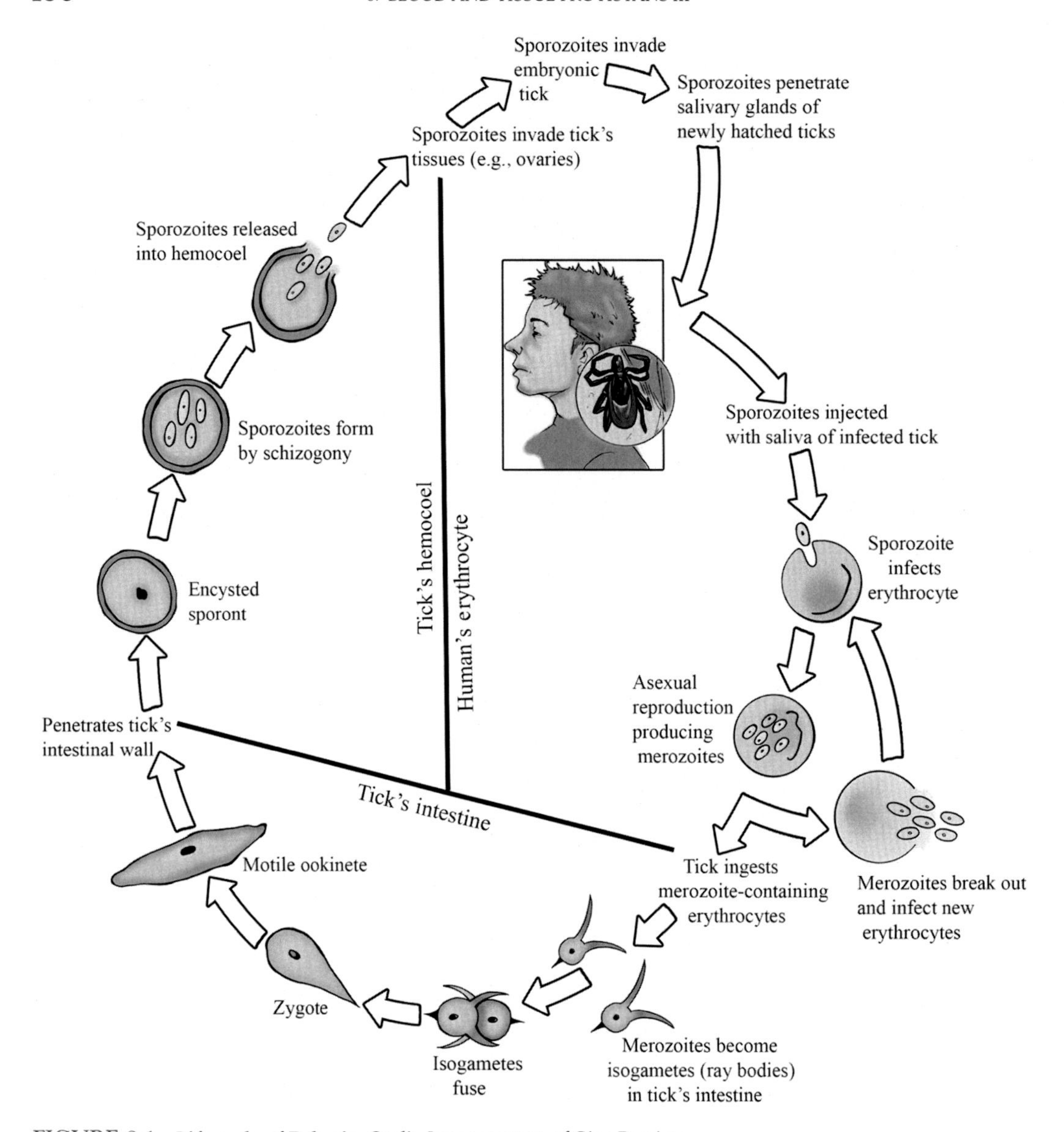

FIGURE 8.1 **Life cycle of Babesia.** *Credit: Image courtesy of Gino Barzizza.*

vertebrate portion of the life cycle. The vermicle enters the host erythrocyte where it develops into a trophozoite, rapidly increases in size, and undergoes binary fission producing numerous merozoites. The merozoites erupt from the infected erythrocyte and enter other erythrocytes where the cycle of growth, division, and re-entry continues, resulting in an extremely large intra-erythrocytic population in a short time.

The tick initially acquires the infection by ingesting infected vertebrate blood. Parasites are released into the tick's intestine and are immediately transformed into motile, polymorphic isogametes. Gametic fusion to form zygotes occurs within 24 h after the blood meal. The zygotes transform into cigar-shaped ookinetes, 8–10-μm long, which penetrate the tick's intestine and become encysted sporonts, each of which grows to about 16 μm in diameter within 2 days. Each sporont nucleus then undergoes schizogony, and the resulting vermicles, each 9–13-μm long, migrate into the hemocoel, whence they invade various tissues of the tick, particularly ovarian. In the ovaries, they undergo several more divisions and invasions within embryonic ticks. Final generations of vermicles eventually migrate to the salivary glands of the newly hatched tick and are injected into the vertebrate host when the young tick takes a blood meal. The passage of *Babesia* from a tick to its progeny in this manner is known as **transovarian transmission**.

Symptomatology and Diagnosis

Human babesiosis can be fatal in immunologically compromised individuals. The disease mimics mild malaria, and, unless there is reason to suspect babesiosis, the erythrocytic stages in blood smears are often mistaken for malarial organisms. The usual manifestation of the infection is basically hemolytic anemia.

Chemotherapy

Since babesiosis can be mistaken for malaria, it is sometimes treated with chloroquine. In spite of some claims of success, there is no real evidence that the drug is effective against this disease. A 7–10-day treatment with clindamycin plus quinine has been used successfully to treat the disease. The procedure, however, is still considered investigational. Since most individuals recover spontaneously, treatment of symptoms may be the best way to manage the disease. For patients in whom the number of parasites becomes life-threatening, exchange transfusion is a viable option.

TOXOPLASMA GONDII

Human toxoplasmosis is caused by a coccidian, *T. gondii*, originally discovered in 1908 in a desert rodent. This potentially perilous parasite is estimated to infect 50% of the population of the United States. Fortunately, most of the infections are asymptomatic, with clinical toxoplasmosis affecting only a limited number of individuals. Occasionally, however, minor epidemics do occur. The principal means of acquiring the infection is either by ingestion of inadequately cooked meat, primarily beef, pork, and lamb, or by contact with feral or domestic cats. Any cat, no matter how well cared for, may carry and pass the infective stage of *Toxoplasma*. Congenital toxoplasmosis is a very serious disease, and for this reason, pregnant women should avoid contact with litter box filler used by cats. Flies and cockroaches have also been implicated as carriers of the infective stages from cat feces to food.

Life Cycle (Fig. 8.2)

Toxoplasma can attack a wide variety of tissue cells but seems to favor muscle, lymph nodes, and intestinal epithelium. Infection of intestinal epithelial cells occurs only in felines, probably the "normal" hosts, and this developmental pathway is termed the **enteric** or **enteroepithelial phase**. It is during this phase that the formation of sporozoite-containing oocysts, the primary source of human infection, occurs. In other hosts, including many species of carnivores, insectivores, and primates, there is only the **extraintestinal** or **tissue phase**. Ingestion of a sporulated oocyst is the precursor for either developmental phase. Each oocyst, measuring 10–13 μm by 9–11 μm, contains two sporocysts, each of which, in turn, contains four sporozoites. The sporozoites are released from the oocyst in the lumen of the host's small intestine. In cats, some sporozoites penetrate intestinal epithelial cells to begin the enteric phase, while others penetrate the mucosa and develop in cells of underlying tissues, including lymph nodes and leukocytes.

In the enteric phase, the sporozoites enter the host cell, are enclosed in a parasitophorous vacuole, become trophozoites, and reproduce by endodyogeny and/or endopolyogeny. The number of asexual cycles varies according to the physiological condition of the feline host, but about 2–40 merozoites (Fig. 8.3) arise from each trophozoite. Three to 15 days after infection, some of the merozoites enter new host cells and develop into either microgametocytes (males) or macrogametocytes (females). About 2%–4% of the gametocytic population is microgametocytes, and each produces about 12 microgametes. Fertilization is intracellular, the microgametes bursting from their host cell to invade other host cells containing macrogametes. Following fertilization, the resulting zygote develops into an oocyst that breaks out of the cell into the lumen of the feline intestine to be passed out with feces. Within 2–5 days, in the presence of oxygen, the oocyst undergoes sporogony, forming two sporocysts, each containing four sporozoites. The feline enteric phase, then, proceeds through three stages: asexual reproduction producing merozoites, gamogony producing gametes, and sporogony producing oocysts containing the infective sporozoites.

In the cat, extraintestinal and enteric developments, which follow entirely different patterns, can occur simultaneously. In mammals other than cats, including humans, only the extraintestinal phase occurs.

In extraintestinal development, there are two avenues for infection: infective oocysts may be ingested, in which case sporozoites invade cells other than those of the intestinal epithelium; or tissue infected with pseudocysts may be ingested, whereupon liberated **bradyzoites** can invade cells. In either case, the intestinal wall is penetrated and the parasites are engulfed by macrophages and transported throughout the body. In the macrophage, the organisms, protected from the lysosomal activity of the cell, form rapidly dividing merozoites called **tachyzoites**, which measure 7 by 2 μm. In acute infections, 8–16 tachyzoites are produced within a parasitophorous vacuole in the host cell, eventually causing the cell to disintegrate, probably by pressure, releasing the parasites to invade new cells. An accumulation of tachyzoites in a host cell is known as a **group**. Many types of cells are vulnerable to infection. As the disease becomes chronic, parasites infecting cells of the brain, heart, and skeletal muscle reproduce more slowly than during the acute phase. At this time, they are designated as **bradyzoites** and accumulate in

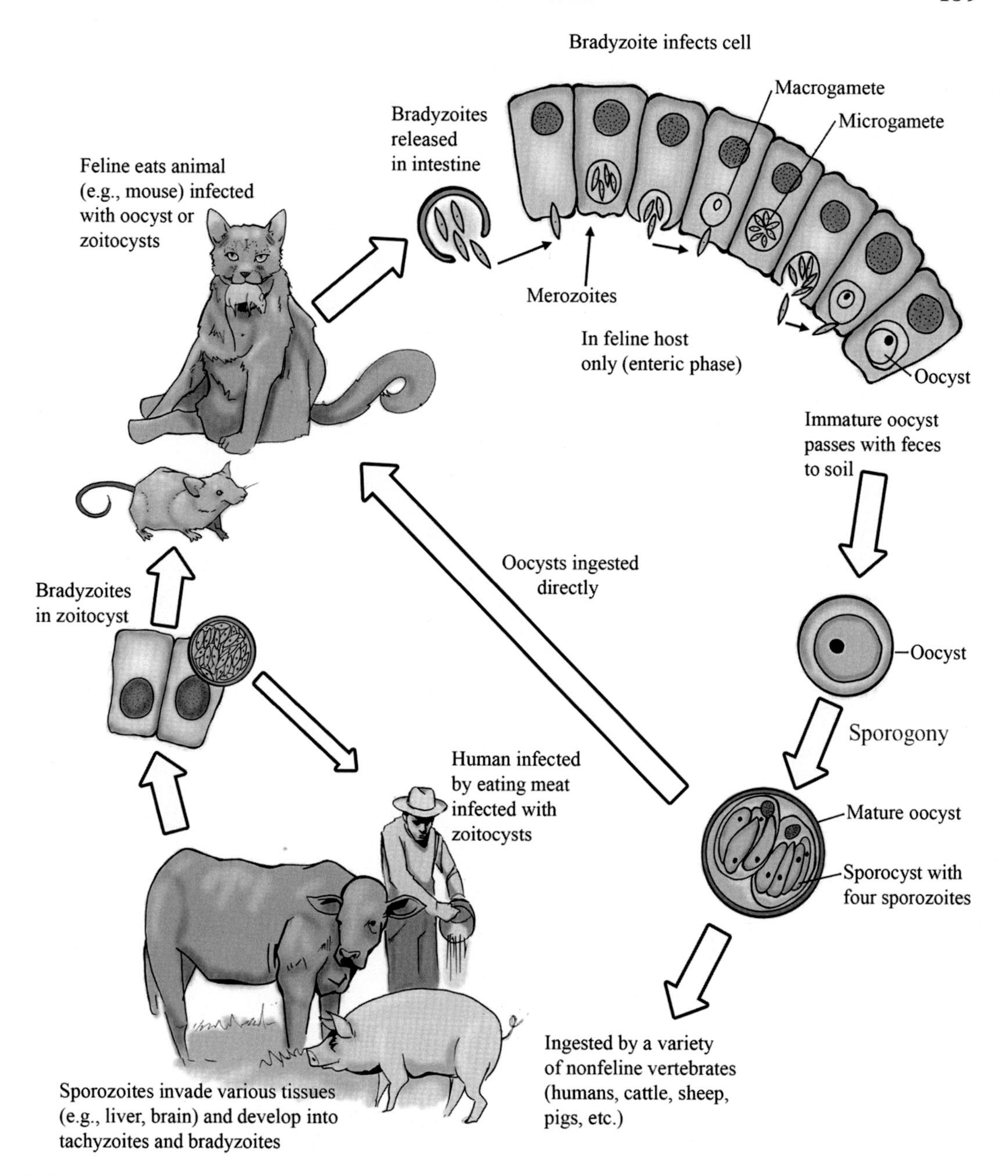

FIGURE 8.2 **Life cycle of *Toxoplasma*.** *Credit: Image courtesy of Gino Barzizza.*

large numbers within an infected cell. Gradually, thick walls develop around the masses of bradyzoites to form **zoitocysts** or **tissue cysts**, which may persist for months or even years, especially in nerve tissue. Therefore, all animals except cats can be considered paratenic hosts since the parasite never completes its sexual phases in those creatures.

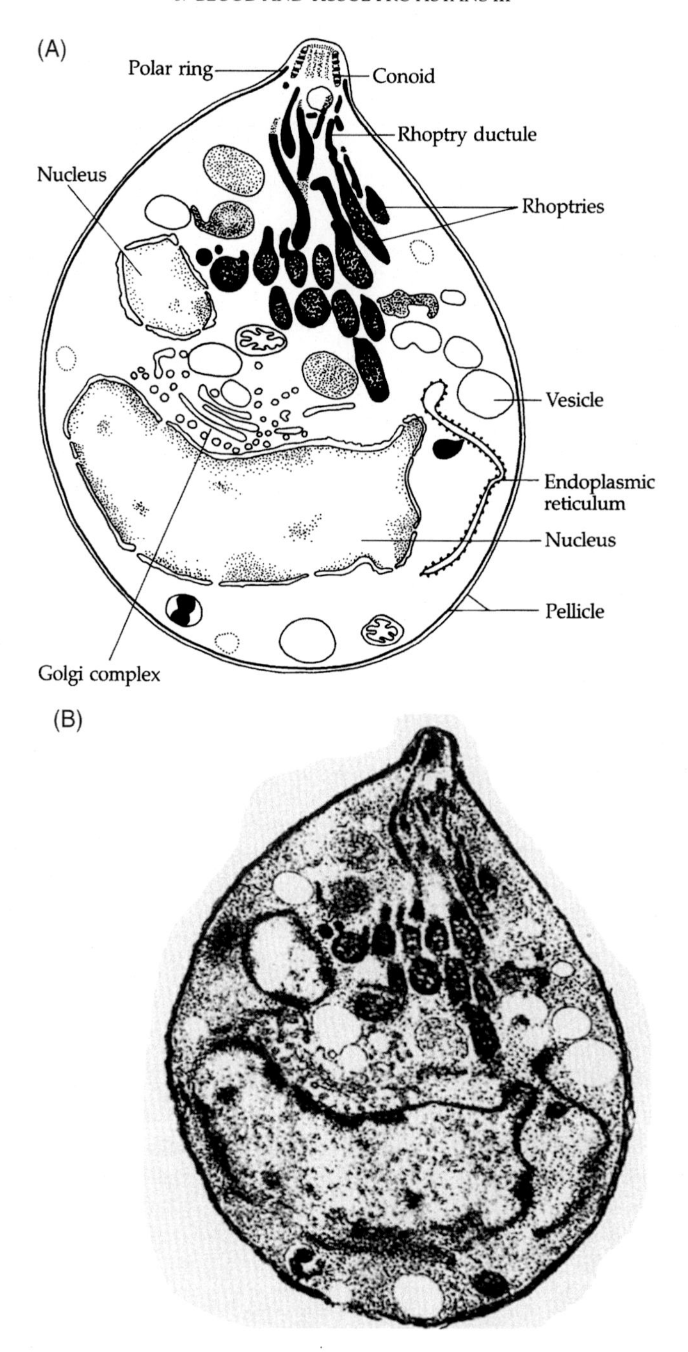

FIGURE 8.3 **Apical complex.** (A) Drawing of merozoite of *Toxoplasma gondii* showing some constituents of the apical complex. (B) Transmission electron micrograph from which (A) was drawn.

Zoitocyst formation coincides with the development of immunity in the host. This immunity, involving both humoral and cellular reactions, is usually permanent and precludes establishment of a new infection. If immunity wanes, it is restored by release of bradyzoites from cysts, restimulating the host immune system.

Epidemiology

T. gondii is cosmopolitan in distribution. All mammals, including humans, are capable of transmitting toxoplasmosis transplacentally. Sporulated oocysts, tachyzoites, and bradyzoites all serve as infective agents. Sources of infection vary, ranging from direct contamination, as from handling cat litter, to ingestion of inadequately cooked meat or raw milk.

Immunologic surveys reveal that humans throughout the world carry antibodies to *Toxoplasma*; however, clinical toxoplasmosis is rare, and infections are generally asymptomatic. Although not all of the influential factors are presently known, it has been established that the following affect the level of pathology: (1) age of the host, with older hosts being more resistant to the disease; (2) virulence of the strain of *T. gondii* involved; (3) natural susceptibility of the host; and (4) degree of acquired immunity of the host.

At least three different strains of *T. gondii* have been identified and studied. They differ primarily in life cycle duration and the number and morphology of merozoites produced. Genomic studies of the three strains infecting host neuroepithelial cells have revealed that they affect gene expression. Strain I largely affects genes related to the host's central nervous system. Strain III influenced those genes whose expression is related to nucleotide metabolism, while strain II infection does not appear to affect any clearly defined set of genes. Strain I is considered the most lethal strain and is most prevalent in congenital infections. Strain II is most prevalent in immunocompromised patients.

Symptomatology and Diagnosis

Symptomatic, or clinical, toxoplasmosis may be classified as acute, subacute, chronic, or congenital. Acute toxoplasmosis in humans is characterized by parasitic invasion of the mesenteric lymph nodes and liver parenchyma. The most common symptom is painful, swollen, lymph glands in the inguinal, cervical, and subclavicular regions, frequently accompanied by fever, headache, anemia, muscle pain, and sometimes pulmonary complications. The tachyzoites proliferate in many tissues and tend to kill host cells rapidly. When cells from sites such as the retina or brain are involved, serious lesions often develop. Subacute toxoplasmosis is merely a prolongation of the acute stage.

Normally, the duration of the chronic stage is limited by the host's immunological system. However, if immunity develops slowly, the course of clinical toxoplasmosis can be protracted. During this period, tachyzoites continue to destroy cells, producing extensive lesions in lung, heart, liver, brain, and eyes. Damage is usually greater to the central nervous system than to non-nerve tissues because of lower immunocompetence in the former. Toxoplasmosis becomes chronic when immunity in the host, accompanied by the formation of zoitocysts, becomes sufficient to suppress tachyzoite proliferation. The zoitocysts may remain intact for years, producing no clinical symptoms. A zoitocyst wall may occasionally rupture, however, releasing bradyzoites, most of which are destroyed by host responses, although some may penetrate cells and form new zoitocysts. Death of bradyzoites elicits a hypersensitive response. In the brain, nodules of glial cells gradually form at the sites of such reactions. In cases in which there are sufficient numbers of such nodules, the victim may develop symptoms of chronic encephalitis, sometimes accompanied by spastic paralysis. This is especially true of AIDS (acquired immune deficiency syndrome) patients, in whom *Toxoplasma* can cause severe brain damage. The presence and rupture of pseudocysts in the retina and choroid can lead to blindness. Chronic toxoplasmosis can also cause myocarditis, leading to permanent heart damage and pneumonia.

Congenital toxoplasmosis results from fetal transplacental infection. Such infection may result in stillbirth or a number of severe birth defects. Approximately, 12% of infected infants born alive die shortly after birth, and fewer than 20% of those surviving are normal by age 4. Abnormalities occur in the central nervous system, eyes, and viscera with symptoms such as jaundice, microcephaly, and hydrocephaly appearing at birth or shortly thereafter.

Diagnosis is based, primarily, on serological tests using killed antigens. Demonstration of organisms in mice following inoculation with suspected fluid or biopsied tissue (xenodiagnosis) constitutes positive diagnosis.

Chemotherapy

Oral administration of pyrimethamine, usually accompanied by sulfadiazine, is the treatment of choice at this time. Since pyrimethamine is an antifol and thus can cause folic acid deficiency in the host, supplemental folic acid (e.g., leucovorin, a folic acid analog) may also be added to the regimen as a precautionary measure.

Host Immune Response

Toxoplasmosis is considered an opportunistic disease (p. 30). The disease characteristically displays two stages, an acute stage defined by the early spread of tachyzoites by lymphocytes to the tissues, and a chronic stage characterized by the appearance of cysts. During the acute stage, IL-12 from activated macrophages along with TNF-α induces NK cells to produce IFN-γ. IFN-γ and TNF-α in turn activate anti-*Toxoplasma* activity in macrophages. $CD4^+$ and $CD8^+$ T cells are not involved in the innate immunity response. During the chronic stage, acquired immunity to *T. gondii* exhibits inflammatory characteristics associated with Th1 type responses. In this instance, $CD8^+$ T cells and IFN-γ play major roles.

CRYPTOSPORIDIUM PARVUM

Cryptosporidiosis is a condition caused by the coccidian *Cryptosporidium parvum*. The organisms live in the brush border underneath the cell membrane of the small intestine and respiratory epithelium of a number of mammals, including humans. This site, described as an intracellular/extracellular site, probably allows the organism to evade host immune surveillance. Acquisition is by ingestion of the oocyst usually in contaminated drinking water.

Life Cycle (Fig. 8.4)

The infective oocysts are oval, measuring 4–5-μm wide, and contain four slender sporozoites but no sporocysts. Oocysts are expelled with the feces of a number of infected mammals and, when ingested, sporozoites excyst in the small intestine and attach to the epithelial surfaces of the ileum and colon (Fig. 8.5). Once enclosed in a parasitophorous vacuole formed from the convergence of microvilli of infected cells, each sporozoite becomes a trophozoite (Fig. 8.6), which, in turn, undergoes schizogony to produce eight first-generation merozoites. These erupt from the infected cells into the intestinal lumen. The following sequence of events then transpires (1) attachment of each released first-generation merozoite to an uninfected

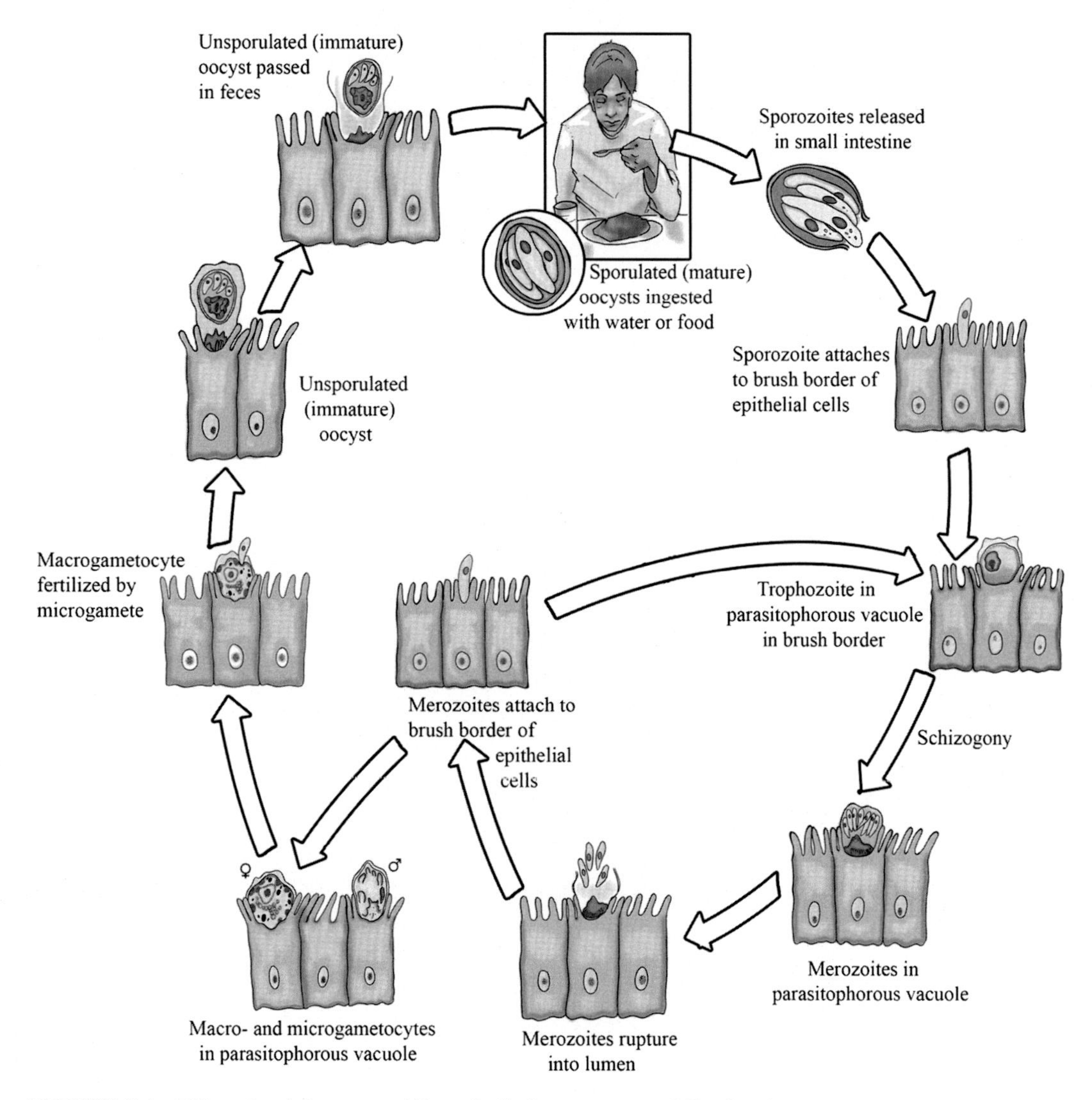

FIGURE 8.4 **Life cycle of *Crpytosporidium*.** *Credit: Image courtesy of Gino Barzizza.*

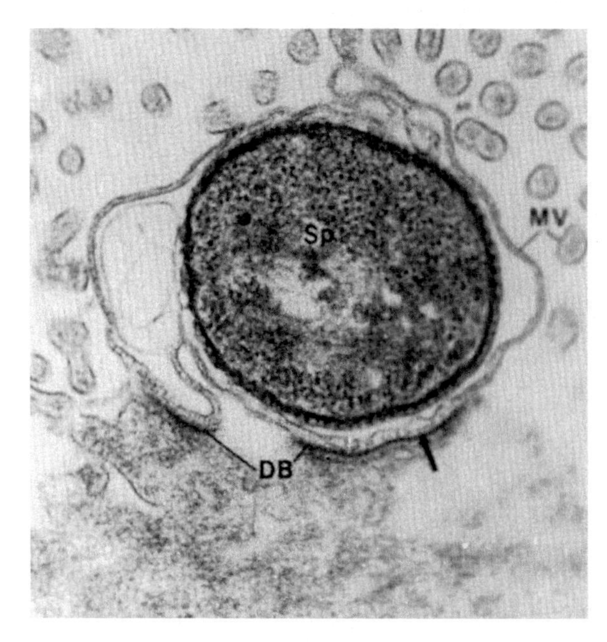

FIGURE 8.5 ***Sporozoite* (sp) of *Cryptosporidium parvum* surrounded by intestinal microvilli (mv).** Note dense band (DB) on the surface of the intestinal cell apparently induced by organism and microfilaments (*arrow*).

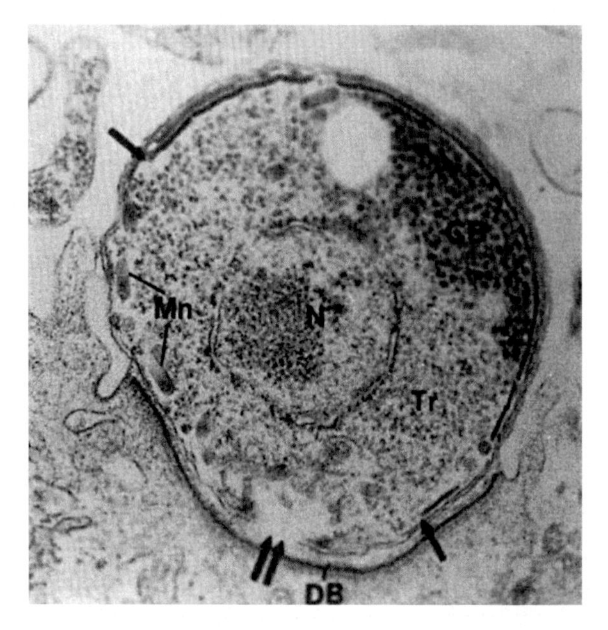

FIGURE 8.6 **Early trophozoite (Tr) of *Cryptosporidium parvum*.** Note the micronemes (Mn), nucleus (N), crystalloid bodies (CB), and dense band (DB) of the trophozoite. *Arrows* point to the parasite's plasma membrane.

epithelial cell surface, (2) envelopment of the merozoite by microvilli, (3) transformation to trophozoite, (4) schizogony to form a second generation of four merozoites, (5) eruption of second-generation merozoites from the infected cell, and (6) incorporation of these merozoites to the surface of still other uninfected epithelial cells. The second-generation merozoites differentiate into microgametocytes and macrogametocytes. The former undergo several divisions producing numerous microgametes, while the latter transform to macrogametes, usually one per parasitophorous vacuole. The microgametes burst from the infected cell and enter cells containing macrogametes. Following fertilization, the resulting zygote differentiates into an unsporulated oocyst, which frees itself from the superficial parasitophorous vacuole and is shed with the feces of the host. The oocyst is infectious upon shedding. Each oocyst sporulates when swallowed by a new host or in the soil.

Epidemiology

Cryptosporidium is cosmopolitan in distribution, occurring in a wide variety of hosts such as primates (including humans), cattle, sheep, rodents, and birds. Although several distinct species of *Cryptosporidium* have been described, studies have shown a surprisingly high degree of cross-infectivity of *Cryptosporidium* recovered from different hosts, with *C. parvum* being the most dominant species in nature. Therefore, a large number of hosts can act as reservoirs. Outbreaks of cryptosporidiosis were reported in 1993 in Milwaukee, WI and Washington, DC. Spring flood water laden with infective oocysts, ostensibly washed downstream in cattle manure, were blamed for overloading the cities' water purification systems. In Washington, drinking water was brought in from Charlottesville, VA for use by athletes participating in a weekend football game to protect them from infection. Lesser outbreaks have been reported from other cities. None of the purification chemicals used in municipal water treatment plants is effective against the oocysts. Filtration is the best means of reducing the number of oocysts in public drinking water. Since the number of asymptomatic individuals who pass oocysts has been found to be higher than previously thought, outbreaks attributed to person-to-person transmission, especially in day-care centers and nursing homes, have been reported.

Symptomatology and Diagnosis

Morphological alteration in the intestinal epithelium of infected individuals includes villous atrophy, mitochondrial changes, and increased lysosomal activity in infected cells. Symptoms vary from none to mild diarrhea to diarrhea with severe cramping, anorexia, nausea, and vomiting. In immunologically competent individuals, the infection is self-limiting, lasting from several days in most patients to several weeks. In immunologically compromised individuals, however, cryptosporidiosis is a chronic disease lasting months or even years. In extreme cases (e.g., AIDS patients), the disease can be extremely severe with patients losing as much as three liters of fluid daily. Mortality in such instances may reach 50%, and secondary extraintestinal complications are common, such as biliary disease and, possibly, pneumonitis from respiratory tree infection. Diagnosis depends upon identification of oocysts in the stool. Histological sections from biopsy of intestinal epithelium showing any of the stages of the organism also constitute positive identification.

Chemotherapy

In immunocompetent patients, no treatment is recommended since the infection is self-limiting. Nitazoxanide has been recommended as treatment in immunocompetent patients with severe diarrhea. For other patients (i.e., immunosuppressed patients), while no chemotherapeutic agents have yet proven effective, the antibiotic agent Azithromycin has been used successfully to treat severe diarrhea in immunosuppressed children. Octreotide also has no effect on the organism but has been used to control diarrhea. Paromomycin has been used with some success to control the organism. While there is still a lack of the usual drug targets used for the control and treatment of other parasitic protists, genomic studies have identified several plant-like and bacterial-like enzymes in *C. parvum* that are being pursued as potential candidates for drug exploration.

Host Immune Response

Knowledge regarding the human immune response to *C. parvum* infection is at best fragmentary. Preliminary investigations on human models reveal that activation of an early immune response by IL-15 appears to be critical for the initial clearance of the parasite. $CD4^+$ T cells and IFN-γ play key roles in the human memory response. Little is known of the roles of β-cell and $CD8^+$ T-cell functions. TGF-β and IL-10 cells are likely involved in anti-inflammatory and healing processes. They may also lead to generation of IgA, which may prevent re-infection.

CYCLOSPORA CAYENTANENSIS

Cyclospora cayentanensis is a relatively new addition to the realm of organisms dangerous to human health. It is now considered an emerging pathogen. Although the organism has been known since 1979, it was first isolated from patients from Peru in 1985. The first outbreak of cyclosporidiosis in the United States was recorded in 1990 in Chicago, IL. Subsequently, at least 1000 cases of the disease have been reported in the United States and numerous incidents have been confirmed from Central and South America, the Caribbean, Asia, and Eastern Europe. In Kathmandu, Nepal, for example, *Cyclospora* has been identified in 11% of individuals with gastrointestinal symptoms. Most infections have been traced to drinking water contaminated with sporulated oocysts. However, an outbreak of cyclosporidiosis in Charleston, South Carolina in 1996, was initially traced to contaminated strawberries and later in contaminated Guatemalan raspberries as well. The infective stage, a sporulated oocyst, measures 8–10 μm in diameter. In the laboratory, sporulation requires 7–13 days at 25–32°C, with each sporulated oocyst enclosing two sporocysts, each containing two sporozoites. Sporozoites measure 1.2-μm wide by 9-μm long. The complete life cycle and epidemiological features still remain largely unknown.

Symptomatology and Diagnosis

Clinical symptoms of cyclosporidiosis resemble those of cryptosporidiosis (i.e., nausea, vomiting, anorexia, weight loss, and explosive watery diarrhea lasting 1–7 weeks). Diagnosis is difficult because *Cyclospora* oocysts recovered from the feces of infected humans are often mistaken for those of *Cryptosporidium*. It is strongly recommended that precise measurements of the oocyst be made to differentiate between *Cyclospora* oocysts and those of *Cryptosporidium*.

Chemotherapy

As in any diarrheic incident, it is important to correct and maintain hydration. *Cyclospora* can be treated successfully in children and adults with trimethoprim–sulfamethoxazole.

ISOSPORA BELLI

Isospora belli, the causative agent of isosporiasis in humans, is endemic in South America, the Caribbean, Africa, and Southeast Asia. The infective stage is the sporulated oocyst containing two sporocysts, each with four sporozoites. The apicomplexan attacks the columnar epithelium of the small intestine, causing diarrhea, which is usually mild in normally healthy patients. However, in immunologically compromised individuals, the infection can be life-threatening, causing high fever and persistent severe diarrhea. In Haiti, there is a 15% prevalence among AIDS patients. A second species, *I. hominis*, has been implicated in some human cases of isosporiasis. Trimethoprim–sulfamethoxazole is an effective treatment.

BLASTOCYSTIS HOMINIS

Blastocystis hominis is an enteric parasite of humans and a wide variety of animals. Its geographic range is global and the organism has been known since the early part of the 20th century. It is the causative agent of traveler's diarrhea, rectal bleeding, fever, and irritable bowel syndrome. The taxonomic status of the organism is still questionable, but based on information derived from sequencing studies from multiple conserved genes, it is considered a polymorphic protozoan in the stramenopile group of protists. Its life cycle includes four stages, a vacuolated stage, most commonly found in stool samples, amoeboid, precystic, and cystic stages. The amoeboid stage reproduces by binary fission, while the cystic stages are considered, by many, the transmissible stages. The cystic stage comprises thin-walled and thick-walled types with the former probably being the autoinfective stage and the latter's role being that of external transmission. *B. hominis* displays extreme genetic diversity. The forms that are infective to humans can be assigned to at least seven zoonotic subtypes based on their genotypes. It is now believed that such genetic diversity is indicative of the pathogenic and non-pathogenic nature of the organism. While no treatment is indicated, metronidazole has proven effective in a number of clinical cases.

Suggested Readings

Allred, D. R. (2003). Babesiosis: persistence in the face of adversity. Trends in Parasitology, *19*, 51–55.

Bern, C., Otega, Y., Checkley, W., Roberts, J. M., Lescano, A. G., Cabrera, L., et al. (2002). Epidemiologic differences between cyclosporiasis and cryptosporidiosis in Peruvian children. Emerging Infectious Diseases, *8*, 581–585.

Clark, D. P., & Sears, C. L. (1996). The pathogenesis of cryptosporidiosis. Parasitology Today, *12*, 221–225.

Ho, A. Y., Lopez, A. S., Eberhart, M. G., Johnson, C. C., & Herwaldt, B. L. (2002). Outbreak of cyclosporiasis associated with imported raspberries. Emerging Infectious Diseases, *8*, 783–788.

Hughes, H. P. A. (1985). Toxoplasmosis—a neglected disease. Parasitology Today, *1*, 41–44.

Jackson, M. H., & Hutchison, W. M. (1989). The prevalence and source of *Toxoplasma* infection in the environment. Advances in Parasitology, *28*, 55–105.

Kjemtrup, A. M., & Conrad, P. A. (2000). Human babesiosis: an emerging tick-borne disease. International Journal of Parasitology, *30*, 1323–1337.

Lindsay, D. S., & Blagburn, B. L. (1994). Biology of mammalian *Isospora*. Parasitology Today, *10*, 214–220.

Luder, C. G. K., Bohne, W., & Soldati, D. (2001). Toxoplasmosis: a persisting challenge. Trends in Parasitology, *17*, 460–463.

Millar, B. C., Finn, M., Xiao, L. H., Lowery, C. J., Dooley, J. S., Rooney, P. J., et al. (2002). *Cryptosporidium* in foodstuffs—an emerging aetiologial route of human foodborne illness. Trends in Food Science and Technology, *13*, 168–187.

Morris, A. M., Beard, C. B., & Huang, L. (2002). Update on the epidemiology and transmission of *Pneumocystis carinii*. Microbes and Infection, *4*, 95–103.

Perkins, M. E. (1992). Rhoptry organelles of apicomplexan parasites. Parasitology Today, *8*, 28–32.

Spielman, A. (1988). Lyme disease and human babesiosis: evidence incriminating vector and reservoir hosts. In P. T. Englund, & A. Sher (Eds.), The biology of parasitism (pp. 147–165). New York, NY: Wiley-Liss.

Ynes, R., Ortega, M. S., Sterling, C. R., Gilman, R. H., Cama, V. A., & Diaz, F. (1993). *Cyclospora* species—a new protozoan pathogen of humans. New England Journal of Medicine, *328*, 1308–1312.

Walzer, P. D. (1999). Immunological features of *Pneumocystis carinii* infection in humans. Clinical and Diagnostic Laboratory Immunology, *6*, 149–155 (1773–1785).

Zipori, T. (1988). Cryptosporidiosis in perspective. Advances in Parasitology, *27*, 63–129.

CHAPTER

9

General Characteristics of the Trematoda

Chapter 9 gives the reader an overview of the trematoda as one of the two groups representative of the flatworm phylum Platyhelminthes discussed in this text. Various aspects of the morphology of adult trematodes are dealt with in detail. The discussion begins with an examination of the unique outer covering of members of this group, the tegument. This is followed by an examination of the diversity of what are considered incomplete digestive tracts. The discussion of the reproductive system follows the same detailed format as presented for the tegument and digestive tracts. Ample illustrations help to augment these descriptions. Lesser detail is presented in the examination of the muscular, nervous, and reproductive systems. Within the context of the reproductive system discussion, a review of trematode generalized life cycle patterns is pictorially presented. An introduction to the general physiology and chemotherapy illustrative of the trematodes is included. The chapter concludes with a brief classification of the group emphasizing only those trematodes that will be discussed in subsequent sections.

The phylum Platyhelminthes includes various dorsoventrally flattened animals commonly known as flatworms (Fig. 9.1). All members are typically bilaterally symmetrical and lack a body cavity. The digestive tract, if present, is incomplete (i.e., the ceca end blindly); therefore, the only opening to the exterior, the mouth, serves for both ingestion and egestion. Skeletal, circulatory, and respiratory systems are usually lacking. The space between the body wall and the internal organs contains connective tissue fibers, muscle, and unattached and fixed cells of various types. The intercellular spaces are filled with body fluids. The fibers, cells, and the spaces between them are referred to collectively as the **parenchyma**.

Four classes make up the phylum. Two of these, Trematoda and Cestoidea, contain flatworms parasitic to humans. One evolutionary scheme for the trematodes proposes that they arose from a stock of free-living flatworms (progenitors of present-day rhabdocoel turbellarians), became intimately associated with mollusks and, ultimately, developed into parasitic forms. Evolutionary divergence within this endoparasitic population gave rise to two groups, designated as subclasses Digenea and Aspidobothrea. The ancestral digeneans proliferated asexually in the mollusk; later adult forms parasitized evolving vertebrates. The ancestral, nonproliferative, aspidogastrean forms, on the other hand, remained within their molluskan hosts through adulthood. All trematodes parasitic to humans belong to the subclass Digenea.

Human Parasitology. http://dx.doi.org/10.1016/B978-0-12-813712-3.00009-6

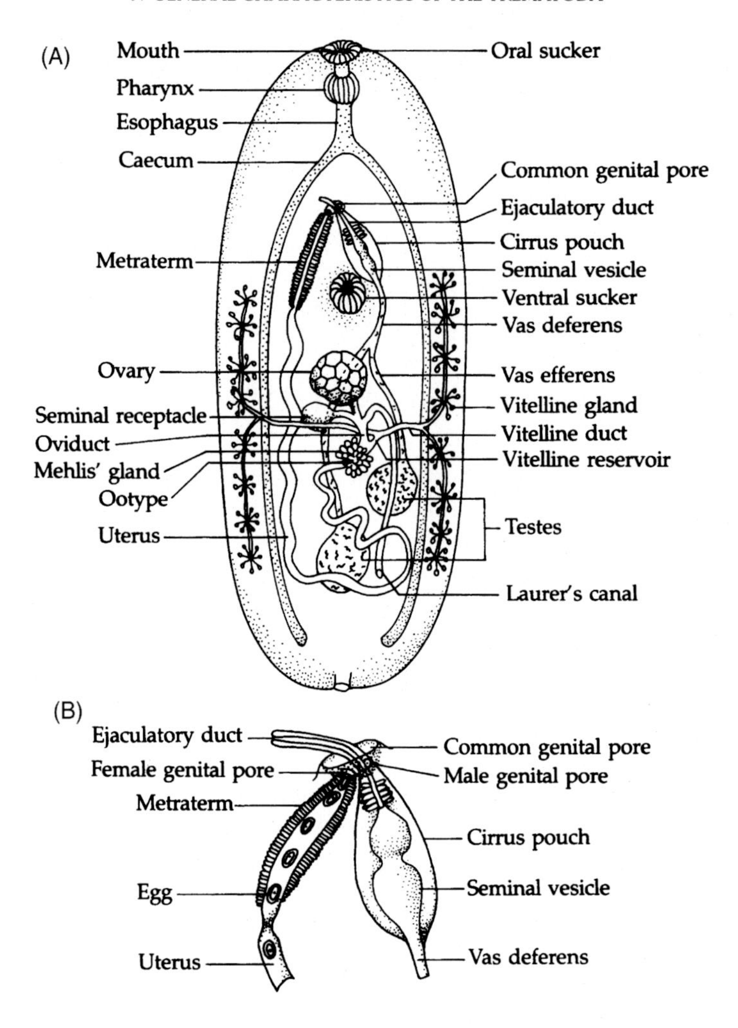

FIGURE 9.1 **Generalized digenetic trematode.** (A) Diagram of entire organism. (B) Detailed diagram of male (*right*) and female (*left*) genitalia.

Digenetic trematodes constitute one of the largest groups of platyhelminths, parasitizing a wide range of invertebrate and vertebrate hosts, including humans. Within human hosts, these worms are found in numerous organs, including the intestine, lungs, liver, and vascular system.

STRUCTURE OF ADULT

Despite superficial differences, the morphology of the various groups of digenetic trematodes is basically uniform. The following description represents a hypothetical composite exemplifying the various anatomical features (Fig. 9.1).

Tegument

Once considered a nonliving, protective "cuticle," the tegument is now recognized as a dynamic, cellular structure. Under light microscopy, it appears as a generally homogeneous layer about 7–16 μm thick (Fig. 9.2). The tegument is a **syncytium** (i.e., a multinucleated tissue with no cell boundaries [Fig. 9.3]). The outer zone of this syncytium, the **distal cytoplasm**, is delineated at its surface by a plasma membrane measuring about 10 nm thick. Associated with the plasma membrane is a surface coat, or **glycocalyx**, that varies in thickness according to species. Surface invaginations, the number and extent of which also vary according to species, serve to increase tegumental surface area, much like microvilli on the surface of human intestinal cells. The ability of the tegument to absorb exogenous molecules is generally proportional to the number and extent of invaginations and the number of mitochondria in the distal cytoplasm. Hydrolytic enzymatic activity in the glycocalyx facilitates the uptake of certain molecules such as sugars and amino acids from the environment. The glycocalyx is also a protective structure, shielding the worm from such hostile environmental influences as antibodies and host digestive enzymes. For example, the presence of acid mucopolysaccharides in the glycocalyx is of particular significance since such molecules are known to inhibit a number of digestive enzymes. Their presence on the body surface may account for the ability of intestinal trematodes to resist host digestive enzymes.

Embedded in the distal cytoplasm of some species are tegumental spines, with bases lying just above the basal plasma membrane of the distal cytoplasm and tips projecting outward but still covered by the surface membrane. Although the function of these spines has not been firmly established, it is speculated that they may serve as ancillary holdfast mechanisms

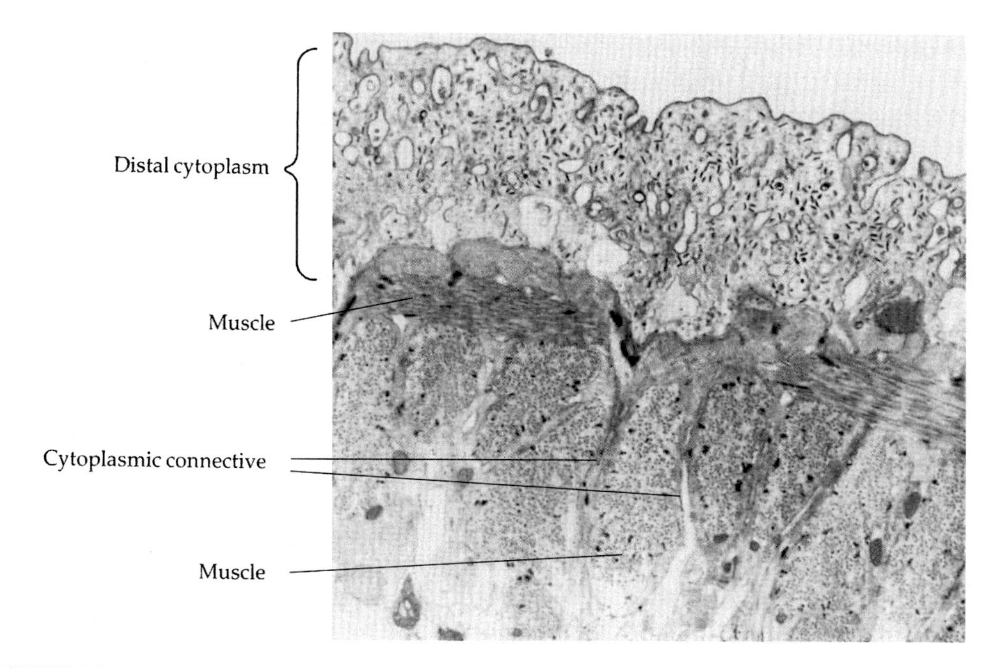

FIGURE 9.2 **Transmission electron micrograph of the tegument of a digenetic trematode.**

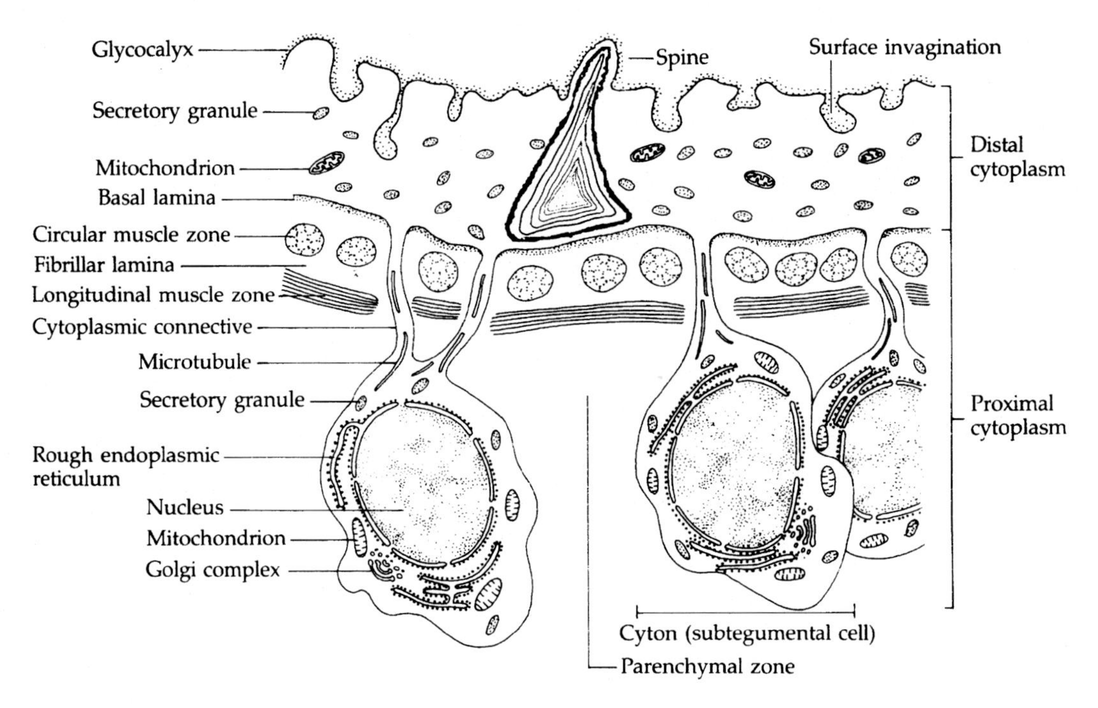

FIGURE 9.3 **Tegument of a digenetic trematode.**

and/or storage sites for certain essential molecules. The matrix of the distal cytoplasm also contains one or two types of secretory vesicles.

The outer, distal cytoplasm is connected to the inner, **proximal cytoplasm (=cyton region)** by cytoplasmic bridges. The proximal cytoplasm contains nuclei, endoplasmic reticular Golgi complexes, glycogen deposits, mitochondria, and various types of vesicles. This region of the tegument is the locus for synthesis of materials for repair and maintenance of components comprising the distal cytoplasm. The vesicles in the distal cytoplasm are packets of substances produced in the proximal cytoplasm that continually maintain the outer plasma membrane and its glycocalyx and assist in the maintenance of the matrix and spines. The translocation of these vesicles from proximal to distal cytoplasm is facilitated by microtubules in the cyton region and in the cytoplasmic bridges.

Digestive Tract

Digenetic trematodes possess incomplete digestive tracts (Fig. 9.1). The anterior mouth, surrounded by a muscular oral sucker, leads into a bulbous, muscular pharynx, in many species via a short prepharynx. The esophagus connects the pharynx and the alimentary tract, the latter bifurcating into two **ceca**. The mouth, pharynx, and esophagus make up the foregut, analogous to that of higher forms. The lining of the foregut is morphologically similar to the general tegument (i.e., it is a syncytium consisting of two cytoplasmic zones, distal and proximal). There are, however, no spines associated with this lining. The foregut is the site of ingestion and assimilation of food, and the ability to accomplish these functions is enhanced by specific modifications of the foregut in these organisms. For instance, in many forms the proximal cytoplasm produces enzymes that are released into the lumen of the foregut where they partially degrade ingested food. In addition, heavily muscularized regions of the foregut, such as the pharynx or, in some species, the esophagus, mechanically break food into smaller particles.

The transition from the tegument-like structure of the foregut to the simple epithelium, or **gastrodermis**, of the two ceca is abrupt and marked by a prominent cell junction (Fig. 9.4). The gastrodermis consists of cells either with or without distinct lateral cell boundaries (i.e., a syncytium). There is no discernible physiological basis for this variation. The ceca are two longitudinal, blind tubes of variable length. In many species, they extend almost to the posterior tip of the body; in others, they may extend no further than a third of the body length. In certain larger digeneans, the ceca exhibit extensive diverticulation.

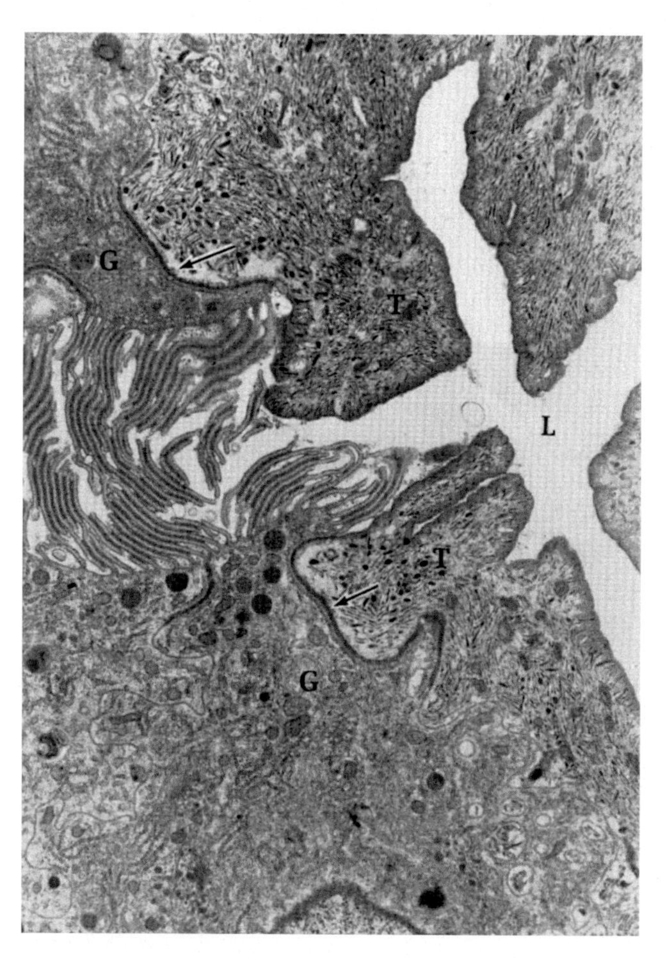

FIGURE 9.4 **Section through junction between tegument (T) and gastrodermis (G).** Note abrupt transition at desmosome region (*arrows*). L denotes the lumen of the foregut.

The gastrodermal surface is bounded by a plasma membrane amplified in either fingerlike microvilli (Fig. 9.5A) or leaflike lamellae (Fig. 9.5B). Digestion and absorption of food occur in the ceca, and such amplifications increase the absorptive surface. The type of amplification is independent of the type of food ingested. For example, some blood-feeding digeneans exhibit a microvillar type, while others display the lamellar type. Associated with the plasma membrane is a prominent glycocalyx that, similar to the tegumental glycocalyx, appears to afford protection as well as aid in the uptake of certain molecules, the latter function abetted by its enzymatic activity. Digestion is typically extracellular, occurring in the gut lumen.

The gastrodermis is highly active in protein synthesis and secretion, for which it possesses abundant rough endoplasmic reticulum, Golgi complexes, numerous mitochondria, and vesicles most of which originate from the Golgi complex. The vesicles contain either material for maintaining the glycocalyx or hydrolytic enzymes that are released into the cecal lumen for the purpose of digestion. The basal plasma membrane of the gastrodermis often contains extensive infoldings that seem to be associated with the organism's ability to transport ions. The basal plasma membrane rests on an extensive **basal lamina** in which are embedded two layers of muscles, one circular, the other longitudinal.

Muscular and Nervous Systems

There are two muscle zones in adult digenetic trematodes. Underlying the tegument is the **subtegumental zone**, which consists of three layers of muscle: longitudinal, circular, and diagonal. Contractile activity by these muscles is usually minimal although certain species exhibit

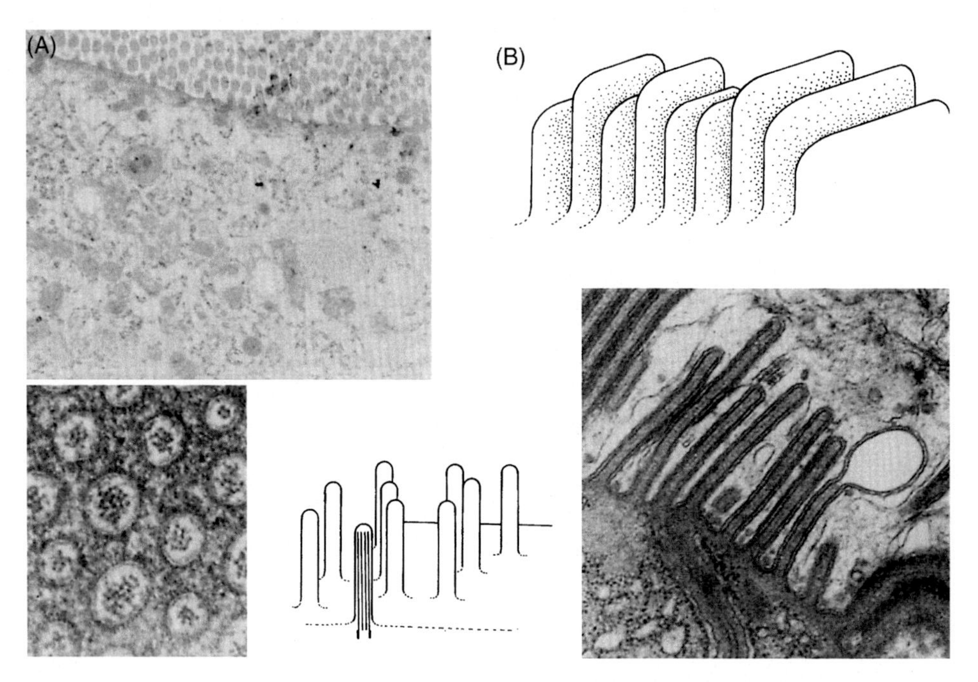

FIGURE 9.5 (A) Microvillar amplification of gastrodermis. Bottom left is a cross section through microvilli. (B) Lamellar amplifications of gastrodermis.

more active subtegumental musculature. The muscle layers are typically more distinct in the anterior part of the body. The orientation of contractile fibers allows the organism to elongate, contract, and/or twist its body in almost any given plane. This zone of smooth muscles contains a nucleated region, or **myoblast**, connected to the bundles of myofibers. The second zone of musculature, the **gastrodermal zone** described in the previous section, helps move food up and down the cecal lumen. Well-developed contractile fibers are also present in the oral and ventral suckers.

The nervous system of adult digeneans (Fig. 9.6) is of the "ladder" type. Three pairs of longitudinal nerve trunks, a prominent ventral pair, a lateral pair, and a dorsal pair, extend posteriorly and anteriorly from two connected dorsal ganglia (=brain) near the pharynx; all three pairs of trunks are interconnected by transverse commissures. Smaller branches, emanating from the brain and longitudinal trunks, supply motor and sensory innervation to the tegument, suckers, reproductive systems, and other organs.

Sensory organs are evident in some tissues of the digenean body, particularly the tegument and gastrodermis. These usually appear as modified cilia projecting from bulbous nerve endings that extend outward from the tegumental surface or from the gastrodermis into the lumen of the digestive tract. They may serve as pressure, rheotactic sensors, or chemoreceptors. In larval stages, the types of sensory organs vary widely, including papillae, pigmented eyespots, uniciliated organs, and organs containing up to six ciliary eyespots. Such diversity undoubtedly aids these free-swimming larvae in locating hosts. Nerve end organs act as chemo-, mechano-, and photoreceptors for the larvae.

Osmoregulatory System

The osmoregulatory system of digenetic trematodes is of the typical protonephridial type, a tubular system closed at one end and open at the other. Currents are produced at the closed ends by **flame cells**, each of which is equipped with a tuft of fused, vigorously beating cilia

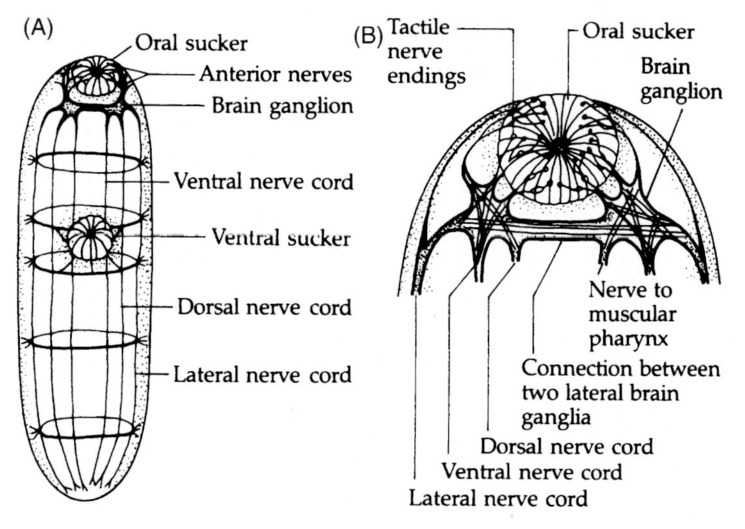

FIGURE 9.6 (A) Nervous system of a digenetic trematode. (B) Innervation of the anterior end and oral sucker of a digenetic trematode.

(Fig. 9.7). The number and arrangement of such cells in digeneans are specific enough to serve as taxonomic indicators of phylogenetic relationships (Fig. 9.8). Each cell opens into a terminal tubule, several of which converge to form larger collecting tubules. The collecting tubules on each side of the body lead posteriorly and empty into a common **excretory bladder** (Fig. 9.9), the duct of which opens to the exterior through the excretory pore located at or near the posterior end of the body.

While it seems likely that this system serves in both osmoregulation and/or excretion, not all details of these functions are clearly defined. In some forms, the structural and biochemical features suggest reabsorptive functions. For instance, microvilli and alkaline phosphatase activity, both characteristic of absorptive epithelium, occur in the walls of the collecting ducts in such forms.

The major nitrogenous waste product in digeneans is ammonia, although other soluble compounds, such as urea and other nitrogenous compounds, also occur. It is not entirely clear how much of each of these compounds is eliminated through the tegument, digestive tract, or excretory system. In a number of species, uric acid forms in the excretory bladder and tubules and is eliminated as insoluble crystals via the excretory pore.

Reproductive Systems

With the exception of schistosomes, digenetic trematodes are hermaphroditic. The male reproductive system may mature prior to the female system, reducing the likelihood of self-fertilization.

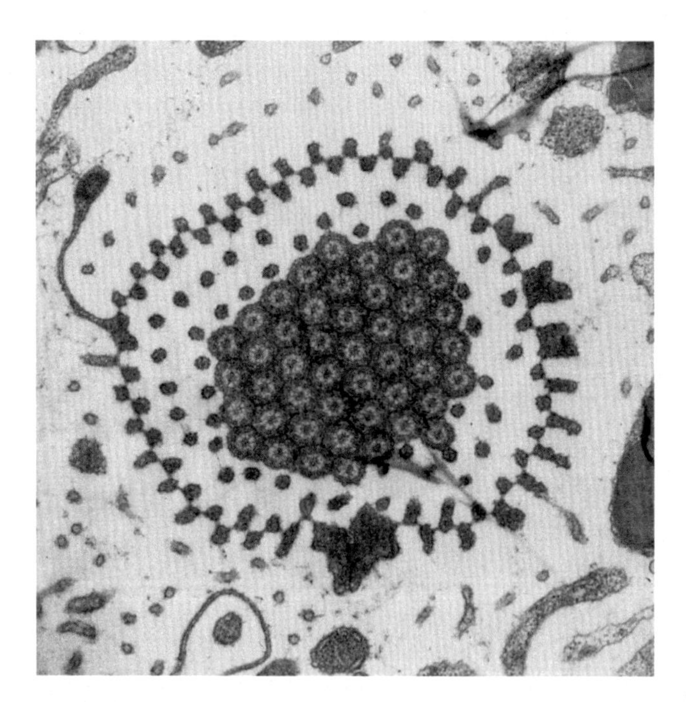

FIGURE 9.7 **Transmission electron micrograph of the ciliary tuft of a trematode flame cell.**

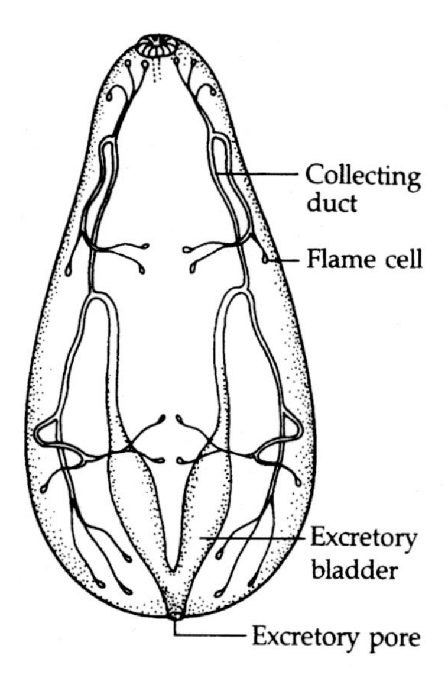

FIGURE 9.8 **Excretory system of heterophyes.**

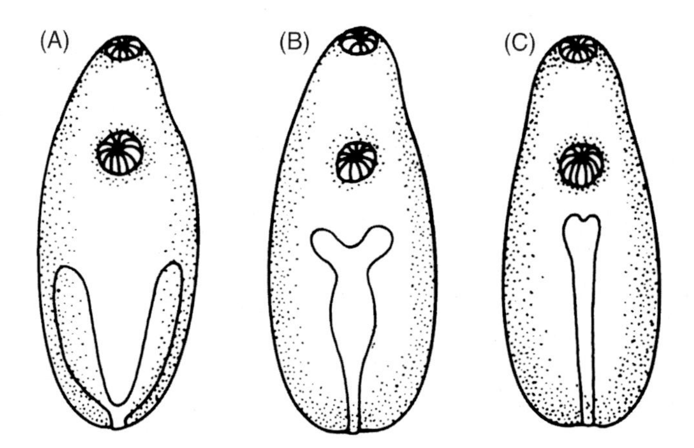

FIGURE 9.9 **Shapes of excretory bladders of digenetic trematodes.** (A) V shaped. (B) Y shaped. (C) I shaped.

Male System

The male reproductive system (Fig. 9.10) generally includes two testes, although schistosomes are multitesticular. The position of the testes in the parenchyma varies according to species, as do shape and orientation to each other. For instance, testes can be located anywhere from the middle to the posterior portion of the body. They can be ovoid, round, smooth, branched, or lobed. They can be tandem, side by side, or diagonal to each other. Such characteristics are useful in the identification of species. Spermatogenesis in the testes produces biflagellated sperm.

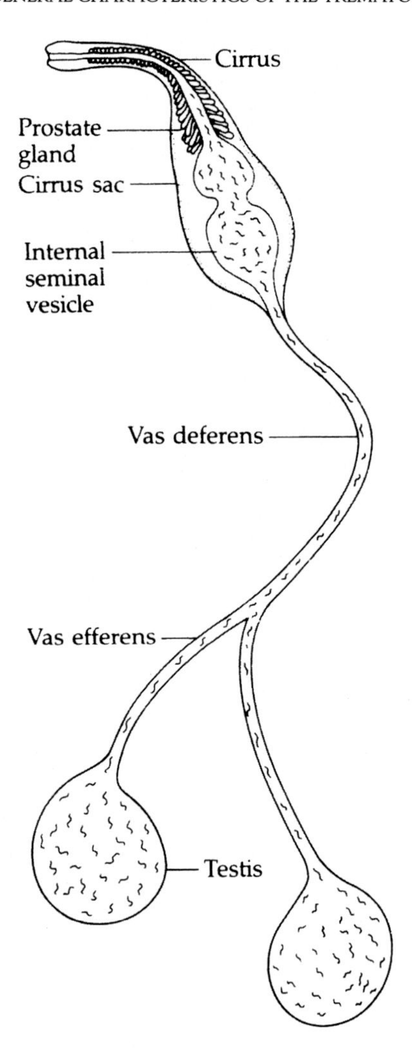

FIGURE 9.10 **Male reproductive system of a digenetic trematode.**

Leading from each testis is a **vas efferens**, each of which unites with others anteriorly to form the common **vas deferens**. Distally, this duct forms the male copulatory organ, or **cirrus**. The cirrus may be surrounded by a **cirrus sac** into which it can invaginate when not everted. Also enclosed by the cirrus sac are the sperm-storing **seminal vesicle** and the **prostate gland**. The eversible cirrus can be protruded to the exterior through a genital pore on the ventral surface of the organism.

Among digeneans, there are certain variations in the components of the vas deferens and their relative positions. One or more of these components may be missing (e.g., the prostate gland and/or the cirrus sac and/or a protrusable cirrus), and the seminal vesicle often varies in size and position. The seminal vesicle, usually enclosed within the cirrus sac, is sometimes located outside the sac, in which case it is designated as an **external seminal vesicle**.

Sperm

A mature sperm of digenetic trematodes and cestodes presents the same morphology with but few exceptions. Basically, it is an elongated cell with two flagella and no discernible external regionalization. An anterior, elongated nucleus and a single mitochondrion are observed in what might be considered the head region and extend posteriorly through a region that may be defined as the middle piece. Two flagella arise from basal bodies located at the junction of the head region and middle piece (Fig. 9.11). The geometry of the microtubules comprising the axonemes is unique in that they display a 9 + 1 arrangement as opposed to the typical 9 + 2 arrangement observed throughout the animal kingdom. The central structure resembles a bull's eye. Each basal body is anchored by a prominent striated rootlet extending anteriad. Underlying the plasma membrane are cortically arranged microtubules. Proximally, the plasma membrane divides forming two short processes each containing the terminal portions of the flagella.

Female System

The female reproductive system consists of a single ovary embedded in the parenchyma, either anterior to, posterior to, or between the testes, depending upon the species. **Ova** (actually secondary oocytes), formed in the ovary, are released via a short **oviduct** and undergo a sequence of events analogous to an assembly line (Fig. 9.12). After leaving the ovary, each ovum passes down the oviduct to a minute chamber, the **ootype**. In this vicinity, the duct of the seminal receptacle, in which sperm deposited earlier are stored, joins the oviduct. Fertilization occurs at this point, whereupon oogenesis is completed and cleavage begins. The ootype is surrounded by **Mehlis' gland** consisting of two groups of unicellular glands, each group characterized by the type of material it secretes into the ootype. One group secretes

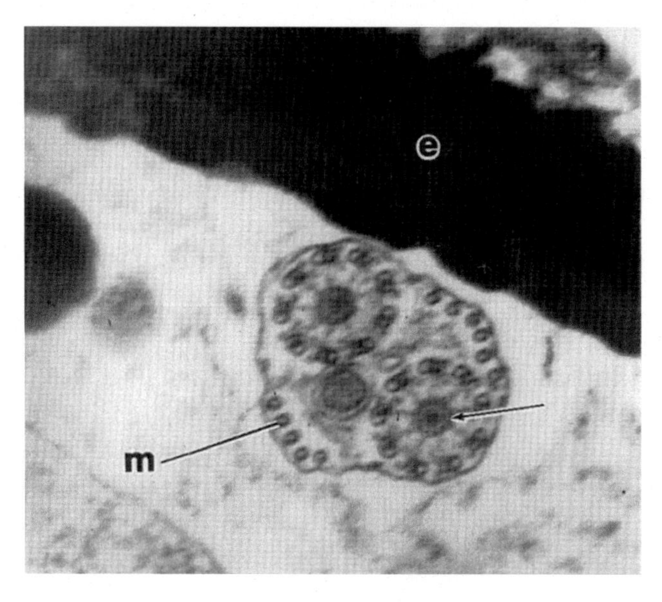

FIGURE 9.11 **Cross section through a sperm of a digenetic trematode.** Note the pattern of nine outer doublers and a single, central microtubule (*arrow*) in each axoneme of the biflagellated sperm. *m*, cortical microtubule; *e*, eggshell.

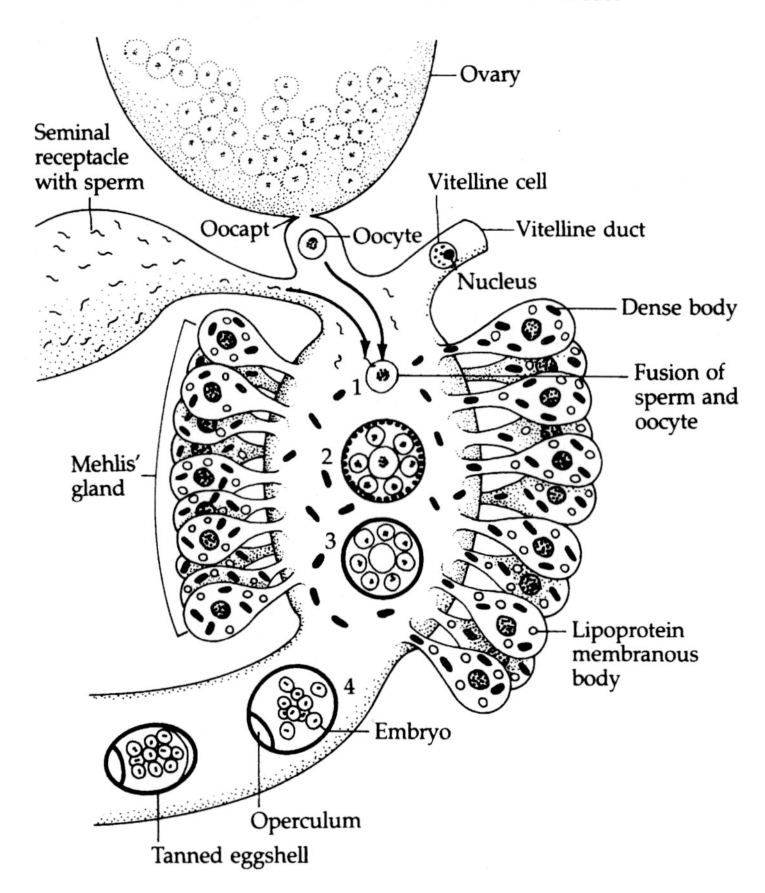

FIGURE 9.12 **Sites of formation of constituents of a digenetic trematode egg (*see text for description*).**

a membranous body, while the other secretes a dense body. Among several functions suggested for these secretions, the most likely is that the membranous body provides a template for the deposition of shell material (Fig. 9.13) and the dense body provides lubrication for the passage of the forming, shelled egg. Other functions postulated for these secretions include activation of sperm, activation of vitelline glands to release shell material, and enhancement of the hardening process of the eggshell. Other glands that communicate with the ootype are the **vitelline glands** (Fig. 9.14). In digeneans, these glands are composed of numerous multicellular clusters. Each cell synthesizes globules, which are then stored in its cytoplasm. As each cell attains a certain level of maturity, it detaches and enters a vitelline ductule. Groups of glands are usually situated bilaterally, although distribution may vary according to species. The smaller vitelline ductules converge to form right and left **vitelline ducts** which, in turn, merge to form the **common vitelline duct** opening into the ootype. In some digeneans, the right and left vitelline ducts merge into a common chamber, the **vitelline reservoir**, which is connected to the ootype by a short duct.

The vitelline gland cells are essential to eggshell formation. They release into the ootype globules, which become aligned against the membranous template derived from Mehlis'

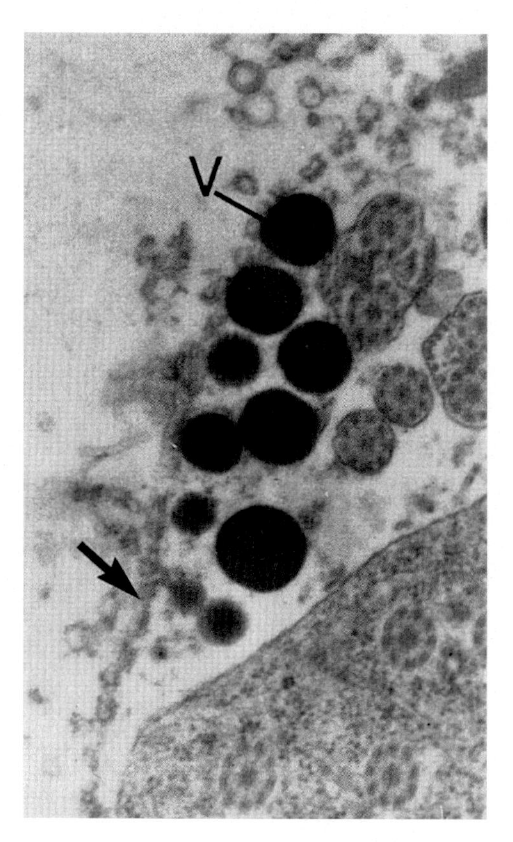

FIGURE 9.13 **Transmission electron micrograph of a forming egg shell of a digenetic trematode.** Vitelline globules (V) line the inner aspect of the lipoprotein membrane (*arrow*) prior to fusion.

gland. These globules coalesce and eventually toughen to form the shell (Fig. 9.13). This toughening process is accomplished by enzymatic cross-linking of proteins in the coalesced globules. In some species, tyrosine residues of the proteins are oxidized into a "tanned" egg-shell. This type of shell is rather hard and brownish in color. In other forms, disulfide links are formed, producing an eggshell more elastic than the tanned form. Cytoplasm of the vitelline cells also provides nourishment for the developing embryo. The uterine lining in some digeneans may also supply essential elements for eggshell formation. The uterus is a long, often convoluted tube through which shelled eggs are transported to the exterior via the genital pore. In some species, a muscular, distal portion of the uterus, a **metraterm**, helps propel the eggs out of the uterus as well as aiding in copulation. In addition to transporting eggs to the exterior, the uterus also permits sperm to move in the opposite direction to the seminal receptacle. During copulation, the cirrus is inserted into the distal end of the uterus; sperm, ejaculated into the metraterm of the uterus, then swim to the seminal receptacle where they are stored. In some species, a **Laurer's canal**, originating on the surface of the ootype, passes to the dorsal surface where it may or may not open to the exterior. This canal may represent a vestigial vagina or it may serve as an outlet for excess sperm and extraneous matter formed during egg formation.

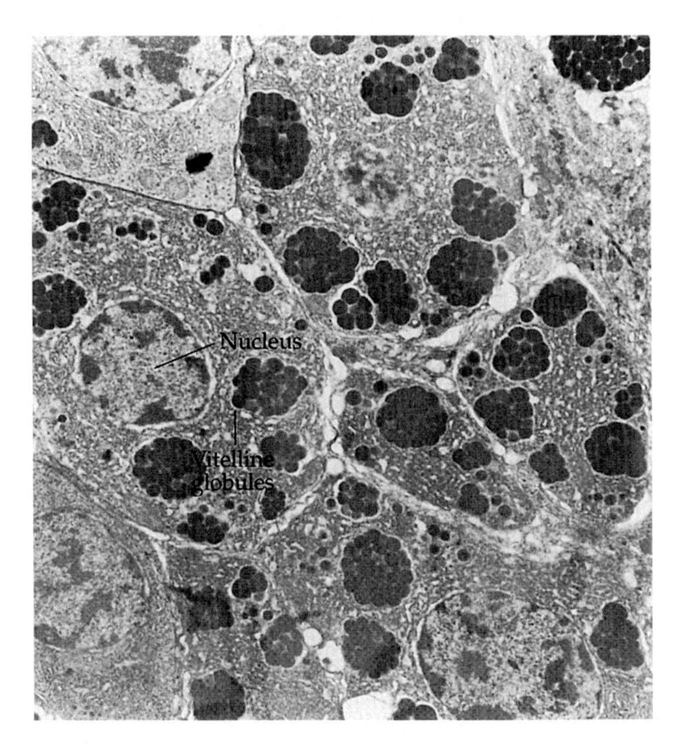

FIGURE 9.14 **Transmission electron micrograph through the vitelline gland of a digenetic trematode.**

The Egg

The ovoid, shelled egg contains vitelline substance, the embryo, ancillary membranes, and other materials (Fig. 9.15). The typical eggshell is equipped at one end with a lid-like structure, the **operculum**, which allows the larva to hatch. The eggshells of human blood flukes possessing no operculum rupture longitudinally when they hatch. Hatching occurs only under precise conditions of temperature, osmolarity, and light. For instance, in schistosomes infecting humans, hatching of embryonated eggs is inhibited by an osmolarity equivalent to 0.85% saline and a temperature of 37°C, thereby reducing the likelihood that the egg will hatch prematurely within the host's body. Most digenean eggs are unembryonated when they pass out of the human host and are unable to hatch without further development. Specific size and structural characteristics of trematode eggs, especially those of medically important species, are useful in diagnosis (Fig. 9.16).

GENERALIZED LIFE CYCLE PATTERNS

The following brief overview of the life cycles of digeneans that infect humans will illustrate the various larval stages and the relationship of each to the succeeding one (Fig. 9.17). Eggs are usually released into the lumen of the host's organ housing the adult worm (gut,

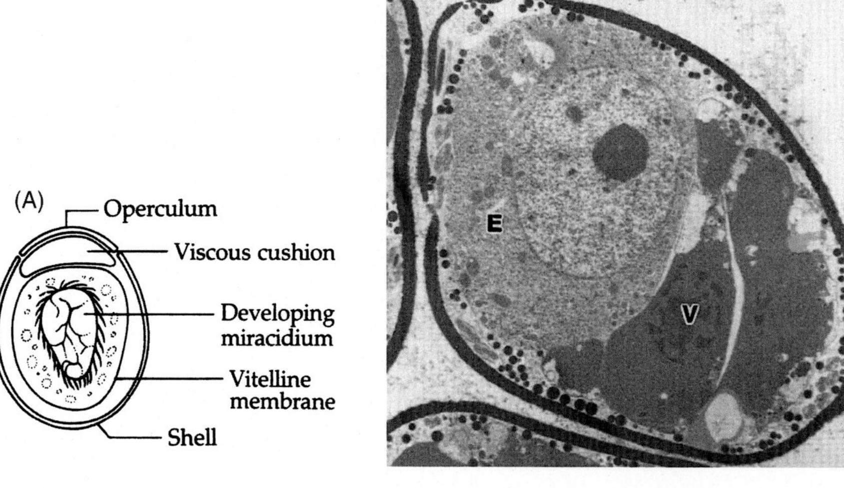

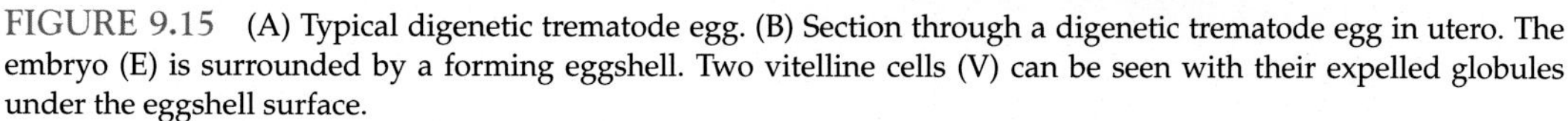
FIGURE 9.15 (A) Typical digenetic trematode egg. (B) Section through a digenetic trematode egg in utero. The embryo (E) is surrounded by a forming eggshell. Two vitelline cells (V) can be seen with their expelled globules under the eggshell surface.

lungs, urinary bladder, etc.) and pass to the exterior through feces, sputum, or urine. After finding its way to water, the egg completes its development, and a free-swimming **miracidium** (Fig. 9.18) hatches. Within 24 h, the miracidium must find and penetrate the integument of a suitable freshwater snail host (the first intermediate host), shedding its ciliated epidermis in the process and metamorphosing into a **primary sporocyst** (Fig. 9.19). This primary sporocyst may produce numerous secondary sporocysts or **primary rediae** asexually (Fig. 9.20). In some forms, the egg must be ingested by the snail (particularly if the snail is terrestrial) before the miracidium hatches.

While the digestive gland of the snail is a common site for further development, the gonad, the mantle, the lymph spaces surrounding the intestine as well as other organs, may also serve for such development. Once established in a suitable location, the sporocyst or redia grows, matures, and continues to proliferate.

Sporocysts are commonly elongate and hollow and contain germ cells, formed in the miracidium, that multiply by mitosis and develop into **germ balls**. A redia differs from a sporocyst in that a redia possesses a functional, sac-like gut and a pharynx.

The redia or secondary sporocyst (depending on species) eventually gives rise to a tailed larva called a **cercaria** (Fig. 9.21). Cercariae that escape from the molluskan host experience only a brief (several hours) free-swimming existence since they do not feed outside the host. In some species, the cercariae may actively penetrate or attach to the surface of a second intermediate host or attach to vegetation; they lose their tails, encyst, and are then known as **metacercariae** (Fig. 9.22). Upon ingestion by a vertebrate definitive host, the encysted metacercaria excysts in the vertebrate small intestine, migrates to the definitive site, and gradually matures into the adult stage. In schistosomes, cercariae penetrate the definitive host directly, thereby foregoing an encysted metacercarial stage.

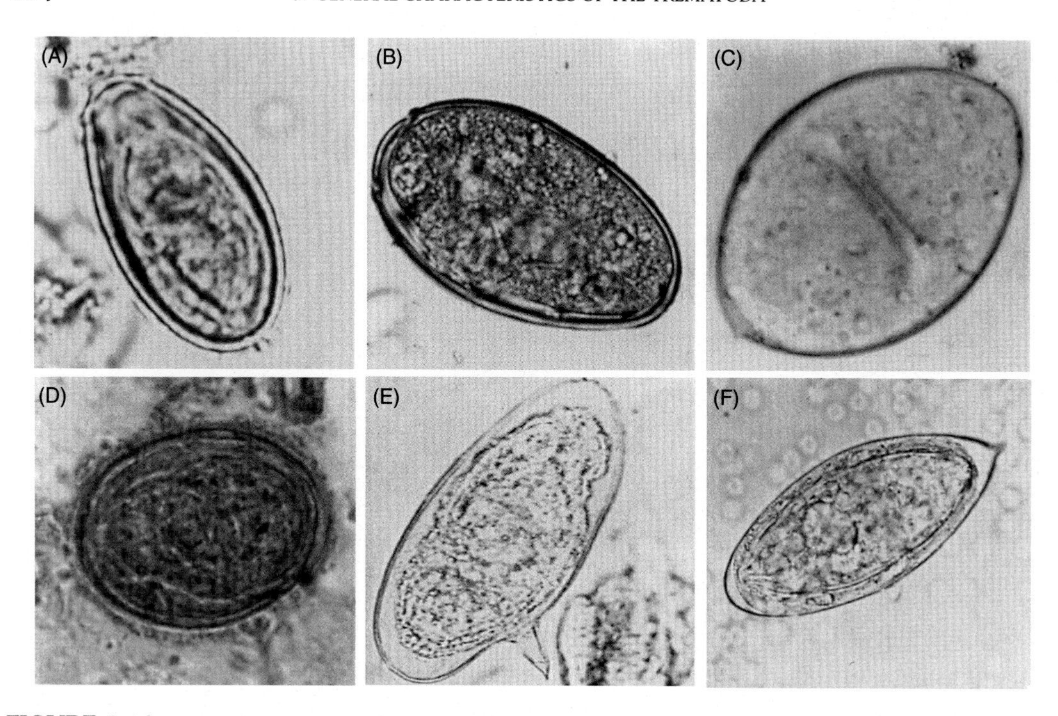

FIGURE 9.16 **Eggs of some digenetic trematodes parasitic in humans.** (A) *Clonorchis sinensis*, 27–35 mm by 12–20 mm, nonshaped, operculum at narrow end. (B) *Paragonimus westermani*, 80–118 mm by 48–60 mm, oval, operculum at flattened end. (C) *Fasciolopsis buski*, 130–140 mm by 80–85 mm, ellipsoidal, inconspicuous operculum. (D) *Schistosoma japonicum*, 7–100 mm by 50–65 mm, round to oval, inconspicuous lateral spine, no operculum. (E) *Schistoma mansoni*, 114–175 mm by 45–68 mm elongate oval, longer lateral spine, no operculum. (F) *Schistoma haematobium*, 112–170 mm by 40–70 mm, spindle shaped, posterior terminal spine, no operculum.

The Miracidium

The miracidium is a ciliated, nonfeeding larva (Fig. 9.18). Under favorable conditions, it escapes from the eggshell, usually through the operculum, into the environment. The miracidium is elongated and covered with flattened, ciliated epidermal plates. At the junctures of adjacent epidermal plates are **cytoplasmic ridges**, which play a prominent role during the miracidium's metamorphosis into the next larval stage (Fig. 9.24). Beneath the epidermal plates are well-developed circular and longitudinal muscles.

At the anterior tip of the miracidium is a flexible **apical papilla** with sensory organs and three secretary glands, the apical gland and two lateral glands, which secrete materials at the tip of the papilla during host penetration (Fig. 9.23). During penetration, the papilla becomes partially invaginated and secretions from the glands are captured in the depression. The papilla thus acts as a suction cup, holding the miracidium to the site of penetration and allowing the secretions to exert both adhesive and lytic actions. In addition to the sensory structures in the papilla, there may be two to three anterior **eyespots**, as well as **lateral papillae**, on each side of the body. The "brain" of the miracidium lies in the parenchyma behind the apical region from

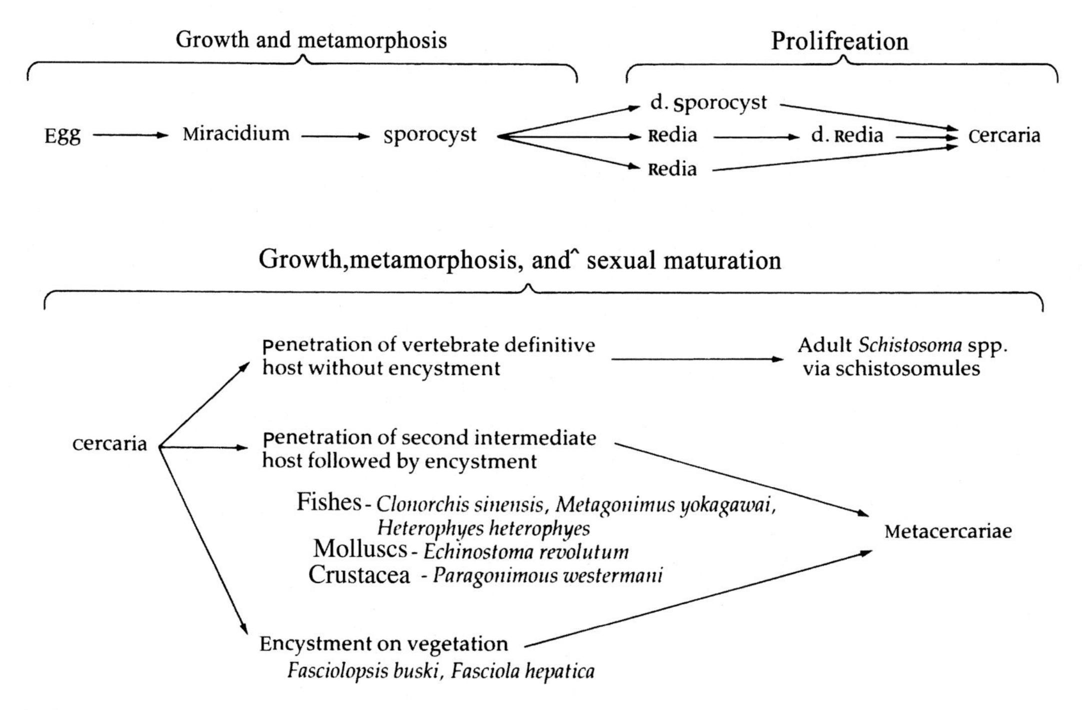

FIGURE 9.17 **Flowchart showing life cycles of trematodes that infect humans.**

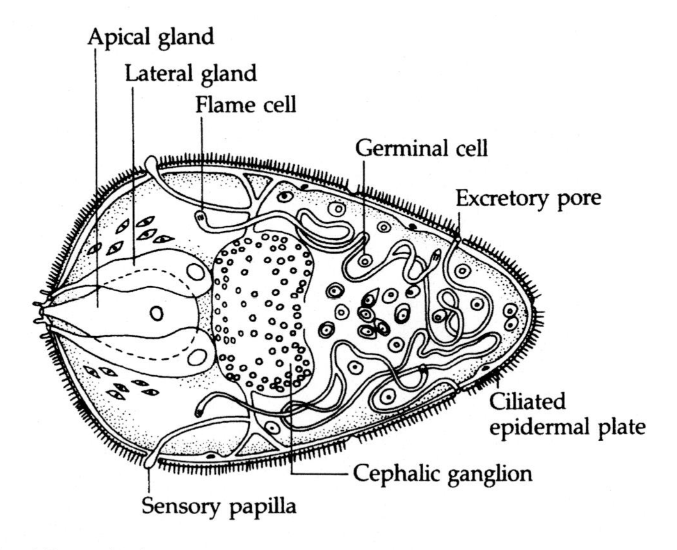

FIGURE 9.18 **Miracidium of *Schistosoma*.**

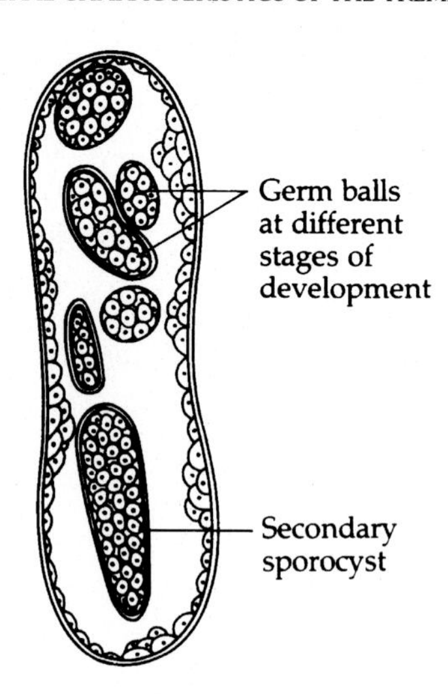

FIGURE 9.19 **Primary sporocyst.**

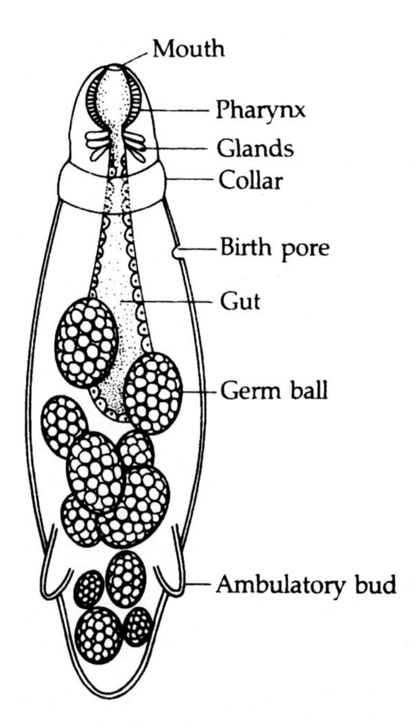

FIGURE 9.20 **Redia of *Fasciola hepatica*.**

which nerve fibers innervate various tissues and organs of the body. The miracidium also has a simple, protonephridial excretory system. Waste-containing body fluids are collected by two or three pairs of flame cells and excreted through two lateral excretory pores. During differentiation of the miracidium, germ cells grow and divide to form germ balls. Each germ ball eventually develops into a distinct, membrane-enclosed entity that is the next larval generation.

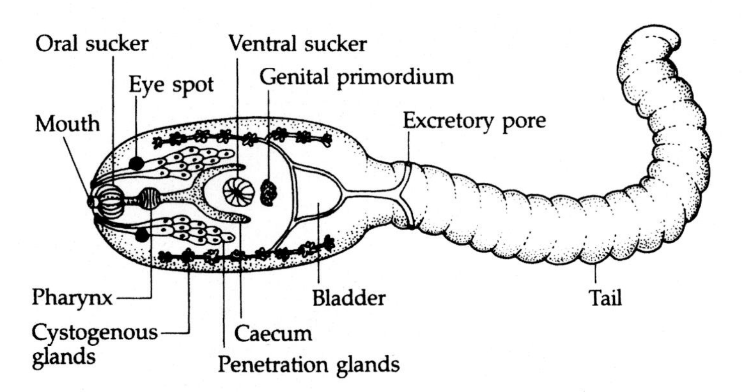

FIGURE 9.21 **Cercaria of *Clonorchis sinensis*.**

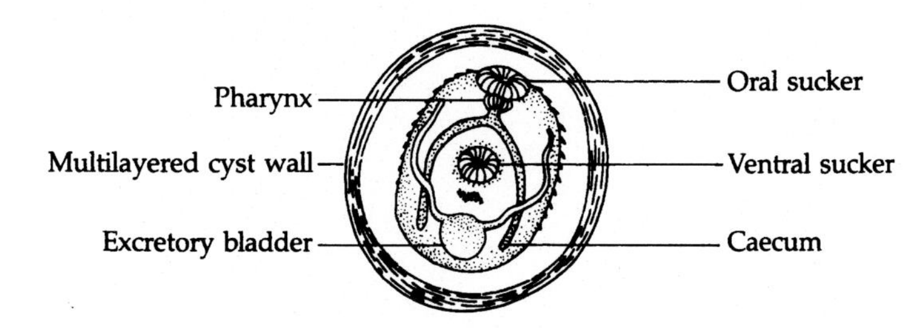

FIGURE 9.22 **Generalized drawing of an encysted metacercaria.**

The Sporocyst

First-generation sporocysts, having differentiated from miracidia, usually accumulate near the site of penetration (mantle and headfoot of the molluskan host), but occasionally they may reach the hemocoel in the digestive gland. In certain species, they congregate along the digestive tract. The sporocyst varies in shape from ovoid to elongate to tubular and on occasion may even be extensively branched (Fig. 9.19).

The thickness of the sporocyst wall, as seen in cross section, varies according to age and species. Its outermost layer, as in all subsequent stages in the life cycle of digeneans, is a syncytial tegument derived from the cytoplasmic ridges of the miracidium. Prior to completing penetration of the first intermediate host, or shortly thereafter, the miracidial epidermal plates are shed. As the miracidium transforms into a sporocyst, the cytoplasmic ridges spread to form the tegument (Fig. 9.24). In most sporocysts, the microvilli produce extensive amplification of the tegumental surface.

Beneath the tegument lies a thin basal lamina in which is embedded a layer of circular muscles. A thin layer of parenchyma underlies this area. These layers form the lining of the fluid-filled **brood chamber**, a cavity containing the germ balls. Sporocysts possess no digestive tracts, and essential nutrients must diffuse across the absorptive tegument. Carbohydrates derived from body fluids of the infected mollusk are the chief source of energy for this stage. Definitive nervous and reproductive systems also are lacking, although flame cells are generally present.

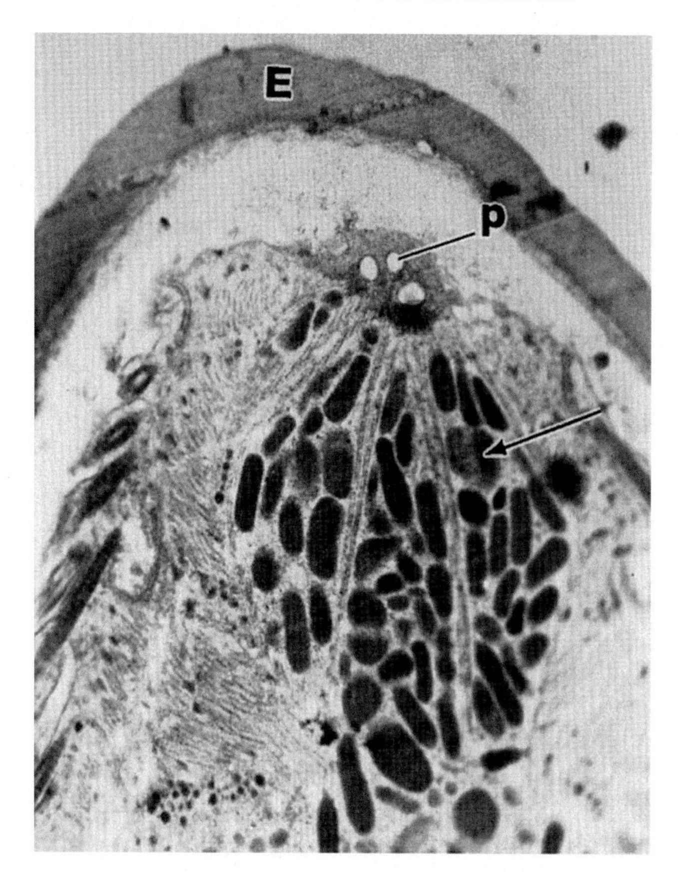

FIGURE 9.23 **Section through a miracidium enclosed by an eggshell (E).** Note the gland openings in the apical papilla (p). The *arrow* points to gland contents.

In some species, germ balls in the brood chamber of mother (or primary) sporocysts differentiate to form daughter (or secondary) sporocysts. In other species, the germ balls differentiate into rediae or directly into cercariae. Daughter sporocysts usually are morphologically similar to primary sporocysts but can be distinguished by their larger size and their common occurrence in deeper body organs of the molluskan host, such as the digestive gland and gonads.

The Redia

Rediae, if present, develop from germ balls in the brood chamber of the primary sporocyst. They eventually escape from the sporocyst through the molluskan tissue and migrate to the digestive gland. Each redia is elongate and normally possesses two or four bud-like, antero- and posterolateral projections, the **ambulatory buds** (or **procruscula**) (Fig. 9.20). As their name implies, the ambulatory buds facilitate movement of the larva through the tissues of the molluskan host. This movement is abetted by contractions of the redial body. Unlike the sporocyst, the redia possesses a digestive tract with an anterior mouth, a muscular pharynx,

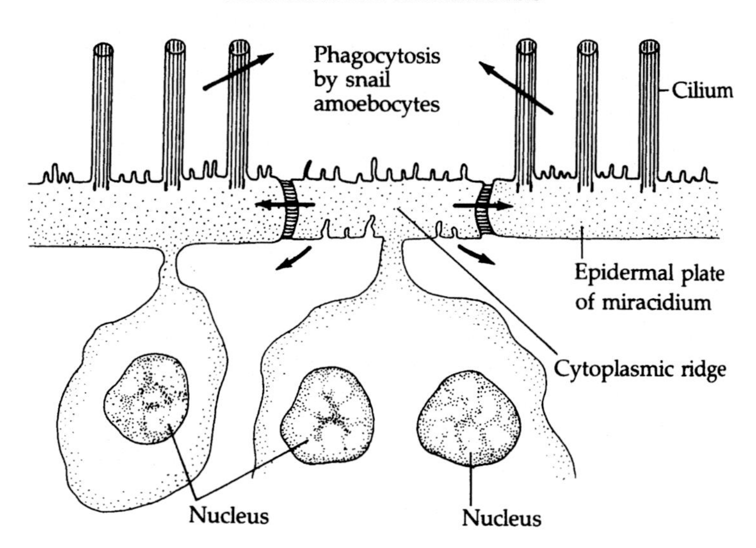

FIGURE 9.24 **Changes in the body of a miracidium after its penetration of a snail.** Epidermal plates are shed and phagocytosed by snail amebocytes. Cytoplasmic ridges expand to form the outer layer (tegument) of the sporocyst.

and an unbranched cecum. As the redia moves through the host's tissues, it actively ingests host cells. Digestion of the cells takes place in the lumen of the cecum. Some rediae augment these intestinal feedings by secreting hydrolytic enzymes to the exterior and lysing surrounding host cells. The resulting molecules are then absorbed through the tegument, which is morphologically similar to that of the sporocyst. The physical effect of rediae upon the molluskan host is considerably more deleterious than that produced by sporocysts.

On each side of the pharynx is a cephalic ganglion from which nerve fibers radiate. Flame cells occur in most rediae, terminating at the bladder(s) in either single or multiple excretory pores. Usually, near the mouth of the redia there is a birth pore leading from the brood chamber. Within the brood chamber of the redia, germ balls differentiate into either secondary (daughter) rediae or the next larval stage, the cercaria.

The Cercaria

Cercariae differentiate from the germ balls in the brood chambers of secondary, tertiary, or subsequent generations of sporocysts or rediae depending upon the life cycle of the particular digenean. A cercaria is usually equipped with a tail, enabling it to swim; a few species, such as the human lung fluke, have minute tails, forcing the cercaria to crawl on a substrate rather than swim. After escaping from the brood chamber through the birth pore, cercariae leave the molluskan host and actively seek the next host or suitable vegetation.

The distribution of internal organs in all cercariae, regardless of species, usually resembles that of the adult worm (Fig. 9.21). The mouth, situated at the anterior end of the body and surrounded by the oral sucker, leads into the foregut and paired ceca. There is a variably positioned ventral sucker whose location remains constant through adulthood. In many cercariae, several types of glands open anteriorly. The name of a gland indicates its assumed function. For example, in schistosome cercariae, a pair of **escape glands** lies near the mouth (Fig. 9.25). The contents of these glands are secreted during the cercaria's emergence from the sporocyst in which it had developed as well as during its exit from the snail host. In liver fluke cercariae, **cystogenous glands** secrete substances to form a cyst wall. Other glands, such as **penetration glands** and **mucoid glands**, may play a role in host penetration and may be further augmented by cuticular stylets capable of puncturing chitin-covered arthropods.

Embedded in the parenchyma, near the ventral sucker, is a genital primordium, a mass of germinal cells that eventually forms the male and female reproductive systems of the adult. Cercariae also have protonephridial (flame cell) excretory systems. But, unlike sporocysts and rediae, cercariae have two lateral collecting tubules that empty into a common, posterior, excretory bladder from which a tube may extend into the tail. The arrangement of flame cells in cercariae is similar to the arrangement in the adult. Thus, if the flame cell pattern in a cercaria is known, it can be used to correlate the larval form with its adult stage.

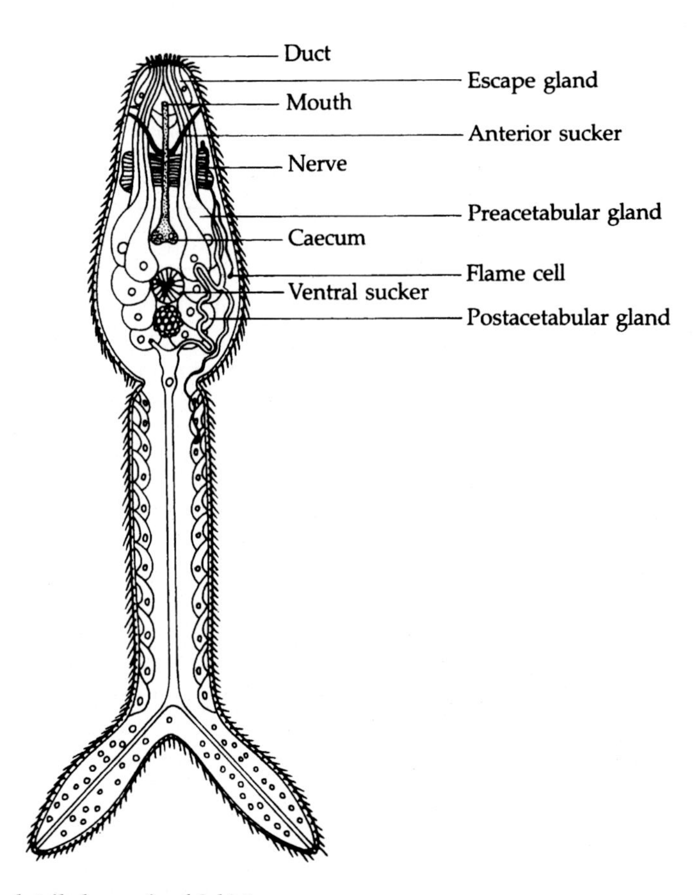

FIGURE 9.25 **Fork-tailed cercaria of *Schistosoma* spp.**

The Metacercaria

The next larval stage of most digeneans is the metacercaria (Fig. 9.22). When the free-swimming (or crawling) cercaria locates a suitable substrate or penetrates a second intermediate host, it sheds its tail and encysts. Within the cyst wall, the metacercaria may grow and develop into a more mature juvenile preadult, or its genital primordium may differentiate into a complete reproductive system that usually is nonfunctional. Many cercarial features (i.e., certain glands and sensory structures) shortly disappear.

Very little is known about the metabolism of metacercariae. It is generally assumed that they subsist primarily, if not exclusively, on stored nutrients. However, while many metacercariae appear metabolically quiescent, some that are encysted in animal tissues grow and may exert a dynamic affect upon their host's metabolism. They commonly induce the host's cellular defenses resulting in the encapsulation of the metacercariae in fibrous connective tissue. All metacercariae, however, undergo one common developmental change: they become infective to their definitive hosts.

GERM CELL CYCLE

Reproduction in the miracidium–sporocyst–redia–cercaria progression is asexual, with the progeny arising through differentiation of germinal cells passed from one generation to the next. This phase is known as the **germ cell cycle**. Only in the adult does sexual reproduction occur. Currently, the most widely accepted concept of intramolluskan reproduction among digeneans is that of **sequential polyembryony**, the production of multiple embryos from the same zygote with no intervening gamete production. There is some evidence of parthenogenesis in sporocysts of several digeneans as well as polyembryonic proliferation.

Of particular importance to those interested in gene expression is the observation that all stages in the digenean life cycle carry identical sets of genes. Yet, the expression of these genes differs among the several stages. It appears that certain genes are activated while others are suppressed at various times during the sequence of larval development. For example, genes responsible for the development of an intramolluskan larva into a redia are suppressed in circumstances in which development into a sporocyst is indicated, and vice versa. What triggers this activation–suppression cycle has yet to be established.

Although the life cycle in a given species almost always follows the same pattern, environmental factors, such as temperature changes, and experimental manipulations, such as transplanting from one mollusk to another, can alter the sequential pattern in some digeneans. These alterations consist primarily of variations in the number and type of intramolluskan generations of larvae.

PHYSIOLOGY

During the past two decades, considerable information has been compiled concerning the biochemistry and physiology of digenetic trematodes. Much of this information has been derived from studies of the adult sheep liver fluke *Fasciola hepatica.* A number of factors are responsible for this emphasis: the adult worms are large and easy to work with, the life cycle can be easily maintained in the laboratory, and the economic and medical importance of this species attracts interest and funding for experimentation. The physiology of human blood flukes, or schistosomes, also has been studied intensively for similar reasons. Unfortunately, physiological and biochemical information on other genera and on larval forms is meager. Much of the following overview of the physiology of digeneans is derived from studies of the genera *Fasciola* and *Schistosoma*.

Substrate-level phosphorylation, via glycolysis, is the main source of energy for these forms, with glycogen and glucose as the principal carbohydrates metabolized. Even in the presence of oxygen, as in the blood vessel environment of schistosomes, glycolysis provides the primary energy supply. Schistosomes cannot de novo synthesize fatty acids, sterols, purines, nine essential amino acids, arginine or tyrosine. They exploit fatty acids and cholesterol from host blood and plasma via many transporters. Genomic studies have shown that the schistosome genome can encode transporters such as fatty acid binding proteins, scavenger receptors, and others. In *Fasciola*, oxygen is used when available, but its contribution toward satisfying the organism's overall energy requirements is difficult to assess. It is debatable whether a functional TCA cycle exists in *Fasciola* or in the schistosomes. If such a cycle does function, its overall role in energy production is probably minimal. Some enzymes usually involved in the TCA cycle actually may, in some instances, serve in other metabolic pathways.

The dependence of digeneans on glycolysis for energy has been useful in devising effective drugs for treatment of patients infected with these parasites. For example, the efficacy of trivalent antimony compounds in treating schistosomiasis derives from their ability to inhibit phosphofructokinase, an important enzyme in the glycolytic pathway. A requisite for any effective drug is that the equivalent host enzyme must not be affected by the same level of drug concentration that affects the parasite.

The miracidia and cercariae of all species studied to date are obligate aerobes, relying on oxidative phosphorylation for energy. Intramolluskan stages, on the other hand, resemble more the adult forms in their dependency upon substrate-level phosphorylation to supply energy-rich compounds.

CHEMOTHERAPY

The most effective chemotherapeutic regimen against most trematodes infecting humans is the oral administration of the drug praziquantel. While the mode of action of the drug is still being investigated, it is known that its primary target is the tegument producing vacuolization and, by affecting calcium ion permeability, causing rapid diffusion of calcium ions resulting in rapid muscle paralysis in the parasite. However, praziquantel is not effective against *F. hepatica*. Apparently, at the dosage used for treatment, the tegument of *F. hepatica* is more resistant to the drug than are other flukes. Whether this is due to differences in permeability of the tegument or to differences in susceptibility of the worm's tissues is not known at this time. What is known is that the amount of the drug required to produce muscle tetany is 100 times greater for *F. hepatica* than that required for other worms. Drug resistance is also a major problem with any therapeutic protocol and resistance to praziquantel is becoming more evident. The search for new drugs is ongoing. Most recently, a one-time dose of triclabendazole, a veterinary fasciolide, has been used successfully. The drug binds to worm tubulin. It is available only through the manufacturer, Victoria Pharmacy in Switzerland.

CLASSIFICATION OF THE TREMATODA

For those readers interested in a more recent treatment of the classification of the flatworms, reference is made to the following publication:

Littlewood, R.A., and Bray, R.A. Interrelationships of the Platyhelminthes. In *The systematic association special volume*. London: Taylor & Francis Publishing Co., Series 60, 2001, pp.356.

Phylum Platyhelminthes

Superclass Neodermata

Ciliated epidermis cast off at end of free-swimming larval stage and replaced by nonciliated, syncytial tegument.

Class Trematoda

All parasitic, mainly in the digestive tracts of all classes of vertebrates.

Subclass Digenea

At least two hosts in life cycle, the first almost always a mollusk; perhaps most diversification in bony marine fish, although many species in all other groups of vertebrates.

Order Strigeata

Family Schistosomatidae

Dioecious adults in vascular system of definitive host; pharynx absent; no second intermediate host in life cycle; cercaria furcocercous with relatively short rami; oral sucker of cercaria replaced by protractile penetration organ; cercarial eyespots either pigmented or not; parasites of fishes, reptiles, birds, and mammals (genus mentioned in text: *Schistosoma*).

Order Echinoformata

Family Fasciolidae

Tegument commonly armed with spines (genera mentioned in text: *Fasciola*, *Fasciolopsis*).

Family Echinostomatidae

Adult and cercaria usually with circumoral collar; parasiote of birds, reptiles, and mammals (genus mentioned in text: *Echinostoma*).

Order Opisthorchiata

Family Opisthorchiidae

Cercariae with well-developed penetration glands; cercarial ventral sucker rudimentary; spinose tegument in adult; parasites of fish, amphibians, reptiles, birds, and mammals; cercarial tail one of several types; cercarial oral sucker protractile (genera mentioned in text: *Clonorchis*, *Opisthorchis*).

Family Heterophyidae

Cirrus pouch absent; vas deferens and oviduct unite distally to form a hermaphroditic duct; cirrus pouch absent; parasites of birds and mammals (genera mentioned in text: *Heterophyes*, *Metagonimus*).

Order Plagiorchiata

Family Troglotrematidae

Adult one of several morphologic types; with or without eyespots; oral sucker usually simple, some with appendages; acetabulum in anterior half of body; testes in posterior half of body; ovary pretesticular; cercaria one of several morphologic types; cercaria usually with eyespots; parasites of fish, amphibians, reptiles, and mammals (genus mentioned in text: *Paragonimus*).

Suggested Readings

Bogitsh, B. J. (1986). An overview of surface specializations in the digenetic trematodes. Hydrobiologia, *132*, 305–310.

Fried, B., & Haseeb, M. A. (1991). Platyhelminthes: aspidogastrea, monogenea, and digenea. In F. W. Harrison, & B. J. Bogitsh (Eds.), Microscopic anatomy of the invertebrates (pp. 141–209). (3). New York, NY: Wiley-Liss.

Galaktionov, K. V., & Dobrovolskij, A. A. (2003). The biology and evolution of trematodes. An essay on the biology, morphoplogy, life cycles, transmissions, and evolution of digenetic trematodes. Dordrecht: Kluwer Academic Publishers 620.

Keiser, J., & Utzinger, J. (2005). Emerging foodborne trematodiasis. Emerging Infectious Diseases, *11*, 1507–1514.

Llewellyn, J. (1965). The evolution of parasitic platyhelminths. In A. Taylor (Ed.), Evolution of Parasites. Oxford, England: Blackwell Scientific Publications.

Paulin, R., & Cribb, T. H. (2002). Trematode life cycles: short is sweet. Trends in Parasitology, *18*, 176–183.

Skelly, P. J., & Wilson, R. A. (2006). Making sense of the schistosome surface. In J. R. Baker, R. Muller, & D. Rollinson (Eds.), Advances in parasitology (pp. 185–284). (63). New York, NY: Academic Press Inc.

Smyth, J. D. (1995). Rare, new and emerging helminth zoonoses. Advances in Parasitology, *36*, 1–45.

CHAPTER

10

Visceral Flukes

Chapter 10 discusses five families of visceral flukes that are found as common infections of various visceral organs of humans. The various sites of infection by the adult worms in the definitive hosts include the lungs, intestines, and liver. Some of the more common liver flukes belonging to the genera *Fasciola* and *Clonorchis* are well illustrated using light micrographs of the adult parasite accompanied by an illustrated life cycle for each fluke described. Discussion of each parasite described includes the life cycle, the epidemiology, the symptomatology and diagnosis, as well as the chemotherapy and the host immune response to the parasite.

Species belonging to eight genera, representing five families, commonly infect various visceral organs of humans. Members of the genera *Fasciola*, *Clonorchis*, and *Opisthorchis* reside in the liver; those of another four, *Fasciolopsis*, *Heterophyes*, *Metagonimus*, and *Echinostoma*, inhabit the small intestine; and several members of the genus *Paragonimus*, notably *P. westermani*, live in the lungs. For the sake of convenience, discussion of the organisms is organized according to site of infection rather than phylogenetic relationships. The classification system at the end of Chapter 9 summarizes phylogenetic affinities.

LIVER FLUKES

Fasciola hepatica

Fasciola hepatica, the sheep liver fluke, is one of the largest digeneans parasitizing humans, measuring 30-mm long by 13-mm wide (Fig. 10.1). In addition to its size, *F. hepatica* can be distinguished from other digeneans by its highly branched testes and intestinal caeca; the short, convoluted uterus; oral and ventral suckers of equal size, the former situated on an anterior prominence called the **cephalic cone**; and vitellaria that extend along the lateral edges of the body to the posterior end. Adult worms live in the bile ducts, gall bladder, and liver tissue of their mammalian hosts.

Life Cycle (Fig. 10.2).

F. hepatica occupies a prominent place in parasitology because its life cycle was the first among digenetic trematodes to be completely elucidated, and that achievement has been the impetus for all subsequent investigations on life histories.

Human Parasitology. http://dx.doi.org/10.1016/B978-0-12-813712-3.00010-2

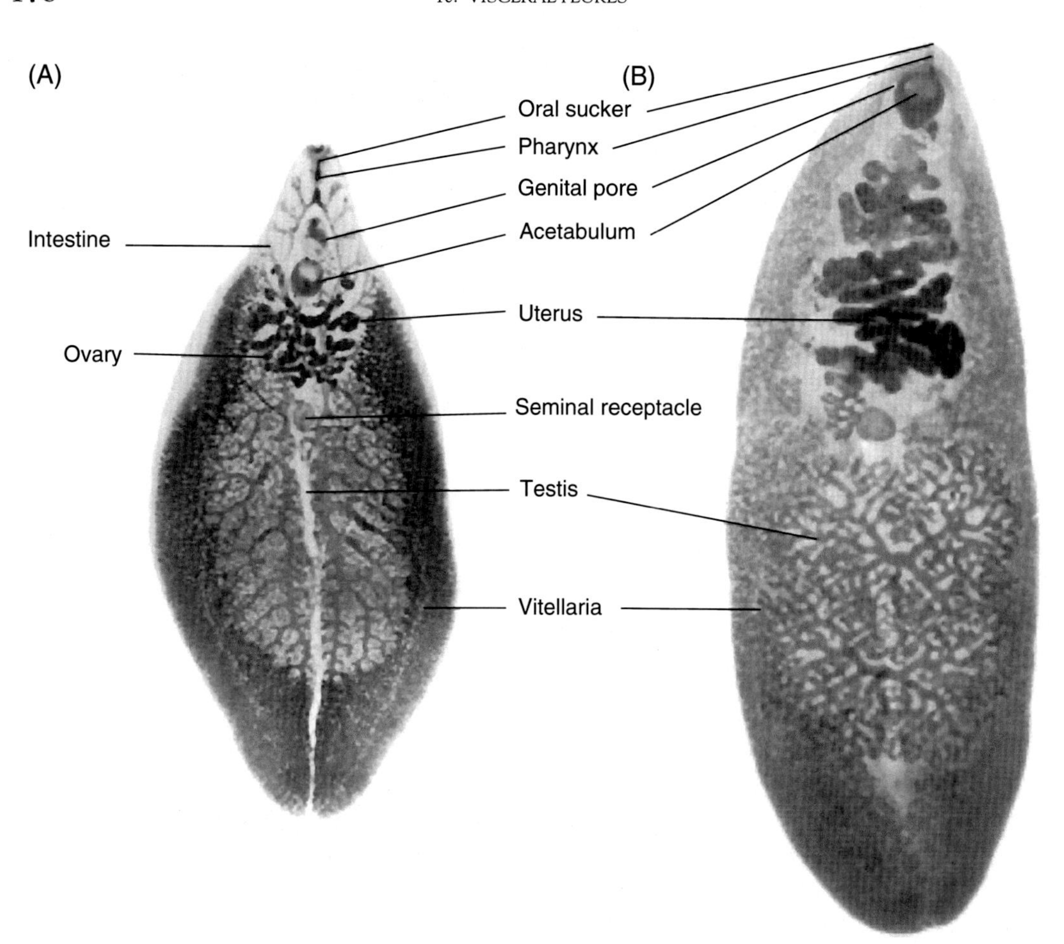

FIGURE 10.1 **Two fasciolid flukes.** (A) *Fasciola hepatica*, the sheep liver fluke. (B) *Fasciolopsis buski*.

The ovoid eggs are relatively large (130–150 μm by 63–90 μm), operculate, and yellowish brown in color. They are expelled before the miracidium is fully developed, pass into the host's alimentary tract via the common bile duct, and eventually reach the exterior with feces, at which time they must encounter fresh water if the cycle is to continue.

After 4–15 days in water at approximately 22°C, the completely developed miracidium escapes when the operculum opens. The miracidium has eyespots and is positively phototaxic. Its survival depends upon its success in locating and penetrating a suitable snail host within 8 h after hatching. Members of the amphibious snail genera *Lymnaea, Succinea, Fossaria*, and *Practicolella* serve as first intermediate hosts. Upon penetration, each miracidium metamorphoses into a sporocyst that gives rise to mother rediae, which, in turn, produce daughter rediae. Germ balls in the brood chamber of daughter rediae develop into cercariae, which emerge from the snail and become free-swimming. Upon reaching aquatic, emergent

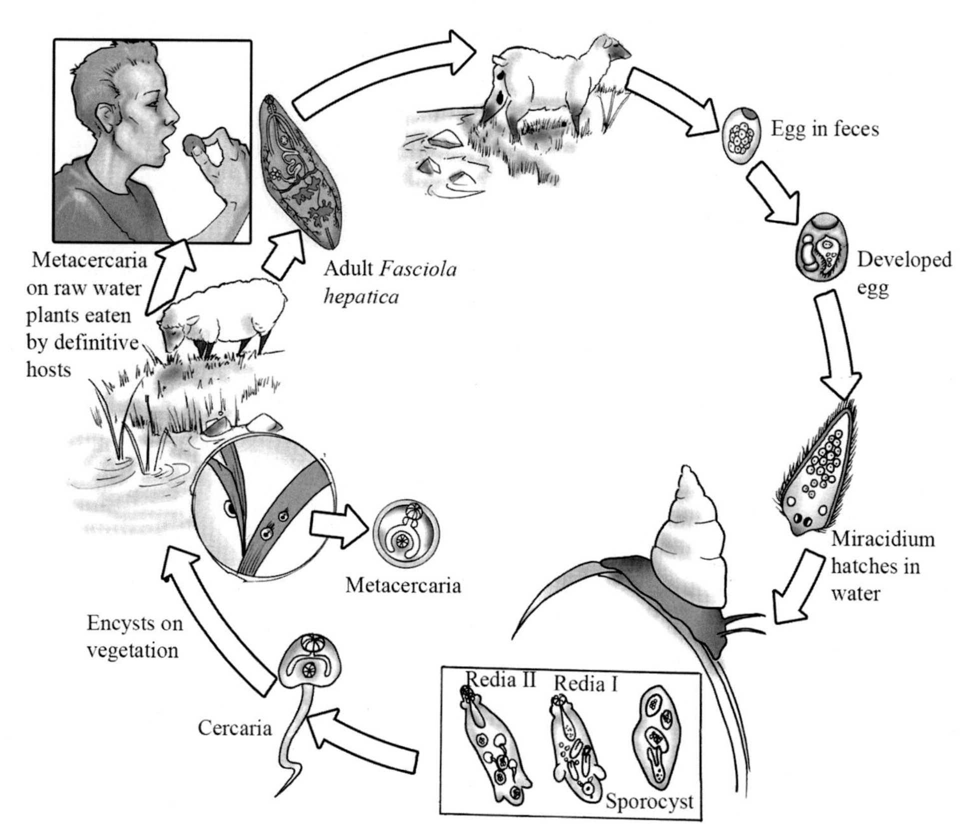

FIGURE 10.2 **Life cycle of *Fasciola hepatica*.** *Credit: Image courtesy of Gino Barzizza.*

vegetation (e.g., grass) or even submerged bark, the cercariae shed their tails and encyst as metacercariae on plants upon which sheep and cattle commonly feed. Humans become infected by eating contaminated watercress and other vegetation.

Metacercariae swallowed by the definitive host excyst in the duodenum, penetrate the intestinal wall, and enter the liver capsule via the body cavity. Migration through the liver parenchyma allows them to consume liver cells and blood before they reach the bile ducts, where they attain sexual maturity in approximately 12 weeks. Adult *F. hepatica* can live up to 11 years.

Epidemiology

Human infection by *F. hepatica* occurs throughout the world and is of increasing importance in some Caribbean islands and South America, as well as southern France, Great Britain, and Algeria. Ecologically, human infection occurs most frequently in sheep- and cattle-raising regions. Such livestock infections can result in heavy economic losses in wool, milk, and meat production. In one abattoir, annual economic losses from infected beef liver were estimated to be in the thousands of dollars.

Symptomatology and Diagnosis

F. hepatica infection in humans, or fascioliasis, is characterized by extensive destruction of liver tissue and bile ducts, hemorrhage, atrophy of portal vessels, and secondary, potentially lethal, pathological conditions. In addition to mechanical damage, the worms may evoke inflammatory reactions when the host becomes sensitized to the worms' metabolic products. Juvenile worms may get lost in the body cavity, encyst in ectopic tissues, and eventually become calcified. Initial symptoms frequently include severe headache, backache, chills, and fever. An enlarged, tender, or cirrhotic liver, accompanied by diarrhea and anemia, indicates advanced infection.

Laboratory diagnosis is based on identification of the characteristic eggs (Fig. 9.16) from patients' feces. Computer tomography (CT) scans are also diagnostically useful, and an enzyme-linked immunosorbent assay (ELISA) is especially effective in diagnosis of extrahepatic infections. Immunological techniques using monoclonal antibodies to identify excretory/secretory antigens have also proven useful.

It is of interest that, when eaten by humans in the Middle East, raw bovine liver harboring *F. hepatica* produces pain, irritation, hoarseness, and coughing due to young worms becoming attached to buccal or pharyngeal membranes. This condition, known as **halzoun**, is more commonly caused by pentastomids, or tongue worms, and leeches acquired in similar manner.

Chemotherapy

Praziquantel, a drug that has proven successful in treating a wide variety of trematode infections, has shown limited efficacy against *Fasciola*. The current recommended treatment for fascioliasis is a one-time oral administration of the tubulin binder triclabendazole or, as an alternative, 10–15 doses of bithionol.

Host Immune Response

Many of the immunological studies pertaining to human fascioliasis have concentrated on identifying those excretory/secretory antigens that are useful for immunodiagnostic techniques. Some of these antigens (e.g., *F. hepatica* proteases) have also proven useful in the development of vaccines for livestock. Immune responses to *F. hepatica* excretory/secretory antigens are of the Th2 type. However, the use of cathepsin L, a protease that has been identified as being secreted by the gastrodermis as the basis for a vaccine, has elicited a Th1 or a mixed Th1/Th2 response. Humans infected with *F. hepatica* also develop specific antibodies of the IgM, IgA, and IgE class.

Clonorchis sinensis

Clonorchis sinensis, the Chinese or Oriental liver fluke, is distributed widely in Russia, Korea, China, and Vietnam and has been estimated to infect at least 15–20 million persons with 200 million people at risk of infection. It is considered the most common human liver fluke in East Asia. It was classified as a group 1 biological carcinogen in 2009.

A distinctive feature of these members of the family Opisthorchiidae is their small suckers. The flukes infect the biliary systems of reptiles, birds, and mammals. Prosobranch snails serve as the molluskan hosts, and freshwater fishes, such as carp, serve as second intermediate hosts.

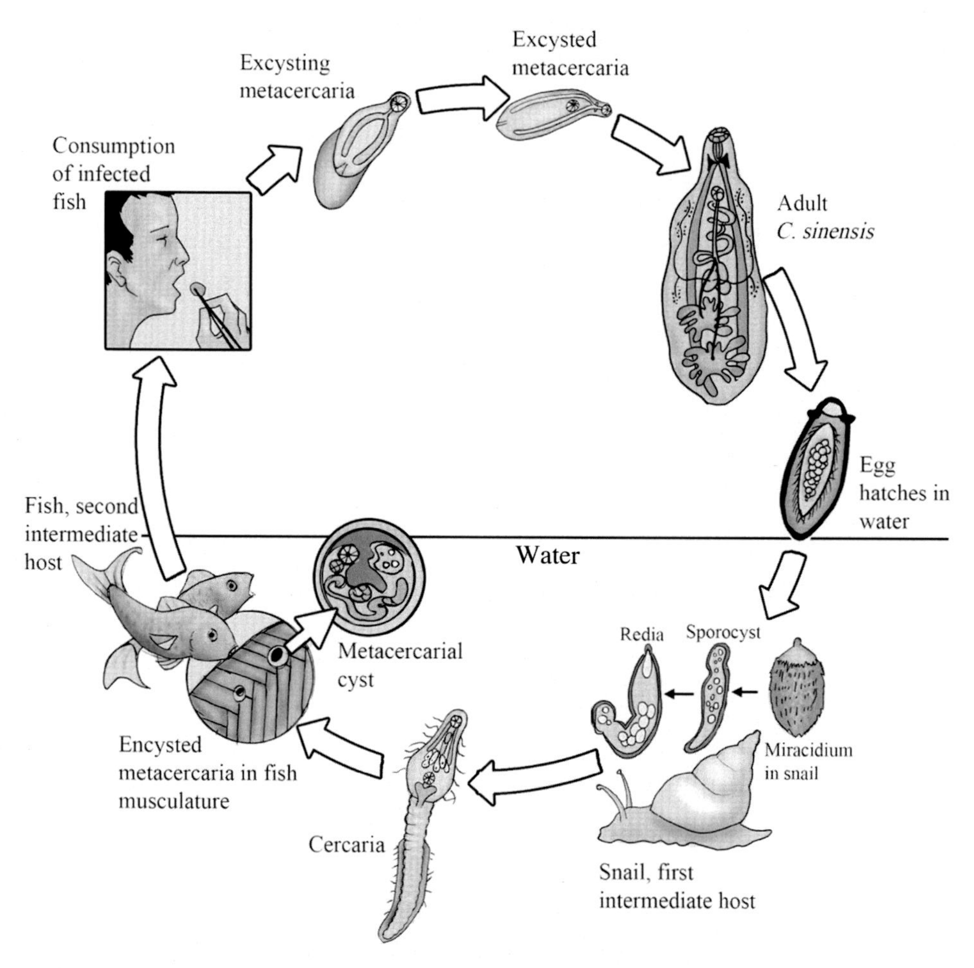

FIGURE 10.3 **Life cycle of *Clonorchis sinensis*.** *Credit: Image courtesy of Gino Barzizza.*

Life Cycle

The adult worm resides in the bile ducts of the human host and varies in size from 12 to 20 mm long and from 3 to 5 mm wide (Fig. 10.3). The body is tapered anteriorly, while the posterior end is somewhat blunt (Fig. 10.4). The poorly developed ventral sucker lies about one-fourth the body length from the anterior end, just behind the common genital pore. The caeca extend to the posterior region of the body. A centrally located ovary lies just anterior to the branched, tandemly arranged testes. The vitellaria are lateral, and a loosely coiled gravid uterus extends from the region of the ovary to the genital pore. The tanned eggs, which measure 29 μm by 16 μm, are operculated, with a ridge or collar at the base of the operculum, giving them an unusual urn shape (Fig. 9.15). There is also a characteristic knob at the abopercular end of the shell.

Eggs containing partially developed embryos are deposited in the biliary ducts and pass out of the host's body with feces. Although the egg contains a fully developed miracidium

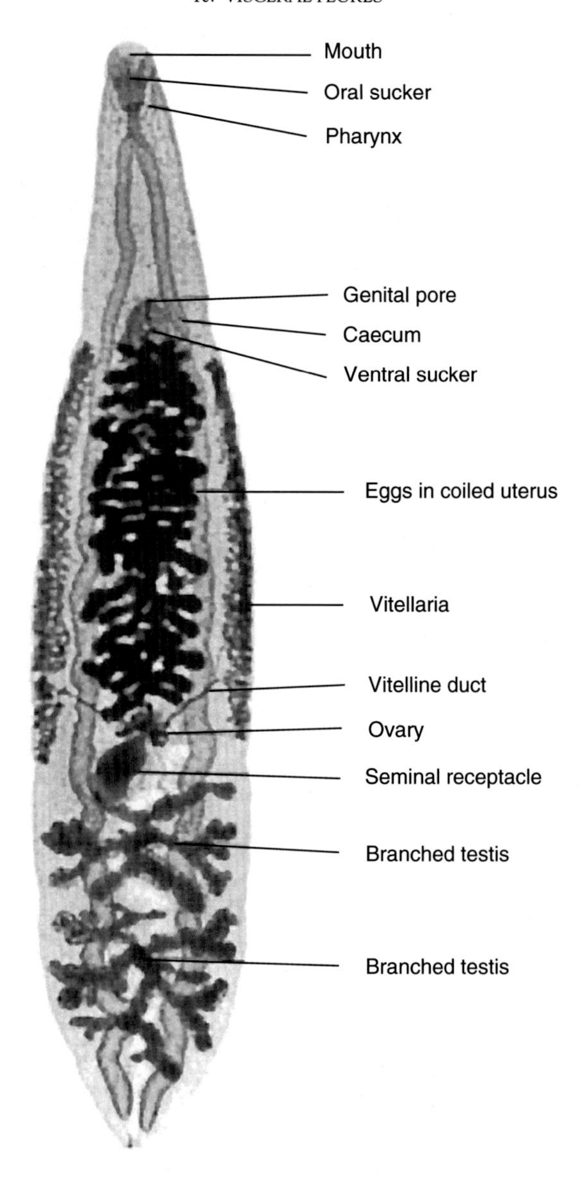

FIGURE 10.4 **The digenetic trematode *Clonorchis sinensis*, the Chinese liver fluke.**

by the time it reaches fresh water, it does not hatch immediately. Instead, hatching is delayed until the egg reaches the digestive tract of a suitable snail intermediate host, one belonging to any of the genera *Parafossarulus, Bulimus, Semisulcospira, Alocinma*, or *Melanoides*. The released miracidium first penetrates the intestinal wall of the snail, enters the hemocoel, and then the digestive gland, where it metamorphoses into a sporocyst. The sporocyst gives rise to rediae, which erupt from the sporocyst and, in turn, produce cercariae. In order to survive, the free-swimming cercariae that emerge from the snail must penetrate the skin, gills, fins, or muscles of freshwater fish hosts within 24–48 h. Each cercaria burrows into one of these tissues, loses its tail, and encysts as a metacercaria. A wide variety of fish species belonging to the family Cyprinidae serve as second intermediate hosts. Humans become infected by eating raw or poorly cooked fish (including steamed, smoked, or pickled). The metacercaria excysts in the duodenum, migrates up the common bile duct and into the biliary ducts where it reaches sexual maturity, feeding continually on the contents of the ducts. Following excystation, the worm reaches sexual maturity in about a month. *C. sinensis* has been known to live up to 25 years in the human host.

Epidemiology

Reservoir hosts, including cats, dogs, tigers, foxes, badgers, and mink, play a significant role in maintaining Oriental fluke populations in endemic areas. The high incidence of mammalian infection, the increase in freshwater fish farming in the Orient, and the practice of eating fish raw have made clonorchiasis a serious problem. Oriental aquaculture ponds are commonly fertilized with human excrement to enhance the growth of vegetation on which the fish feed. In Hong Kong, where fish farming is very common, the prevalence of human clonorchiasis is about 14%; in rural endemic areas, it may reach 80%.

Symptomatology and Diagnosis

Damage to human hosts is most severe in the bile ducts, as manifested by mechanical and toxic irritation. The extent of damage is proportional to the number of worms present, with some infections running into the thousands; over 6000 adult worms have been recovered from a single patient at autopsy. In such extreme cases, liver enlargement, thickening of the bile ducts, fibrosis, and some destruction of liver parenchyma are evident. Unlike *Fasciola*, however, *Clonorchis* does not invade liver tissues and, therefore, does not cause extensive liver necrosis. Intestinal disturbances are also common, but clonorchiasis is rarely fatal except when lowered resistance leaves patients vulnerable to secondary infections.

Positive diagnosis depends on identification of eggs from either feces or biliary drainage, and these must be differentiated from eggs of heterophyids (see below). Clinical examination is warranted whenever there is liver enlargement coupled with a history of residence in endemic areas. Immunodiagnostic techniques using recombinant *C. sinensis* proteins as antigens for detecting *C. sinensis*-specific IgG and IgE antibodies are under investigation.

Chemotherapy

Praziquantel is the chemotherapeutic agent of choice, although tribendimidine is under investigation as a promising alternative.

Opisthorchis felineus and *O. viverrini*

Opisthorchis felineus and *O. viverrini* are parasitic in the bile ducts of fish-eating mammals, including humans (Fig. 10.5). *O. felineus* is most commonly seen in southern, central, and eastern Europe, Turkey, the southern part of Russia, Vietnam, India, and Japan. It is also present in Puerto Rico and possibly other Caribbean islands. In Thailand, Laos, and Southeast Asia, *O. viverrini* occurs in an estimated 1–3 million humans. Patients harboring these worms suffer from diarrhea and thickening and eventual erosion of the bile duct wall.

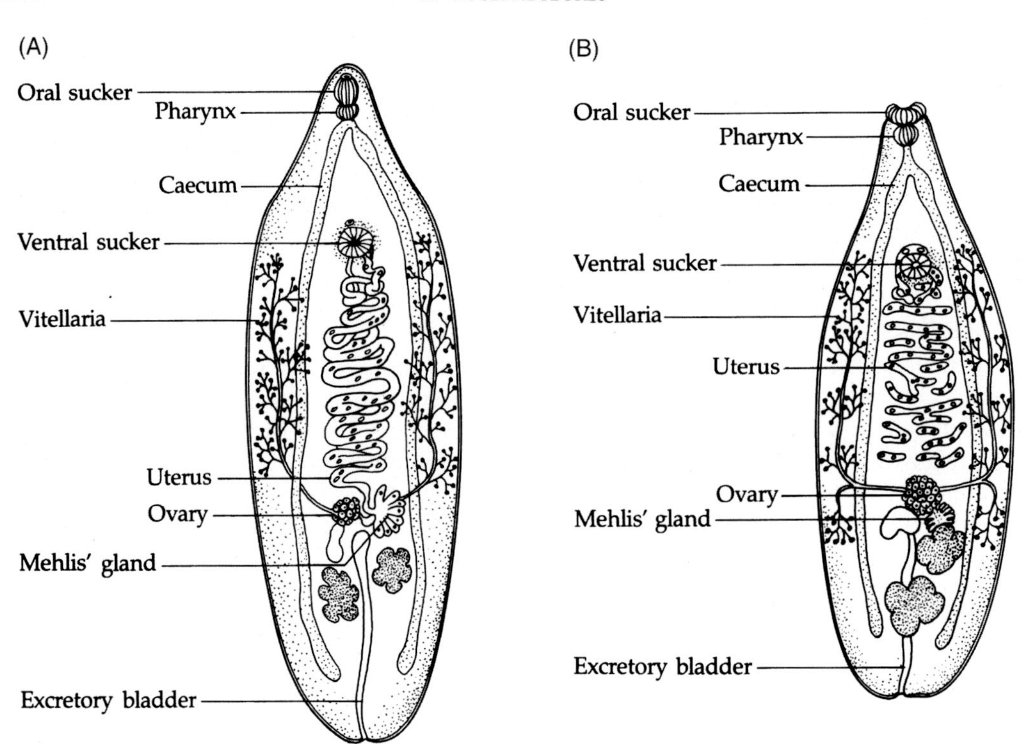

FIGURE 10.5 (A) *Opisthorchis felineus* adult. (B) *O. viverrini* adult.

The life cycles of *O. felineus* and *O. viverrini* are similar to that of *C. sinensis*, employing intermediate hosts of the snail genus *Bithynia* as well as cyprinid fishes, such as the chub or trench. Although infection by these worms is more common in animals, human infection results from eating raw or improperly cooked fish infected with encysted metacercariae. Felines are important reservoir hosts in endemic areas for both *O. felineus* and *O. viverrini*. Clinically, opisthorchiasis is indistinguishable from clonorchiasis and treatment is similar to that for clonorchiasis.

INTESTINAL FLUKES

Fasciolopsis buski

Fasciolopsis buski is the largest digenean infecting humans, reaching a size of 75-mm long and 20-mm wide. It is morphologically similar to *F. hepatica* with a few notable differences (Fig. 10.1). The most obvious of these are that, in *F. buski*, the caeca lack side branches, the ventral sucker is much larger than the oral sucker, and there is no cephalic cone. It differs further from all other members of the family Fasciolidae in that the definitive habitat is the small intestine of humans and pigs rather than the liver.

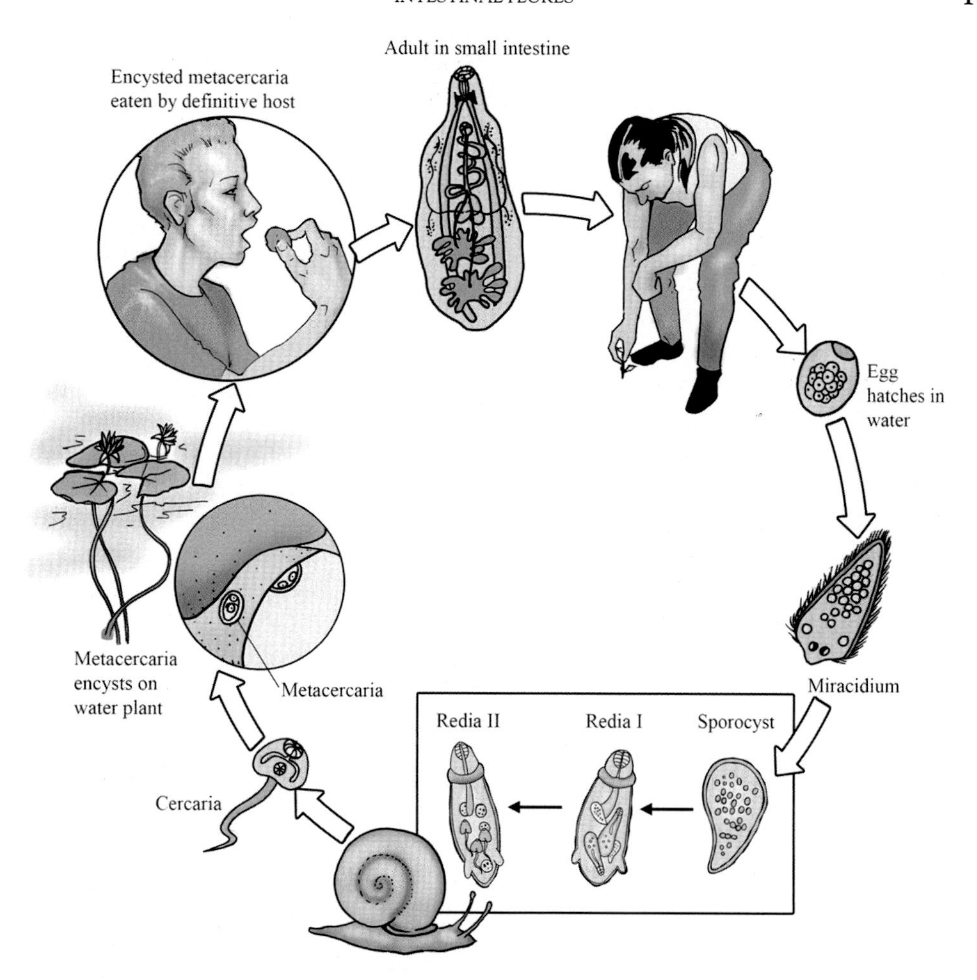

FIGURE 10.6 **Life cycle of *Fasciolopsis buski*.** *Credit: Image courtesy of Gino Barzizza.*

Life Cycle

Adult worms inhabit the duodenal and jejunal regions of the small intestine, either attached by the suckers to the mucosal epithelium or lie buried in the mucus secretions (Fig 10.6). Each deposits about 25,000 eggs daily, which are indistinguishable from those of *F. hepatica*; the eggs are expelled in feces and must reach freshwater in order to continue the cycle. In freshwater, miracidia develop and hatch in 3–7 weeks, depending on temperature. After locating and penetrating a planorbid snail of the genus *Segmentina* or *Hippeutis*, the miracidia metamorphose into sporocysts, which subsequently give rise to two sequential redial generations. Cercariae emerge from the daughter rediae 4–7 weeks after miracidial penetration of the snail, reenter the water, and encyst on freshwater vegetation, most commonly water chestnut, water caltrop, water bamboo, and lotus. Encysted metacercariae are ingested by humans when they eat contaminated raw plants or peel the pods or stalks with their teeth before eating, thereby freeing the encysted metacercaria, which are then swallowed. Metacercariae excyst in the small intestine, attach to the mucosa, and develop to sexual maturity in 25–30 days.

Epidemiology

Human infection by *F. buski* occurs throughout central and south China, Taiwan, Laos, Vietnam, Cambodia, India, Korea, and Indonesia. The infection is usually acquired by ingestion of encysted metacercariae from fresh plants grown in ponds fertilized with human or swine excrement. Drying or cooking the plants before eating kills metacercariae. Hogs are the most important animal reservoirs for *F. buski*.

Symptomatology and Diagnosis

The worms feed actively not only on the host's intestinal contents but also on the superficial mucosa, causing inflammation, ulceration, and abscesses at sites of attachment. Diarrhea, nausea, and intestinal pain are common, especially in the morning hours. Intestinal edema occurs in severe infections. Eating may relieve the abdominal distress unless the food happens to consist of infested aquatic plants, in which case symptoms will eventually be exacerbated. The large size of the worms may also lead to intestinal obstruction. Reaction to the worms' metabolites can produce such clinical symptoms as general leukocytosis, anemia, and eosinophilia. Patients purged of the worms usually recover completely, although advanced, severe infections can be fatal. When clinical symptoms appear in an endemic area, diagnosis must be confirmed by fecal examination for eggs, or, occasionally, by retrieval of whole worms vomited or passed in feces.

Echinostoma trivolvis

Several members of the genus *Echinostoma* and related genera occasionally infect humans as well as other mammals. Adult echinostomes, while varying greatly in size, are easily identified by the collar of spines along the dorsal and lateral sides of the head. In general appearance (Fig. 10.7), the adult worm is elongated, with a relatively large ventral sucker situated immediately behind the head. The testes lie in tandem in the posterior portion of the body; the ovary is anterior to the testes, and the short uterus consists only of an ascending limb terminating at the genital pore anterior to the ventral sucker. Large, operculate eggs measure 90–126-μm long by 54–71-μm wide, with only a few in the uterus at any given time. *Echinostoma trivolvis* is the prototype for echinostome species infecting humans, essential differences consisting primarily of the number and arrangement of collar spines.

Life Cycle

The life cycle of *E. trivolvis* is typical of most echinostomes. Operculated eggs are passed from the definitive host with feces and must reach freshwater for the cycle to continue. The enclosed miracidium is at a very early stage of development when the egg is deposited and requires 2–5 weeks to reach maturity, after which it hatches and penetrates a snail of one of the genera *Lymnaea*, *Physa*, or *Bithynia*. Various freshwater pelecypods and gastropods serve as hosts for other echinostomes. A single sporocyst generation and two redial generations develop in the molluskan host. Free-swimming cercariae escape from daughter rediae, enter the water, and penetrate and encyst in a variety of aquatic animals including mollusks, some

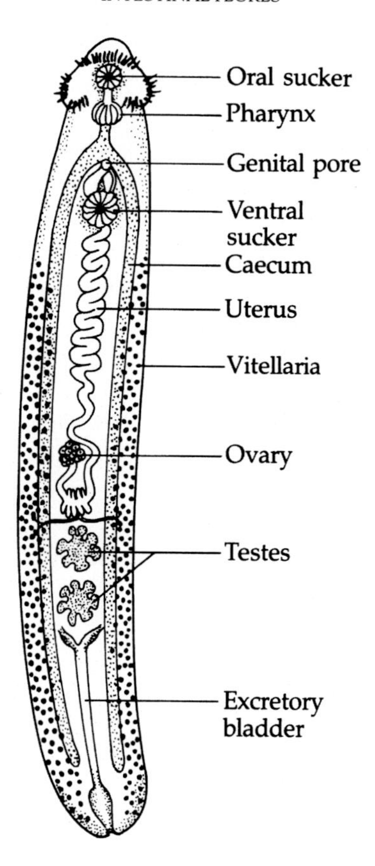

FIGURE 10.7 ***Echinostoma revolutum*** **adult.**

of which may serve as first intermediate hosts, or, at times, on aquatic vegetation. The life cycle is completed when the definitive host ingests encysted metacercariae, which excyst and develop to sexual maturity in the small intestine.

Epidemiology

While as many as 15 species of *Echinostoma* have been reported in humans, most are incidental parasites. *E. trivolvis* and *E. revolutum,* for instance, are commonly parasites of birds and mammals in the United States and Europe, respectively. Human infections of *E. ilocanum* are most frequently reported from the Philippines, China, Taiwan, and Indonesia. Infection occurs when the infected second intermediate host is eaten either raw or improperly cooked. Because of the variety and number of potential intermediate and definitive hosts, it is impossible to control the parasite, but human infections can be prevented if food is cooked adequately.

Symptomatology and Diagnosis

Echinostomiasis in humans is usually a minor affliction, often causing nothing more serious than diarrhea. In severe infections, the spinose collar may cause ulceration of the intestinal

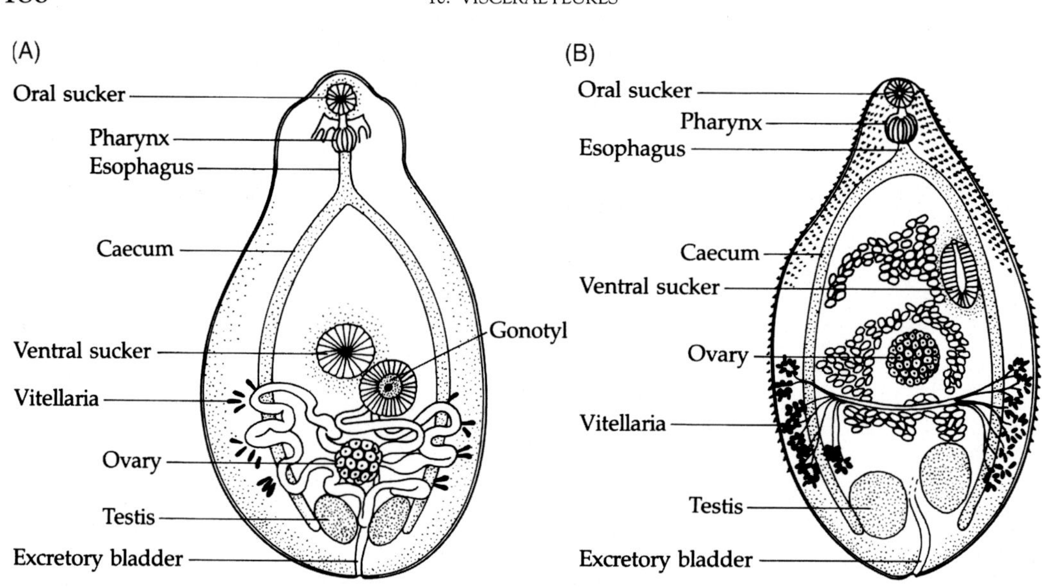

FIGURE 10.8 (A) *Heterophyes heterophyes* adult. (B) *Metagonimus yokogawai* adult.

mucosa. Children sometimes experience abdominal pain, diarrhea, anemia, and/or edema. The principal diagnostic technique, identification of eggs from feces, is facilitated by a number of distinctive features of echinostome eggs, namely, their dark brownish color and the very immature larvae, even uncleaved zygotes, that are unlike those of other intestinal trematodes.

Heterophyes heterophyes and *Metagonimus yokagawai*

Heterophyes heterophyes and *Metagonimus yokagawai* belong to the family Heterophyidae. Measuring 1.4-mm long and 0.5-mm wide, they are among the smallest digeneans infecting humans (Fig. 10.8). The tegument of these pyriform flukes contains scalelike spines. In *H. heterophyes*, the genital pore, situated posterolateral to the prominent ventral sucker, is surrounded by a genital sucker or **gonotyl**. In *M. yokagawai*, the ventral sucker and gonotyl are fused, and the complex is displaced to the left side of the body. The reproductive system is located in the posterior half of the body with the testes lying side by side; the ovary is medial, just anterior to the testes, with lateral, follicular vitelline glands restricted to the posterior third of the body. The gravid uterus loops between the long intestinal caeca, terminating at the gonotyl. The eggs of both species resemble those of *C. sinensis* except for their indistinct opercular shoulders and the absence of an abopercular knob.

Life Cycle

H. heterophyes and *M. yokagawai* commonly inhabit the mid-region of the small intestine, and their life cycles are almost identical. Human infection by *H. heterophyes* occurs in Asia, Egypt, and Hawaii while *M. yokagawai* is the most common intestinal fluke of humans in the Far East, Spain, and the Balkan countries. Eggs containing fully developed miracidia pass out of the human host in feces and hatch only when ingested by a suitable molluskan first intermediate host. In *H. heterophyes*, this is a freshwater or brackish water snail belonging to one of the genera *Pirenella* (in Egypt), *Cerithidia* (in Japan), or *Tarebia* (in Hawaii); *M. yokagawai* infects members of the snail genus *Semisulcospira*. The hatched miracidium penetrates the intestine of the snail and transforms into a sporocyst in the digestive gland. Two generations of rediae follow the sporocyst with daughter rediae giving rise to cercariae that escape to the external environment, penetrate the musculature of any of a number of food fishes, and encyst as metacercariae. One of the principal fishes used by *H. heterophyes* as the second intermediate host is the mullet, *Mugil cephalus*, which can harbor several thousand metacercariae. Salmonoid fishes commonly serve as second intermediate hosts for *M. yokagawai* metacercariae. Human infection results from consumption of raw or improperly cooked fish. The metacercariae excyst in the duodenum, migrate to the jejunum, and attain sexual maturity in about a week.

Epidemiology

In addition to *H. heterophyes* and *M. yokagawai*, at least 14 other heterophyids have been reported in humans. There is an unusually high incidence of infection by *H. heterophyes* in Egypt, especially in parts of the lower Nile valley. Poor sanitation practices by local fishermen, boatmen, and other residents continually pollute the water with eggs. One of the principal food fishes in the region is the mullet, and parasitic infection results from eating fresh mullet, either incompletely cooked or poorly pickled. Human infection with *H. heterophyes* is also common in Japan, central and south China, Korea, Taiwan, Greece, Israel, and Hawaii. Infection with *M. yokagawai* occurs when the infected second intermediate host, such as a salmonoid fish, is consumed raw or improperly processed. A variety of fish-eating mammals, including cats and dogs, serve as reservoirs for these parasites.

Symptomatology and Diagnosis

The pathology, symptomatology, and diagnosis in cases of infection by these two digeneans are very similar. Adult worms often produce little distress to the patient, but more severe infections may elicit inflammatory reactions at sites of contact as well as diarrhea and abdominal pain. Eosinophilia is also common but without anemia. Adult worms sometimes erode the mucosa and deposit eggs that may infiltrate the lymphatics or venules. The eggs are then carried to various parts of the body where they may cause granulomatous responses in such organs as the heart or brain. *Heterophyid myocarditis* sometimes precipitates fatal heart attacks, and neurological complications also have been reported.

Diagnosis depends on positive identification of eggs from feces. Because of the great degree of similarity, care must be exercised to differentiate the eggs from those of other heterophyids and of opisthorchids.

LUNG FLUKES

Paragonimus westermani

Paragonimus westermani, the Oriental lung fluke, belongs to the family Troglotrematidae and is one of several digeneans of the same genus that infect the human respiratory tract. The first report of human infection was from Taiwan during the latter part of the 19th century.

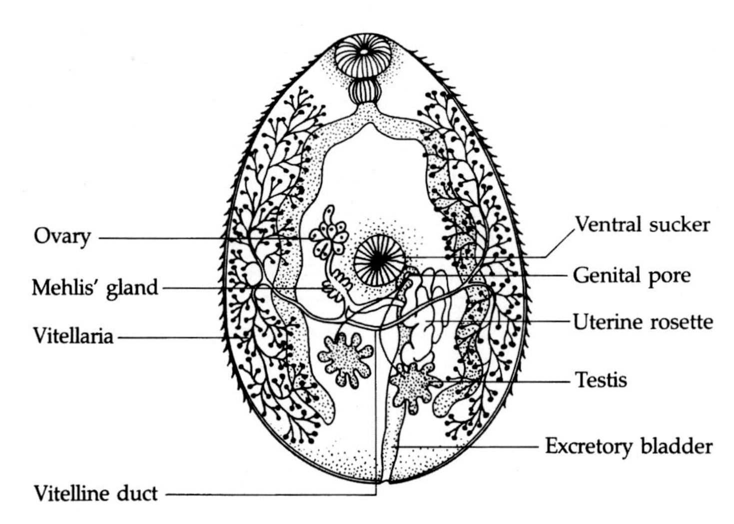

FIGURE 10.9 ***Paragonimus westermani* adult.**

Following that initial account, numerous other infections were quickly diagnosed in eastern Asia where the condition remains prevalent today.

The thick-bodied, reddish-brown adult worm measures 7.5–12-mm long and 4–6-mm wide (Fig. 10.9). The male reproductive system consists of two irregularly lobed testes situated side by side about two-thirds down the length of the body. The lobed ovary, anterior to the right testis, is connected via the oviduct to the uterus, a tightly coiled rosette that lies anterior to the left testis at the same level as the ovary. Vitellaria extend bilaterally along the length of the body, lateral to and paralleling the caeca. A medially located ventral sucker lies between the ovary and the uterus. Brownish, operculated eggs (Fig. 9.15), smaller than but similar to those of *F. hepatica*, are released through the genital pore situated in the center of the uterine rosette.

Life Cycle

Paired adult *P. westermani* are usually found encapsulated in the bronchioles of the victim's lungs. Commonly, eggs containing uncleaved embryos or zygotes are coughed up and expelled with sputum when the capsules enclosing the adult worms rupture (Fig 10.10). However, some eggs may be swallowed with sputum, pass through the digestive system, and be expelled with feces; others become trapped in the surrounding lung tissue and produce bronchial abscesses. Once the egg reaches water, several weeks are required for the miracidium to develop. Fully developed miracidia hatch spontaneously, and each must then locate and penetrate a suitable snail host of the genera *Semisulcospira*, *Tarebia*, or *Brotia* within 24 h or perish. Following penetration of the snail host and metamorphosis into a sporocyst, two redial generations are produced in the digestive gland. Cercariae, formed within the brood chambers of the daughter rediae, emerge from the snail tissue into the surrounding water approximately 11 weeks after the snail is infected. The cercariae possess knoblike tails useless for swimming; instead, the cercariae crawl over solid surfaces until they encounter suitable crustaceans, such as freshwater crabs and crayfish. Using a sharply

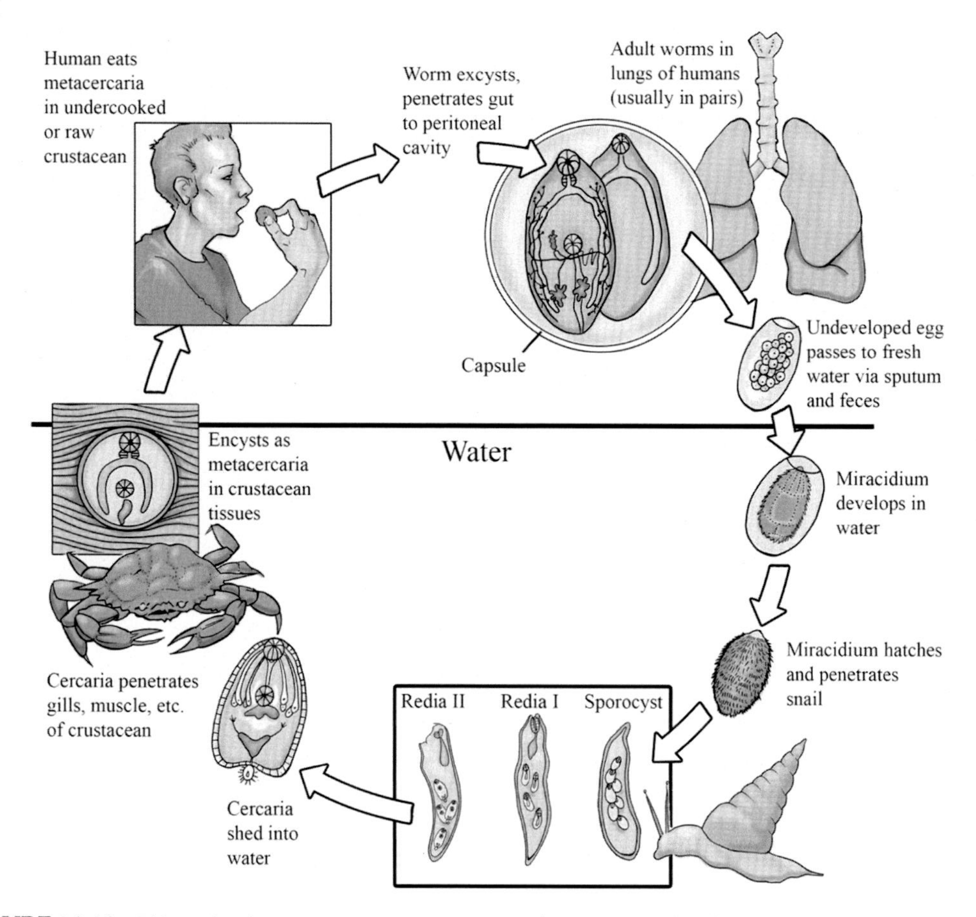

FIGURE 10.10 **Life cycle of *Paragonimus westermani*.** *Credit: Image courtesy of Gino Barzizza.*

pointed, cuticular stylet, they can penetrate the crustacean's exoskeleton at various vulnerable sites. There is evidence that the crustacean second intermediate host may also acquire infection by eating infected snails. Once inside the host, the cercariae encyst in muscles, gills, and viscera (Fig. 10.11) where they develop into metacercariae. The encysted metacercaria is not folded over ventrally, as are most encysted metacercariae, but lies in an extended position within the cyst wall.

Humans acquire infection by eating freshwater crustaceans raw, inadequately pickled, or incompletely cooked. The metacercaria excysts in the small intestine and partially penetrates the intestinal wall. Young adults remain at this site for several days before entering the coelom. Then, traversing the diaphragm and pleura, they enter the peribronchiolar tissues of the lungs, where they become encapsulated in pairs by host connective tissue and develop to sexual maturity within 8–12 weeks. During migration, young adult worms often become lodged in other organs, producing ectopic lesions before succumbing to host reactions.

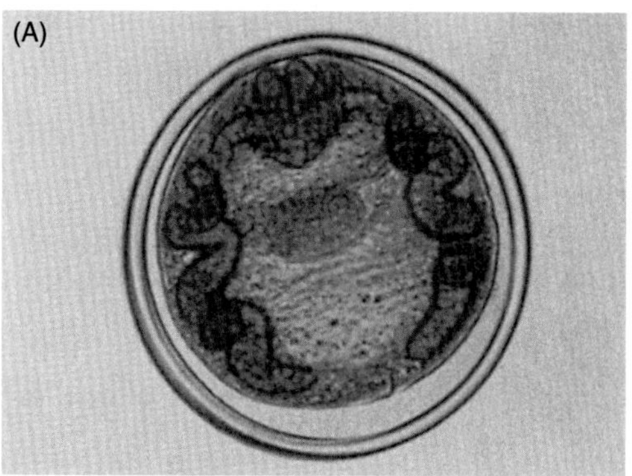

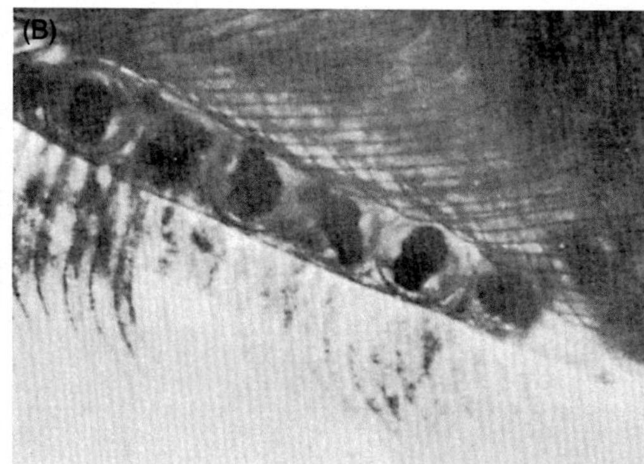

FIGURE 10.11 (A) Encysted metacercaria of *P aragonimus westermani.* (B) Several metacercariae encysted in gill filament of crab host.

Epidemiology

While *P. westermani* is worldwide in distribution, human infections are confined mainly to such Oriental countries as Japan, South Korea, Thailand, Taiwan, China, and the Philippines. A number of animals, including dogs, cats, some rodents, and pigs, can serve as reservoirs. As noted previously, infection, in humans as well as reservoir hosts, results from consumption of infected freshwater crustaceans raw, pickled, or undercooked. The pickling process coagulates muscle protein, giving the meat the appearance of being cooked and, therefore, harmless while, actually, it has no effect upon the encysted metacercariae. Metacercariae dislodged from the crustacean during the cleaning process may also adhere to utensils, which then become a source of infection to food handlers. Humans may also become infected by consuming the juices obtained from crushed crabs, a medicinal practice common in parts of the Far East.

Other members of the genus *Paragonimus*, *P. ohirai* and *P. iloktsuenensis*, infect the lungs of humans and other mammals in the Far East, while *P. kellicotti* has been reported infecting the lungs of humans in the United States. The life cycles are similar to *P. westermani* except that the second intermediate hosts are estuarine crabs (Far East) and crayfish (United States) that inhabit the brackish waters at the mouths of rivers, primarily in Japan, and freshwater streams in the United States.

Symptomatology and Diagnosis

P. westermani adults and eggs stimulate formation of connective tissue capsules in the host, both in the lungs and at ectopic sites. In addition to adult worms, the capsules contain eggs and infiltrated host cells in a hemorrhagic, semifluid mass. The capsules often ulcerate, giving the lungs a peppered appearance. Early symptoms include a cough producing blood-tinged sputum, pulmonary pain, and even pleurisy. A low-grade fever usually accompanies these symptoms. At present, paragonimiasis is difficult to distinguish from other pulmonary disorders such as pneumonia and tuberculosis.

Encysted worms may be found at such ectopic sites as the abdominal wall, lymph nodes, heart, and portions of the nervous system. Infection of the abdominal wall may produce abdominal pain, diarrhea, and bleeding. In the brain, infection may produce a variety of neurological symptoms such as epilepsy and paralysis. Fatalities have been recorded from cardiac involvement as well as from heavy pulmonary infections.

Identification of eggs from sputum, pleural aspirate, or feces is the most reliable diagnostic procedure. Patients from endemic areas who show such symptoms as pulmonary distress, blood-tinged sputum, and eosinophilia should be examined carefully. For ectopic infections, immunological tests with antigens derived from *Paragonimus* have proven useful.

Host Immune Response

Few investigations are available regarding the host responses to *P. westermani*. As is the case with the aforementioned visceral inhabiting trematodes, most research has been aimed at defining antigens that can be used in immunodiagnostic procedures. The tissue migratory stages of *P. westermani* activate host peritoneal macrophages and cause an increase in host IgE. Migrating metacercariae have been shown to secrete proteases that reduce host eosinophil-associated tissue inflammation. No changes in $CD4^+$ T cells and $CD8^+$ T cells have been observed in infected hosts.

Suggested Readings

Hong, S. J., Yun Kim, T., Gan, X. X., Shen, L. Y., Sukontason, K., & Kang, S. Y. (2002). *Clonorchis sinensis*: glutathione S-transferase as a serodiagnostic antigen for detecting IgG and IgE antibodies. *Experimental Parasitology, 101*, 231–233.

Koniya, Y. (1966). *Clonorchis* and clonorchiasis. *Advances in Parasitology, 4*, 53–106.

Murrell, K. D., Cross, J. H., & Chongsuphajaisiddhi, T. (1996). The importance of food-borne parasitic zoonoses. *Parasitology Today, 12*, 171–173.

Park, G. M. (2007). Genetic comparisons of liver flukes, *Clonorchis sinensis* and *Opisthorchis viverrini*, based on rDNA and mtDNA gene sequences. *Parasitology Research, 100*, 351–357.

Rim, H. -J., Farag, H. F., Sommani, S., & Cross, J. H. (1994). Food-borne trematodes: ignored or emerging? *Parasitology Today, 10*, 207–209.

Yokagawa, M. (1969). *Paragonimus* and paragonimiasis. *Advances in Parasitology, 7*, 375–387.

CHAPTER

11

Blood Flukes

Chapter 11 deals with those trematodes that cause the condition known as schistosomiasis or snail fever or bilharziasis. The disease is caused primarily by three members belonging to the genus *Schistosoma, Schistosoma mansoni, Schistosoma haematobium*, and *Schistosoma japonicum*. The impact of the disease on world history is briefly discussed. Of particular historical interest is the effect schistosomiasis had on ancient Egypt and the Middle East in biblical times and how this disease still has an impact in modern times. The morphology and composite life cycle of the three species with emphasis on the variations that exist between them is presented and illustrated. The symptomatology of the disease is described. Emphasis is placed on the hosts' immune responses and the ability of the adult worms to evade these responses. A brief discussion of the epidemiology of the three species follows. The latest chemotherapeutic regimens are listed, and how the use of genomics to identify new targets for drug and vaccine intervention is presented. The chapter concludes with a discussion of those schistosome species that cause the dermatological condition known as swimmer's itch.

The human disease complex known as **schistosomiasis** is also referred to as **bilharziasis** or **snail fever**. It is caused primarily by three members of the genus *Schistosoma* (family Schistosomatidae): *S. haematobium*, *S. mansoni*, and *S. japonicum*. There are several other species that infect humans, but they are rare. Infections number in excess of 250 million people in 76 countries and, in spite of efforts to control this disease, the level of incidence has shown no significant decrease. In the People's Republic of China, a recent estimate indicated at least 15 million cases of schistosomiasis japonica, representing the single most serious disease in that country. As a result of concerted control measures, incidence in China has been somewhat reduced. Egypt has one of the most heavily infected populations in the world, since not only is *S. haematobium* endemic to that country, but *S. mansoni* also occurs with great frequency. Among the inhabitants of some endemic areas of the Nile valley, the infection rate exceeds 80%. Other areas of high incidence include tropical and subtropical Africa, parts of South America, and several of the Caribbean islands (Fig. 11.1).

While it was not until 1852 that the young German parasitologist Theodor Bilharz, working in Egypt, discovered one of the parasites (*S. haematobium*) responsible for urinary schistosomiasis, there are recorded accounts of the disease dating from pharaonic times. In the Ebers papyrus

Human Parasitology. http://dx.doi.org/10.1016/B978-0-12-813712-3.00011-4

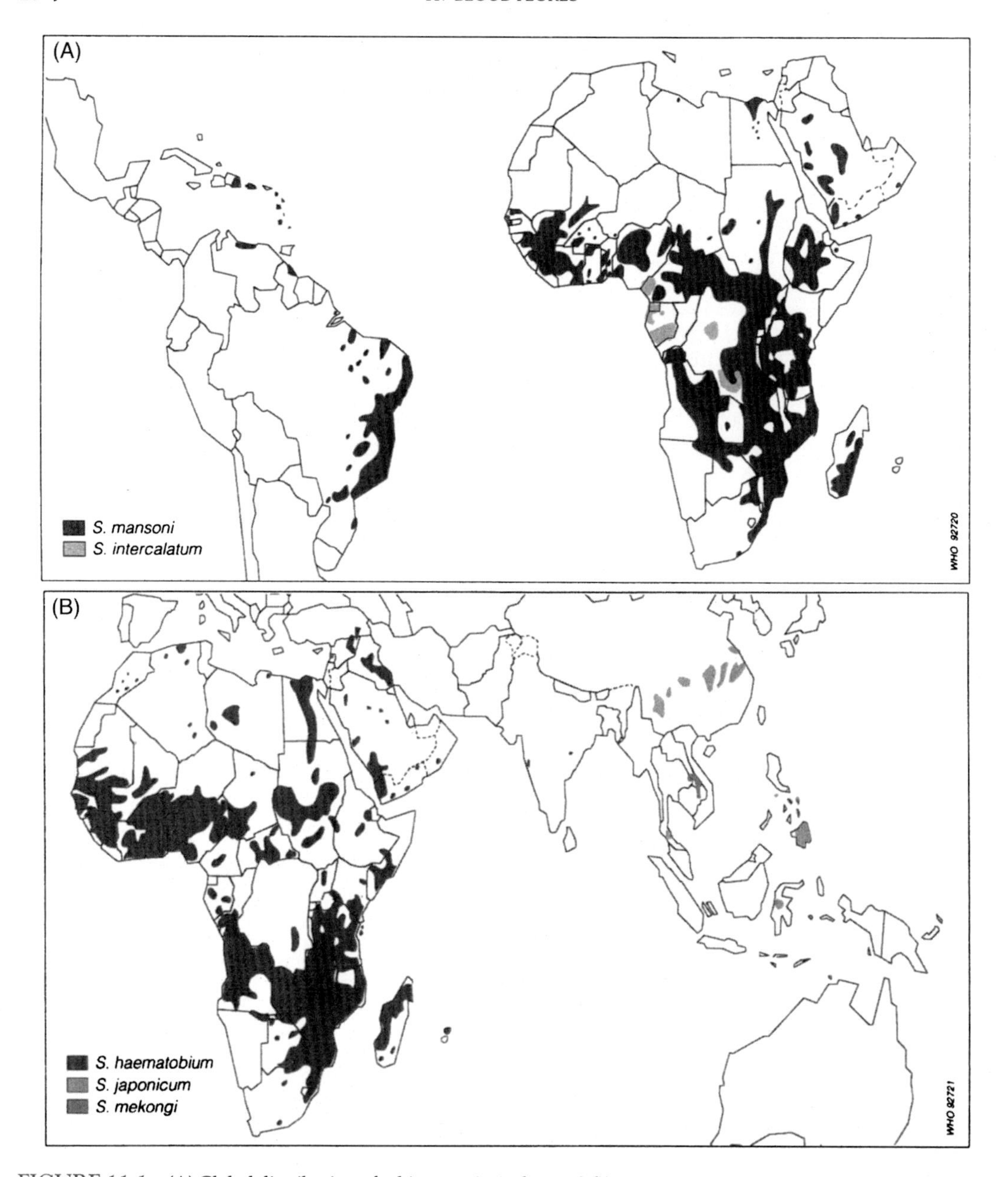

FIGURE 11.1 (A) Global distribution of schistosomiasis due to *Schistosoma mansoni* and *S. intercalatum*. (B) Global distribution of schistosomiasis due to *Schistosoma haematobium*, *S. japonicum*, and *S. mekongi*.

from 1500 BCE, there is a reference to treatment of hematuria (bloody urine) and calcified eggs of *S. haematobium* have been found in the viscera of Egyptian mummies dating from 1200 BCE.

Fossilized bulinid snails that may have served as intermediate hosts for *S. haematobium* have been unearthed in the ancient biblical city of Jericho. An interesting hypothesis,

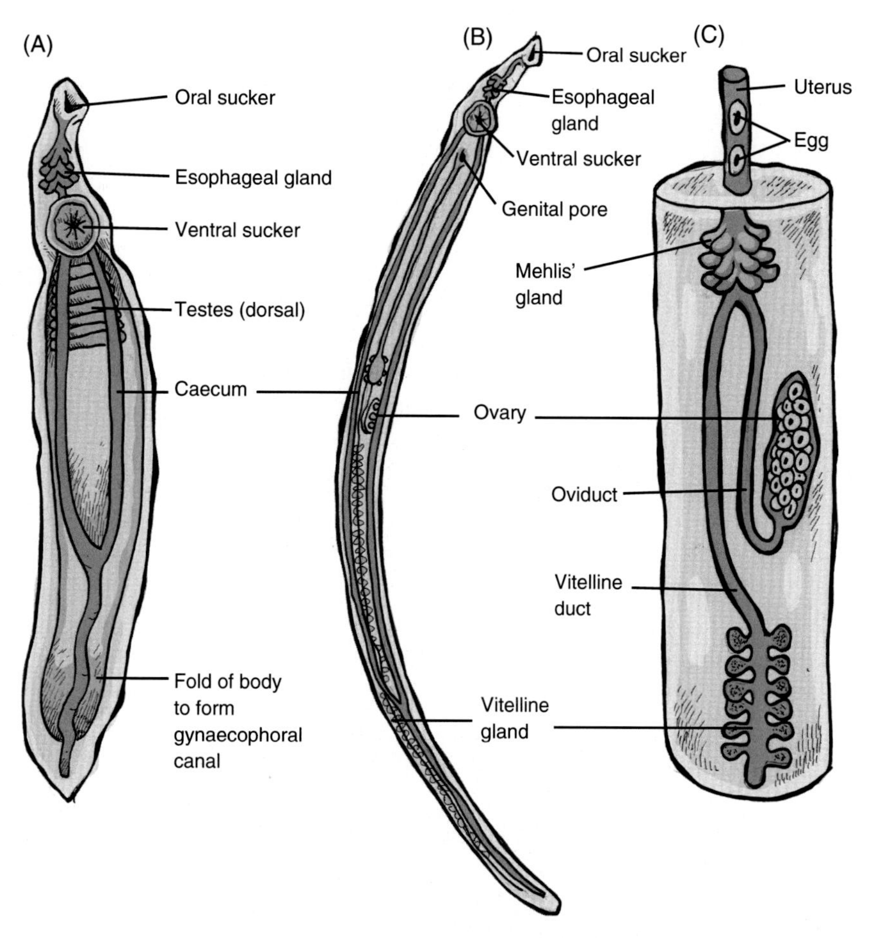

FIGURE 11.2 Generalized adult schistosomes (A) Male. (B) Female. (C) Female reproductive organs. *Credit: Image courtesy of Gino Barzizza.*

based on this discovery, is that the city's well was infested with infected snails, producing a high incidence of schistosomiasis among the citizenry. Too debilitated by the disease to defend their city or repair its decaying walls, they were easily defeated by Joshua's army. Without knowing the cause of this heinous disease, Joshua, in order to prevent its spread, destroyed Jericho and proclaimed a curse upon any who would rebuild it, thus precluding subsequent repopulation. The city remained deserted for more than 500 years. Centuries of recurring drought apparently destroyed the snails, and the city became, and remains, free of the parasite.

The French invasion of Egypt during the latter part of the 18th century was probably one of the first large-scale contacts people of the Western world had with schistosomiasis. Besides amoebic dysentery, the French had to contend with two other diseases previously unknown to them, hematuria and the eye disease known as trachoma. The former, as we know now, is caused by *S. haematobium*, while the latter is caused by a fly-transmitted microorganism (see p. 345). Both diseases remain firmly entrenched in Egypt.

MORPHOLOGY

Figure 11.2 depicts adult male and female worms. The mouth of the adult schistosome is surrounded by an oral sucker, and a ventral sucker is located immediately posterior to the level of bifurcation of the gut. While no pharynx is present, there is an esophagus with prominent **esophageal glands**. The paired caeca reunite posteriorly, forming a single caecum that extends the remaining length of the body. Schistosomes are unique among digeneans in being dioecious and sexually dimorphic. The adult male is more robust than the female and possesses a ventral fold or groove called the **gynaecophoric canal**. The female, longer and more slender than the male, is held in this canal, permitting almost continuous mating (Figs. 11.3 and 11.4).

The male possesses from 5 to 9 testes, and the male genital pore opens ventrally, immediately posterior to the ventral sucker. There is no cirrus. In the female, the position of the single ovary varies according to species, and the uterus may be long or short, depending on the position of the ovary relative to the female genital pore.

LIFE CYCLE

The life cycles (Fig. 11.5) of the three species are virtually identical and will be so treated. Individual differences will be noted in the section following description of the life cycle.

FIGURE 11.3 **Scanning electron micrograph of adult male schistosome showing mouth and ventral sucker.** Note female worm in gynaecophoric canal. *Source: National Cancer Institute.*

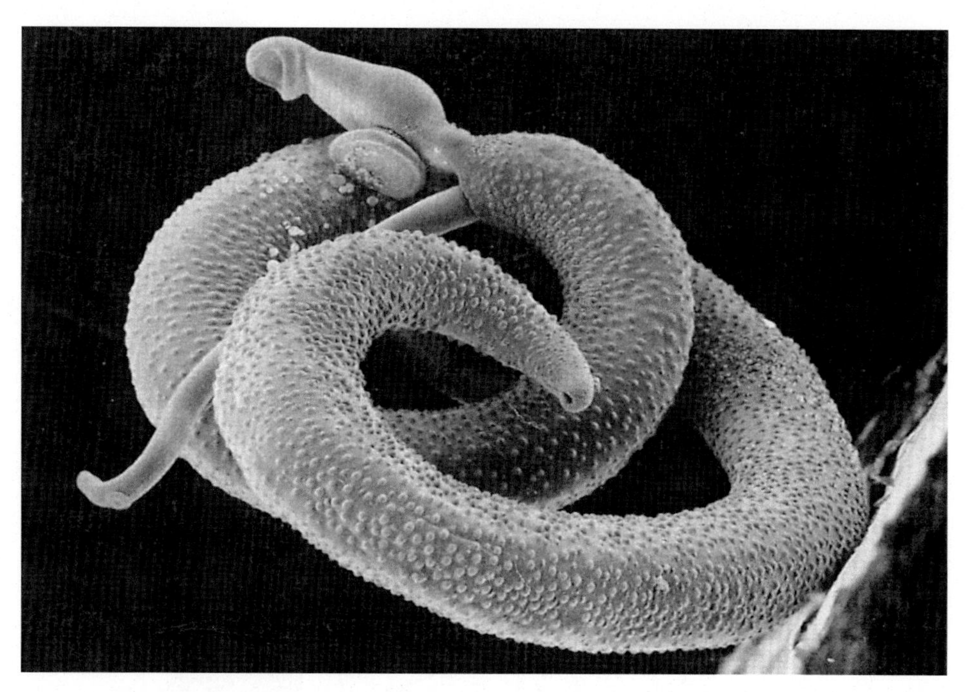

FIGURE 11.4 **Scanning electron micrograph of *Schistosoma mansoni* adult worms in copula.**

Adult schistosomes reside in mesenteric veins that drain the intestine (*S. mansoni* and *S. japonicum*) or in vesicular veins serving the urinary bladder (*S. haematobium*). In single-sex infections, the sexual organs of female worms are underdeveloped, leading to the hypothesis that one or more male factors are essential for complete maturation of the female.

The female usually migrates to smaller venules before depositing eggs. The morphology of the egg is distinctive in each species and serves as a diagnostic criterion (Fig. 9.16). The enclosed miracidium is poorly developed at time of oviposition, but is well formed before it reaches the lumen of the infected organ. The egg must penetrate the venule endothelium and then traverse the intervening tissues and mucosal lining before entering the lumen of the gut or the bladder to escape to the outside. The method by which the egg passes through the tissues remains speculative but probably involves the secretion of hydrolytic enzymes emitted through the porous shell. The secretions induce an immune response from the host which, in turn, produces a granulomatous response. The granuloma consists of a number of motile cells (e.g., macrophages, eosinophils) which aids in the movement of the granuloma through the venule endothelium to the lumen of the intestine or bladder. The granuloma is dispersed releasing the egg to the lumen of the infected organ. The process is obviously inefficient, since only about one-third of the eggs produced reach the exterior; the remaining eggs are either trapped in the urinary bladder or intestinal walls or are swept back by the blood flow to become lodged in ectopic sites such as the liver (Fig. 11.6) and, occasionally, the spleen and other tissues. After reaching the intestinal lumen, the egg passes to the exterior in either feces or urine.

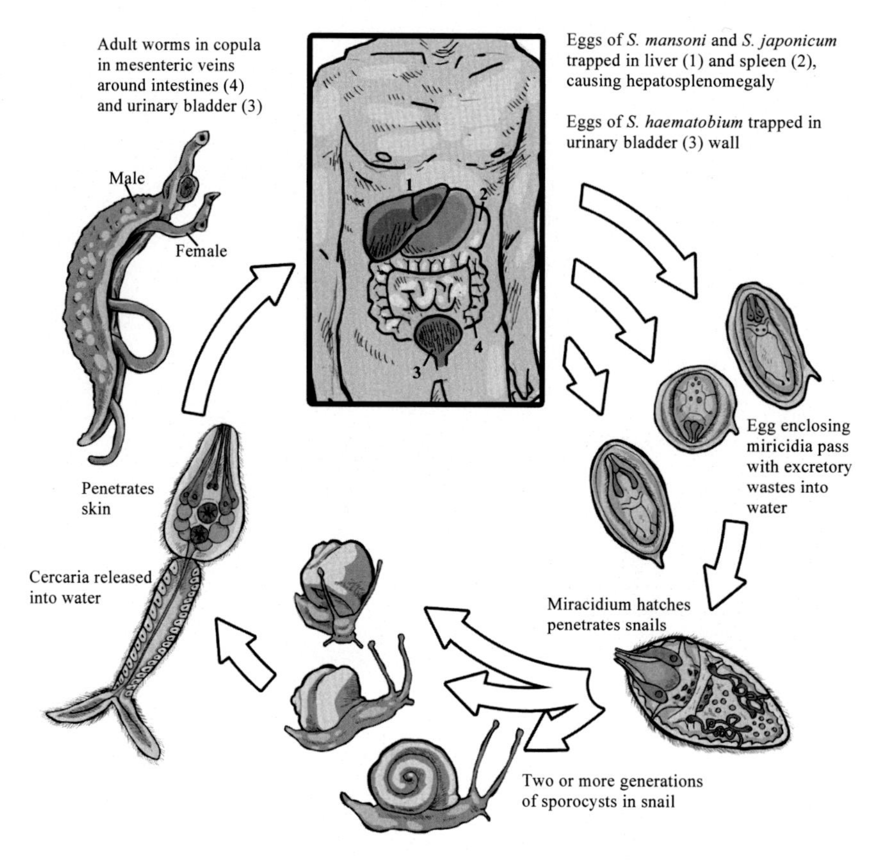

FIGURE 11.5 **Life cycles of Schistosoma spp.** *Credit: Image courtesy of Gino Barzizza.*

Upon reaching fresh water, the egg escapes the inhibitory osmolarity of the host's body fluids, thereby activating the miracidium to hatch. Since schistosome eggs have no operculum, hatching occurs through a rupture of the eggshell along a line known as the **suture**.

The free-swimming miracidium (Fig. 11.7) must penetrate a suitable snail intermediate host within a few hours after hatching or it dies. After penetration, the miracidium transforms into a sporocyst in the head foot of the snail. A second generation of migratory sporocysts is produced (see Fig. 11.8) which move to the digestive gland or gonads where it either reproduce additional generations of sporocysts or give rise to the cercarial generation. The cercariae leave the sporocyst in which they have developed via a birth pore and pass through the tissues of the snail to the exterior. This passage is facilitated by secretions from a pair of **escape glands** located in the cephalic region of the cercaria (Fig. 11.9A).

Actively swimming cercariae possess distinctive forked tails and move in a figure-eight pattern characteristic of schistosomes. They may swim upward to the surface of the water and then sink slowly toward the bottom, or they may adhere to the surface film and come to rest awaiting contact with their next host. The cercariae are stimulated to attach and penetrate by the secretions of the mammalian skin. In fresh water, a mucoid surface coat protects the free-swimming cercariae from the hypoosmolarity of the environment. Cercariae have five

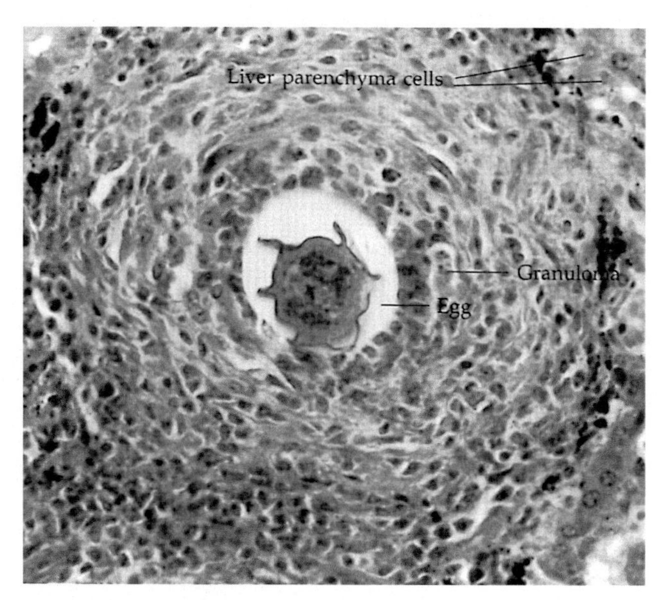

FIGURE 11.6 ***Schistosoma mansoni*** **egg in liver granuloma.**

pairs of unicellular glands. Two of these, the **preacetabular glands**, are anterior to the ventral sucker, while the other three pairs, the **postacetabular glands**, lie behind the ventral sucker. Each gland cell is equipped with a duct that empties separately at the anterior margin of the oral sucker (Fig. 11.9B).

Once the cercaria enters the skin, it burrows to the peripheral capillary bed or enters the lymphatic system; in either case, the worm migrates to the right side of the heart and then enters the lungs. During the penetration process, three significant morphological changes occur in the cercaria: the tail is lost, the surface coat is lost, and the contents of the penetration glands are spent. Following these changes, the transformed cercaria is called a **schistosomule** (Fig. 11.10).

Schistosomules appear in pulmonary capillaries by the 3rd day postpenetration. On day 4, these juveniles begin feeding on host erythrocytes, initiating a period of rapid growth and development (Fig. 11.11). The period spent in the host's lungs varies even within the same schistosome species. After a week to 10 days, the schistosomules move through the pulmonary vein to the left side of the heart and then into the systemic circulation. Approximately 3-weeks postpenetration, the worms reach the hepatic portal veins, where they reach sexual maturity and mate after 40 days. Males with females enclosed in their gynaecophoric canals then migrate against the portal flow to venules at the definitive sites. The sex of the worms is genetically determined at the time of fertilization.

VARIATIONS

While the morphology and life cycles of the three major schistosomes are basically similar, there are certain clinical differences that are useful in diagnosis.

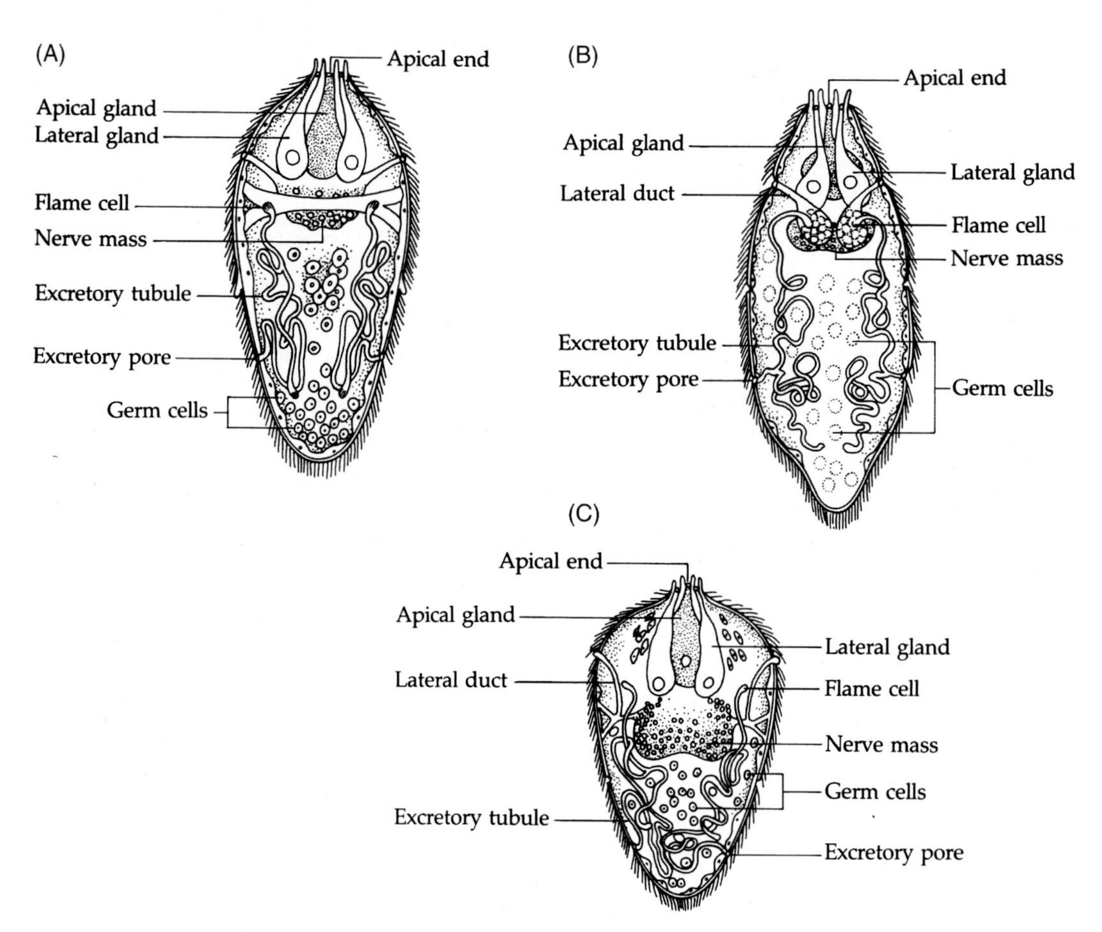

FIGURE 11.7 **Miracidia of three major schistosomes that infect humans.** (A) *Schistosoma haematobium.* (B) *S. mansoni.* (C) *S. japonicum.*

Schistosoma Haematobium

In India and Portugal, intermediate hosts for *S. haematobium* belong to the snail genera *Ferrissia* and *Planorbarius*, respectively; in all other major endemic areas, from north to south Africa (particularly the Nile valley), central and west Africa, and a number of countries in the Middle East, several species of the snail genus *Bulinus* serve as intermediate hosts. The male worm may attain a length of 15 mm, while females may reach 20 mm. There are 4 or 5 testes in the male; in the female, the single ovary is situated at about the midpoint of the body. The tegument of the male has many knoblike tubercles on the dorsal surface, while the tegument of the female is smooth. In both sexes, the caeca reunite posteriorly at a point about two-thirds of the way down the length of the body.

Females deposit about 30 eggs daily, each egg measuring 112–170 μm long by 40–70 μm wide. The *S. haematobium* egg is readily identifiable by its small, distinct terminal spine (Fig. 9.16). Intramolluskan development requires approximately 4–6 weeks after penetration by the miracidium. The prepatent period for *S. haematobium* in the human host is 10–12 weeks.

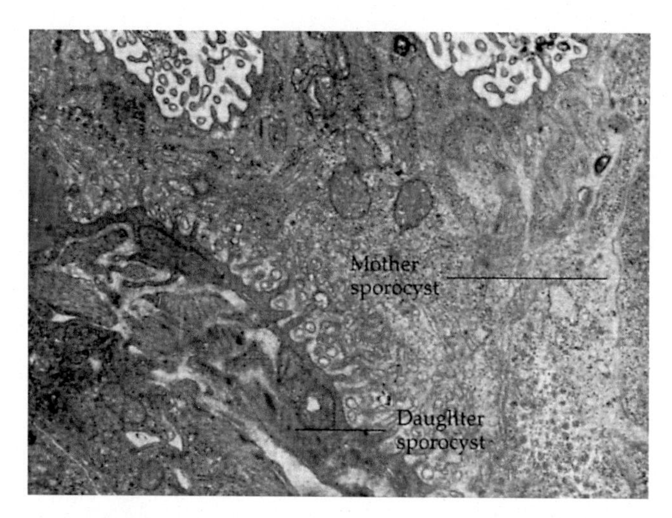

FIGURE 11.8 **Transmission electron micrograph of the tegument of a mother sporocyst with enclosed daughter sporocyst of *Schistosoma mansoni*.**

Schistosoma Mansoni

This species occurs widely throughout Africa and South America, especially in Brazil, Venezuela, Surinam, and Guyana, and on several Caribbean islands, including Puerto Rico, St. Lucia, Martinique, and Guadalupe. It may have been brought to the Western Hemisphere during the time of the African slave trade when a number of susceptible snail hosts were introduced, possibly in casks of drinking water accompanying the infected slaves.

The male worm measures up to 10 mm in length; the female, up to 14 mm. As in *S. haematobium*, the tegument of the male has tubercles on the dorsal surface, while that of the female is smooth. The male has 6–9 testes, while in the female a single ovary is situated in the anterior half of the body. Female worms deposit 190–300 eggs daily, each measuring 114–175 μm long by 45–68 μm wide and bearing a prominent, lateral spine (Fig. 9.16). Many species of the genus *Biomphalaria* are suitable snail hosts in endemic areas throughout the world except in Brazil where members of the snail genus *Tropicorbis* serve as intermediate hosts. Intramolluskan development requires 3–4 weeks, and the prepatent period in humans is 7–8 weeks.

Schistosoma Japonicum

This schistosome is found in the far Eastern countries of China, Japan, and the Philippines, Taiwan, and Indonesia. The snail host belongs to the genus *Oncomelania*. Adult worms are the largest schistosomes infecting humans, with males attaining a length of 20 mm and females 26 mm. The tegumental surface of both sexes is smooth. The male has 7 testes, while the female has a single ovary lying in the posterior half of the body. The relatively long uterus may contain as many as 300 eggs at any given time. The female is a prodigious egg producer, capable of depositing 3500 eggs per day, and, unlike those of *S. haematobium* and *S. mansoni*, the eggs are deposited in clusters, rather than singly. The eggs of *S. japonicum* are the smallest of the three species, measuring 70–100 μm long by 50–65 μm wide, and may bear a minute, lateral spine (Fig. 9.16). Intramolluskan development requires 4–9 weeks, while the prepatent period in the human is 5–6 weeks.

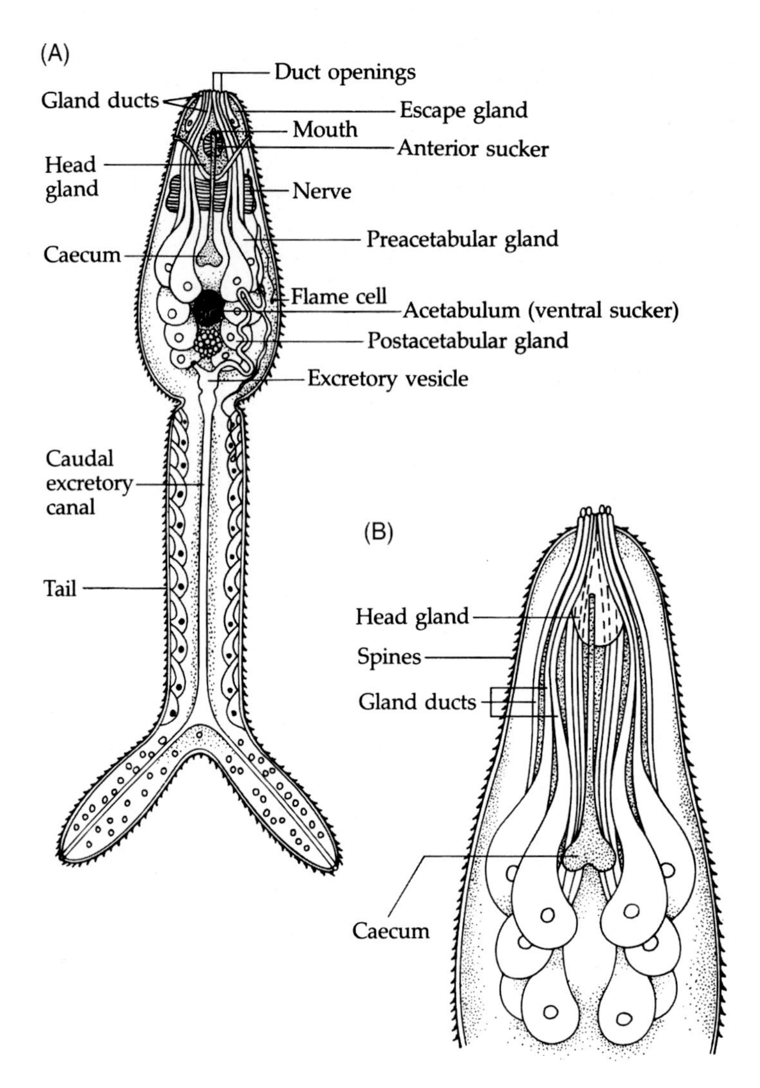

FIGURE 11.9 **Cercaria of a schistosome.** (A) Entire cercaria. (B) Anterior end showing head gland.

SYMPTOMATOLOGY AND DIAGNOSIS

The first symptom of schistosomiasis is a localized dermatitis, often appearing after cercariae penetrate the skin. The characteristic itching and local edema usually disappear after 4 days. Following skin penetration, the symptoms of human schistosomiasis appear in three phases. The first is the migration phase, characterized by toxic reactions and pulmonary congestion accompanied by fever. This phase may last 4–10 weeks, during which the worms migrate from the lungs to the liver where they reach sexual maturity and mate. Mated pairs of *S. mansoni* and *S. japonicum* then migrate to the mesenteric veins via the hepatic portal system. The second phase is considerably longer, lasting 2 months to several years. Characteristic

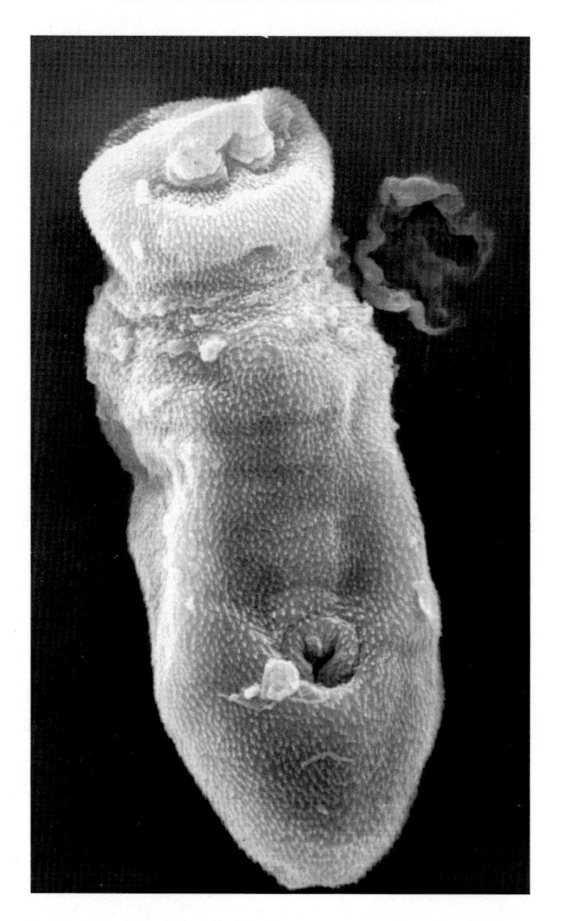

FIGURE 11.10 **Scanning electron micrograph of *Schistosoma mansoni* schistosomule.**

symptoms, such as bloody stools (*S. mansoni* and *S. japonicum*) and hematuria (*S. haematobium*), are caused by the passage of eggs through the intestinal and urinary bladder walls. Pathological alteration of these organs accompanies these incursions. The last phase, the most serious, is characterized by severe intestinal, renal, and hepatic pathology, caused primarily by the reaction of the host to the schistosome eggs.

Hepatosplenomegaly (enlargement of the liver and spleen) is a common symptom of advanced schistosomiasis. Eggs trapped in the walls of the intestine and urinary bladder as well as in ectopic regions, notably the liver and spleen, elicit inflammatory reactions due to leukocytic and fibroblastic infiltration, which produce cirrhosis, anemia, etc. Eventually, a **granuloma** (pseudotubercle) forms around each egg or cluster of eggs (Fig. 11.6). Small abscesses, accompanied by occlusion of small blood vessels, lead to necrosis and ulceration. The most severe consequences result from an increase in portal blood pressure as the liver becomes fibrotic and filled with blood. Fluid accumulation in the peritoneal cavity and the formation of new blood vessels bypassing infected organs such as the liver are usually associated with these changes. The latter are subject to bursting, which may lead to life-threatening bleeding. In endemic areas, where reinfection is common, repeated penetration of the intestinal and urinary bladder walls by migrating eggs results in excessive scar tissue formation, which impedes the normal functioning of these organs. For example, *S. haematobium* infections alter absorptive properties of the urinary bladder wall, predisposing it to malignancy. In Egypt, for example, *S. haematobium* infection is commonly associated with bladder cancer among male agricultural workers. Because formation of scar tissue also blocks the migration of eggs through infected organs, more eggs are swept back to other sites, producing organ enlargement (e.g., hepatosplenomegaly).

The surest means of diagnosis is finding and identifying characteristic eggs in excreta or in tissue biopsies, particularly rectal biopsies. In chronic cases in which very few, if any, eggs are passed, biopsies can be used advantageously, but these are expensive and require the services of trained specialists. The development of inexpensive but reliable diagnostic procedures adaptable to primitive conditions is clearly needed. Currently, the most promising diagnostic method utilizes immunodiagnostic techniques; however, positive results from these tests should be confirmed by identification of eggs since false positives sometimes result from concomitant infections with other parasites or exposure to various animal schistosome cercariae. The latter sometimes produces a severe dermatitis called **swimmer's itch** (Fig. 11.12).

CHEMOTHERAPY

No reliable prophylactic regimen is presently available other than the observance of proper hygiene and sanitation procedures, avoidance of cercaria-infested waters, and prevention of water contamination by human excreta. The chemotherapeutic agent recommended for all species of human schistosomes is praziquantel (see p. 172). In the case of schistosomes, this drug, by disrupting the integrity of the tegument, apparently exposes otherwise inaccessible antigens as targets for host antibodies. Unfortunately, administration of praziquantel does not prevent reinfection and it also displays varying degrees of activity among different stages of the worm's life cycle (e.g., the effects of the drug are not discernible until egg

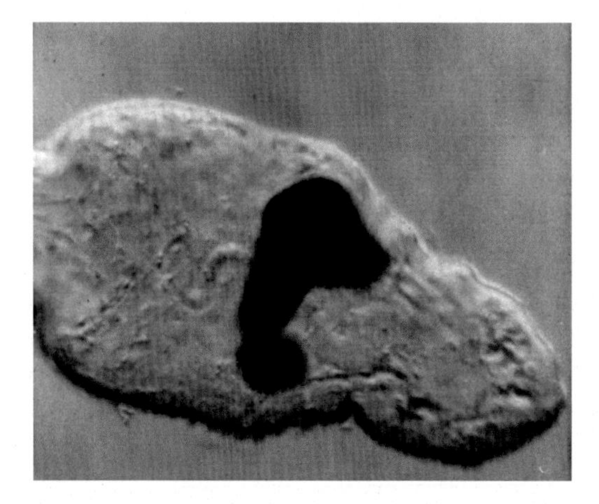

FIGURE 11.11 **Light micrograph of a feeding schistosomule.** Note the black pigment derived from host hemoglobin in the digestive tract.

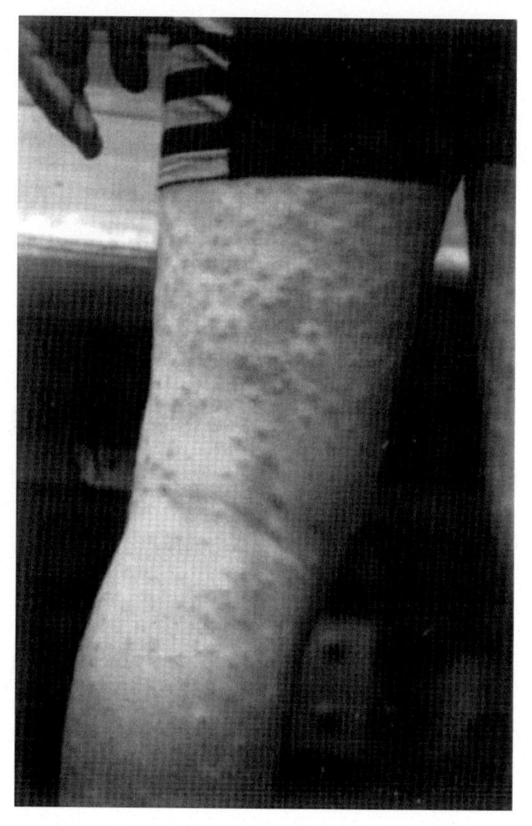

FIGURE 11.12 **Swimmer's itch.** Note the inflammatory reaction on the leg.

production). Evidence has also surfaced that repeated treatment is causing the development of drug resistance. As we have noted, drug resistance is a continual problem in the study of host/parasite relationships and resulting parasite-caused diseases.

In a quest for alternative drugs, oxamniquine was introduced in 1972. However, its use is limited since it is only effective against *S. mansoni* infections. Its usefulness, therefore, is confined to the Americas (e.g., Brazil) where *S. mansoni* is the only schistosome causing disease. The preference for the more expensive oxamniquine in Brazil is primarily due to the fact that the medical community is concerned that praziquantel pills are too large for children to swallow. Oxamniquine action is similar to that of praziquantel, and it also has presented resistant strains. Oxamniquine activity relies on a protein (Smp) translated from the parasite's chromosome 6 locus (QTL). Smp has been identified as a sulfotransferase which converts the drug to its active form. In drug resistant schistosomes, a mutation altering the binding site of the enzyme annuls activation of the drug and its usefulness in treatment of *S. mansoni* infections. As noted previously (see page 9), finding a molecular cause of drug resistance is a major step in finding a solution to the problem by either modulating treatment at the population level or establishing new drugs.

Research efforts to find new targets for drug intervention, to improve existing drugs, and to synthesize new and more effective ones must continue, at least until a vaccine becomes available. For instance, the antimalarial drug, artemisinin and its derivatives, has shown anti-schistosomicidal affects against all life cycle stages.

According to the World Health Organization, the key to eventual schistosomiasis control lies in a four-pronged attack: population-based chemotherapy; drug administration to infected individuals, use of molluskicides; introduction of biological controls, such as carnivorous snails and fish; and education of the population.

GENOMIC STUDIES

Since 2015, the completely sequenced genomes of approximately 2 dozen helminths have been published with *S. mansoni* being one of the first of the flatworms, so analyzed. The progress in using genomic technology related to the schistosomes is lagging the advances that have been made related to the Protista (e.g., *Plasmodium* spp.). One of the major factors for this discrepancy is the size of the schistosome genome, the genome of *S. mansoni* is at least 10 times larger than that of *P. falciparum*.

One of the techniques that is being utilized to circumvent this difficulty is **RNA interference** (**RNAi**). This procedure uses double-stranded RNA molecules to silence gene expression by neutralizing mRNA molecules. The technique has been used in the study of parasites either to determine gene function or to target proteins that are involved in a parasite's physiology. In so doing, viable targets may be exposed for drug or vaccine intervention. For instance, in schistosomes, RNAi has been used to inhibit the expression of the parasite's enzyme cathepsin B which is essential for normal growth. RNAi has also been used to inhibit the expression of the gynecophoral canal protein. This protein, which peaks at 28 days postinfection in male worms, is an important component in the mating process.

The gene that codes for the protein SminAct is expressed in the vitelline glands and reproductive organs of female schistosome worms and is involved with pairing of male and female worms as well as with egg formation. Since eggs are one of the primary causes of pathology in schistosome infections, inhibition of SminAct, causing abortion in eggs, would provide a means of reducing the pathology.

HOST IMMUNE RESPONSE

Immunity to human schistosomiasis is not totally understood. It is known that some animals experimentally infected with schistosomes normally infective to humans develop immunity to these parasites. The life span of adult schistosomes in the human host can be more than 30 years. However, worms do not accumulate with age even after continual exposure to the infective stage. Instead, in endemic areas a distinct peak of infection intensity occurs during early teenage years, after which the number of worms are reduced without being completely eliminated. As noted earlier, during much of their lifetime the worms are prolific egg producers, and this output over such a long period elicits a wide range of immune responses, both humoral and cell mediated. In fact, such responses can be correlated with various parasitic stages in the human host. The first, or skin penetration stage, is characterized by an antibody reaction to the penetrating cercaria's surface coat and penetration gland secretions, both released into the host's tissues. During the second, or early development stage, there is a response caused by the tegumental changes in the migrating schistosomule accompanied by a marked increase in Th1-like responses. The third, or adult worm and egg stages, precipitates a response to immunogens released from the adult worm's intestine, tegument, excretory system, and egg production. These activities are marked by a decrease in Th1 responses and the development of a strong Th2 response induced primarily by the eggs. The reaction to eggs is in response to immunogens released from either migrating or trapped eggs. The immunogens are primarily glycosylated macromolecules associated with the egg surface or those that diffuse through micropores in the eggshell. While some of these macromolecules facilitate migration of the eggs through the tissues, they also elicit a granulomatous response to eggs trapped in the tissues, causing pronounced pathological changes. The granulomatous response, which is part of the chronic phase of the disease, is orchestrated by $CD4^+$ T-cell activity induced by the egg antigens. During this phase, modulation of the Th2 response occurs with a concomitant reduction in the size of newly formed granulomas and an increase in fibrosis under the influence of interferon-γ (IFN-γ) and interleukin-13 (IL-13). Fibrogenesis is promoted by the Th2 cytokine IL-13 as a consequence of prolonged Th2 responses. However, the Th1 proinflammatory responses persist throughout the chronic phase. The pathology of schistosomiasis in humans is firmly associated with severe periportal fibrosis rather than with a large number of egg granulomas.

The question may well be asked: "If worms produce immunogens to which the host responds, how do the worms evade this response?" While there is not yet a complete answer to this question, several significant factors provide clues. Perhaps the most remarkable of these is the ability of the worm to acquire protective host antigens (e.g., host proteins) on its surface. The antigens afford protection by disguising the worm's surface so that it escapes detection by the host immune mediators. Antigen acquisition apparently begins during the early schistosomule stage and may be associated with the initiation of feeding on host blood. One such antigen has been identified as the soluble serum protein beta-2 macroglobulin. Beta-2 macroglobulin binds to the parasite's surface protein, paramyosin, which is responsible for bonding host serum proteins. Logically, therefore, an effective vaccine must target the larval stage before it attains this protection.

OTHER SCHISTOSOMES

Since the mid-20th century, new discoveries of metazoan parasites pathogenic to humans have been rare. There were, for many years, pockets of what were thought to be *S. japonicum* infections in Southeast Asia, especially in parts of Laos and Cambodia. However, American involvement in Southeast Asia during the Vietnam War prompted a reevaluation of the causative organism for schistosomiasis in that region. As a result, that parasite is now considered to be a separate species, *Schistosoma mekongi*. While it closely resembles *S. japonicum* in pathology and morphology, there are significant differences: the molluskan intermediate host is a minute snail, *Tricula aperta*, the eggs are smaller, and the prepatent period is a week longer than for *S. japonicum*.

Another schistosome species, *Schistosoma intercalatum*, is known to cause human schistosomiasis in Cameroon and Zaire in Africa. In Cameroon, *Bulinus forskalii* serves as the molluskan intermediate host; in Zaire, *Bulinus globosus* does so. Because of the terminal spine on its eggs, *S. intercalatum*, normally a blood fluke of cattle, is generally considered more closely related to *S. haematobium*. However, the eggs, like those of *S. mansoni*, are voided in feces rather than urine. *S. intercalatum* is also less pathogenic to humans than *S. haematobium*.

SWIMMER'S ITCH

An interesting phenomenon of schistosome biology is cercarial dermatitis, or **swimmer's itch**. While the condition is not life threatening, it can have a negative impact on the economy of regions where outbreaks occur, especially those popular with tourists. A number of lake and seashore resorts in MI, MN, NJ, NE, NC, WI, MA, CT, and Canada have suffered economic losses when outbreaks have driven vacationers away. The condition is caused when cercariae of blood flukes that normally parasitize aquatic birds and mammals penetrate human skin, sensitizing points of entry and causing pustules and an itchy rash. Since humans are not suitable definitive hosts for these flukes, the cercariae do not normally enter the blood stream and mature. Instead, after penetrating the skin, they are destroyed by the victim's immune responses. Allergenic substances released from dead and dying cercariae produce a localized inflammatory reaction (Fig. 11.12).

In freshwater lakes of North America, cercariae of the genera *Trichobilharzia*, *Gigantobilharzia*, and *Bilharziella*, which normally infect birds, are the common dermatitis-producing schistosomes, while the mammal parasite, *Heterobilharzia*, is the culprit in Gulf Coast states. One of the most common causative agents of marine swimmer's itch on both the east and west coasts of North America is *Microbilharzia variglandis*, a blood fluke of sea gulls, the cercariae of which develop in the mudflat snail, *Ilyanassa obsoleta*. Other marine genera implicated in swimmer's itch are *Austrobilharzia* and *Ornithobilharzia*, which use members of the snail genera *Littorina* and *Batillaria*, respectively, as intermediate hosts. Swimmer's itch is by no means confined to North America; outbreaks have also been reported in Asia, Africa, Europe, and the Middle East.

Suggested Readings

Basch, P. F., & Samuelson, J. (1990). Cell biology of schistosomes. I. In D. J. Wyler (Ed.), *Modern parasite biology. Cellular, immunological, and molecular aspects ultrastructure and transformations* (pp. 91–106). New York: W. H. Freeman.

Bogitsh, B. J. (1995). The feeding of A type red blood cells *in vitro* and the ability of *Schistosoma mansoni* schistosomules to acquire A epitopes on their surface. *Journal of Parasitology, 79*, 946–948.

Butterworth, A. E. (1990). Immunology of schistosomiasis. In D. J. Wyler (Ed.), *Modern parasite biology. Cellular, immunological, and molecular aspects* (pp. 262–288). New York: W. H. Freeman.

Engels, D., Citsulo, L., Montresor, A., & Savioli, L. (2002). The global epidemiological situation of schistosomiasis and new approaches to control and research. *Acta Tropica, 82*, 139–146.

Foster, L. A., & Bogitsh, B. J. (1986). Utilization of the heme moiety of hemoglobin by *Schistosoma mansoni*. *Journal of Parasitology, 72*, 669–676.

Jordan, P., & Webbe, G. (1982). *Schistosomiasis: epidemiology, treatment and control*. London: Heinemann Medical Books Ltd, Wm.

Loker, E. S. (1983). A comparative study of the life-histories of mammalian schistosomes. *Parasitology, 87*, 343–369.

Pearce, E. J., & MacDonald, A. S. (2002). The immunobiology of schistosomiasis. *Nature Reviews Immunology, 2*, 499–511.

Popiel, I. (1986). The reproductive biology of schistosomes. *Parasitology Today, 2*, 10–15.

Redman, C. A., Robertson, A., Fallon, P. G., Modha, J., Kusel, J. R., Doenhof, M. J., et al. (1996). An urgent and exciting challenge. Praziquantel. *Parasitology Today, 12*, 14–20.

Rollinson, D., & Simpson, A. J. G. (1987). *The biology of schistosomes: from genes to latrines*. New York: Academic Press.

Smith, K. (2009). Sequencing unlocks secrets of blood parasites. *Nature News, 10*, 1038.

Stirewalt, M. A. (1974). *Schistosoma mansoni*: cercaria to schistosomule. *Advances in Parasitology, 12*, 115–182.

Todd, C. W., & Colley, D. G. (2002). Practical and ethical issues in the development of a vaccine against schistosomiasis mansoni. *American Journal of Tropical Medicine and Hygiene, 66*, 348–358.

CHAPTER

12

General Characteristics of the Cestoidea

This chapter describes the basic structural characteristics and physiology of the Cestoidea. Cestoidea is a class of parasitic flatworms, commonly called tapeworms, of the phylum Platyhelminthes. These parasites live in the digestive tract of vertebrates as adults, and often in the bodies of various animals as juveniles. A little over 1000 species have been described and all vertebrates can be infected by at least one species of tapeworm. Humans can be infected after consuming the parasite in underprepared meat such as pork (*Taenia solium*), beef (*Taenia saginata*), and fish (*Diphyllobothrium*). In addition, food prepared in conditions of poor hygiene can also lead to infection (*Hymenolepis* or *Echinococcus*). The general life cycles of these parasites is also described.

The Cestoidea, or tapeworms, being a class of parasitic flatworms, possess all the characteristics of the phylum Platyhelminthes presented in Chapter 9. The most striking difference between members of this class and the class Trematoda is that tapeworms lack a mouth and digestive tract.

The Cestoidea are the most highly specialized flatworm parasites known. Adults of this class are endoparasitic in the alimentary tract and associated ducts of various vertebrates, including humans; the larvae, on the other hand, infect both vertebrates and invertebrates. The life cycle requires one or two intermediate hosts, in each of which the tapeworm undergoes a specific developmental phase.

Tapeworms were first written by Hippocrates and Aristotle about 300 BC and are among the first known parasites of humans recorded. There is much speculation about the origin and phylogeny of tapeworms. One evolutionary scheme proposes that they arose from a stock of aquatic, free-living, bottom-dwelling protomonogeneans that, in turn, evolved from a rhabdocoel-like ancestor similar to the ancestral form suggested for digenetic trematodes. The immediate ancestors of modern tapeworms evolved adhesive organs that enabled them to become attached to, and subsequently ectoparasitic upon, bottom-dwelling vertebrates. Some of these ectoparasitic forms migrated internally to the gut of these vertebrates and became endoparasitic. To survive in that hostile environment, these organisms evolved protective modifications, such as a glycocalyx on the body surface and resistant, quinone-tanned eggshells. Such modifications enabled them to resist the actions of the hosts' digestive enzymes. They also underwent physiological adaptations that allowed them to survive in an environment with reduced oxygen tension. At least one branch of these essentially monozoic

Human Parasitology. http://dx.doi.org/10.1016/B978-0-12-813712-3.00012-6

animals evolved additional modifications, such as the loss of the gut, development of anterior attachment organs, and duplication of reproductive systems, the last feature leading eventually to segmentation of the body. Later modifications, as the group became increasingly diverse, included adoption of intermediate hosts and the appearance of **apolysis**, the release of gravid (egg filled) body segments from the chain to the exterior. A causal relationship appears certain between the development of apolysis and the loss of capacity to form resistant eggshells by tanning, hence, loss of protection from host digestive enzymes, although there is a "chicken or egg" kind of question as to which occurred first.

MORPHOLOGY

The body of the typical adult tapeworm consists of three distinct regions: **scolex**, **neck**, and **strobila** (Fig. 12.1). The scolex, located at the anterior end, is the attachment terminal, the morphology and dimensions of which are key features in identification of these worms. The neck, an unsegmented, poorly differentiated region immediately posterior to the scolex, is generally the narrowest part of the worm. It is from the neck region that new segments, or **proglottids**, differentiate. As new proglottids are formed in the neck region, they push the older ones progressively posteriad, creating a chain of proglottids, the strobila, that constitutes the third body region. The asexual process of forming segments is termed **strobilation** (Fig. 12.2). As each proglottid is shifted posteriad, its sexual reproductive system matures progressively; hence, the anteriormost proglottids have the least developed reproductive systems, while the more posteriorly the proglottids are located, the higher their level of development. This progressive maturity of the reproductive systems permits a loose subdivision of the strobila into regions of **immature**, **mature**, and **gravid** proglottids (Fig. 12.2). The reproductive organs in immature proglottids are visible but nonfunctional while those in mature proglottids are fully functional. At the posterior end of the strobila are the gravid (egg filled) proglottids. Often, the reproductive organs in gravid proglottids have atrophied. In **apolytic** species, gravid proglottids detach from the strobila and exit the body of the host with fecal wastes. In **anapolytic** species, eggs are released through a uterine (or genital) pore directly into the host's intestine and, subsequently, also are discharged to the exterior in feces. Most anapolytic tapeworms produce protective, tanned eggshells.

Tegument

The tegument of tapeworms (Fig. 12.3) is essentially similar to that of digeneans with a few notable differences. The surface of the tapeworm tegument bears specialized microvilli, known as **microthrices** (singular, **microthrix**), which project from the outer, limiting membrane of the tegument. The dimensions of these projections vary according to species and location on the strobila. Unlike typical microvilli, each microthrix includes an electron-dense, apical tip separated from the more basal region by a multilaminar plate. These tips, applied to the host's intestinal epithelium, not only provide resistance to the peristaltic movement of the intestine but also with each movement of the worm, agitate intestinal fluids in the immediate microhabitat, thus increasing accessibility of nutrient materials as well as flushing away waste products. Covering the entire surface of the tegument is a layer of carbohydrate-containing

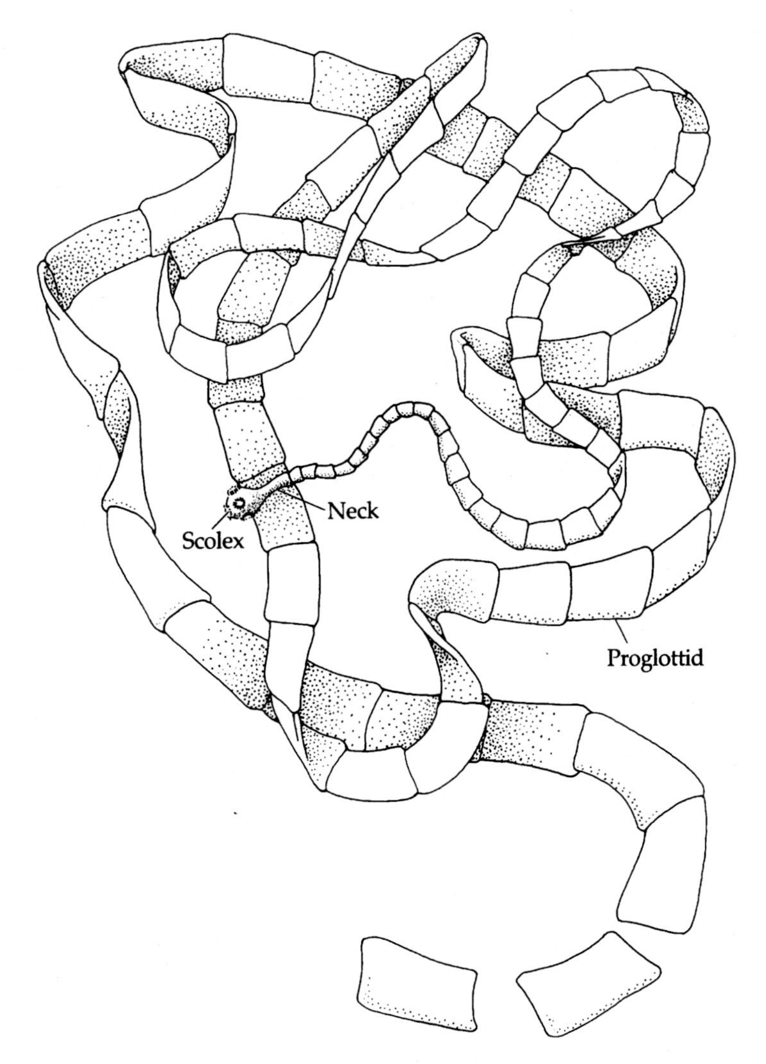

FIGURE 12.1 **The three major regions of a generalized eucestode.** Note the scolex and neck regions. The remaining region, made up of many proglottids, is the strobila.

macromolecules, the **glycocalyx**, which serves several important functions, among which are protecting the parasite from host digestive enzymes, enhancing nutrient absorption, and maintaining the parasite's surface membrane.

As in digeneans, the tegumental syncytium of tapeworms consists of two cytoplasmic regions, **distal** and **proximal**. The distal cytoplasm is replete with mitochondria (usually aligned in a broad, basal band), as well as several types of vesicles and scattered membranes. Glycogen granules are also present in this region in some species. The vesicles arise in the nucleated, proximal cytoplasm, or **cyton**, sunk deep in the parenchyma. The cyton contains Golgi complexes, mitochondria, rough endoplasmic reticulum, and other organelles involved in protein synthesis and packaging. The mechanisms involved in protein synthesis and

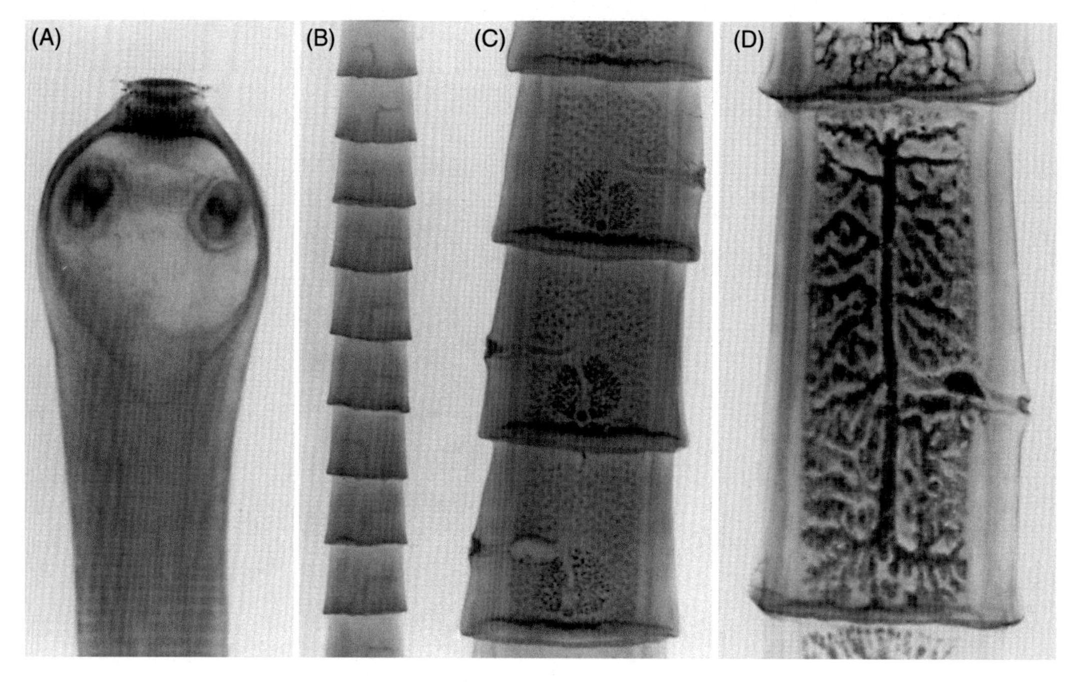

FIGURE 12.2 **Regions along the length of a cestode, Taenia solium.** (A) Scolex and neck region. (B) Immature proglottids. (C) Mature proglottids. (D) Gravid proglottid.

packaging maintain the distal cytoplasmic components, that is, materials so synthesized are translocated, via cytoplasmic connectives abetted by microtubules, to the distal cytoplasm where they maintain the glycocalyx, membranes, microthrices, etc. Underlying the distal cytoplasm are two layers of muscles, collectively known as the **tegumental musculature**, consisting of an outer layer with its contractile fibrils oriented in a circular pattern and an inner layer with contractile fibrils oriented longitudinally.

Parenchyma

The space enclosed by the tegument—except for the portion occupied by reproductive organs, osmoregulatory structures, muscle fibers, and nervous tissue—is filled with a spongy tissue known as the **parenchyma**. In live tapeworms, fluid fills the spaces between parenchymal cells. Parenchymal cells are the primary sites for synthesis and storage of glycogen. There is speculation that a single category of cells, the myoblasts, gives rise to both the parenchyma and the musculature of most tapeworms.

Parenchymal Muscles

Parenchymal musculature, as distinguished from tegumental musculature, is unique to eucestodes. Bipolar muscle cells and fibers embedded in the parenchyma form a broad band that encircles each proglottid about midway between the outer surface and the central axis and divides the parenchyma into an outer **cortical** region and an inner **medullary** region. In addition to the dominant, longitudinally aligned, contractile myofibers that help stabilize the strobila against peristalsis in the host intestine, circular myofibers are also present.

Scolex

To facilitate attachment to the host's intestinal wall, tapeworms utilize several types of structures on their scolices, the most common of which are suckers. Muscles in the scolex make possible the holdfast action of this organ. The musculature of the scolex consists of sets

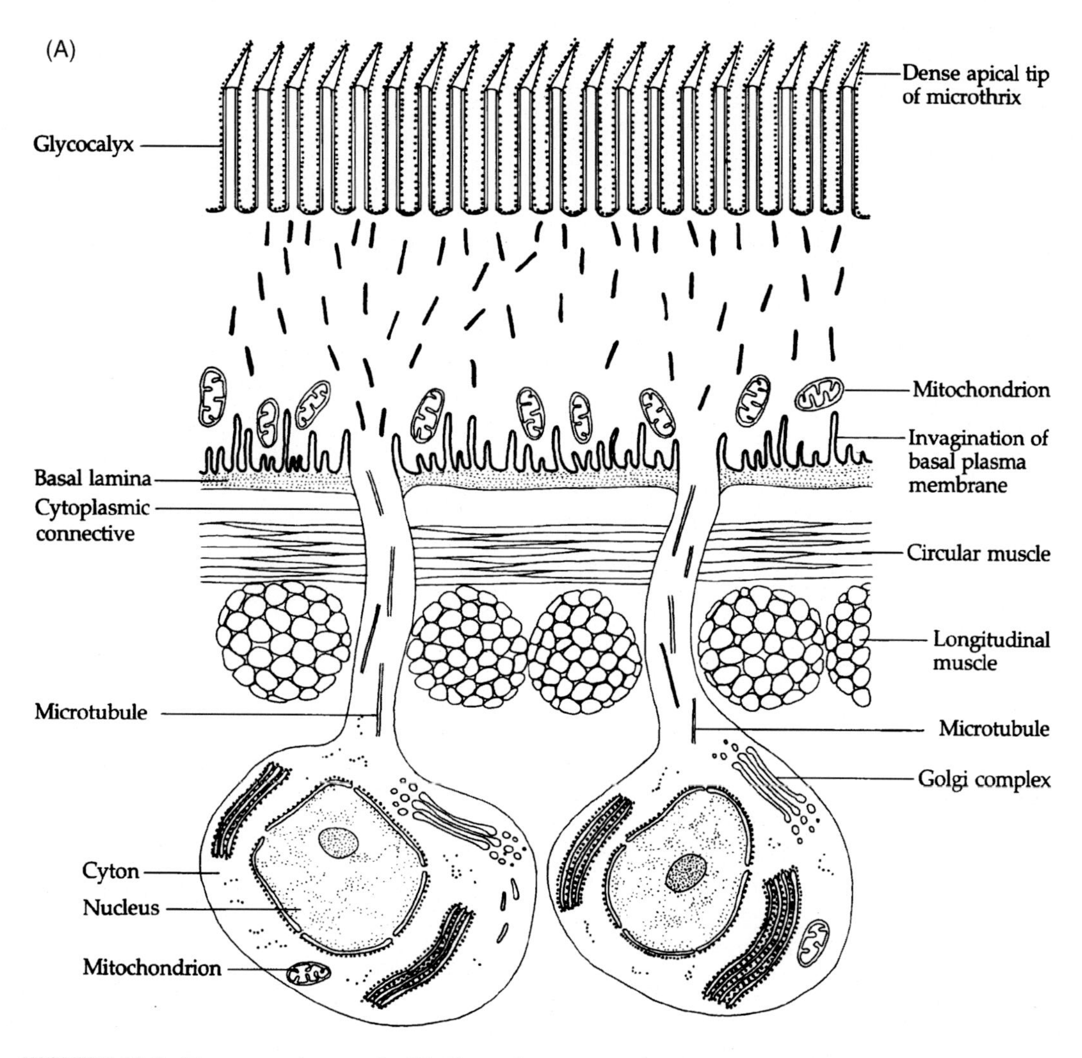

FIGURE 12.3 **Tegument of a cestode.** (A) The surface is covered with microthrices, each ending in a thickened, spinelike cap. As usual among parasitic flatworms, the cell bodies are secretory and produce surface coast constituents among other material. (B) Transmission electron micrograph of the tegumental surface showing microthrices.

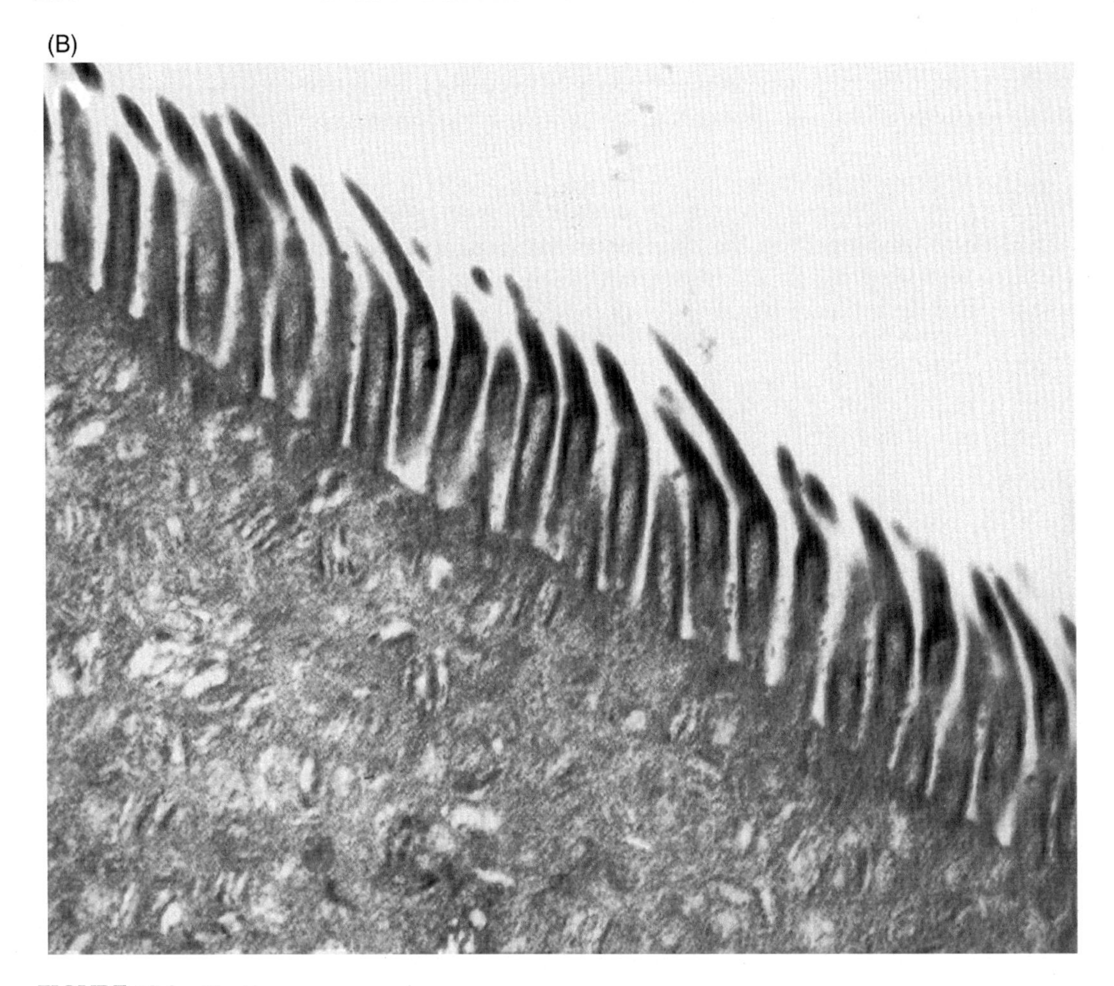

FIGURE 12.3 (*Cont.*).

of crisscrossing fibers attached to the inner surfaces of the suckers, enabling them to contract. Scolices of tapeworms that infect humans are categorized as either **acetabulate** or **bothriate**, depending on the type of sucker (Fig. 12.4).

An acetabulate scolex is characterized by the presence of four muscular cups sunk into the equatorial surface of the scolex (Fig. 12.4A). These cups are radially arranged equidistant from each other. While the rim of each cup is usually round, it may be oval or even slitlike in some species, and it may be flush with the surface or project beyond it. Each cup is covered by a thin layer of tegument continuous with that covering the remaining body of the worm. In addition to the muscular cups, there may be accessory holdfast structures, such as hooks, to help anchor the scolex to the host's intestinal wall, in which case the scolex is called an **armed scolex**. Such hooks usually are grouped at the apical end of the scolex on a protrusible **rostellum** (Fig. 12.4C). The presence, number, size, and shape of the hooks are of taxonomic importance.

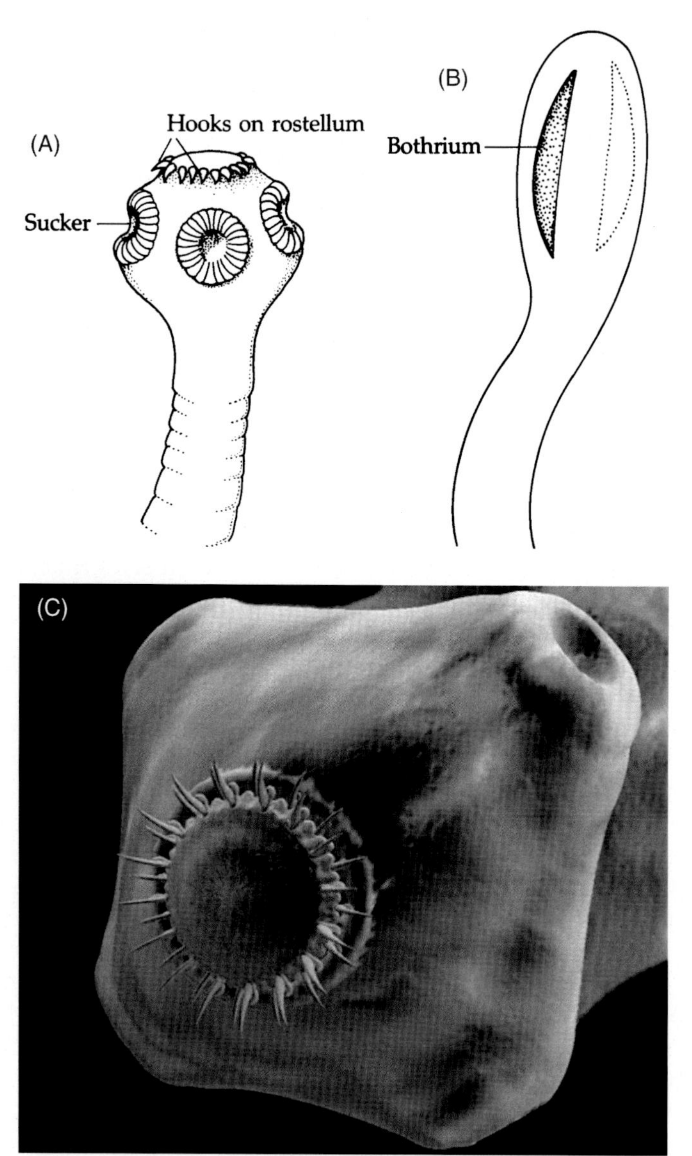

FIGURE 12.4 **Types of scolices found on tapeworms that infect humans.** (A) Acetabulate, showing three of the four suckers and an armed rostellum. (B) Bothriate, showing one of the two bothridial grooves. (C) SEM of *Taenia solium* scolex showing rostellar hooks.

A bothriate scolex is characterized by the presence of two, or rarely four or six, longitudinally arranged, shallow depressions called **bothria** (singular, **bothrium**) (Fig. 12.4B).

Various types of glandular secretions are associated with the scolices of many tapeworms. The function of these secretions has not been established with certainty, although it has been speculated that they are proteolytic, adhesive, and/or stimulatory, depending upon the species.

Calcareous Corpuscles

Large numbers of concretions, known as calcareous corpuscles, occur in the parenchyma of numerous cestode species as well as some trematodes. These spherical bodies, most noticeable in larval forms, consist of organic and inorganic components. The organic portion is composed of DNA, RNA, proteins, glycogen, glycosaminoglycans, and alkaline phosphatase; the inorganic portion is made up primarily of calcium, magnesium, phosphorus, and trace metals. While the functions of these inclusions remain unclear, among the theories advanced have been that they may act as buffers against anaerobically produced acids, serve as reservoirs for inorganic ions required during development, act as enzyme activators, or are a form of excretory product of metabolism.

Osmoregulatory System

The cestode osmoregulatory–excretory system is essentially the same as the flame cell, protonephritic type found in digeneans. In most cases, it serves to maintain within the worm an optimal hydrostatic pressure for extensory movements of the strobila and scolex. Some tapeworms, however, such as *Hymenolepis diminuta*, are known conformers, that is, they lack the ability to regulate osmotic pressure and therefore adapt or "conform" to the prevailing environmental conditions. In these species, the system appears to be strictly excretory. The morphology of the system varies somewhat among the different taxa, but sufficient similarity exists to justify the following generalized description.

The osmoregulatory–excretory system consists of two components: **collecting canals** and **flame cells**. Four laterally aligned collecting canals, two dorsal and two ventral, extend the entire length of the strobila (Fig. 12.5). All four canals lie just inside the medullary margin of the parenchyma, and a single transverse canal connects the ventral canals at the posterior end of each proglottid. The ventral canals carry fluid away from the scolex, the dorsal canals toward it. In some tapeworms, the four longitudinal canals are linked within the scolex by either a network of canals or a single ring vessel; in others, the dorsal and ventral canals on each side are linked by a simple connection in the region of the scolex, with no apparent exchange between the two sides.

In the terminal proglottid of young worms, there is an excretory vesicle into which the ventral canals empty. However, in older tapeworms that have sloughed the original posteriormost proglottid, the posterior ends of the ventral canals open independently to the exterior. Flame cells, usually arranged in groups of four, are associated with the ventral canals. Fluid collected by the flame cells passes through secondary tubules into the main canals. Analysis of fluid within the osmoregulatory system of certain species of tapeworm has revealed that it consists primarily of glucose, soluble proteins, lactic acid, urea, and ammonia. Reabsorption of essential molecules in this system has not been verified.

Nervous System

The cestode nervous system is relatively complex (Fig. 12.6). The "brain," located in the scolex, is a rectangular or circular arrangement of nerve tissue varying in complexity from a simple ganglion to a combination of several ganglia and commissures (Fig. 12.6). It gives rise to a system of short anterior and posterior nerves that richly supply various portions of the

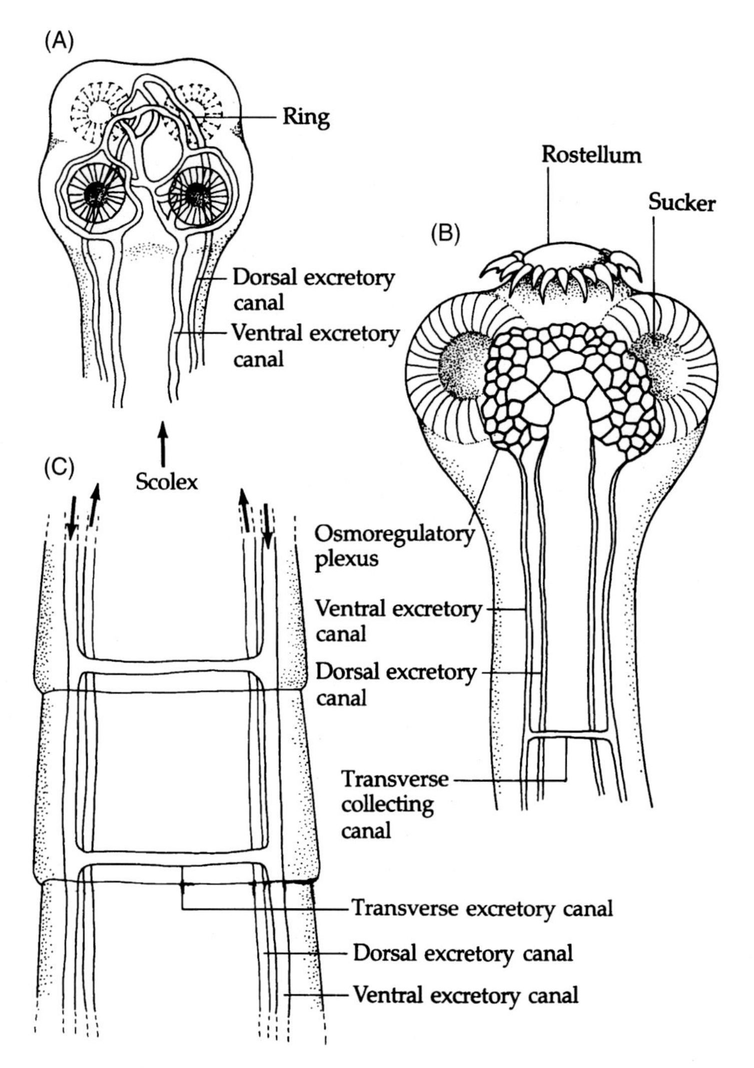

FIGURE 12.5 **Morphology of the osmoregulatory–excretory system of cestodes.** (A) Scolex of *Proteocephalus* sp. showing single-ring type of connection of the osmoregulatory canals. (B) Scolex of *Taenia* sp. showing network type of osmoregulatory plexus. (C) Proglottids showing longitudinal collecting canals. *Arrows* show direction of flow.

scolex with motor fibers and receive sensory fibers from rostellum, suckers, and tegument. Several pairs of longitudinal nerve cords extend posteriorly from this "brain" along the length of the strobila, lateral to the osmoregulatory canals. The cords are connected in each proglottid by cross-connectives, producing a ladderlike appearance. Small motor nerves emanating from the cords and cross-connectives innervate the reproductive organs and musculature, and small sensory nerves supplying the tegument merge with the cords and connectives. Certain organs of both the scolex and the proglottids, such as parts of the reproductive system and suckers, are more extensively innervated than others.

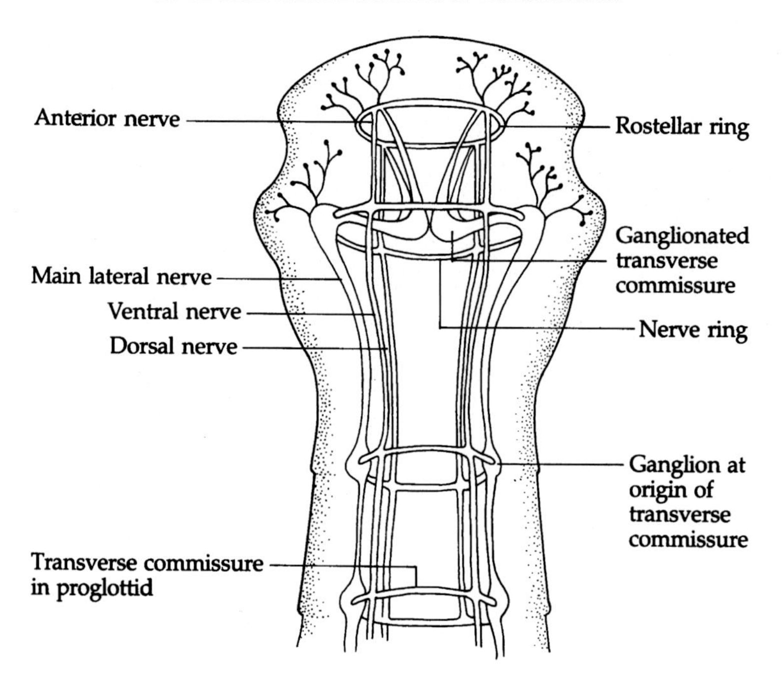

FIGURE 12.6 **Cestode nervous system.**

Reproductive Systems

The general pattern of the reproductive system of cestodes resembles that of digenetic trematodes except for the **cul-de-sac uterus** in some forms (cyclophyllideans), the presence of a separate vaginal canal, and often a laterally situated genital pore. A generalized description of the cestode reproductive system follows, with specific variations noted.

Male System

The male reproductive system consists of from one to many testes embedded in the medullary parenchyma of each proglottid (Fig. 12.7). Emanating from each testis is a single vas efferens; in cases of multiple testes, the vasa efferentia unite to form a common vas deferens, which is usually coiled. The distal portion of the vas deferens is modified as a muscular **cirrus**, usually enclosed within a **cirrus sac**. In some species, the cirrus is equipped with spines that hold the organ in place during copulation. The cirrus everts through the male genital pore, which then opens into the common **genital atrium**.

In most species there is an enlarged area of the vas deferens, the **seminal vesicle**, for storage of sperm. When located within the cirrus sac, it is designated an **internal seminal vesicle**; located outside the sac, it is termed an **external seminal vesicle**. Some species possess both.

Female System

Ova are produced in a single, sometimes bilobed ovary (Fig. 12.8). Following fertilization in the proximal portion of the oviduct, the resulting zygote passes into a region of the

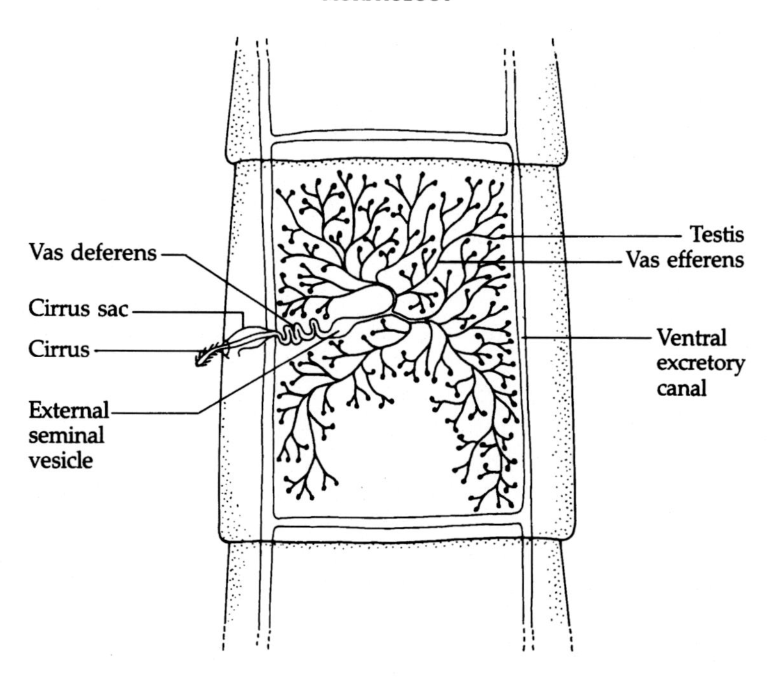

FIGURE 12.7 **Male reproductive system of a typical eucestode.**

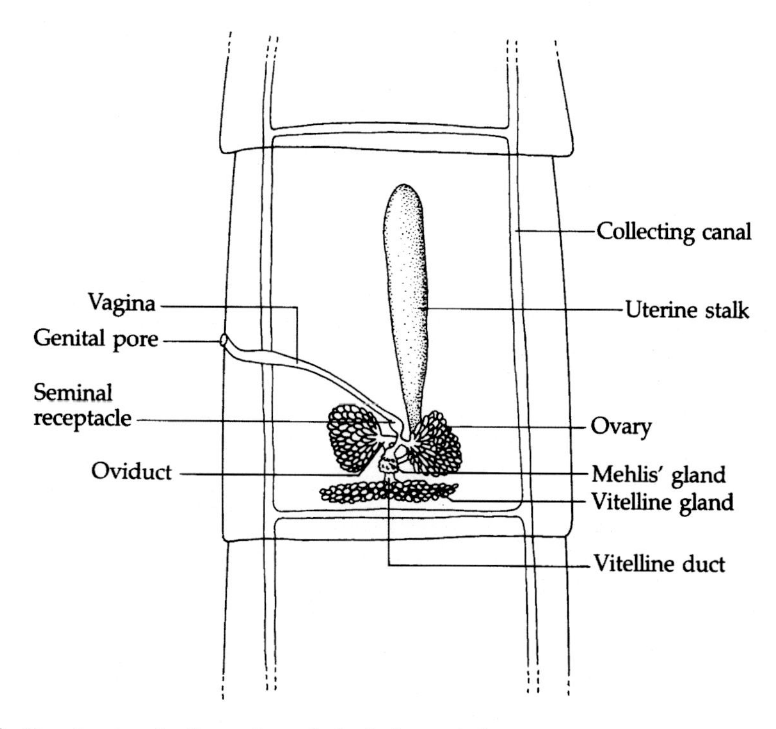

FIGURE 12.8 **Female reproductive system of a typical eucestode.**

oviduct, the **ootype**, equipped with structures involved in eggshell formation similar to those found in digeneans. A **Mehlis' gland** surrounds the ootype and secretes into it material essential to formation of the eggshell; a single, common **vitelline duct** enters the oviduct in the vicinity of the ootype. As in digeneans, the common vitelline duct is formed by the union of many **primary vitelline ducts** arising from vitelline glands, which vary in size and location according to species. Vitelline glands (collectively designated as the **vitellaria**) may form a compact body or consist of numerous follicles scattered throughout the medullary region of the parenchyma. With few exceptions, secretions of the vitelline glands contain shell precursors and provide nourishment for the developing larva. The uterine wall may also contribute materially to extraembryonic membranes and capsules.

The **vagina**, a tubular organ that joins the oviduct at the level of Mehlis' gland, provides a passage for sperm between the genital atrium and the oviduct. Fertilization occurs in the region where the vagina and oviduct join. Sperm are stored in an enlargement of the vagina known as the **seminal receptacle**. The oviduct continues as the uterus, which, in some tapeworms such as the anapolytic members of the order Pseudophyllidea, opens to the outside of the proglottid through a **uterine pore**. Eggs are produced continuously and are expelled through this opening. In other species, including members of the order Cyclophyllidea, the uterus is a blind sac in which developing eggs accumulate; the uterus becomes distended with eggs, filling the medullary region of the proglottid. The gravid proglottid eventually becomes detached from the strobila and is discharged from the host. In some tapeworms (*Dipylidium*, etc.), there is a modification of uterus–egg interaction in which a much reduced uterus, upon receiving a specific number of eggs, begins pinching off **egg capsules**, which eventually fill the medullary region of the gravid proglottid.

During copulation, the cirrus of one proglottid may be inserted in the vagina of another proglottid of the same, or another, worm. Cross-fertilization between two worms is advantageous, at least periodically, to insure vitality and prevent the development of deleterious features due to excessive self-breeding.

The Egg

The morphology of tapeworm eggs is important for species identification. The following is a brief general description of the basic parts of a typical egg (Fig. 12.9). The **oncosphere**, containing three pairs of hooks, is encased in an **inner envelope** that is surrounded by another

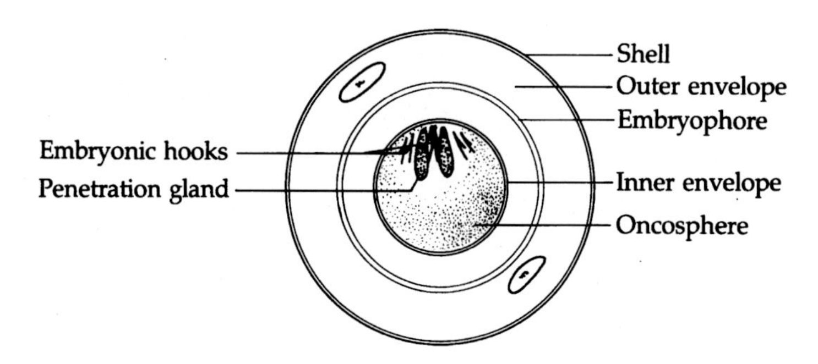

FIGURE 12.9 **General structure of a cestode egg and enclosed oncosphere.**

membranous structure, the **embryophore**. A cellular zone known as the **outer envelope** lies between the embryophore and the **shell** (or **capsule**), usually the outermost covering of the egg. Tapeworm eggs exhibit certain variations within this basic pattern and are classified into four types: pseudophyllidean, dipylidean, taenioid, and stilesian. Of these, all but the last are represented among tapeworms infecting humans.

The pseudophyllidean egg (Fig. 12.10A), of which the eggs of *Diphyllobothrium latum* are representative, is most similar morphologically and developmentally to those of digenetic trematodes. The fully developed egg has a thick, quinone-tanned shell, usually with a lid-like operculum at one end. Numerous vitelline cells are associated with the zygote, providing stored food for subsequent development. The zygote develops into an oncosphere, which is covered by a ciliated embryophore enabling it to swim upon hatching. This ciliated form of the organism is called a **coracidium** (plural, **coracidia**).

The dipylidean egg, seen in the genera *Dipylidium* and *Hymenolepis*, possesses a thin shell, a thin, nonciliated embryophore, and a relatively thick outer envelope (Fig. 12.10B). In the taenioid egg, characteristic of members of the genera *Taenia* and *Echinococcus*, the shell and outer envelope are lacking and the thick nonciliated embryophore constitutes the outermost covering (Fig. 12.10C). In *Dipylidium* and taenioid eggs, in contrast to pseudophyllidean eggs, only one or very few vitelline cells are associated with the zygote (Fig. 12.11).

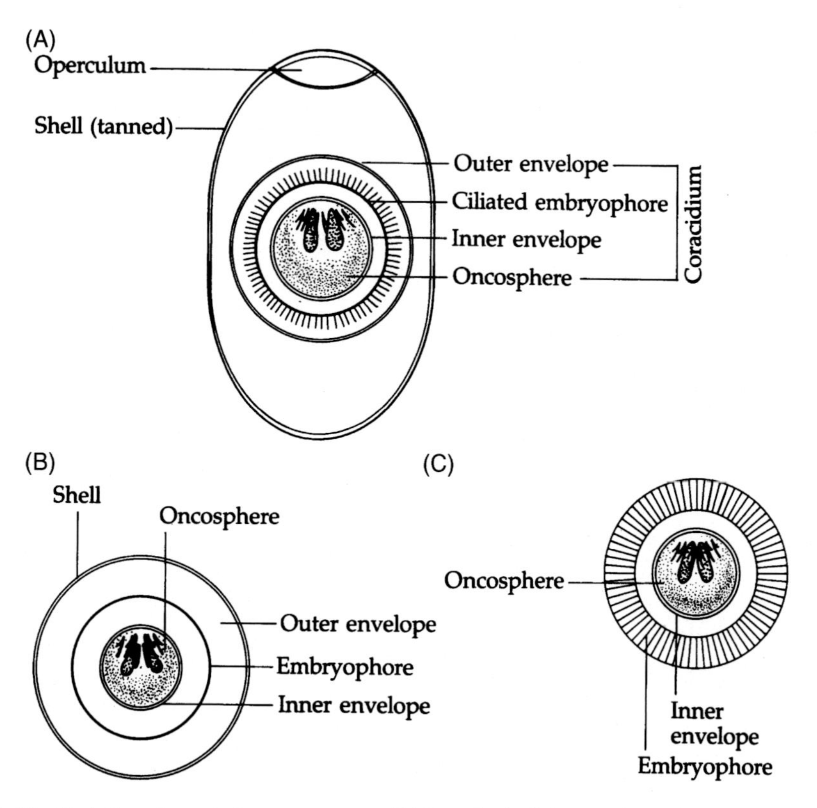

FIGURE 12.10 **Variations in cestode egg structure.** (A) Pseudophyllidean. (B) Dipylidean. (C) Taenioid.

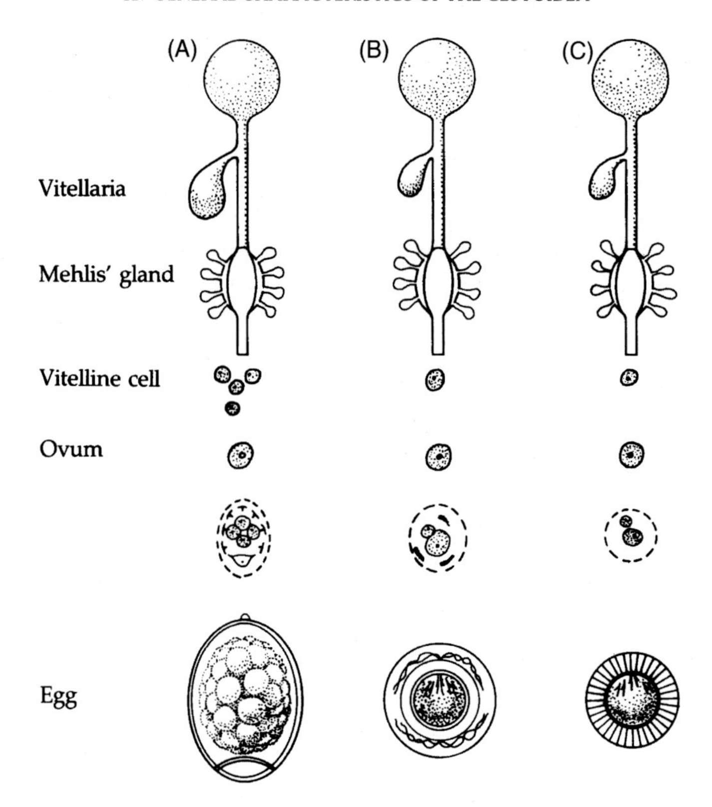

FIGURE 12.11 **The three types of egg-forming tapeworms that infect humans.** (A) Pseudophyllidean, for example, in *Diphyllobothrium latum*. (B) Dipylidean, for example, in *Hymenolepis nana*. (C) Taenioid, for example, in *Taenia solium*.

LIFE CYCLE PATTERNS

Tapeworms that infect humans display one of two basic life cycle patterns, one typical of members of the order Pseudophyllidea, the other of members of the order Cyclophyllidea (Fig. 12.12).

Pseudophyllidean Pattern

In members of the order Pseudophyllidea, coracidia-containing eggs leave the host with feces into water. The coracidium escapes from the eggshell through the operculum and swims for a brief time by means of its ciliated embryophore. Survival of the organism depends upon ingestion of the coracidium by an aquatic arthropod, the first intermediate host, within which the embryo sheds its ciliated embryophore and metamorphoses into a globular **procercoid** in the hemocoel. During this stage of development, the oncosphere hooks are retained, albeit nonfunctionally, in a tail-like structure called the **cercomer**. When the first intermediate host is ingested by a second intermediate host, usually a fish, the procercoid migrates via the peritoneal cavity to any of several parts of the body, principally the musculature, where it grows and develops into a solid, vermiform **plerocercoid** that

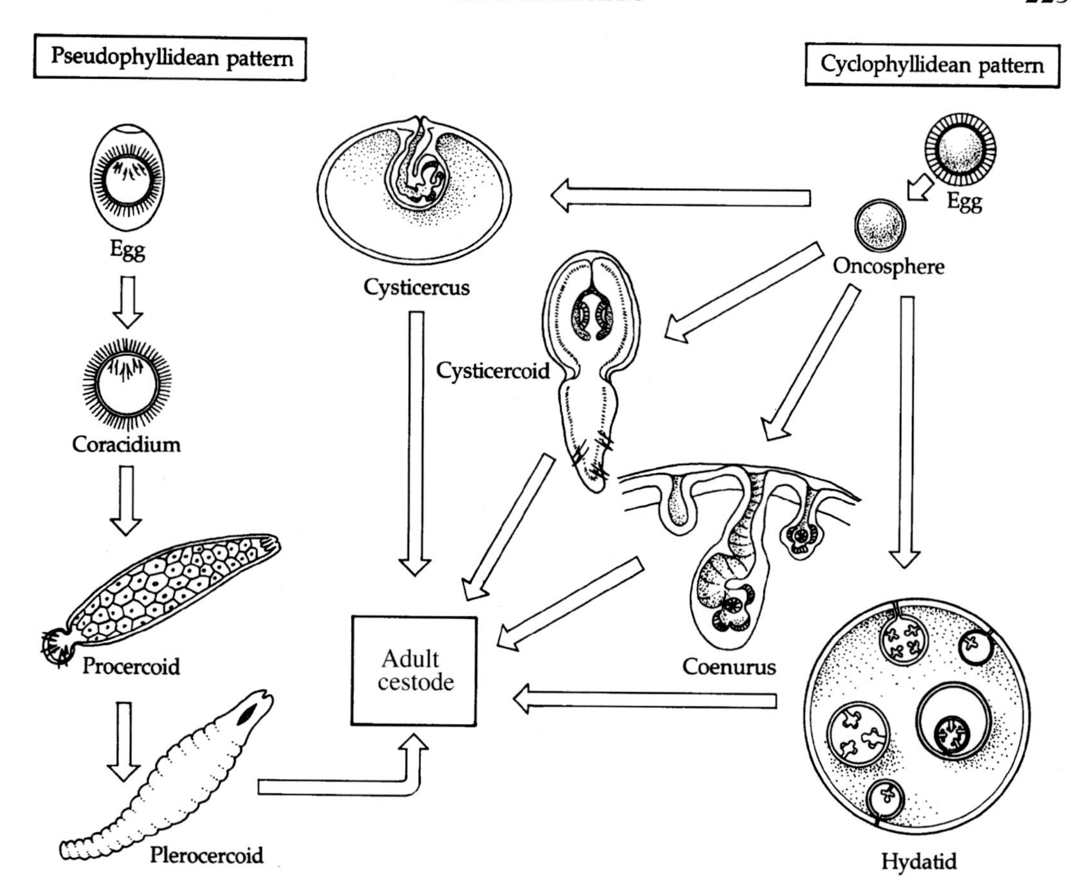

FIGURE 12.12 **Life cycle patterns of tapeworms that infect humans.**

manifests the beginning of strobilation and a developing adult scolex. The plerocercoid is infective to humans; when ingested, it attaches to the wall of the small intestine where strobilation occurs. All the aforementioned stages possess penetration glands that secrete lytic enzymes to aid in penetration of and migration among the various host tissues and organs.

Cyclophyllidean Pattern

Adapted as it is to terrestrial hosts, the cyclophyllidean hexacanth embryo, or oncosphere, lacks a ciliated embryophore and must remain passive until the egg is ingested by a vertebrate or invertebrate intermediate host. In species that normally utilize an invertebrate intermediate host, usually an arthropod, the oncosphere, upon hatching in the digestive tract, employs its six hooks and its penetration glands to enter the hemocoel, where it metamorphoses into a **cysticercoid**. This form is solid-bodied and possesses a fully developed acetabulate scolex. It is surrounded by several layers of cystic tissue and has a prominent **cercomer** equipped with hooks. The cystic layers and the cercomer are digested away in the digestive tract of a definitive host, freeing the scolex and neck to begin strobilation.

In species that utilize vertebrate intermediate hosts, the oncosphere, after ingestion, penetrates the intestinal lining and enters a venule. It is carried by the circulating blood to several areas of the body where it develops into a **cysticercus** with an acetabulate scolex invaginated into a fluid-filled vesicle or bladder; hence, the common name **bladderworm**. Two other forms that follow this developmental pattern are the **coenurus** and the **hydatid** cysts. In the former, the wall of the bladder develops several invaginated scolices; in the latter, secondary cysts are formed as invaginations on the walls. These second-generation cysts are called **brood capsules** since they give rise to scolices. When ingested by a suitable definitive host each of these scolices can develop into an adult worm.

In some tapeworms, certain immature stages, such as the cysticercus, coenurus, and hydatid cyst of some cyclophyllideans and the plerocercoid of several pseudophyllidean tapeworms, are capable of developing in extraintestinal tissues of humans (see Chapter 14).

PHYSIOLOGY

The adaptive morphology of adult tapeworms and the environment in which they dwell are probably the most significant factors influencing their physiology. Lacking a digestive tract, these worms must derive all nutrient molecules from the host or from their microhabitat and such molecules must cross the tegument. The environment in which tapeworms reside, the small intestine, is one of very low oxygen tension, necessitating anaerobic metabolism. The methods by which nutrients cross the tegument include active transport, facilitated diffusion, and simple diffusion. The most important nutrient molecule is glucose, which, after polymerization within the parasite, is stored as glycogen usually in the parenchyma and interstitial fluid. The only other major, transported carbohydrate is galactose. Considering the intestinal environment, it is not surprising that most energy derives from substrate level phosphorylation via glycolysis. Some tapeworms possess a mammalian-style electron transport system, but its role in energy production is minor. The sites on the tegument over which various molecules are transported vary according to which of the three major types of molecules is being transported. For instance, sites differ for the transport of carbohydrates, amino acids, and purines and pyrimidines. Most adult tapeworms also absorb lipids, probably by simple diffusion.

Metabolic rates differ in different parts of the strobila. The neck and immature proglottids have a much higher rate of metabolism than mature and gravid proglottids, reflecting the high energy requirements for new proglottid formation and organ development. Most of the energy requirement in mature proglottids is for egg production.

CHEMOTHERAPY

In most instances, adult tapeworms have little visible effect upon their hosts except in heavy infections, which may result in anemia, weight loss, and various secondary manifestations. The treatment of choice for all tapeworms infecting the small intestine of humans is essentially the same, namely, oral administration of the drug niclosamide, which disrupts proglottids and interferes with the worm's substrate phosphorylation processes, depriving it of required

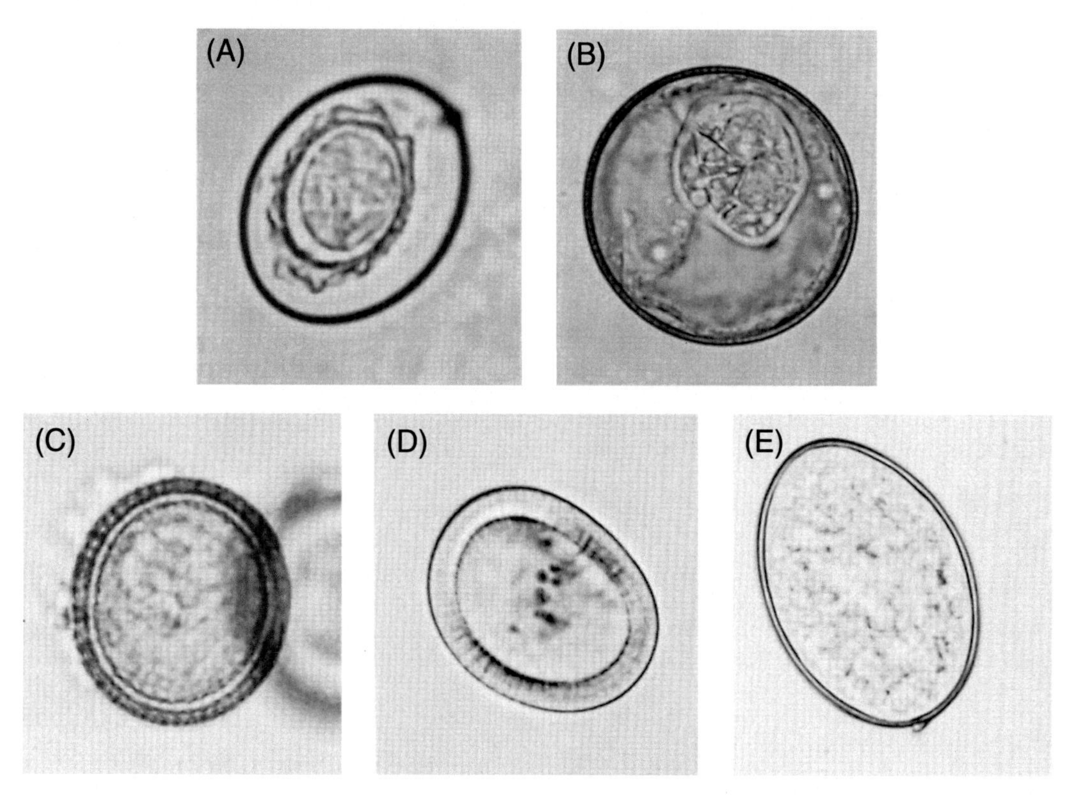

FIGURE 12.13 **Some cestode eggs.** (A) *Hymenolepis nana*: the dwarf tapeworm of humans, rats, and mice. (B) *Hymenolepis diminuta*: the rat tapeworm. (C) *Taenia* spp. (D) *Taenia pisiformis*: the dog and cat tapeworm. (E) *Diphyllobothrium latum*: the broadfish tapeworm.

ATP. Two other drugs, quinacrine hydrochloride and aminocrine, have also proven effective in treating tapeworm infections. Praziquantel is an excellent broad-spectrum anthelmintic and, like niclosamide, causes disruption of proglottids. As in the case of schistosomes, additional effects of praziquantel upon tapeworms are vacuolization of the tegument and rapid paralysis of the worm's musculature. These drugs should be used cautiously in the treatment of *T. solium* infections, since cysticercosis can result from autoinfection by eggs released from disrupted proglottids (see Chapter 14). Five to 6 weeks after treatment, the patient's feces should be reexamined for the reappearance of eggs in case the scolex was retained and developed into a "new" worm (Fig. 12.13).

HOST IMMUNE RESPONSE

Information relative to the factors involved in the immune responses elicited by adult tapeworms is still incomplete and sketchy. Much of the research has centered on the immune effects that extraintestinal tapeworms have upon their respective hosts (see p. 255). Generally, adult tapeworm infections induce a chronic Th1/Th2 response. Response to early infections brings about a transient Th1 response, which is replaced in part or totally by a Th2 response as the infection progresses. The Th1 response, characterized by increased levels of IFN-γ and IL-2, is believed to be correlated with protective immunity, while the Th2 response, characterized by the release of IL-4 and B-cell activation, is thought to be related to susceptibility.

CLASSIFICATION OF THE CLASS CESTOIDEA

Subclass Cercomeromorphae

Posterior end of larva, the cercomer, armed with hooks. All parasitic, being common in all classes of vertebrates except Cyclostomata; intermediate host required for almost all species.

Order Cyclophyllidea

Scolex usually with four suckers; rostellum present or absent, armed or not; neck present or absent; strobila usually with distinct segmentation; monoecious (or rarely, dioecious); genital pores lateral (ventral in Mesocestoididae); vitelline gland compact, single (double in Mesocestoididae), posterior to ovary (anterior or beneath ovary in Tetrabothriidae); uterine pore absent; parasitic in amphibians, reptiles, birds, and mammals (genera mentioned in text: *Hymenolepis*, *Echinococcus*, *Taenia*, *Dipylidium*).

Order Diphyllobothriidea

Scolex with two bothria, with or without hooks; neck present or absent; strobila variable, proglottids anapolytic (senile proglottid detached after it sheds enclosed eggs); genital pores lateral, dorsal, or ventral; testes numerous; ovary posterior; vitellaria follicular as in Trypanorhyncha, occasionally in lateral fields but not interrupted by interproglottidal boundaries; uterine pore present, dorsal or ventral; egg usually operculate, containing coracidium; parasitic in fishes, amphibians, reptiles, birds, and mammals (genus mentioned in text: *Diphyllobothrium*).

Suggested Readings

Coil, W. H. (1991). Platyhelminthes: cestoidea. In F. W. Harrison, & B. J. Bogitsh (Eds.), Microscopic anatomy of the invertebrates (pp. 211–283). (3). New York: Wiley-Liss.

Kuperman, B. I., & Davydov, V. G. (1982). The fine structure of glands in oncospheres, procercoids and plerocercoids of Pseudophyllidea (Cestoda). International Journal of Parasitology, *12*, 135–144.

Lumsden, R. D. (1975). Surface ultrastructure and cytochemistry of parasitic helminths. Experimental Parasitology, *37*, 267–339.

Pappas, P. W., & Read, C. P. (1975). Membrane transport in helminth parasites: a review. Experimental Parasitology, *37*, 469–530.

Read, C. P. (1959). The role of carbohydrates in the biology of cestodes: VIII. Experimental Parasitology, *8*, 365–382.

Schmidt, G. D. (1985). Handbook of tapeworm identification. Boca Raton, FL: CRC Press.

Smyth, J. D. (1969). The physiology of cestodes. San Francisco, CA: W. H. Freeman.

CHAPTER

13

Intestinal Tapeworms

This chapter is the introductory chapter on the tapeworms that considered the overall morphology, life cycle patterns, physiology, and chemotherapy of the group in general. In this chapter, six adult tapeworms are considered in detail. Two others are considered in a subsequent chapter since only the larvae of these forms infect humans. As the chapter title states, all adult tapeworms inhabit the intestinal tracts of their human hosts. *Diplyllobothrium latum* is a pseudophyllidean tapeworm and is discussed first. Its morphology and life cycle are illustrated. The cyclophyllideans, represented by two members of the genus *Taenia*, *T. saginata* and *T. solium*, are discussed relative to their life cycles, symptomatology and diagnosis, and epidemiology. This discussion is followed by consideration of two members of the genus *Hymenolepis*, *H. nana* and *H. diminuta* and, lastly, *Dipylidium caninum*. The latter forms are also discussed within the categories of life cycles, symptomatology and diagnosis, and epidemiology.

Adult tapeworms belonging to six species, representing two orders, infect the human intestinal tract. A single representative of the order Diphyllobothriidea, *D. latum*, is included in this group, while the remaining species, representing three families, belong to the order Cyclophyllidea. The family Taeniidae is represented by *T. solium* and *T. saginata*; Hymenolepididae, by *H. nana* and *H. diminuta*; and Dilepididae, by *D. caninum*. At least, two other members of the family Taeniidae, *Echinococcus granulosus* and *Echinococcus multilocularis*, are of medical importance to humans; however, since only the larvae infect humans, they are discussed in Chapter 14.

DIPHYLLOBOTHRIUM LATUM

D. latum, the broadfish tapeworm, parasitizes several larger mammals, including humans. It is most prevalent in Scandinavia, the former Soviet Union, and parts of temperate South America. Its common name is derived from the fact that the proglottids are wider than they are long.

The adult *D. latum* reproductive system possesses, on the midventral surface of each proglottid, a common genital atrium into which male and female genital pores open (Fig. 13.1). The bilobed ovary lies in the posterior portion of the proglottid. The oviduct, arising from the ovary, continues anteriad as a coiled uterus opening to the exterior through the midventral **uterine pore**. Eggs enclosed in tanned eggshells are expelled via the uterine pore.

Human Parasitology. http://dx.doi.org/10.1016/B978-0-12-813712-3.00013-8

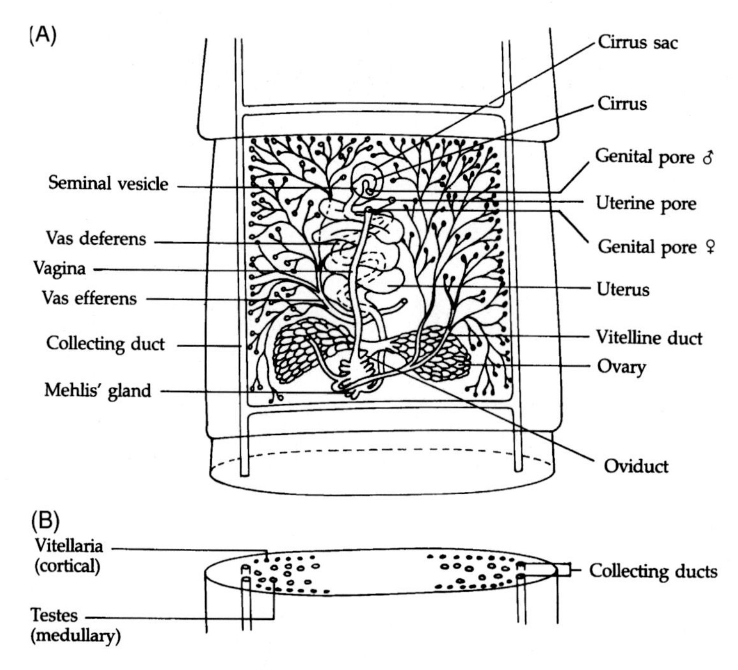

FIGURE 13.1 **Mature proglottid of *Diphyllobothrium latum*.** (A) Ventral view. (B) Cross section.

The follicular cells, which comprise the vitellaria, are scattered throughout the cortical fields of the proglottid. Numerous testes are chiefly medullary in their distribution except for those occupying an area along the midline of each proglottid. Sperm enter the female pore and pass down the vagina to the oviduct where fertilization occurs.

Adult worms may attain a length of 10 m, may be 10–20 mm wide, and may consist of more than 3000 proglottids, making *D. latum* the largest tapeworm found in humans. The extraordinary size of this tapeworm is partially due to anapolysis, the retention of terminal proglottids; approximately 80% of the proglottids are either mature or nearly so. The size of the vertebrate host may also influence the size of the parasite. For example, one of the largest recorded *D. latum*, 12 m long, was recovered at autopsy from a bear from Yellowstone National Park.

Life Cycle

The adult worm attaches to the mucosal lining of the ileum or, sometimes, the jejunum by both bothria (Fig. 13.2). Ovoid, operculated eggs are released from the uterine pore on the ventral surface of the proglottid. At the time of oviposition, the enclosed hexacanth embryo is immature; the eggs must lie dormant in the water for approximately 8–12 days or longer in order for embryonic development to be completed. Typical of the Pseudophyllidea, the hexacanth embryo is sheathed in a ciliated embryophore and is called a **coracidium**.

Within 24 h after hatching, the motile coracidium must be ingested by a freshwater copepod belonging either to the genus *Cyclops* or *Diaptomus*, or it will perish. In the digestive tract of the copepod, the ciliated embryophore is shed and the naked hexacanth larva, by means of

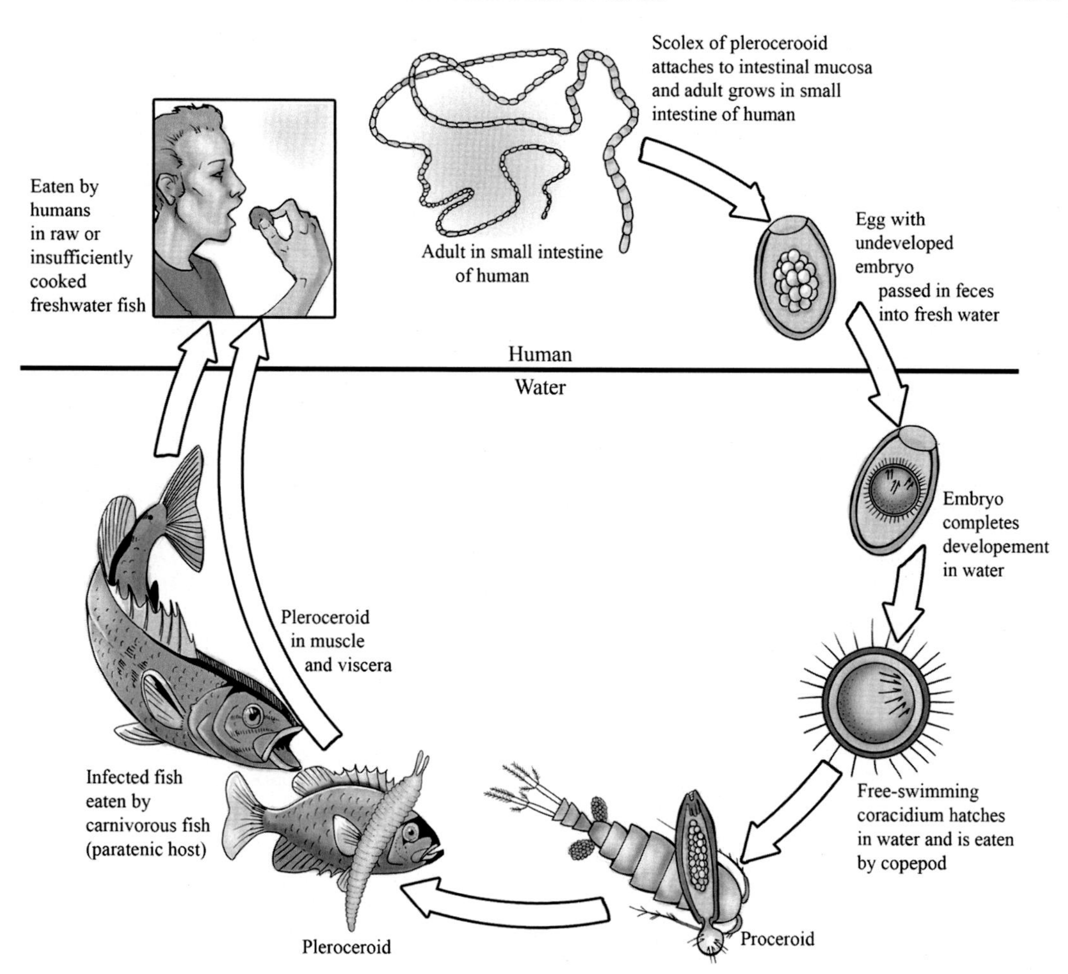

FIGURE 13.2 **Life cycle of *Diphyllobothrium latum*.** *Credit: Image courtesy of Gino Barzizza.*

its hooks and secretions, bores through the intestinal wall into the hemocoel. In 14–18 days, the hexacanth embryos, usually only one or two per infected copepod, metamorphose into elongated, globular **procercoid** larvae, each measuring about 500 μm in length. The prominent cercomer, containing the six hooks, projects posteriorly.

When the infected copepod is ingested by a suitable plankton-feeding freshwater fish, the procercoid larva penetrates the intestinal wall and migrates to the musculature. In 7–30 days, it develops into a long (2–4 cm) pseudosegmented **plerocercoid** larva with a fully developed scolex at one end. Numerous plerocercoids may be present in the musculature of a single fish host.

Unlike most pseudophyllidean plerocercoids, those of *D. latum* are coiled and, at times, encapsulated or, more commonly, lie free in the muscle tissue. Host reaction, such as encapsulation, depends upon the site of plerocercoid invasion. Plerocercoids infiltrating the muscles of the body wall are rarely encapsulated; however, among those settling in or on the viscera, encapsulation is common. Definitive hosts, including humans, acquire infection through ingestion of plerocercoids in undercooked, steamed, smoked, pickled, or raw fish. Upon entering the small intestine of the definitive host, the plerocercoid attaches to the mucosa and grows at the rate of about 30 proglottids a day and, although reaching full sexual maturity in 3–5 weeks, continues to add proglottids for an extended period.

Epidemiology

Human infection with *D. latum* is primarily, although not exclusively, limited to areas where fresh fish are a dietary staple or where the cleaning and handling of fish takes place. In addition to being ingested with raw or improperly cooked fish, plerocercoids that cling to the hands of fish cleaners may be accidentally ingested. Incidence is relatively high, for example, among the residents of Finland and some Baltic communities. A number of freshwater fish from cold-water regions, including such prized food fish as pike, salmon, trout, and whitefish, can serve as second intermediate hosts. In North America, 50–70% of northern and wall-eyed pike found in certain small lakes in the northern United States and Canada harbor plerocercoids of *D. latum.* The parasite is also found in Swiss lakes, the subalpine regions, the basin of the Danube River, the Middle East, Japan, Chile, Argentina, Peru, and Australia. *Diplyllobothrium nihonkaiense* is the primary cause of human diphyllobothriasis in Japan. It is indistinguishable from *D. latum* except by genomic studies. It is commonly found in salmon in Japanese waters.

Although a number of fish-eating mammals harbor the tapeworm, they are responsible, at worst, for the spread of the parasite in areas devoid of human inhabitants. Humans, on the other hand, through inadequate sanitation and the practice of eating raw or improperly cooked fish, coupled with the presence of suitable intermediate hosts, are responsible for establishing and maintaining endemicity in the human population. The eating of raw salmon in sushi restaurants has also led to infections in the United States. Furthermore, the increased incidence of infected fish in the United States can be traced directly to the practice of dumping untreated sewage into lakes and streams.

Symptomatology and Diagnosis

Rarely is more than a single worm found in an infected human, and many victims display few, if any, symptoms. Others complain of abdominal pain, weight loss, weakness, and nervous disorders. Many of these vague symptoms are attributable to the patient's reaction to the parasite's metabolic wastes, to degenerating proglottids, or to irritation of the intestinal mucosa; in some cases, they may also be a psychosomatic manifestation after the patient learns of the presence of the worm.

Occasionally, the worm is found in the upper portions of the jejunum, in which case it can compete successfully with the host for ingested vitamin B_{12} occasionally being responsible for provoking megaloblastic anemia. Since this vitamin is important in the synthesis of hemoglobin, patients deprived of it develop an anemia similar to pernicious anemia. In endemic areas such as Finland 5–10 of every 10,000 individuals infected with *D. latum* suffer from this type of anemia. If the worm is forced to retreat further down the intestine, by chemotherapy, for instance, the anemia subsides.

Laboratory diagnosis consists of identifying eggs (Fig. 12.13) or the characteristic broad proglottids from feces or vomitus.

TAENIA SOLIUM

This parasite, commonly referred to as the "human pork tapeworm," is common among humans in areas where raw or inadequately cooked pork is regularly consumed. The adult worm usually measures from 180 to 400 cm (sometimes up to 800 cm) long and comprises 800–900 proglottids. The small scolex (Fig. 13.3A), measuring about 1 mm in diameter, is armed with two circles of 22–32 rostellar hooks. These hooks are of two sizes, long (180 μm) and short (130 μm), alternating in the two circular rows.

The mature proglottid (Fig. 13.4) is squarish in outline with the common genital pore situated on the lateral edge about halfway down. The pores on successive proglottids may be on alternate sides or may be positioned unilaterally. The testes are scattered throughout the medullary region of the proglottid. The ovary consists of two prominent lobes and one small, central lobe. The vitellaria are compact, as is characteristic of cyclophyllideans, and are located in the basal portion of the proglottid just posterior to the ovary. The oviduct arises at the junction of the three ovarial lobes and continues anteriad as the *cul-de-sac* **uterine stalk**. As eggs are produced, they are "pushed up" into the stalk, which forms lateral branches as the number of eggs increases. The number of lateral branches serves as a tool for distinguishing *T. solium* from *T. saginata*; *T. solium* has 7–12 lateral branches, while *T. saginata* has more than 12 (Fig. 13.3B).

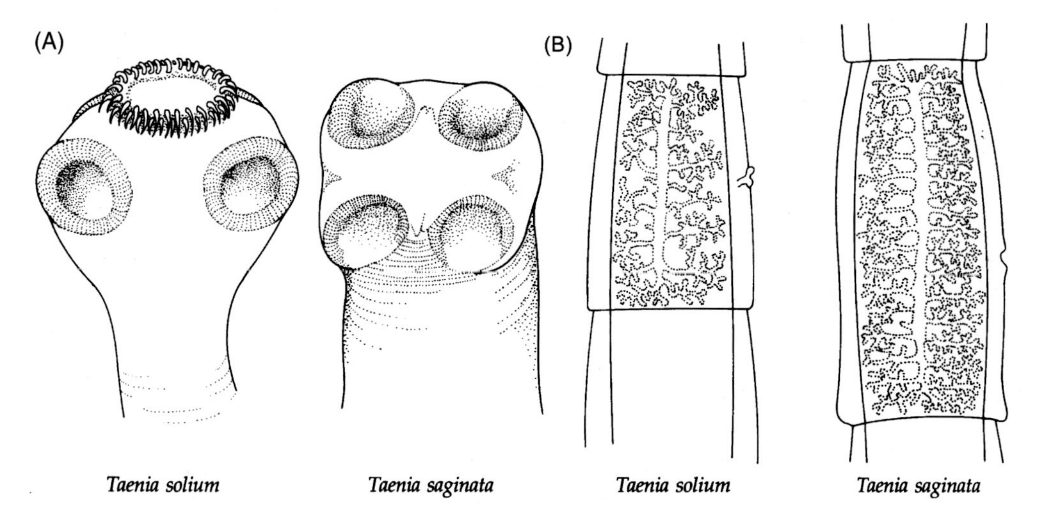

FIGURE 13.3 (A) Armed scolex of *Taenia solium* (*left*) and unarmed scolex of *T. saginata* (*right*). (B) Gravid proglottids of *Taenia solium* (*left*) and *T. saginata* (*right*).

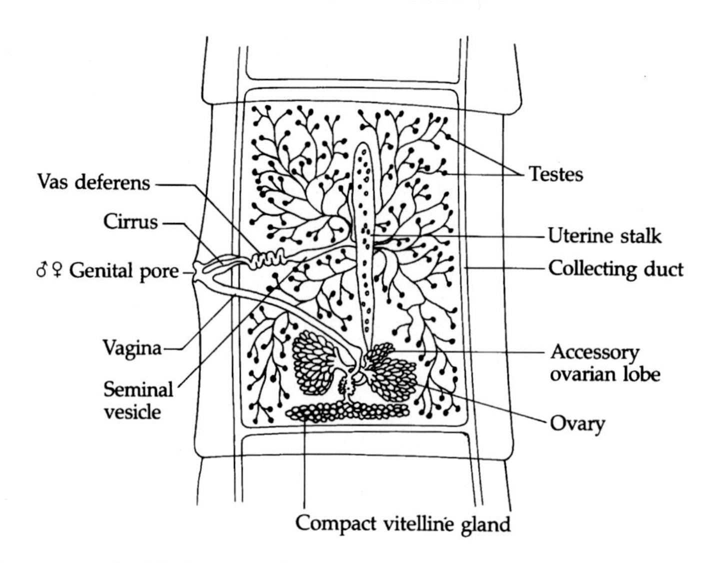

FIGURE 13.4 **Mature proglottid of *Taenia solium*.**

Life Cycle

Groups of 5 or 6 gravid proglottids, each containing thousands of eggs, exit the human host daily (Fig. 13.5). Eggs of *T. solium* are indistinguishable from those of *T. saginata*. The outer egg-covering is radially striated and covers the oncosphere. The proglottid may rupture either in the host's intestine or after it leaves the host. When eggs are ingested by pigs, the liberated oncospheres, using their hooks and penetration gland secretions, penetrate the intestinal wall, enter the circulatory system, and are carried by blood or lymph to muscles, viscera, and other organs, where they develop into cysticerci. Each white, ovoid, fluid-filled cycticercus, formerly termed *Cysticercus cellulosae*, measures 6–18 mm in length and contains a single, invaginated scolex. When infected pork is consumed by a human, the scolex evaginates and attaches to the jejunal wall where the parasite develops to sexual maturity in 2–3 months. Humans are the only known natural definitive hosts.

Epidemiology

The prevalence of pork tapeworm infection in humans varies by region. The very low incidence in the United States can be attributed to the isolation of pigs from human feces. Religious dietary proscriptions forbidding pork consumption by adherents of Islam and Judaism render human infection very rare in Muslim and Jewish communities. It is, however, common in other parts of Africa, India, China, several countries in South and Central America, and Mexico. Conversely, the beef tapeworm is comparatively rare in Hindu (India) populations, where cows are regarded with reverence and rarely eaten by humans.

Symptomatology and Diagnosis

Usually only a single adult tapeworm infects a human. The armed scolex may cause irritation of the mucosal lining, and there have been cases in which the scolex perforated the intestine leading to peritonitis. However, the greatest hazard to human health associated with

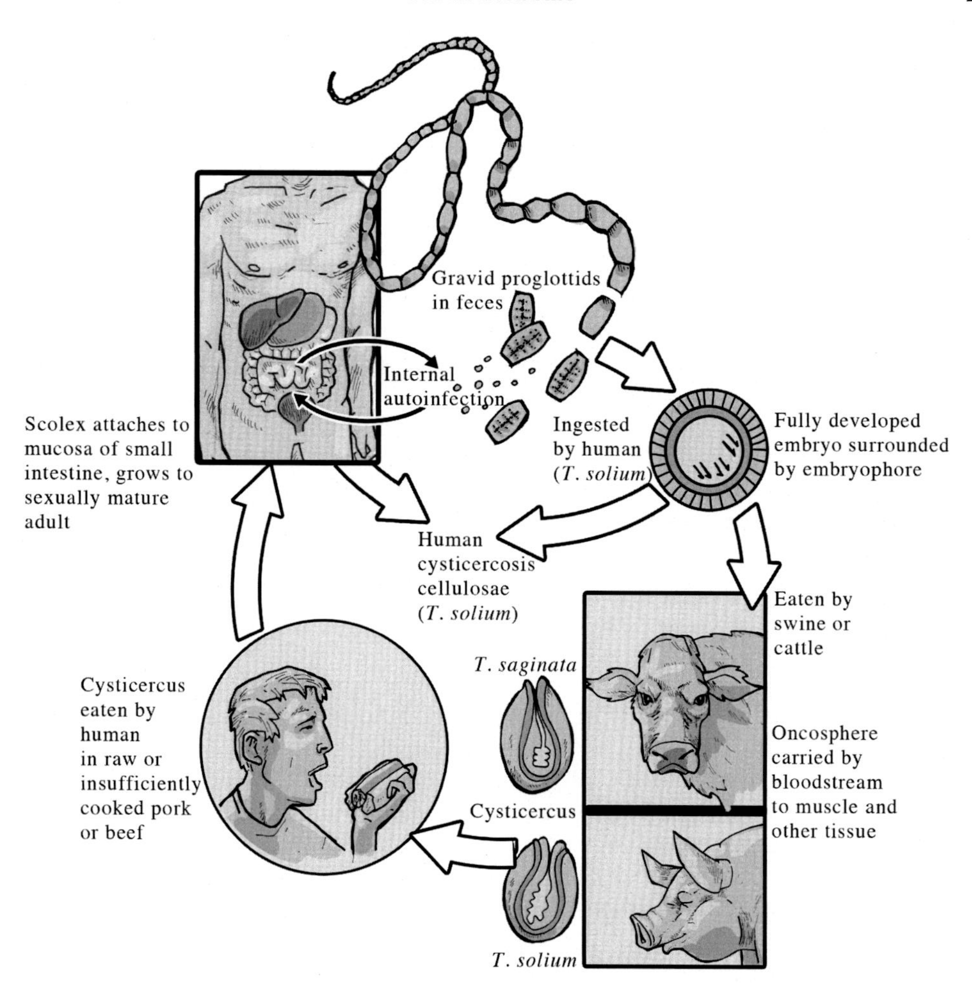

FIGURE 13.5 **Life cycles of *Taenia solium* and *T. saginata*.** *Credit: Image courtesy of Gino Barzizza.*

this parasite is infection with the cysticercus causing the potentially serious disease, **human cysticercosis** (see Chapter 14).

Identification of proglottids in feces is the most reliable method of diagnosis. Since most taenioid eggs are morphologically indistinguishable (Fig. 12.13), positive diagnosis is established by examination of gravid proglottids to determine the number of main lateral uterine branches (7–12 in *T. solium*) (Fig. 13.3B). The morphology of the scolex, particularly the rostellum, is also useful in diagnosis; *T. saginata* has no rostellum and its scolex bears no hooks, making it easily distinguishable from *T. solium*, which has an armed rostellum.

TAENIA SAGINATA

T. saginata is the most common of the large tapeworms of humans. Morphologically, the adult worm resembles *T. solium*. Usually 35–60 cm long, specimens as long as 225 cm have been reported. The strobila comprises approximately 1000 proglottids. The scolex is unarmed, having neither hooks nor rostellum (Fig. 13.3A). The morphology of mature proglottids in the two species differs primarily in that *T. saginata* has a bilobed ovary and about twice as many testes as *T. solium*. In addition, as previously noted, the gravid uterus of *T. saginata* has in excess of 12 main lateral branches (Fig. 13.3B).

Life Cycle

The life cycle of *T. saginata* strongly resembles that of *T. solium*. Adults of both species reside in the jejunum of humans, and gravid proglottids of *T. saginata* detach singly from the strobila and pass to the outside with feces (Fig. 13.5). The eggs of *T. saginata*, indistinguishable from those of *T. solium*, are ingested by a suitable intermediate host, such as cattle or other ungulates. The liberated oncosphere penetrates the intestinal wall and is carried by the lymphatic or blood circulatory system to intramuscular connective tissue where it develops into a cysticercus known as *Cysticercus bovis*. Humans become infected by ingesting cysticerci in undercooked or raw beef, particularly in muscles of the head and heart. Following evagination of the scolex and subsequent attachment to the jejunal wall, the worm develops to sexual maturity in 8–10 weeks.

Epidemiology

T. saginata is distributed throughout the world. Humans acquire infection by eating raw or insufficiently cooked beef infected with the cysticerci, as in dishes such as steak tartare. Cattle acquire *Cysticercus bovis* by grazing in fields upon which human excrement has been deposited either through fertilization with "night soil" or from poor sanitation. Pastures flooded by rivers and creeks contaminated with human excrement provide another source of infection for cattle. Under such conditions, eggs may remain viable for 2 months or longer. Thorough cooking of beef at 57 °C until the reddish color disappears or freezing at −10 °C for 5 days effectively destroys infective cysticerci.

Symptomatology and Diagnosis

Saginata taeniasis (=taeniosis) in humans is often characterized by such symptoms as abdominal pain, greatly diminished appetite, and weight loss. These symptoms are especially common in patients already debilitated by malnutrition or other illness. Unlike victims of *T. solium* infection, *T. saginata* victims rarely develop cysticercosis, and the prognosis is generally good. Diagnostic procedures are the same as those for *T. solium*.

HYMENOLEPIS NANA

Known as the dwarf tapeworm of mice and humans, the adult of this species is the smallest of the tapeworms infecting humans. It ranges from 7 to 50 mm in length and may consist of as many as 200 proglottids. The acetabulate scolex has a retractible rostellum armed with a single circle of small hooks. The mature proglottid (Fig. 13.6), approximately four times as broad

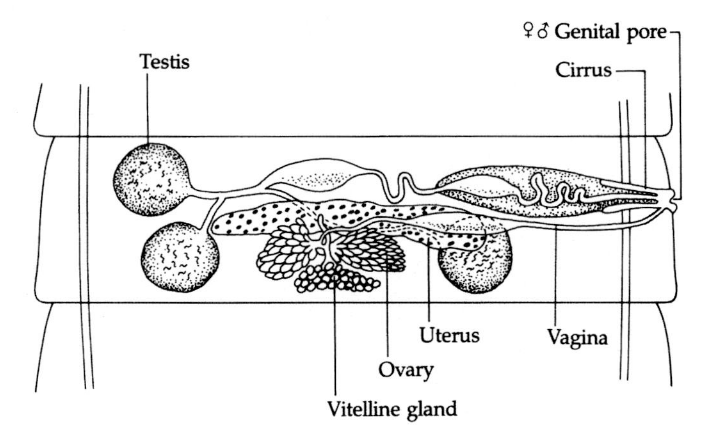

FIGURE 13.6 **Mature proglottid of *Hymenolepis nana*.**

as it is long, has a single, common genital atrium on the left margin. The male reproductive system consists of three spherical testes, one situated near the genital pore and separated from the other two by the bilobed ovary. The medullary region of the gravid proglottid is entirely occupied by a sacculate uterus containing up.

Because *H. nana*, unlike *H. diminuta* (see below), possesses an armed rostellum, among other, less conspicuous differences, some consider it to represent a different genus, *Vampirolepis*, hence *V. nana*.

Life Cycle

The life cycle of *H. nana* is of particular biological significance: it represents a modification of the typical cyclophyllidean life cycle pattern in that the parasite requires only one host to complete its development (Fig. 13.7). Natural definitive hosts, in addition to humans, are rodents, particularly mice and rats. Gravid proglottids from adult worms rupture, releasing oncosphere-containing eggs into the host's intestine to be eliminated with feces. The morphology of the egg, infective upon release, is characteristic of hymenolepid eggs. The oncosphere is enclosed in a thin shell and an embryophore with two polar thickenings, from each of which extend four to eight filaments. Upon being ingested by a new host, the oncosphere, freed in the small intestine from its encapsulating membranes, penetrates a villus into the lamina propria. There, about 4 days later, it becomes a modified cysticercoid larva known as a **cercocystis**. The cercocystis erupts from the villus into the lumen of the small intestine, attaches itself to the mucosal lining, and develops into a sexually mature adult in about 30 days. In the case of rodents, an insect, such as the flour beetle, may serve as an intermediate host. In this case, when the insect host is ingested by a rodent, or even accidentally by a human, the cysticercoid attaches to the intestinal wall and develops to sexual maturity. Autoinfection can exacerbate the condition by increasing the number of worms; eggs released from gravid proglottids, instead of passing to the exterior to infect new hosts, hatch in the small intestine and re-infect the same host. The freed oncospheres penetrate villi and repeat the cycle.

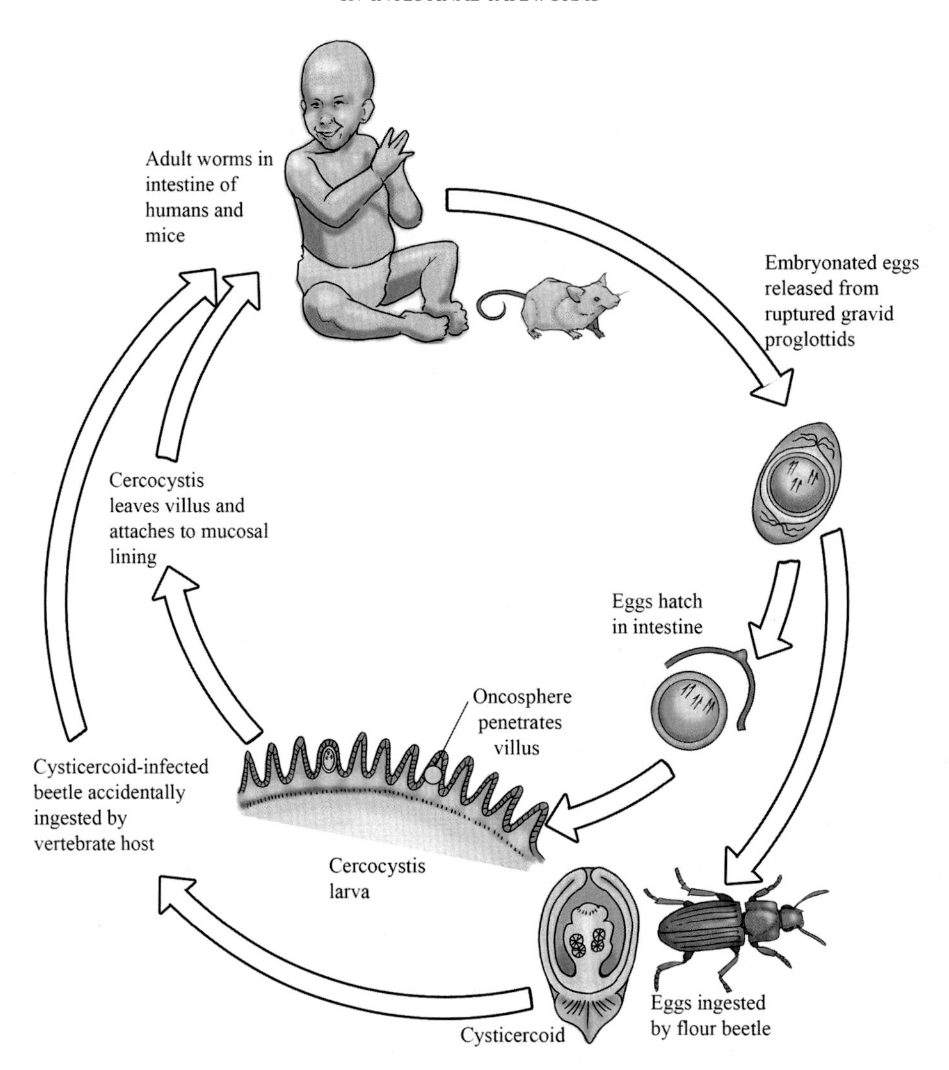

FIGURE 13.7 **Life cycle of *Hymenolepis nana*.** *Credit: Image courtesy of Gino Barzizza.*

Due to certain physiological variations, many authorities recognize two subspecies of *H. nana*: *H. nana nana*, which infects humans, and *H. nana fraterna*, which infects rodents. The life cycle of the rodent subspecies includes fleas and beetles as intermediate hosts in which the infective cysticercoid larva develops. Limited host cross-infectivity does occur.

Epidemiology

H. nana is cosmopolitan in distribution and is possibly the most common cestode parasite of humans in the world, especially among children. Worldwide prevalence ranges from less than 1% in the United States to about 9% in Argentina, with an average worldwide prevalence of 4%. The usual mode of transmission in humans is hand-to-mouth. Although transmission through ingestion of contaminated food is possible, it is rare since infective eggs are very susceptible to such environmental conditions as heat and desiccation. Infection may also be acquired by accidental ingestion of cysticerci-infected insects. The nature of the life cycle (i.e., no essential intermediate host and a high likelihood of autoinfection) renders prevention difficult. Teaching proper personal hygiene to children is perhaps the most effective preventive measure.

Symptomatology and Diagnosis

Since it is possible for a human victim to harbor massive numbers of these parasites, damage to the intestinal mucosa may be sufficient to produce enteritis. Most infections, however, are light and virtually symptomless, although autoinfection can lead to heavy worm burdens, particularly in children and immunosuppressed patients. In adult patients, the infection is usually self-limiting. In children with a moderate parasite burden, there may be loss of appetite, diarrhea, some abdominal pain, and dizziness. In many patients, a low humoral immune response is detectable. Diagnosis is by identification of eggs in feces (Fig. 12.13).

HYMENOLEPIS DIMINUTA

H. diminuta, a common parasite of rats throughout the world, occasionally parasitizes humans. *H. diminuta* exhibits a typical two-host life cycle, utilizing a grain-ingesting insect, such as a flour beetle, as intermediate host. A single worm may reach a length of 90 cm. The scolex in this species is unarmed, and the width of each proglottid is greater than its length. The morphology of proglottids is markedly similar to that of *H. nana* (Fig. 13.6). Of diagnostic relevance, the eggs are usually yellowish-brown and spherical. Unlike that of *H. nana*, the embryophore in these eggs does not bear conspicuous knobs and filaments at the poles. Insects are infected when they consume rodent feces containing either gravid proglottids, which have become detached from the strobila, or eggs. The oncosphere, utilizing its hooks and secretions from the penetration glands, penetrates the intestinal wall of the insect and enters the hemocoel where it develops to the cysticercoid stage. The most common intermediate hosts are grain beetles belonging to the genera *Tribolium* and *Tenebrio*, although cockroaches and fleas are also known to harbor infective cysticercoids. Humans acquire infection by eating cereals, dried fruits, and other similar foods contaminated with infected insects. The cysticercoid, once ingested by the definitive host, is freed from the insect's tissues in the small intestine. The scolex escapes from the cysticercoid tissue via a birth pore, attaches to the mucosa of the small intestine, and becomes sexually mature within 25 days. When human infection occurs, children are the most common victims and may suffer abdominal pain, diarrhea, insomnia, and convulsions. With a life cycle easily maintainable in the laboratory, *H. diminuta* has been a choice parasite for experimental studies for decades. It is perhaps the most studied of all tapeworms.

DIPYLIDIUM CANINUM

D. caninum is a common parasite of dogs, cats, and humans, especially children, throughout the world. The parasite can attain a length of 30 cm and possesses on its scolex a conical, retractible rostellum with one to eight (commonly four to six) rows of hooks (Fig. 13.8). This tapeworm is easily recognizable because each proglottid has two sets of reproductive organs with a genital atrium on each lateral edge. The short inconspicuous uterus atrophies early and, as eggs are produced, they are enclosed in **egg capsules** each containing 8–25 eggs. The medullary region of a typical gravid proglottid is packed with hundreds of egg capsules.

Life Cycle

The adult tapeworm lives in the small intestine of the definitive host where gravid proglottids, 12 mm long and 3 mm wide, separate from the strobila in groups of 2 or 3. The proglottids are capable of moving upon a substrate and can either creep out of the anus or be passed with feces. Eggs and capsules are ingested by larvae of fleas belonging to the genera *Pulex* and *Ctenocephalides* or by the dog louse *Trichodectes canis*. The oncosphere hatches in the gut of the arthropod, burrows through the wall, and develops into a cysticercoid in the hemocoel when

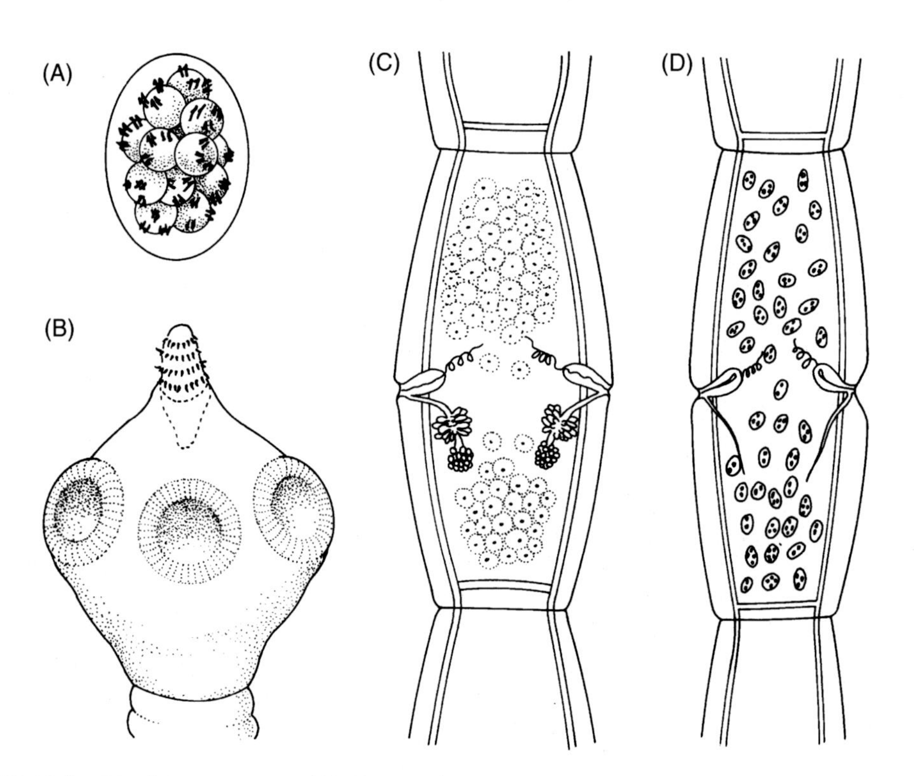

FIGURE 13.8 ***Dipylidium caninum***. (A) Cluster of eggs in a uterine ball. (B) Scolex with armed rostellum. (C) Mature proglottid with two sets of reproductive organs. (D) Gravid proglottid filled with uterine balls.

the flea or louse metamorphoses to the parasitic adult stage. When the infected arthropod is ingested by a suitable definitive host, the cysticercoid is liberated in the small intestine and develops to sexual maturity in about 20 days.

Epidemiology

Most human infections are in children younger than 8 years old, with a high percentage falling in the under-6-months age group. This probably is attributable to the fact that a high percentage of dogs are infected, many of which undoubtedly are pets. Transmission to humans usually results from accidental ingestion of infected fleas or lice or from allowing dogs and cats to lick ("kiss") the mouths of children immediately after the pet has bitten an infected arthropod.

Symptomatology and Diagnosis

It is rare for humans to harbor more than a single parasite, and symptoms are seldom apparent. Diagnosis is confirmed by discovery of characteristic proglottids or eggs in the feces.

Suggested Readings

Arai, H. P. (Ed.). (1980). Biology of the rat tapeworm, Hymenolepis diminuta. New York, NY: Academic Press.

Arme, C., & Pappas, P. W. (Eds.). (1983). The biology of the Eucestoda. New York: Academic Press.

Hernandez-Mendoza, L., Molinari, J. L., Garrido, E., Cortés, I., Solano, S., Miranda, E., et al. (2005). The implantation of *Taenia solium* metacestodes in mice induces down-modulation of T-cell proliferation and cytokine production. Parasitology Research, *95*, 256–265.

Pawlowski, Z., & Schultz, M. G. (1972). Taeniasis and cysticercosis (*Taenia saginata*). Advances in Parasitology, *10*, 269–343.

Peduzzi, R., & Boucher-Rodoni, R. (2001). Resurgence of human bothriocephalosis (*Diphyllobothrium latum*) in the subalpine region. Journal of Limnology, *60*, 41–44.

Tanowitz, H. B., & Wittner, M. (1991). Tapeworm infections. In G. T. Strickland (Ed.), Hunter's tropical medicine (7th ed., pp. 834–859). Philadelphia, PA: W.B. Saunders Company.

von Bonsdorff, B. (1956). *Diphyllobothrium latum* as a cause of pernicious anemia. Experimental Parasitology, *5*, 201–230.

CHAPTER

14

Extraintestinal Tapeworms

This chapter deals with those tapeworm larvae that infect humans and are potentially highly pathogenic to humans. Included in this category are several species of *Diphyllobothrium*, *Taenia solium*, *Echinococcus granulosus*, and *Echinococcus multilocularis*. Other species that may occasionally infect humans are also examined. Human sparganosis is initially discussed followed by a consideration of human cysticercosis and human hydatidosis. The diseases caused by these larvae are discussed under the categories of life cycles, symptomatology and diagnosis, chemotherapy, host immune response, and prevention. Special consideration is placed on the epidemiology of these conditions, such as the use of poultices and the partnerships that humans have with infected domestic animals. Ample light microscope and line drawing illustrations are presented that augment the discussions.

This chapter deals with tapeworms having larvae that are potentially highly pathogenic to humans. Included in this category are several species of *Diphyllobothrium* and related pseudophyllidean cestodes, *T. solium*, and at least two members of the genus *Echinococcus*, namely, *E. granulosus* and *E. multilocularis*. The larvae of two other species, *Hymenolepis nana* and *Taenia multiceps*, sometimes infect humans. However, since human infections are either relatively rare (*T. multiceps*) or, in the larval stage, produce no serious symptoms (*H. nana*) (see p. 237), these species are not considered in this section.

The human disease known a sparganosis is caused by plerocercoid larvae of any of several pseudophyllideans. Human cysticercosis is caused by the cysticercus of *T. solium*. Human hydatidosis results from infection with hydatid cysts of *E. granulosus* or multilocular cysts of *E. multilocularis*.

HUMAN SPARGANOSIS

The plerocercoid larvae of several pseudophyllidean tapeworms are capable of infecting tissues of both humans and other vertebrates (Fig. 14.1). One of the more common of these belongs to the genus *Spirometra*, most commonly *S. mansonoides*, members of which use various carnivores other than humans as definitive hosts. The plerocercoid larva of *Spirometra* was placed originally in the genus *Sparganum* before the association between larva and adult worm was established. The term **sparganosis**, signifying an infection with the plerocercoid larva, stems, therefore, from the generic name that is no longer valid.

Human Parasitology. http://dx.doi.org/10.1016/B978-0-12-813712-3.00014-X

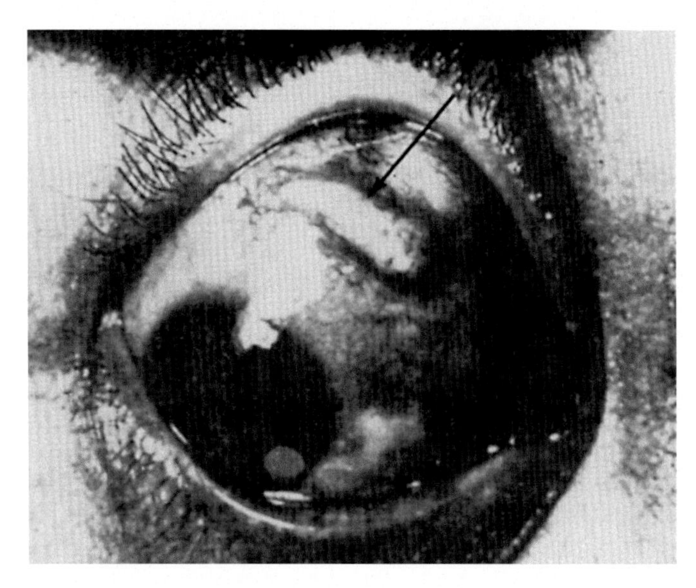

FIGURE 14.1 **Human patient with plerocercoid in conjunctiva.**

Life Cycle

Among various pseudophyllidean tapeworms with plerocercoids that can produce sparganosis in humans, the life cycle of *D. latum* is typical (Fig. 13.2). However, it should be emphasized that ingestion of *D. latum* plerocercoids by humans will also eventually produce the adult tapeworms.

Epidemiology

Humans become infected in several ways. A common vehicle is drinking water contaminated with copepods harboring procercoids. Procercoids released in the human gut penetrate the intestinal wall and migrate to various tissues where they develop into plerocercoids. Backpackers drinking from stagnant pools or lakes are vulnerable to infection in this manner, especially in temperate regions of the world where these parasites are prevalent.

Another way by which humans become infected is through ingestion of insufficiently cooked muscle of fishes, amphibians, reptiles, birds, and such mammals as bears, wild boars, and pigs. Plerocercoids of pseudophyllideans other than *D. latum*, ingested when infected muscle is eaten, are freed in the intestine and migrate to various tissues, most commonly the areas around the eyes, muscles, viscera of the thorax, and subcutaneous regions of the thorax, abdomen, and thighs.

Humans also acquire infection through the use of raw meat poultices (e.g., as a treatment for a black eye where active plerocercoids from infected meat may crawl into the orbit and become established). Similar cases of human ocular sparganosis have been reported from the Far East following treatment of skin ulcers or eye inflammations with poultices made from various infected animals, particularly snakes. In these countries, the most common plerocercoid causing human sparganosis is that of *Diphyllobothrium erinacei*, a tapeworm of carnivores, while in North America, it is that of *S. mansonoides*, a tapeworm of cats. Another species of *Spirometra* that may cause sparganosis in humans displays asexual proliferation of the scolex in the plerocercoid stage; hence its species name *Spirometra proliferatum*.

Because practices such as those described above are widespread in the Far East, human sparganosis is common there. However, the disease also occurs, albeit less frequently, in Europe, Australia, Africa, and North and South America.

Symptomatology and Diagnosis

As noted above, plerocercoids locate in many parts of the body. The movements and secretions of living plerocercoids can induce localized inflammatory reactions; dead and degenerating larvae sometimes cause edema of the surrounding tissue. Chills and fever may accompany infections. Eye infections, particularly common in the Far East, produce conjunctivitis and swelling. In general, severity of infection is determined by the number and location of the larvae and how quickly and completely the patient can be rid of them. Detection of larvae in the host's tissues constitutes diagnosis.

Chemotherapy and Treatment

Surgical removal of the larva is the most dependable treatment. The only widely used drugs to treat tapeworm cysts are the benzimidazoles. Praziquantel, as a supplementary drug, has shown promising results in hastening recovery.

Prevention

A major preventive measure is the thorough cooking of freshwater fishes prior to human consumption. Application of animal flesh to human skin should be avoided and questionable water, especially in endemic areas, should be boiled or filtered before drinking.

Host Immune Response

Many aspects of the effects of sparganosis on the immunity of the mammalian host remain unexplored. In mice, experimentally infected with the plerocercoid larvae of *S. mansonoides*, serum IL-1 and TNF-α levels peaked at week 6 postinfection. Investigators took these findings to deduce that a Th2 response is initiated that regulated the pro-inflammatory cytokines in mouse sparganosis.

HUMAN CYSTICERCOSIS

The fully developed cysticercus of *T. solium* is oval, about 0.5 cm or wider, and usually enclosed by a capsule of host connective tissue (Fig. 14.2). In areas such as certain parts of the brain, the host capsule may be absent, and the larva may attain a diameter of several centimeters. The ability of the cysticerci of *T. solium* to develop in practically any organ of the body and the severity of the resulting pathology render it one of the most pathogenic species of tapeworms infecting humans (Fig. 14.3).

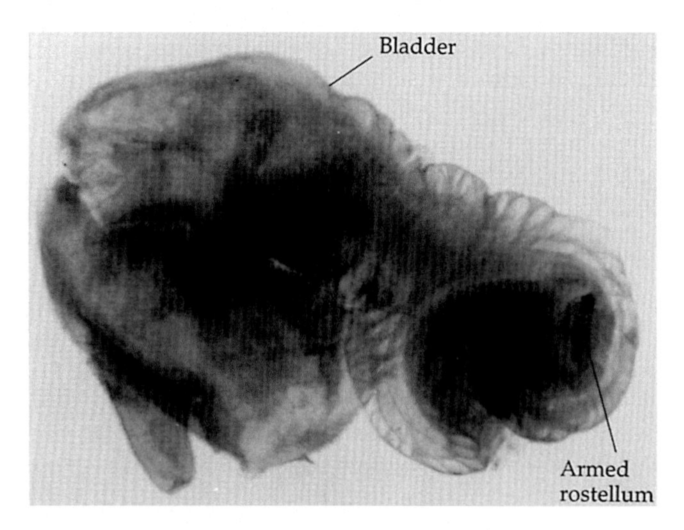

FIGURE 14.2 **Cysticercus of *Taenia* sp.**

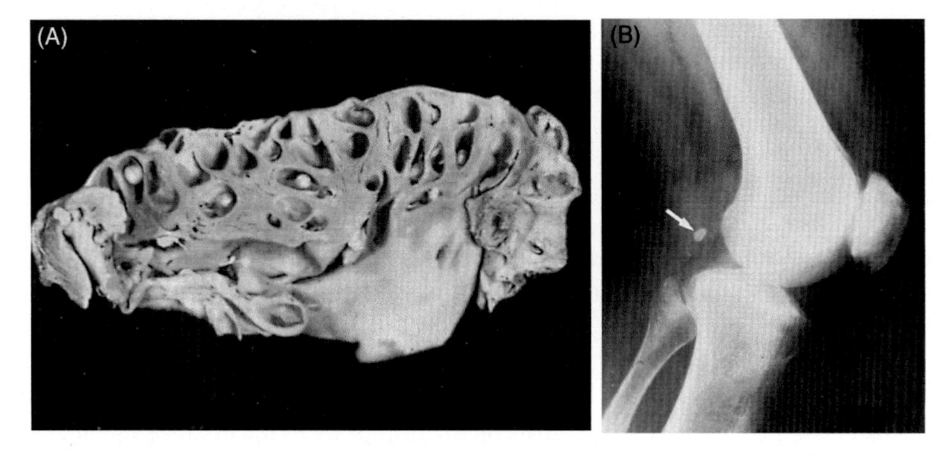

FIGURE 14.3 **Human cysticercosis.** (A) Heart containing cysticerci of *Taenia solium*. (B) *Taenia solium* cysticercus close to bone (*arrow*).

Life Cycle

See Fig. 13.5.

Epidemiology

Prevalence of human cysticercosis predictably parallels the incidence of the adult worm. It is quite common in Latin America (particularly in Mexico, where approximately 3 million people are infected), Africa, Indonesia, India, and China. In the United States, its highest prevalence is in southern California, mainly among migrant workers and recent immigrants from Mexico and countries of Central and South America. In humans, for an unexplained reason, males seem more vulnerable to infection than females. Human infection with cysticerci occurs in several ways, perhaps the most common of which is direct ingestion of eggs. This may result from hand-to-mouth self-infection, from eating food contaminated with eggs during unsanitary food handling practices, or from consuming food or water contaminated with feces containing eggs. Internal autoinfection, whereby eggs are swept back into the stomach by reverse peristalsis, is another method of human infection, although of little epidemiological importance since only about 25% of patients with cysticercosis also harbor the adult tapeworm. Eggs, whether ingested or swept back by reverse peristalsis, pass through the stomach and hatch in the small intestine. The escaping oncospheres penetrate the intestinal wall, enter the circulatory system, and are dispersed throughout the body.

Symptomatology and Diagnosis

While the most common sites for infection are the skeletal muscles and brain, cysticerci may be found in almost any tissue of the body, including the eyes, lungs, and subcutaneous tissues. Cysts are well tolerated in muscles and subcutaneous tissues although heavy infections can produce muscle spasms, weakness, and general malaise. Developing cysts elicit a host inflammatory response resulting in fibrous encapsulation although, as noted earlier, such a capsule may not be formed when the cyst invades parts of the brain. Calcification of the cyst may occur after 1 year, after which time the disease may become asymptomatic. The most serious symptoms appear from 5 to 10 years after infection as a result of dead and dying cysticerci. The degenerating parasite tissues and associated fluid also elicit a host inflammatory reaction that can be very severe, even fatal.

In addition to eliciting host responses, cysts developing in the central nervous system, sense organs, or heart can exert mechanical pressure and cause severe neurological symptoms. Violent headache, convulsion, local paralysis, vomiting, and optic disturbances are common and are sometimes severe enough to be fatal. In non-Muslim developing countries, it accounts for more than one-third of adult onset epilepsy cases. Cysticercosis also accounts for 2% of neurologic/neurosurgical admissions in Southern California and more than 1000 cases per year in the United States.

Clinical diagnosis can be made by linking symptoms such as certain nervous disorders (e.g., the late onset of epileptic convulsions) to a history of residence in an endemic area. X-ray examination of infected muscles or central nervous system may confirm diagnosis by revealing calcified cysts. Computer-assisted tomography scans and magnetic resonance imaging have been used to diagnose cysticerci in the brain. Examination of cerebrospinal fluid and serum by ELISA is also useful in diagnosis of patients with cerebral cysticercosis.

Chemotherapy and Treatment

Surgery is the recommended and most reliable treatment for cysticerci in the fluid-filled spaces of the body. However, in heavy infections of the central nervous system, while removal of some of the cysts has proven helpful, surgery is not generally beneficial. Praziquantel or albendazole with certain steroids (e.g., dexamethasone) are effective in reducing edema and alleviating some of the symptoms of cerebral cysticercosis. Symptomatic treatment with anticonvulsant drugs is also recommended as initial therapy of parenchymal disease. Care must be taken during treatment for adult worms to avoid causing severe vomiting, which may induce reverse peristalsis.

Host Immune Response

One of the more critical results of infection with the cysticercus larva of *T. solium* is the disease neurocysticercosis resulting from the migration of the oncosphere larva to the brain. At this site, it develops into a cysticercus larva or **metacestode**. As the parasite dies, it deteriorates toward a final calcified stage. While the larva is alive, it has the ability to suppress local immune responses. It is not clear how this occurs although there are several hypotheses. One such hypothesis is that the parasite may mask its outer membrane with host-derived proteins. A second possibility is that the parasite may show molecular mimicry by synthesizing proteins that resemble those of the host. Once the parasite begins to die, a localized host response is elicited. Initially, a Th1 response occurs with the production of IFN-γ, IL-4, and IL-18 cytokines, which eventually develops into a chronic mixed Th1 and Th2 profile associated with mature granuloma formation. A similar phenomenon is observed in other helminthic infections such as schistosomiasis (p. 210) where an initial Th1 response is followed by a mixed Th1 and Th2 response with accompanying granuloma formation. The granuloma associated with fibrosis appears to have good and bad repercussions for neurocysticercosis patients. On one hand, it protects adjacent neural tissue from injury; on the other hand, it irreversibly damages the nervous tissue that immediately surrounds the cysticercus.

Prevention

The most effective preventive measures include strict attention to personal hygiene, sexual habits, and environmental sanitation. Removal of adult worms from the patient once infection has been ascertained is important in preventing autoinfection. Visitors to endemic areas should observe such preventive measures as the boiling of drinking water and avoiding salads and raw fruits and vegetables without rinds.

HUMAN HYDATIDOSIS

The two most important species of *Echinococcus* responsible for human hydatidosis are *E. granulosus*, which causes cystic echinococcosis, and *E. multilocularis*, which causes alveolar echinococcosis. Adults of the genus *Echinococcus* are among the smallest tapeworms, measuring 2–8 mm long with strobilae consisting of three or, rarely, four proglottids (Fig. 14.4). Usually, one immature, one mature, and one gravid proglottid make up the strobila. The scolex bears a rostellum with a double row of 28–50 (usually 30–36) hooks, four prominent suckers, and a short neck region. Adult worms inhabit the small intestines of a wide variety of canines and, occasionally, felines. In heavy infections, it is not unusual to find hundreds of worms attached to the mucosa of a dog's small intestine.

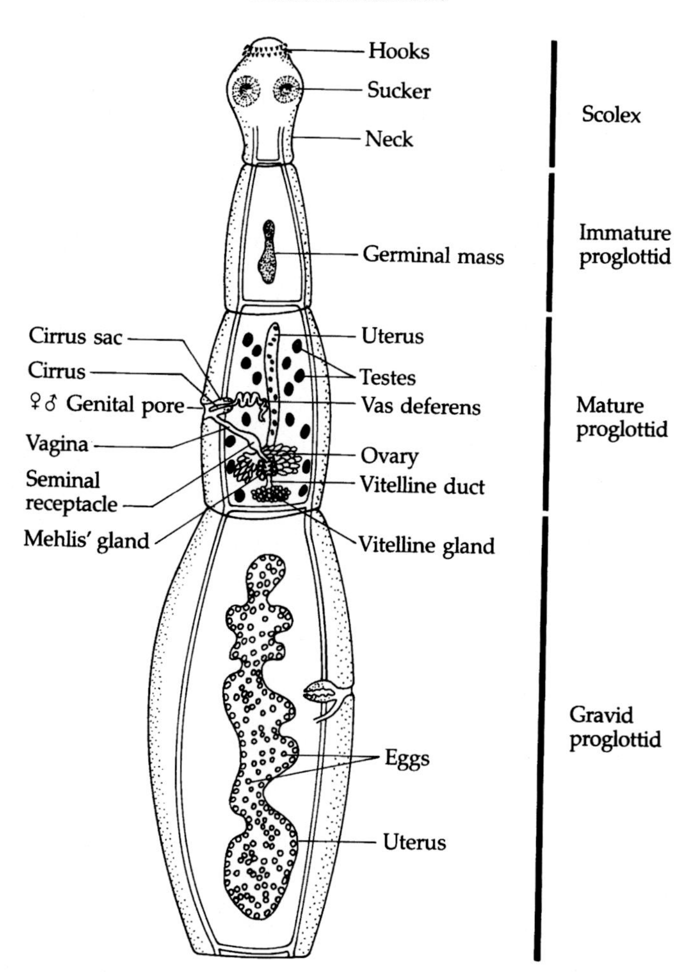

FIGURE 14.4 **Morphology of adult *Echinococcus granulosus*.**

Life Cycle

Eggs, measuring 30-by-38 μm, reach the exterior by host elimination of gravid proglottids with feces (Fig. 14.5). Released when proglottids disintegrate, the eggs are morphologically indistinguishable from other taeniid ova and each contains a fully developed oncosphere. The eggs gain entry into the intermediate host with water or forage contaminated with egg-containing feces. The usual intermediate host for *E. granulosus* is sheep, but cattle and other herbivores, as well as pigs, can also serve in this capacity. Microtine rodents serve as intermediate hosts for *E. multilocularis*, while cats, dogs, and foxes that prey on rodents harbor the adults. Usually, human infection occurs when eggs are ingested as a result of intimate contact with dogs, notably when dogs are allowed to lick human faces after grooming themselves. Humans can also ingest eggs by putting contaminated fingers into the mouth or by eating raw plants contaminated with feces from foxes, cats, or dogs.

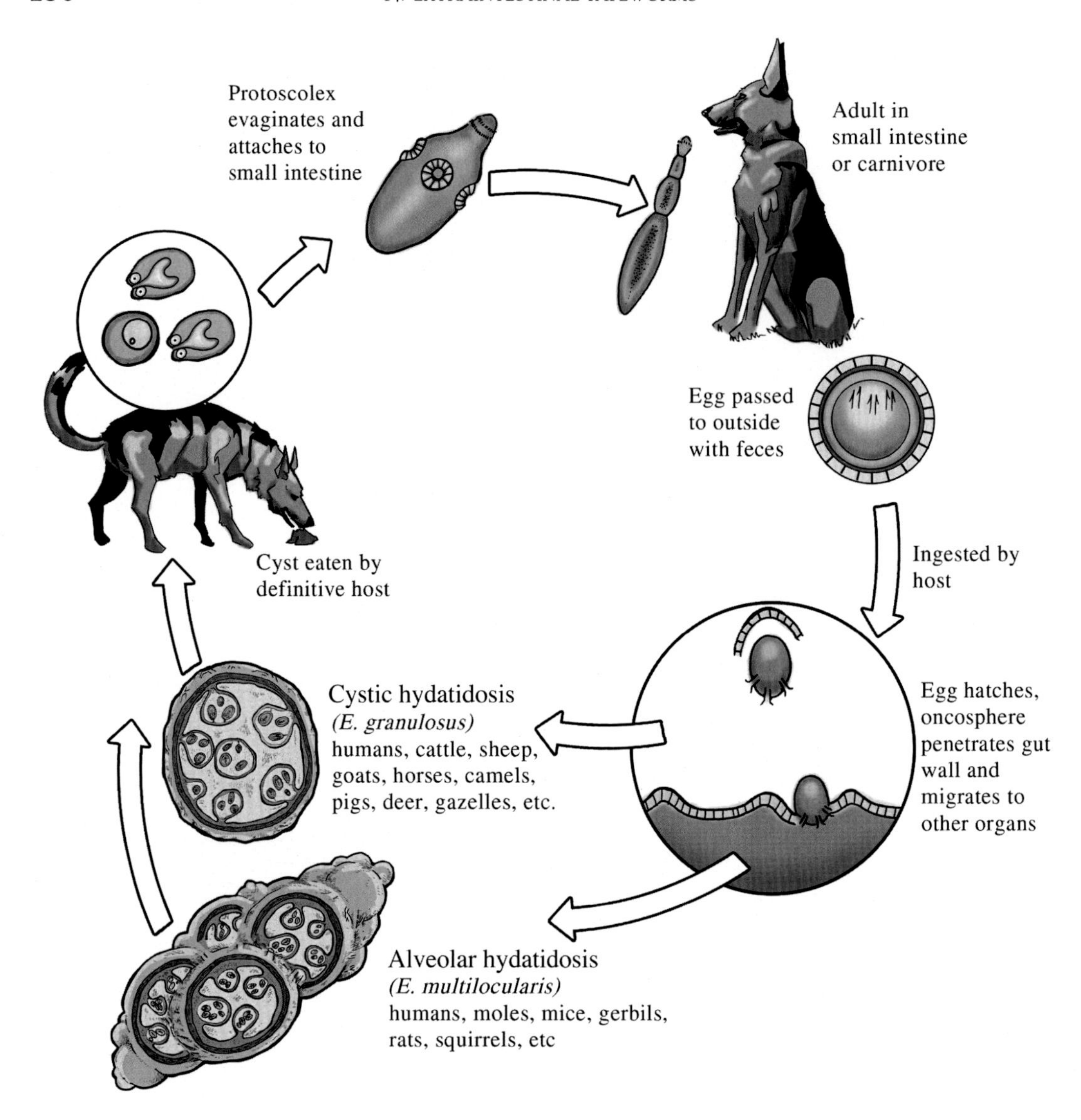

FIGURE 14.5 **Life cycle of Echinococcus.** *Credit: Image courtesy of Gino Barzizza.*

Once swallowed, eggs pass through the stomach and hatch in the small intestine. The freed oncospheres penetrate the intestinal wall, enter the mesenteric venules, and become lodged in capillary beds of various visceral organs as **metacestodes**. In humans, the developing hydatid cysts favor the liver, although other tissues, such as lungs, kidneys, spleen, heart, muscles, brain, and bone marrow, may be invaded. The hydatid cyst grows slowly, reaching a diameter of 10 mm in 5 months. However, it is not unusual for the cyst ultimately to reach the size of an orange or a small grapefruit and contain a considerable amount of fluid. Within the fully formed cyst, minute larvae with inverted scolices develop. Since these immature,

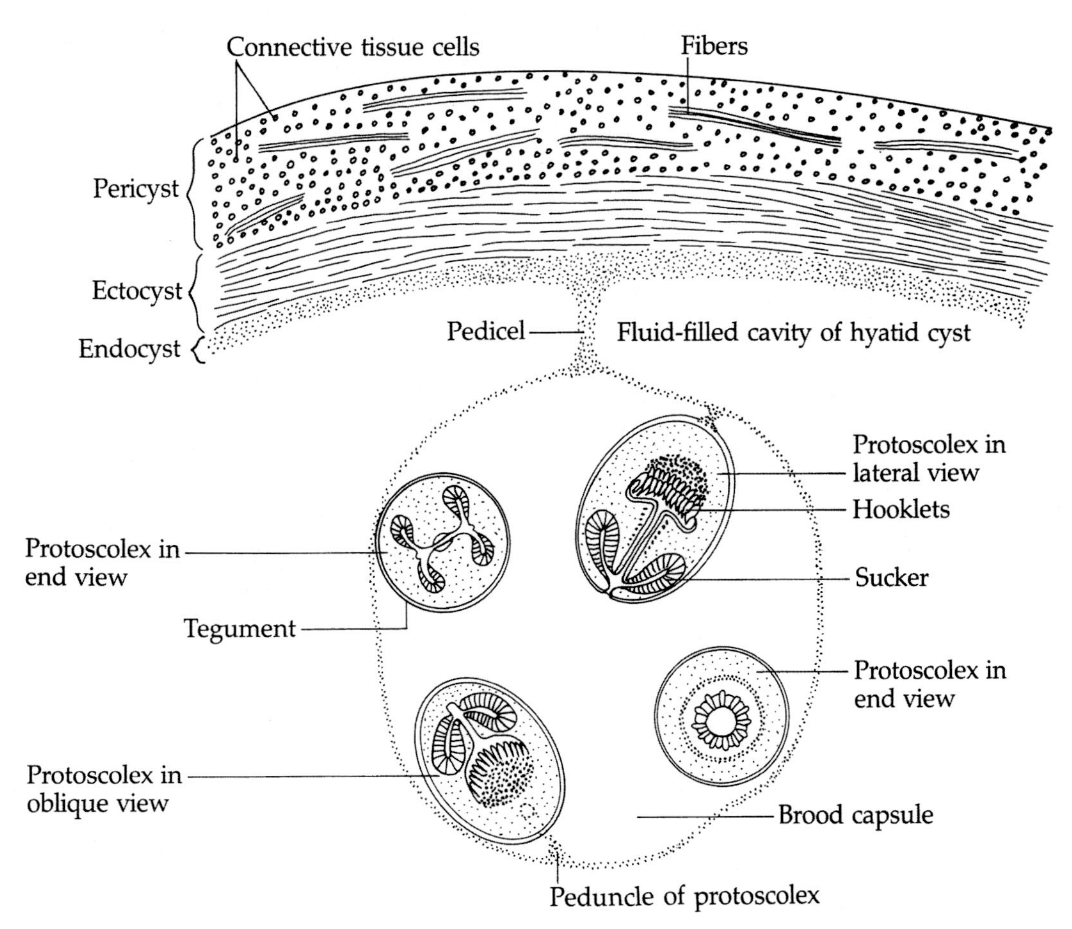

FIGURE 14.6 **A section through part of a unilocular hydatid cyst.**

four-suckered scolices lack individual bladders, they are called **protoscolices** (singular, **protoscolex**), not bladder worms or cysticerci, each of which contains a single scolex within a fluid-filled cyst.

In humans and some domestic animals, formation of hydatid cysts represents a dead end for the parasite. However, many wild animals, such as infected rabbits and squirrels, are potential intermediate hosts since the cysts are ingested when predators feed upon such animals. Upon reaching the small intestine of the definitive sylvatic host (e.g., the predator), each protoscolex has the potential for developing into an adult worm. The average life span of an adult worm is approximately 5 months, although some may survive as long as a year.

Hydatid cysts found in humans fall into three categories: (1) unilocular, (2) osseous, and (3) alveolar, the last representing a developmental stage of *E. multilocularis*. Of the three, the unilocular cyst is the most common and least pathogenic, while the alveolar is the most dangerous.

The diameter of the unilocular cyst may reach 20 cm or more in humans, although its usual diameter ranges from 1 to 7 cm. At maturity, the cyst wall consists of two layers: a thick, laminated, noncellular, outer tegument known as the **ectocyst** and an inner germinal epithelium that produces the protoscolices and is known as the **endocyst** (Fig. 14.6). Brood capsules attached to the germinal epithelium by a stalk, the **pedicel**, extend into the fluid-filled cavity of the cyst. In large cysts, these capsules may rupture and the freed protoscolices, which sink to the bottom of the bladder, are commonly known as **hydatid sand**. Each brood capsule contains 10–30 protoscolices. Second-generation daughter cysts often form within the mother cyst. These are replicas of the mother cyst and produce their own generation of protoscolices. Daughter cysts may, in turn, produce still another generation of cysts. It is not surprising that the average fertile primary cyst is estimated to contain more than 2 million protoscolices. If a cyst ruptures within a host, each liberated protoscolex can produce a daughter cyst. Whether the protoscolex itself develops into a cyst or whether small bits of germinal tissue cling to it and generate a new cyst is conjectural.

Osseous cysts are most commonly found in the ribs, vertebrae, and upper portions of long bones. These cysts usually develop in the marrow cavities. They are much smaller than unilocular cysts and contain little or no fluid and no protoscolices.

The outer membrane of the alveolar cyst is very thin, laminated, and difficult to separate from surrounding tissues. Connective tissue septa divide the cyst into numerous irregular compartments, or alveoli, which are filled with a jellylike material. Alveolar cysts are found most commonly in the liver, where they tend to proliferate by evagination of the thin cyst wall. In humans, the cysts are usually sterile, lacking protoscolices.

Epidemiology

Human hydatidosis is a zoonotic disease that results from intimate contact with dogs (Fig. 14.7). The percentage of infected dogs worldwide where dogs are used to herd domestic animals such as sheep may run as high as 50%, while the prevalence of hydatid cysts may be as high as 30% in sheep and cattle and 10% in hogs. Incidence among humans in Greece, Romania, Spain, Cyprus, Algeria, some Balkan countries, Argentina, Uruguay, and Chile is relatively high due to the close association with working dogs such as sheep dogs. Well-organized prevention programs have reduced the incidence in Australia, New Zealand, and Tasmania. Scattered cases in the human population have been reported in the United States, especially in the lower Mississippi River Valley, Utah, Arizona, and parts of California. In the Canadian Arctic, moose and caribou, rather than sheep, serve as intermediate hosts and are the main source of echinococcosis in the native Inuit. Isolated areas of heavy incidence have also been reported in the United Kingdom, especially in Wales and in some islands off the Scottish coast. Hydatid disease originating in the Middle East is spreading rapidly in Europe, especially in France and parts of Germany.

In nature, the carnivore–herbivore, predator–prey relationship, as in that between the wolf and the moose, wolf and reindeer, or dingo and wallaby, enables *E. granulosus* to complete its life cycle. This condition is known as **sylvatic echinococcosis**. Humans are seldom involved in this type of cycle. *E. granulosus* consists of a number of different genetic strains. It is interesting to note that different strains are adapted to different intermediate hosts, and this fact is important in the epidemiology of the parasite. For example, in Europe strains in which horses and pigs serve as intermediate hosts do not infect humans; however, strains that use sheep and cattle are infective to humans.

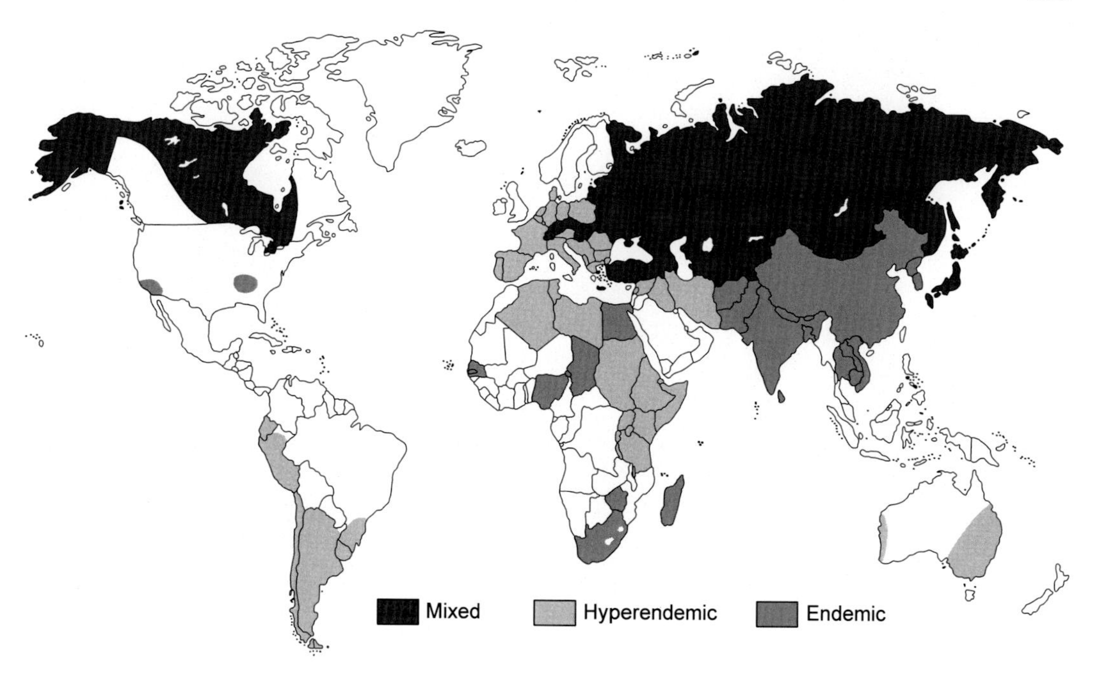

FIGURE 14.7 **Global distribution of hydatidosis.**

Certain unique, ethnic customs promote human infection. For example, members of certain Kenyan cultures, where incidence of human hydatidosis is among the world's highest, utilize dogs not only to herd livestock but also to act as "nurse dogs." In this capacity, the dogs protect babies and clean them after they defecate or vomit by licking their buttocks or faces.

The infection rate is also high among leather tanners in Lebanon since dog feces is an ingredient of the tanning fluid used there. During the preparation process, the tanners' fingers become contaminated, and they accidentally ingest eggs when they put their unwashed fingers into their mouths. It is estimated that there are more than 3 million cases of human cystic echinococcosis in the world.

Human alveolar echinococcosis is common in such countries of the Northern Hemisphere as Japan, China, the former USSR, western Alaska, and central Europe. Approximately 0.3–0.5 million cases are reported, all of which are in the Northern Hemisphere. Recently, human cases have also been documented in Iran, China, India, and the United States, countries previously free of the disease. In the Midwestern United States, a cat–rodent life cycle has been reported for *E. multilocularis* and foxes are significant reservoir hosts in North America and Europe.

Symptomatology and Diagnosis

The presence of unilocular cysts elicits a host inflammatory reaction that results in encapsulation of the cyst. The primary pathology of the unilocular cyst is impairment of organs from mechanical pressure (Fig. 14.8). Increased pressure resulting from cyst growth may

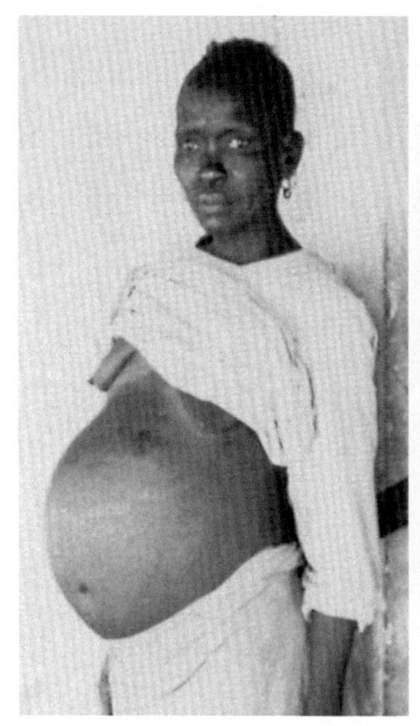

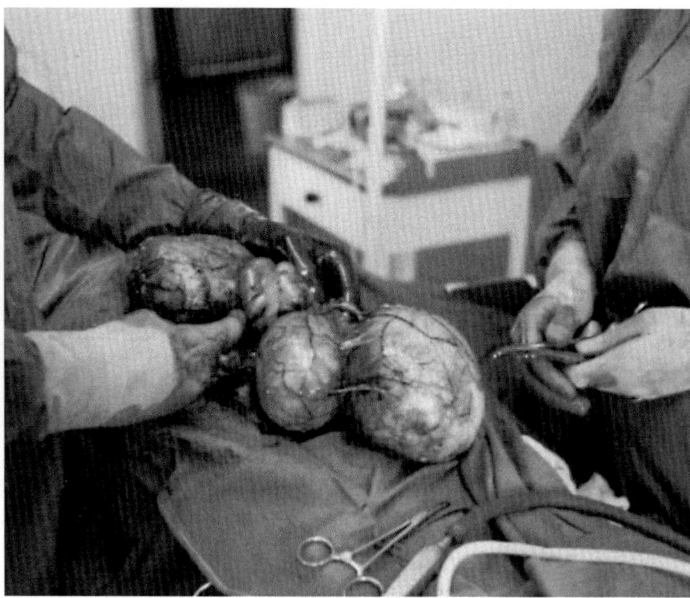

FIGURE 14.8 Human hydatidosis (*left*). Native Kenyan with disease (*right*). Removal of hydatid cysts by surgical means.

cause surrounding tissues to atrophy. The symptoms, therefore, are not unlike those caused by a slow-growing tumor, varying according to the tissues affected. It may take many years for symptoms to appear. For instance, while the liver is the most commonly affected organ, symptoms such as jaundice may take as long as 20 years to emerge. Pulmonary infections, characterized by a cough accompanied by allergic reactions, also are common. The brain, kidneys, spleen, and vertebral column may also be invaded and, over a protracted period, symptoms ranging from seizures to kidney dysfunction appear.

Protoscolices, freed by the rupture of cysts, enter the circulatory system and are transported to tissues throughout the body where they produce secondary echinococcosis. This condition, which may not appear for 2–8 years, is far more serious than the primary infection. The rupture of cysts also releases hydatid fluid, which sometimes causes severe allergic reactions. If a significant amount of fluid enters the bloodstream, it can precipitate anaphylactic shock.

The alveolar cyst is usually proliferative. While growth occurs at its periphery, its center may become calcified. Such cysts commonly occur in the liver and are often mistaken for hepatic carcinoma. They are difficult to extirpate, and the condition is usually fatal within 10 years. The osseous cyst, because of its location, is also difficult to remove. In severe cases, there is often necrosis of diaphyses of long bones, spontaneous fracture, and distortion of cancellous tissue.

Diagnosis of hydatidosis is based upon a number of criteria, such as symptoms (hepatic hypertrophy, etc.), history of residence in an endemic area, and close contact with dogs. X-ray examination is especially useful for revealing calcified cysts and various ultrasound procedures may locate noncalcified cysts. Laboratory diagnosis, which might be more aptly termed "postsurgical confirmation" of hydatidosis, involves detection of protoscolices. Serological tests are diagnostically useful, the direct hemagglutination test being one of the most commonly used. The intracutaneous test (Casoni's intradermal test) is sufficiently sensitive to be helpful in screening for human hydatidosis. However, negative test results are more significant than positive test results since there is about an 18% incidence of false positives.

Chemotherapy and Treatment

Surgery remains the preferred treatment for unilocular hydatidosis. Following drainage of the fluid from the cyst, replacement with 10% formalin for 5 min kills the protoscolices and the germinal epithelium. In any surgical procedure for cyst removal, care is taken to avoid rupturing the cyst. Allergic reactions resulting from stimulation by alveolar fluid respond best to treatment with antihistamines or epinephrine. Most recently, the benzimidazoles (especially albendazole) have been used successfully to reduce the size of both unilocular and alveolar cysts. It is anticipated that, at least in some cases, chemotherapy may eventually replace surgery. A relatively new, nonsurgical approach introduced in Italy involves treatment with albendazole for 12 h, aspiration of some cyst fluid, injection of cyst with 95% ethanol, and subsequent aspiration of the alcoholic solution. Several species of *Echinococcus*, among them *E. granulosus*, display a variety of genetic strains differing developmentally and physiologically. Such variations must be considered when chemotherapy is prescribed since one regimen may not yield uniform results against all strains.

Host Immune Response

Recent reports reveal that *Echinococcus* metacestode antigens promote the secretion of the chemokine IL-8 and IL-10 but reduce the secretion of IL-1 and IL-18. The latter cytokines normally downregulate Th2 responses. It is believed that this early diminution of IL-1 and IL-18 favors Th2 responses thereby reducing pro-inflammatory responses in infected patients, thus preventing overwhelming and pathogenic inflammation. The strong IL-10 response inhibits T-lymphocyte infiltration, preventing them from participating in the effector phase of the cellular immune response.

Prevention

Significant inroads toward prevention of human hydatidosis can be made by reducing contact between dogs and intermediate hosts such as sheep, hogs, and rodents, and by educating the public to the danger of intimate contact with dogs, especially in endemic areas. As added measures, dogs should be treated regularly with anthelmintics and kept away from slaughter houses. Refuse from slaughter houses should be disinfected.

Suggested Readings

Craig, P. S., Rogan, M. T., & Allan, J. C. (1996). Detection, screening and community epidemiology of taeniid cestode zoonoses: cystic echinococcosis, alveolar echinococcosis and neurocysticercosis cases. Advances in Parasitology, *38*, 170–250.

Deplazes, P., Hegglin, D., Gloor, S., & Romig, T. (2004). Wilderness in the city: the urbanization of *Echinococcus multilocularis*. Trends in Parasitology, *20*, 77–84.

Eckert, J., Pawlowski, Z., Dar, F. K., Vuitton, D. A., Kern, P., & Savioli, L. (1995). Medical aspects of echinococcosis. Parasitology Today, *11*, 273–276.

Eger, A., Kirch, A., Manfras, B., Kern, P., Schulz-Key, H., & Soboslay, P. T. (2003). Pro-inflammatory (IL-∃, IL-18) cytokines and IL-8 chemokine release by PBMC in response to *Echinocrroccus multilocularis* meacestode vesicles. Parasite Immunology, *25*, 103–105.

Garcia, H. H., Gonzalez, A. E., Gilman, R. H., Bernal, T., Rodriguez, S., Pretell, E. J., et al. (2002). Circulating parasite antigen in patients with hydrocephalus secondary to neurocysticercosis. American Journal of Tropical Medicine and Hygiene, *66*, 427–430.

Gemmel, M. A. (1977). Experimental epidemiology of hydatidosis and cysticercosis. Advances in Parasitology, *15*, 311–369.

Gottstein, B. (1992). *Echinococcus multilocularis* infection: immunology and immunodiagnosis. Advances in Parasitology, *31*, 321–380.

McManus, D. P., & Smyth, J. D. (1986). Hydatidosis: changing concepts in epidemiology and speciation. Parasitology Today, *2*, 163–167.

Schantz, P. M. (2002). Taenia solium cysticercosis: an overview of global distribution and transmission. *Taenia solium cysticercosis, from basic to clinical science*. New York, NY: CABI Publishing 63-74.

Zhang, W., Ross, A. G., & McManus, D. P. (2008). Mechanisms of immunity in hydatid disease: implications for vaccine development. The Journal of Immunology, *181*, 6679–6685.

CHAPTER

15

General Characteristics of the Nematoda

Chapter 15 opens with a brief discussion of the complexities involved in determining which classification scheme for the roundworms one should use to best illustrate the relationships within this complex group. Accordingly, we have chosen the simplest scheme by placing those roundworms that parasitize humans in the phylum Nematoda. As in previous chapters, the introductory chapter to this group starts with a complete study of the morphology of adult worms. A discussion of the female and male reproductive systems is followed by a detailed presentation of the various larval forms resulting from molting sequences. Ample line drawings clearly illustrate these various forms. The chapter concludes with an overview of the physiology and classification of those nematodes to be studied in subsequent chapters.

After more than a century of debate, the placement of nematodes, or roundworms, in the phylogenetic scheme remains unresolved. Considered by some to constitute an independent phylum, Nematoda (or Nemata), they are regarded by others as members, together with such groups as the Rotifera and the Nematomorpha, of the phylum Aschelminthes. According to still another system of classification, the Nematoda and the Nematomorpha are designated as separate classes of the phylum Nemathelminthes. In the scheme followed herein, roundworms are assigned to a separate phylum, Nematoda.

Parasitic nematodes are of great importance to biologists because they are abundant and widespread, frequently occur as endoparasites infecting a wide variety of invertebrate as well as vertebrate hosts, and often have a serious impact upon human health.

Nematodes that parasitize humans are assigned to either the Class Secernentea (=Class Phasmidia) or the Class Adenophorea (=Class Aphasmidia), distinguishable primarily by the presence or absence of minute sensory structures, known as **phasmids**, on the body surface. Most nematodes that infect humans belong to the class Secernentea.

While the origin of nematodes is obscure, there is marked similarity of structure and of elements of the life cycle among these organisms, whether free-living or parasitic. This consistency argues for "ancestral uniformity," descendance from common ancestors. Since most nematodes are not host specific, they are probably largely independent of the evolution of their hosts, using them merely as vehicles for their own evolution. Life history studies have fostered the view that parasitic nematodes evolved at various times from free-living soil forms, with the free-living ancestor initially utilizing a host for transportation or

Human Parasitology. http://dx.doi.org/10.1016/B978-0-12-813712-3.00015-1

protection only. Movements of such hosts isolated these nematode populations, preventing interbreeding with members of their free-living counterparts, thus providing the basis for speciation.

Significant preadaptation of the third-stage larva (see p. 270) to osmoregulation equips certain nematodes to utilize several environments during their life cycle. Such a third-stage larva is characteristic of all secernentean (phasmidian) parasites of animals and serves as a transfer stage from one environment to another. The third-stage larva of aphasmidians, on the other hand, is less diversified and plays no such role.

STRUCTURE OF THE ADULT

Nematodes are generally elongate, cylindrical, and tapered at both ends. The basic body design is a tube within a tube, the outer tube being the body wall and underlying muscles and the inner tube the digestive tract (Fig. 15.1). Between the tubes is the fluid-filled pseudocoelom in which the reproductive system and other structures are found. In certain species the body is almost uniformly cylindrical and is extremely thin. Sexual dimorphism is evident; at the curved, posterior end of the male, there is a copulatory organ as well as other specialized organs such as alae and papillae. Males are usually smaller than females.

Parasitic nematodes vary widely in size according to species. While some are microscopic and others may reach more than a meter in length, most are between 1 mm and 15 cm long. Nematodes are colorless and vary from translucent (smaller nematodes) to opaque (larger nematodes) when examined alive. It is not uncommon for some to absorb colored matter from surrounding host tissues or fluids.

Cuticle

An elastic **cuticle** covers the body surface of nematodes (Fig. 15.2). The presence of enzymes in the cuticle indicates that it is metabolically active, not an inert covering. Although the cuticle is generally smooth, various structures such as spines, bristles, warts, punctuations, papillae, striations, and ridges may be present on it. Some of these specialized structures are sensory, and some aid in locomotion; their arrangement and position are of taxonomic importance.

The cuticle not only covers the entire external surface, but also lines the buccal cavity, esophagus (=pharynx), rectum, cloaca, vagina, and excretory pore. It consists of four basic layers: the **epicuticle**, the **exocuticle**, the **mesocuticle**, and the **endocuticle** (Fig. 15.3).

The epicuticle is a relatively thin layer and is a consistent component of all nematode cuticles. Typically, it is trilaminate with a glycocalyx. Its function is largely unknown, although it is believed to act, at least in part, as a protective barrier.

The exocuticle is usually composed of two distinct sublayers: the relatively homogeneous **external exocuticle**, with no visible substructure, and the radially striated **internal exocuticle**.

The mesocuticle is the most diverse of the cuticular layers. It commonly consists of obliquely oriented, collagenous, fibrous sublayers that vary in number and in angular relationship to each other. The ability of the mesocuticular fiber sublayers to shift their angles

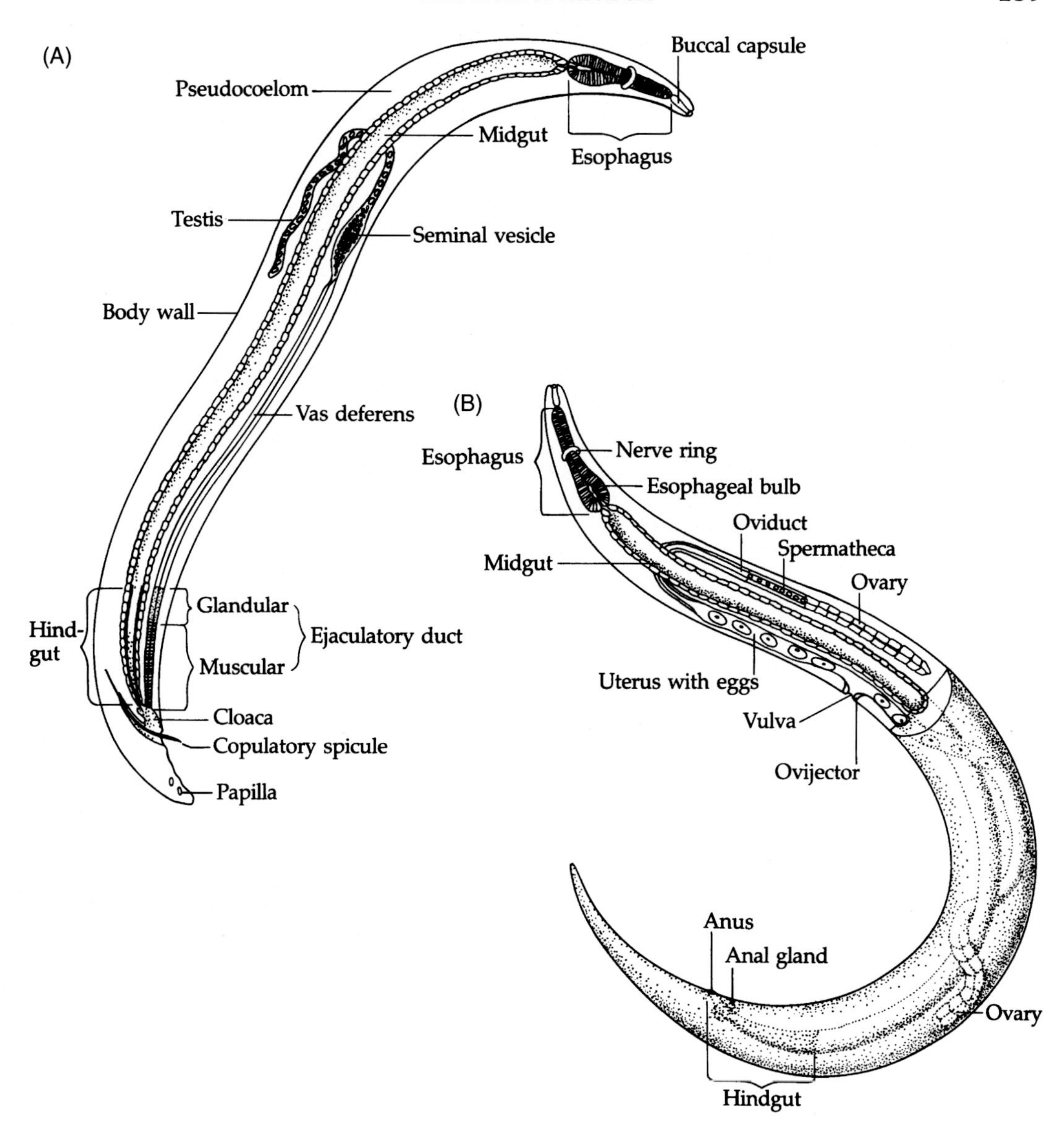

FIGURE 15.1 **Morphology of a generalized nematode.** (A) Male. (B) Female.

of orientation provides flexibility to the cuticle. In some nematodes, the thickness of the mesocuticle is directly proportional to the age of the worm.

The endocuticle is the innermost layer of the cuticle. It is also fibrous, but the orientation of fibers is not as distinct as in the mesocuticle. Often, the pattern is disorganized and with a great deal of overlapping. A basal lamina separates the cuticle from the underlying hypodermis.

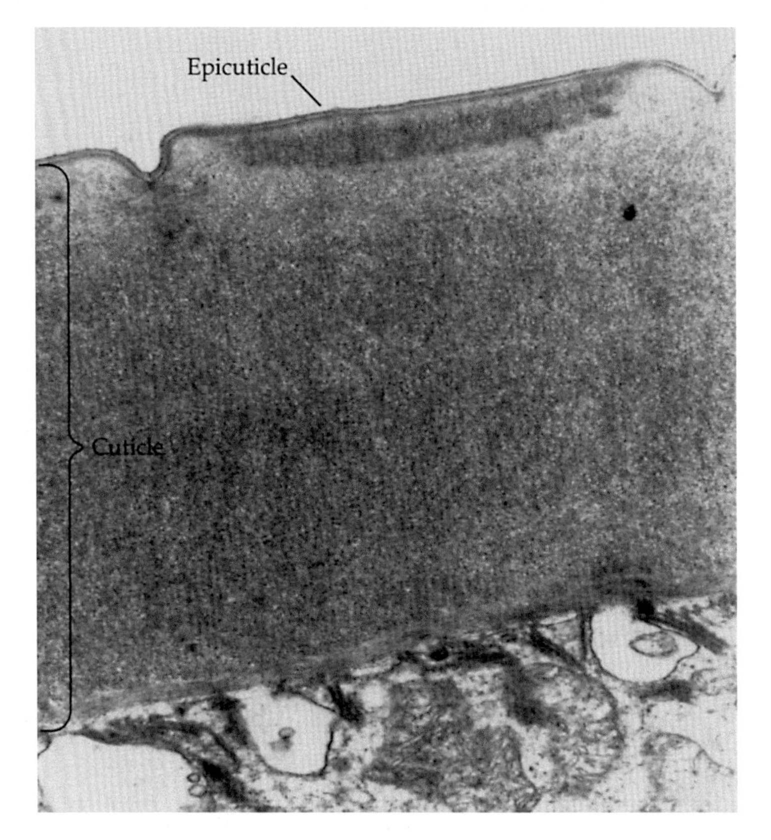

FIGURE 15.2 **Transmission electron micrograph of the cuticle of a nematode.**

Hypodermis

Beneath the basal lamina lies the thin, cellular (in adenophoreans), or syncytial (in secernenteans) hypodermis. A major function of the hypodermis is formation of the cuticle. The hypodermis protrudes into the pseudocoelom along the middorsal, midventral, and lateral lines to form the longitudinal **hypodermal cords**. These partially divide the pseudocoel into quadrants. Hypodermal organelles, such as nuclei and mitochondria, are confined to the cords. The lateral cords are the largest and contain the primary excretory canals when these are present while the dorsal and ventral cords contain longitudinal nerve trunks (Fig. 15.4).

Musculature

Within and closely associated with the hypodermis are one or more layers of longitudinally arranged muscle cells, the **somatic musculature**. Collectively, the cuticle, hypodermis, and somatic musculature make up the body wall. A convenient classification system to describe muscle cell arrangement has been devised based upon the number of rows of muscle cells per quadrant (Fig. 15.5). According to this system, an arrangement of multiple, longitudinal rows of muscle cells in each quadrant is termed polymyarian; one with no more than two rows of

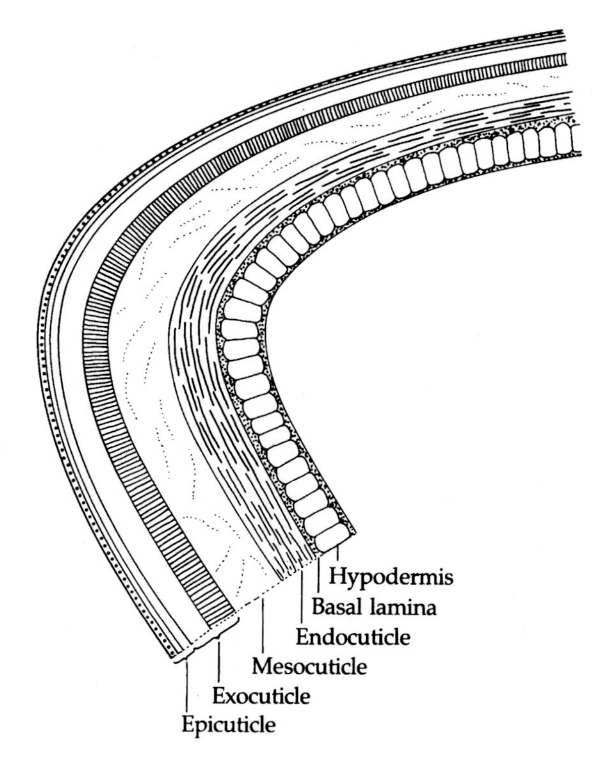

FIGURE 15.3 **Layers of the nematode cuticle.**

cells is designated **holomyarian**; and one with two to five rows is called **meromyarian**. Each muscle cell is comprised of a contractile portion containing myofibrils and a noncontractile portion, in which are found the various organelles, such as the nucleus, mitochondria, ribosomes, and endoplasmic reticulum, as well as stores of glycogen and lipid (Fig. 15.6). Sensory processes usually extend from the noncontractile portion of each cell to the longitudinal nerve trunks. The somatic musculature connects to the cuticle by fibers that originate in the contractile portion of each cell, pass through the basal lamina, and attach to the endocuticle.

Digestive Tract

The digestive tract of nematodes is complete (Fig. 15.1). It consists of an anterior mouth, a gut that comprises three major regions, a cloaca, and a subterminal vent. The major regions of the gut are the foregut, midgut, and hindgut, each displaying a certain degree of specialization.

Foregut

The cuticle-lined foregut begins at the mouth, which, in many species, opens into a **buccal capsule** and continues as the esophagus. When present, the buccal capsule may contain ridges, rods, and plates for maintaining its shape as well as spears, stylets, or teeth for attachment to or penetration of the host or for acquiring food.

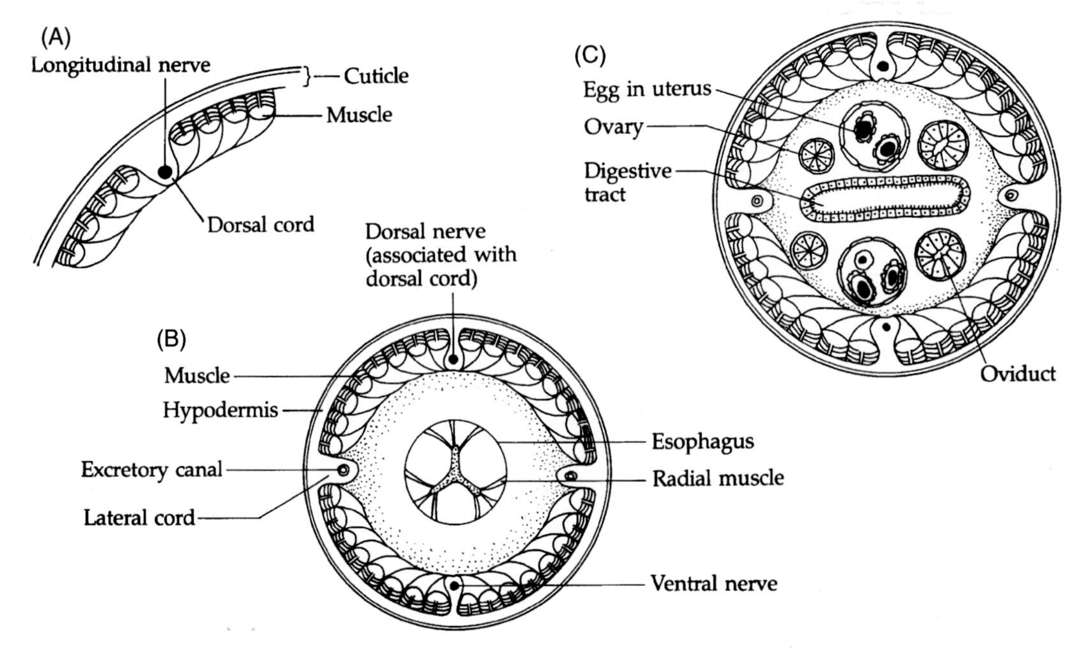

FIGURE 15.4 **Nematode morphology.** (A) Portion of the dorsal body wall showing the relationship of the dorsal cord to the cuticle. (B) Cross-section through the esophageal region. (C) Cross-section through the midgut region.

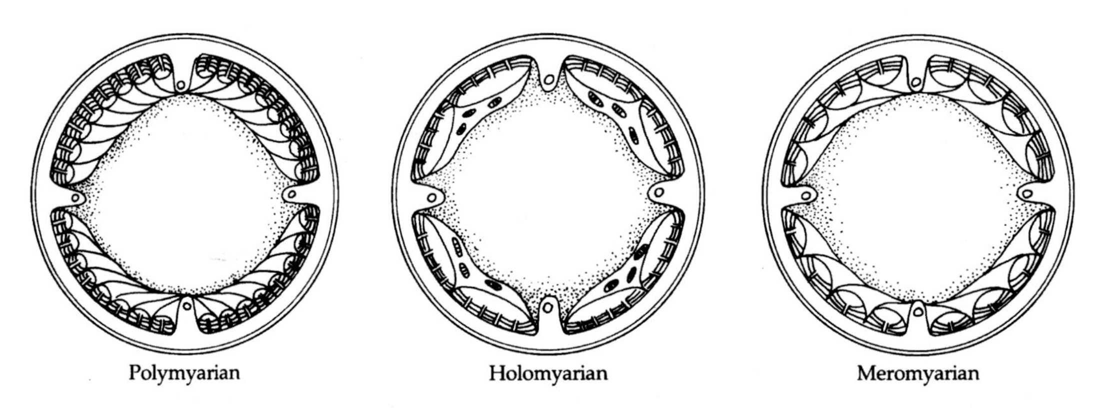

FIGURE 15.5 **Muscle arrangements in nematodes.**

The buccal capsule, or the mouth if a capsule is absent, leads into the esophagus, an elongate structure of varying length and complexity. The lumen of the esophagus is characteristically triradiate in cross-section and is lined with cuticle. The structure of the esophagus varies within the phylum, but its similarity among members of each taxon makes it an important taxonomic feature (Fig. 15.7). It may be completely muscular or completely glandular, or the anterior half may be glandular and the posterior half muscular. Esophageal action is often

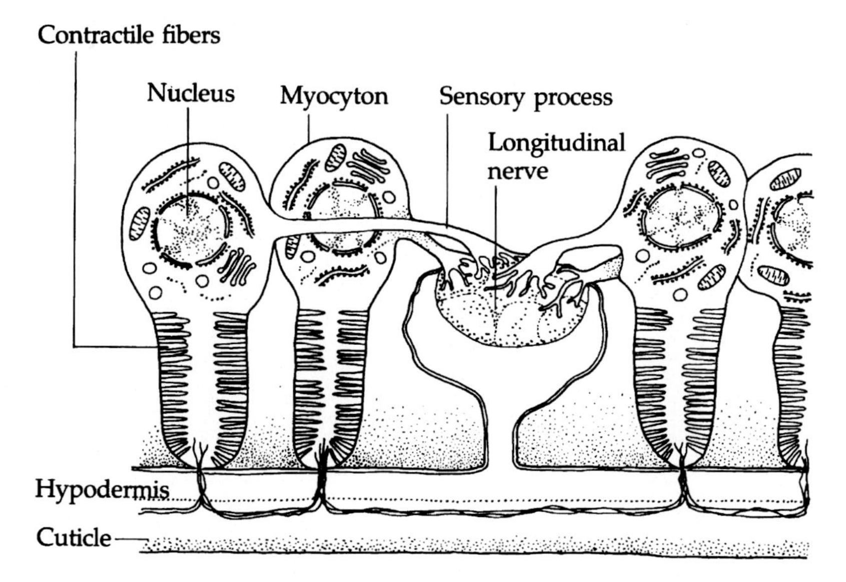

FIGURE 15.6 **Arms of four myocytons forming junctions with a nerve.**

enhanced by one or more muscular enlargements called **bulbs**. Some nematodes display a glandular portion of the esophagus which may consist of a few cells, **stichocytes**, which may comprise a prominent column, the **stichosome**, extending the length the esophagus. The individual stichocytes communicate with the lumen of the esophagus by small ducts and secrete a number of digestive enzymes, including amylase, proteases, and cellulases. In a number of species, these enzymes initiate the digestive process, which is continued in the midgut until digestion is complete. Generally, nutrients are ingested and processed in the nematode foregut for eventual digestion and absorption in the midgut.

Midgut

The esophagus empties into the midgut (or intestine) through a junction called the **esophago-intestinal valve**. The midgut is a straight tube lined with a single layer of cells bearing microvilli and a prominent glycocalyx. In smaller nematodes, the number of cells making up the midgut is fixed for each species making nematodes useful in certain developmental studies. The cellular layer rests on a basal lamina of connective tissue fibers and myofibers connected to the body wall by muscular extensions. The midgut is nonmuscular, the food being moved posteriorly by the muscular activity of the foregut and overall body movements. The form of digestion varies among nematodes. In some, digestion is extracellular; in others, both intercellular and intracellular.

Hindgut

In females, the midgut empties into the cuticle-lined hindgut, or rectum, a short, flattened tube joining the midgut and the anus. In males, the posterior most portion of the hindgut receives the products of the reproductive system via the vas deferens and is therefore called a **cloaca**.

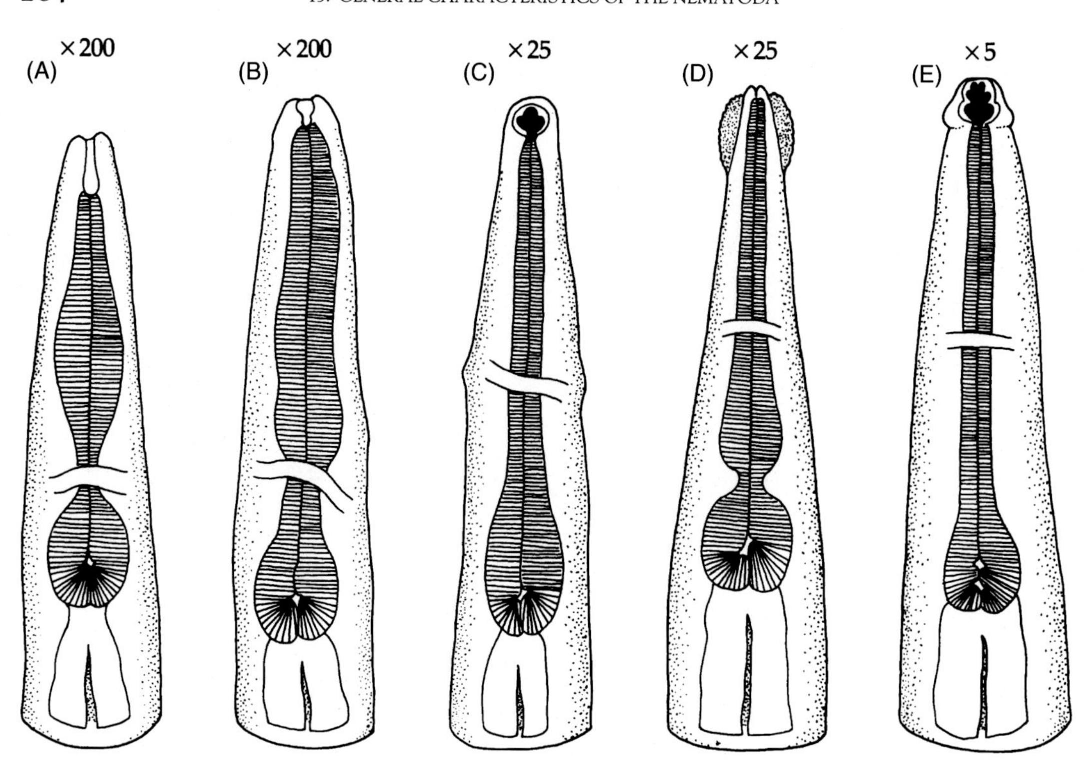

FIGURE 15.7 **Variations in foregut of some nematodes.** (A) *Rhabditis hominus*. (B) *Strongyloides stercoralis*. (C) *Ancylostoma duodenale*. (D) *Enterobius vermicularis*. (E) *Ascaris lumbricoides*.

Nervous System

There are two major nerve centers in nematodes (Fig. 15.8). One, the **circumesophageal commissure**, or **nerve ring**, surrounds the esophagus. In at least one species, the commissure consists of four nerve cells and numerous supporting cells. Associated with the commissure are various ganglia from which longitudinal nerves emanate. The anterior longitudinal nerves innervate the anterior sense organs, such as oral papillae and amphids. The posterior longitudinal nerves, embedded in the dorsal and ventral hypodermal cords, innervate organs in the posterior regions of the body. The ventral longitudinal nerve is the largest nerve in the nematode body. It passes posteriorly as a chain of ganglia. The most posterior chain branches and continues dorsally from the hypodermal cord into the pseudocoelom where it encircles the rectum to form the second nerve center, the **rectal commissure**. Peripheral nerves branch from the main longitudinal trunks and supply sensory organs, such as the phasmids, in the cuticle.

Parasitic nematodes possess both mechano- and chemoreceptors. Located around the mouth are papillae of two main types: **labial papillae** on the lips surrounding the mouth and **cephalic papillae** behind the lips (Fig. 15.9). Papillae are mechanoreceptors and are innervated by **papillary nerves** derived from the circumesophageal commissure. Other papillae may be found at different levels of the nematode body. For example, **caudal papillae**, observed in many male nematodes, aid in copulation. **Amphids** are chemoreceptors located

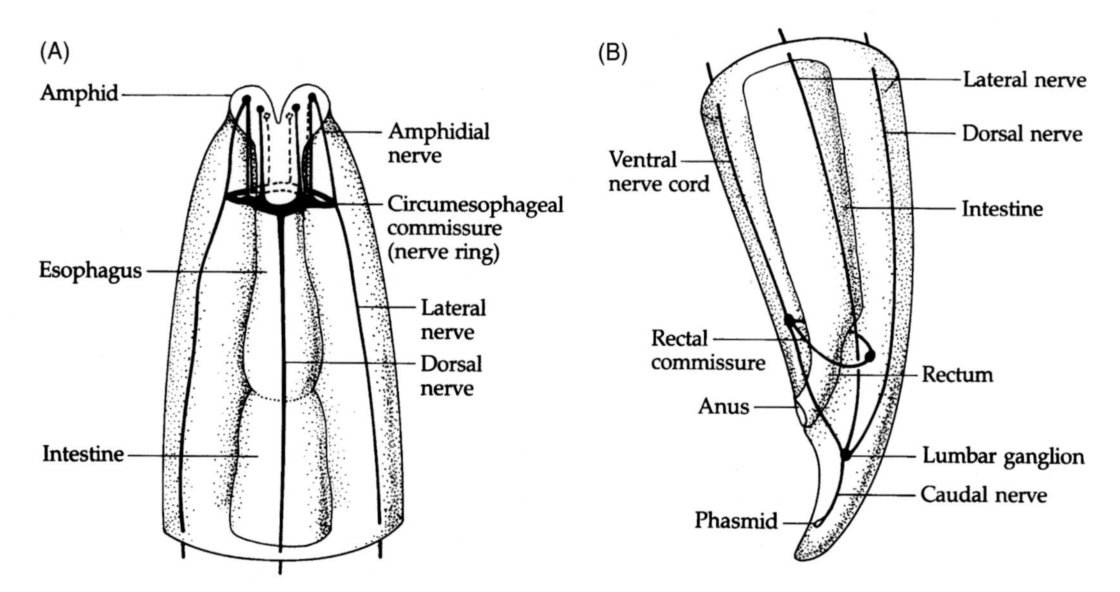

FIGURE 15.8 **Nematode nervous system.** (A) Anterior end. (B) Posterior end.

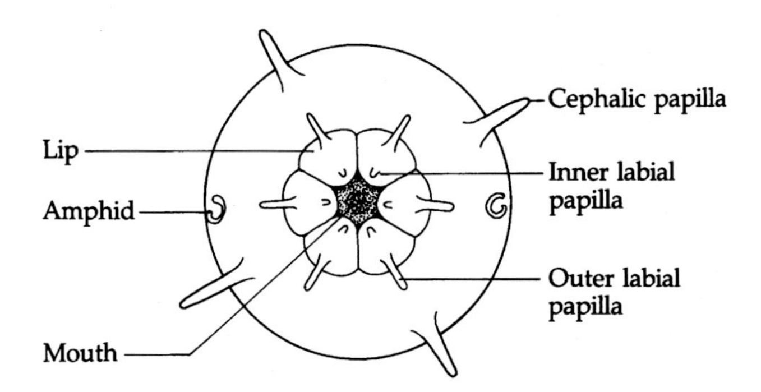

FIGURE 15.9 **Labial and cephalic papillae.** En face view of nematode showing relationship of mouth, lips, amphids, and papillae.

in shallow, anterior depressions or pits at the same level of the body as the cephalic papillae. The sensory endings are modified cilia innervated by **amphidial nerves,** which are also associated with the circumesophageal commissure. **Phasmids** comprise another set of chemoreceptors that appear near the posterior end of many parasitic species as a pair of cuticle-lined organs. While morphologically resembling amphids, phasmids bear a unicellular gland opening into the depression in addition to the sensory nerve endings.

Excretory System

The excretory system of nematodes, when present, is unique. The basic component(s) is(are) one or two **renettes,** large unicellular glands that empty through an excretory pore (Fig. 15.10). The renettes and the excretory pore are usually located anteriorly at approximately the level of

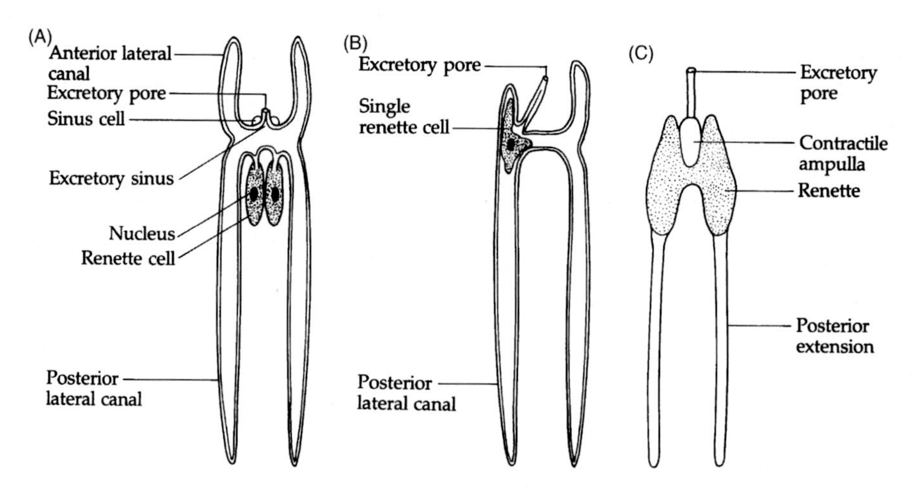

FIGURE 15.10 **Nematode excretory systems.** (A) Rhabditoid type. (B) Ascaroid type. (C) Juvenile Ancylostoma.

the circumesophageal commissure. Most frequently, renettes are associated with longitudinal excretory canals that course the length of the nematode body in the lateral hypodermis. In some genera they empty independently to the exterior through an excretory pore, or two renettes join to form an H configuration, with a crossbar that connects with a common contractile ampulla. Pulsation causes expulsion of the excreta via a duct leading to the excretory pore. In another variation the two renettes join anteriorly. In this case, excreta are emptied via a common duct through the excretory pore.

It has not been shown conclusively that this system serves as the primary vehicle for excretion. Indeed, there is strong evidence that the digestive tract is the principal excretory organ and that the system described above is chiefly osmoregulatory, with merely ancillary excretory and secretory functions.

Reproductive Systems

Although some of the monoecious species are self-fertilizing hermaphrodites and others, such as *Strongyloides stercoralis*, are **protandrogonous** (see p. 301), nematodes are usually dioecious.

Male System

While there is most often a single testis, two are not uncommon (Fig. 15.1). Tubular and usually convoluted and/or recurved, testes can be classified according to the location of their respective **germinal zones**, or regions of sperm formation. In the **telogonic** type, spermatogonial divisions occur at the blind end of the elongate testis, with the remaining portion of the testis making up the **growth zone**; in the **hologonic** type, the germinal zone extends the entire length of the testis. The **vas deferens** (sperm duct), a slender tube continuous at its proximal end with the testis, extends distally to the cloaca. Two specializations of the vas deferens are evident before it enters the cloaca. These are the **seminal vesicle**, in which sperm are stored, and the **ejaculatory duct**. In certain species, numerous unicellular **prostate glands** are present along the length of the ejaculatory duct.

Male nematodes are equipped with one or, more commonly, two **copulatory spicules** (Fig. 15.11). These cuticular structures, which usually resemble slightly curved, pointed blades, are encased within their respective spicule pouches located laterally in the cloacal wall. Each spicule contains a cytoplasmic core formed by cells lining the pouch. The spicules aid during copulation by keeping the female vulva open, thus facilitating the entry of sperm into the female reproductive tract.

In addition to spicules, other accessory structures may be present, such as a sclerotized **spicule guide** or **gubernaculum**. This structure, located along the dorsal wall of the spicule pouch, typically has inwardly curved margins and serves to guide spicules when they are extended.

Nematode sperm have no flagella or acrosomes (Fig. 15.12). Sperm can be classified into several morphological types ranging from small, rounded cells that move by pseudopodia and display distinct anterior and posterior cytoplasmic areas to those with distinct heads and cytoplasmic extensions resembling nonmotile tails. Nematode sperm do not have nuclear

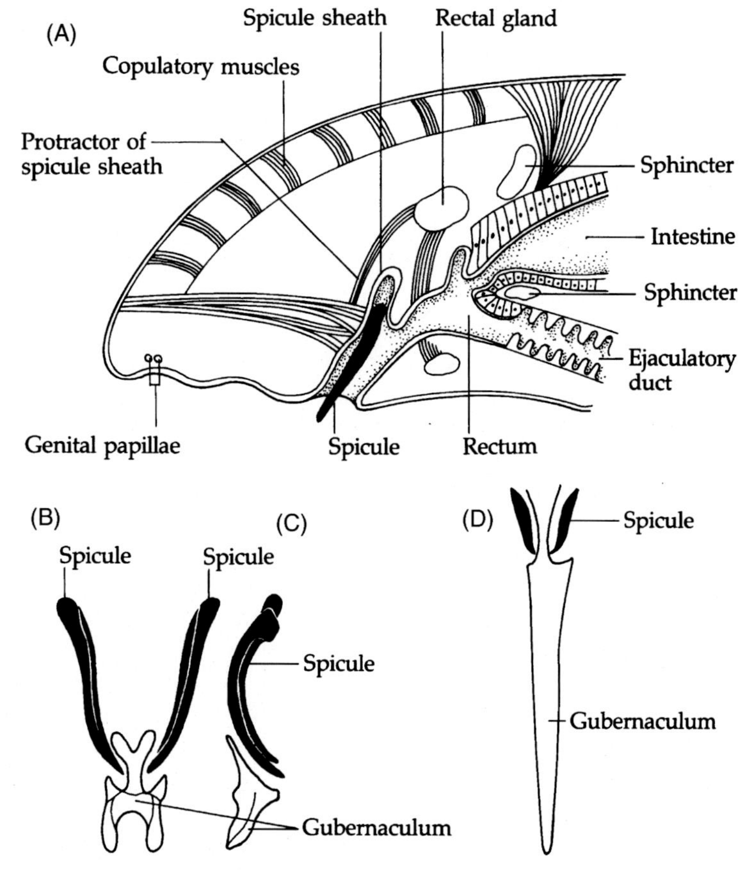

FIGURE 15.11 **Specializations of male reproductive systems of nematodes.** (A) Posterior portion showing relationship of spicules to digestive tract. (B–D) Various relationships of spicules to gubernaculums.

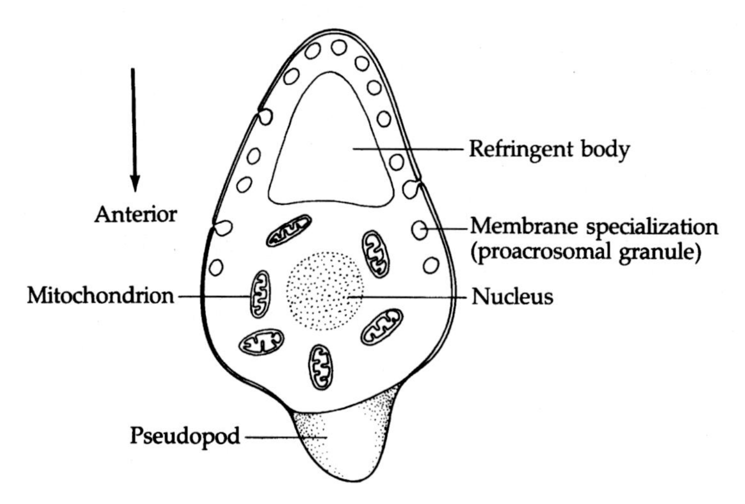

FIGURE 15.12 **A generalized diagram of a nematode sperm.**

envelopes. In some species sperm become activated only after being introduced into the female reproductive tract.

Female System

Female nematodes are usually **didelphic** [i.e., equipped with two cylindrical ovaries and uteri (Fig. 15.1)]. **Monodelphic** species, with one ovary and one uterus, occur less frequently, and, rarely, there are **polydelphic** species with multiple ovaries and uteri. The uteri in didelphic and polydelphic species unite to form a common **vagina** that opens through a gonopore, or **vulva**, usually located near midbody. The ovary, a solid cord of cells attached to a central **rachis**, constitutes the first element in the linearly arranged female reproductive system. Oogonia are produced at the proximal end of the ovary known as the **germinal zone**. As the oogonia develop into oocytes, they move distally along the rachis into the **growth zone**. Approaching the oviduct, oocytes detach from the rachis and pass distally to a portion of the oviduct called the **spermatheca** where sperm are stored. Initiation of meiosis and shell formation begins almost immediately after penetration by a sperm. The developing "egg" is moved down the tract by a combination of uterine peristalsis and hydrostatic pressure. The usually muscular, distal portion of the uterus, the **ovijector**, acts in conjunction with muscles of the vulva to expel ripe eggs.

Upon oviposition, eggs of parasitic nematodes ordinarily consist of three enveloping layers enclosing an embryo (Fig. 15.13) that may range from a few blastomeres to a completely formed larva. Immediately following sperm penetration, the oocyte secretes a fertilization membrane, which gradually thickens to form the chitinous shell. The inner membrane, the lipid layer, is formed by the zygote. As eggs pass down the uterus, a proteinaceous layer is sometimes secreted by the uterine wall and deposited on the shell surface. This layer may be rough-textured (*Ascaris*) or smooth (*Trichuris*).

Eggs of parasitic nematodes may hatch either within the host or in the external environment. The eggs of many nematodes hatch only after ingestion by a host in which case host carbon dioxide tension, salts, pH, and temperature may act as hatching stimuli. These conditions

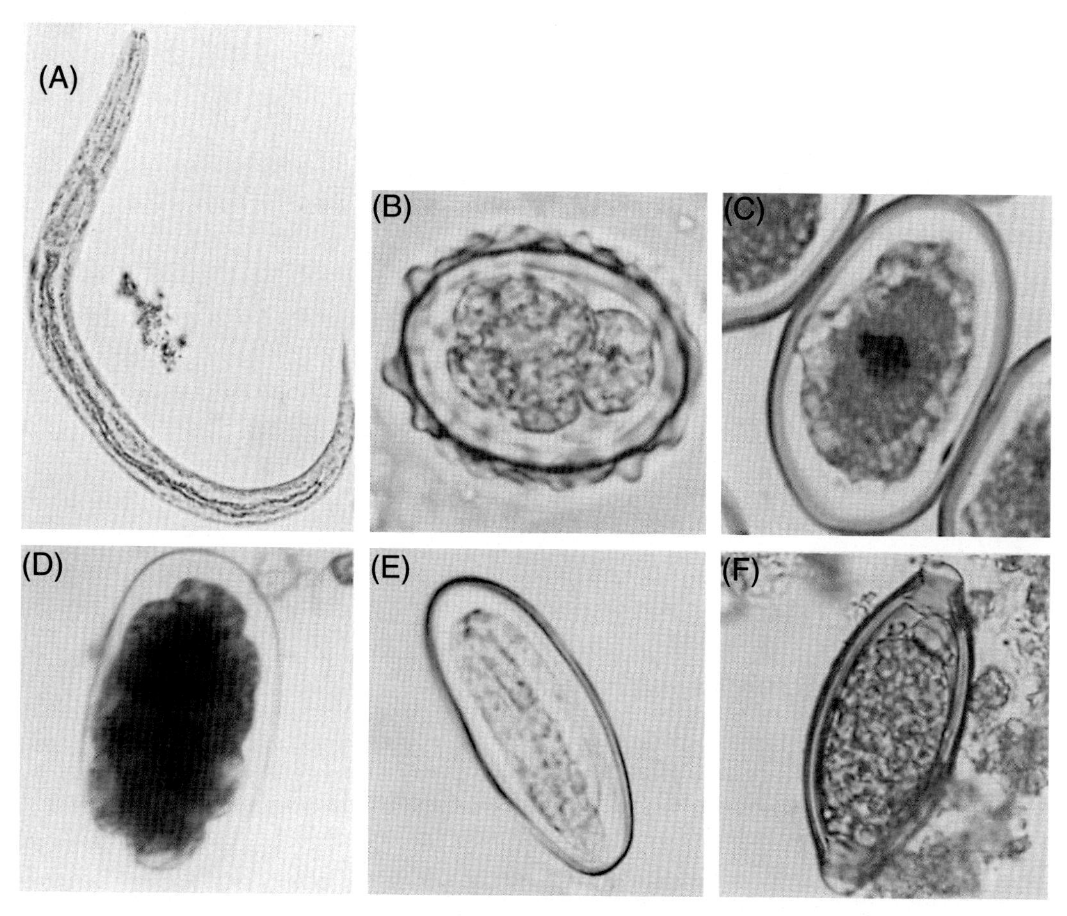

FIGURE 15.13 **Some nematode eggs and larvae.** (A) Strongyloides stercoralis rhabditiform larva. (B) Ascaris lumbricoides normal fertilized egg with developing larva. (C) *Ascaris lumbricoides* unfertilized egg. (D) Hookworm egg. (E) *Enterobius vermicularis* egg. (F) *Trichuris trichiura* egg.

stimulate the enclosed larva to initiate the secretion of enzymes to partially digest the enveloping membranes. When hatching of eggs occurs in the external environment, a first-stage larva (see the "Larval Forms" section) usually emerges. This process is controlled partially by such ambient factors as temperature, moisture, and oxygen tension and partially by the maturity of the larva. An egg will hatch only when external conditions are favorable, thus ensuring that the emerging larva does not enter an unduly harsh environment.

Molting

Nematodes undergo four molts (Fig. 15.14), each of which involves (1) formation of a new cuticle, (2) loosening of the old cuticle, (3) rupturing of the old cuticle, and (4) escape of the larva. This sequence of events is controlled by **exsheathing fluid** secreted by the larva. This fluid digests the cuticle at specific sites on the inner surface, causing it to loosen. Its ability to form new cuticle in the hypodermis before shedding the old one allows the nematode to

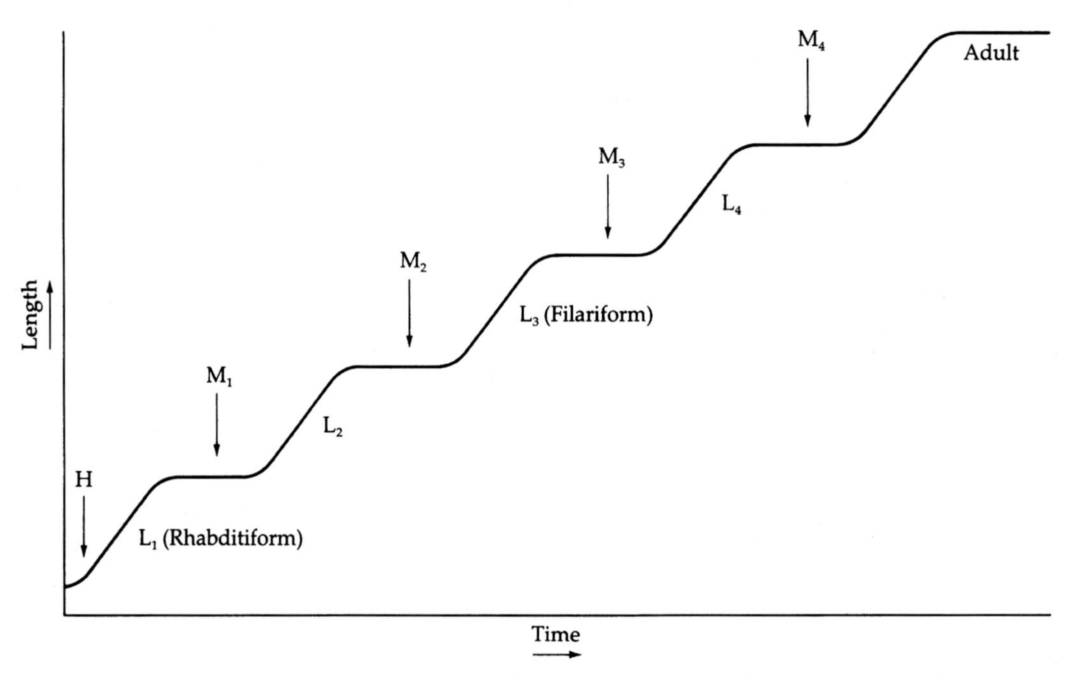

FIGURE 15.14 **Nematode growth pattern.** *H*, hatch; *M*, molt; *L*, larva.

develop continuously between molts; however, growth occurs most rapidly just after molting. This pattern of development strongly resembles that of arthropods.

In some nematodes, there is a lag phase at some stage of development, during which a phase of the life cycle is temporarily arrested. Known as **hypobiosis**, this phenomenon is thought to be an adaptation that allows the larva to withstand adverse environmental conditions while awaiting access to a new host. Renewal of the life cycle following such interruption depends upon stimuli that accompany events such as penetration of host skin or being swallowed by the host. In some species, hypobiosis may occur in the definitive host.

Larval Forms

Larval stages (Fig. 15.15A–D) preceding each of the four molts in the life cycle of parasitic nematodes are generally referred to, respectively, as first-, second-, third-, and fourth-stage larvae (i.e., L_1, L_2, L_3, L_4), the first-stage larva being the stage prior to the first molt. However, various other designations also are used for specific nematode larval forms as follows.

Rhabditiform Larva

The first-stage larvae of such parasitic species as *Strongyloides* and hookworms are called **rhabditiform larvae**. The esophagus of this small larva is joined to a terminal esophageal bulb by a narrow isthmus.

Filariform Larva

After molting twice, the rhabditiform larvae of *Strongyloides* and hookworms normally retain the remnants of their last cuticle and become ensheathed, third-stage or **filariform larvae**, in

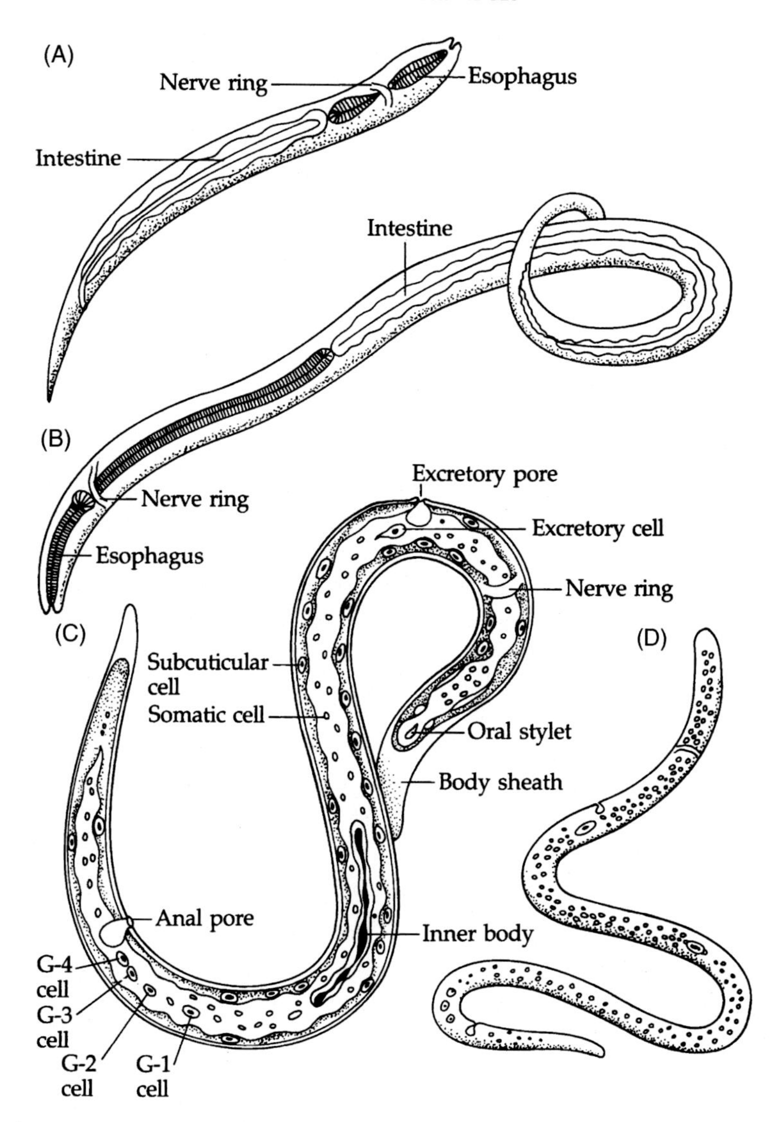

FIGURE 15.15 **Nematode larvae.** (A) Rhabditiform larva. (B) Filariform larva. (C) Sheathed filariform larva of *Wuchereria*. (D) Unsheathed microfilaria of *Onchocerca*.

which the esophagus is typically elongate and cylindrical with no terminal bulb. The filariform larva is usually the stage infective to the definitive host.

Microfilaria

The prelarvae or advanced embryos of filarial nematodes such as *Wuchereria bancrofti* and *Loa loa* are known as **microfilariae**. The larval body surface is covered by a thin layer of flattened epidermal cells. The primordia of various adult structures are visible within the pseudocoelom in the form of a conspicuous cord of nucleated cytoplasm that extends the length of the body and represents the developing digestive tract. This larva, generally found in circulating blood and cutaneous tissues, is between 0.2 and 0.4 mm in length. Unlike tissue-dwelling microfilariae (see p. 283), those that inhabit host blood are usually surrounded by a thin, cuticular sheath.

Physiology

Parasitic nematodes derive much of their energy from the metabolism of glucose. Glycogen reserve is stored primarily in the hypodermis, the intestine, the noncontractile parts of the muscles, and parts of the reproductive system. It is difficult to establish with certainty whether adult intestinal nematodes are exclusively aerobic or anaerobic in their usage of carbohydrate or whether free-living larvae are invariably aerobic. Experimental data strongly suggest the importance of determining not only whether each species uses oxygen but also the manner in which it is utilized. For instance, intestinal nematodes belonging to the genus *Ascaris* can use oxygen if available; however, it serves only as a terminal electron acceptor in a system independent of an electron transport system or a functional Krebs' cycle. On the other hand, both larvae and adults of *Trichinella spiralis* have a functional TCA cycle and an electron transport system with oxygen being the final electron acceptor. Thus, the former organism generates ATP by means of substrate-level phosphorylation, while the latter organisms do so primarily through oxidative phosphorylation. The eggs and first- and second-stage larvae of *Ascaris* spp. exhibit aerobic metabolism with a functional Krebs' cycle. Indeed, optimal larval development in most species that have been studied is dependent upon relatively high concentrations of oxygen.

Certain relatively large nematodes possess some sort of oxygen-binding system, including utilization of the respiratory pigments myoglobin and hemoglobin. Such pigments are usually found in the pseudocoelomic fluid and the hypodermis.

CLASSIFICATION OF THE NEMATODA

The following system of classification of the nematoda is the classical morphology-based system. The following reference is presented for those readers who might be interested in the molecular-based system.

DeLey, P. & Blaxter, M. (2004). A new system for nematoda: combining morphological characters with molecular trees, and translating clades into ranks and taxa. *Nematology Monographs and Perspectives*, 2, 633–653.

Phylum Nematoda

Bilaterally symmetrical unsegmented pseudocoelomates; body generally elongate, cylindrical, covered by cuticle; mouth terminal surrounded by lips; sexes separate; anterior body characteristically with 16 setiform or papilliform sensory organs and two amphids (chemoreceptors); digestive tract complete, with subterminal anus; excretory system, when present, empties through anterior, ventromedian pore; body musculature limited to longitudinally oriented muscles; no respiratory or circulatory systems; eggs with determinate cleavage, oviparous or ovoviviparous; stages in life cycle are egg, four juvenile (larval) stages, and adult.

Class Adenophorea

Amphids postlabial, variable in shape (porelike, pocketlike, circular, or spiral); cephalic sensory organs setiform to papilloid, postlabial, and/or labial; setae and hypodermal glands present; papillae usually present on body; hypodermal cells uninucleate; cuticle usually smooth, but transverse or longitudinal striations sometimes present; excretory organ, if present, single-celled, ventral, without collecting tubules; caudal glands (three) usually present (absent in most members of Dorylaimida, Mermithida, and Trichocephalida); usually two testes in males, with single, ventral series of papilloid or tuboid preanal supplements; male tail rarely with caudal alae; parasitic species associated with invertebrates, vertebrates, and plants; many free-living; includes most marine nematodes.

ORDER TRICHOCEPHALIDA

With protrusible axial spear in early larval stages; amphids adjacent to lip region; posterior esophageal glands in one or two rows along esophageal lumen, not enclosed by stichosome; stichosome and individual gland openings posterior to nerve ring; males and females with single gonad; germinal zones of male and female gonads extend entire length and form a serial germinal area on one side or around gonoduct; males with one or no spicule; eggs operculate; life cycle either direct (often requiring cannibalism) or indirect (involving arthropod or annelid intermediate host); adults parasitic in vertebrates.

SUPERFAMILY TRICHUROIDEA

Stichosome of adults as single row on each side of esophagus (two rows in early larval development); body divided into elongate, narrow anterior end with esophagus with stichosome, and posterior half with reproductive system beginning at esophagointestinal junction; bacillary band (glandular and nonglandular cells of unknown function) occurs laterally; glandular tissue emptying to exterior through cuticular pores; males and females with single gonad, reflexid; males with single spicule; eggs operculate; females oviparous, males small and degenerate in some species, in uterus of female; parasitic in humans and other mammals; life cycle direct or indirect. (Genus mentioned in text: *Trichuris*.)

SUPERFAMILY TRICHINELLOIDEA

Stichosome as single, short row of stichocytes; body not distinctly divided into two regions; no bacillary band present; female genital pore opening far anterior, in region of stichosome; ovary posterior to stichosome; males with single testis but no spicule; females viviparous. (Genus mentioned in text: *Trichinella*.)

Class Secernentea

Amphids usually opening to exterior through pores located dorsolaterally on lateral lips or anterior extremity (in some species the amphidial apertures are oval, cleftlike, slitlike, or located postlabially); cephalic sensory organs are situated on lips and are porelike or papilliform, generally 16 in number arranged in two circles (a circumoral circle of 6 and an outer circle of 10), may be reduced in some species; caudal phasmids present; hypodermis uninucleate or multinucleate; cuticle from two to four layers, almost always transversely striated, laterally modified into a "wing" area marked by longitudinal striae or ridges, generally raised slightly above body contour; lateral alae may extend out a distance equal to body diameter;

esophagus of most species with three esophageal glands, one dorsal (opening in anterior half of body) and two subventral (opening in posterior half of body); excretory system emptying ventromedially through cuticularized duct on one or both sides of body; somatic setae or papillae absent on females; caudal papillae may occur on males; male preanal supplements paired, often elaborate; some males with medioventral preanal supplementary papillae; males commonly with caudal alae (known as copulatory bursa).

SUBCLASS RHABDITIDA Esophagus of juveniles divided into corpus, isthmus, and valved post-corporal bulb; lumen of esophageal bulb expanded into trilobed reservoir lined with cuticle; buccal cavity (stoma) without movable armature and composed of two parts (cheilostome and esophastome), each possibly subdivided into two or more sections; males generally with well developed bursae supported by cuticular rays or papillae.

ORDER RHABDITIA

Number of lips varies from six to none (6, 3, 2, 0); buccal cavity generally tubular but may be separated into five or more sections; esophagus divided into corpus, isthmus, and bulb; terminal excretory duct lined with cuticle with paired, lateral collecting tubules running posteriorly; females with one or two ovaries; intestinal cells uni-, bi-, or tetranucleate; caudal alae (copulatory bursa), if present, contain papillae rather than supporting rays; parasites of invertebrates and vertebrates.

SUBORDER RHABDITINA Buccal cavity usually cylindrical, without distinct separation, generally two or more times as long as wide; lips usually distinct, with cephalic sensory papillae and porelike amphids; esophagus divided into corpus (procorpus and metacorpus) and postcorpus (isthmus and valved bulb); females with one or two ovaries; males generally with paired spicules and gubernaculum; caudal alae (copulatory bursa) common (absent in some families); parasites of invertebrates and vertebrates.

SUPERFAMILY RHABDITOIDCA Well-developed, cylindrical buccal cavity (stoma); lips vary from two to six; esophagus, at least in larvae, include muscular posterior bulb with rhabditoid valve; caudal alae of males supported by five to nine papilloid supplements; parasites of invertebrates and vertebrates. (Genus mentioned in text: *Strongyloidcs*.)

ORDER STRONGYLIDA

Labial region of three or six lips or replaced by corona radiata; stoma well developed or rudimentary (never collapsed and unobtrusive); esophagus of juveniles typically rhabditiform (corpus, isthmus, bulb); esophageal bulb contains typical trilobed rhabdiform valve; esophagus of adults cylindrical to clavate; excretory system includes paired lateral canals and paired subventral glands; females with one or two ovaries and heavily muscular uterus; males with muscular copulatory bursa; with paired genital papillae; males with paired, equal spicules; adults parasitic in vertebrates.

SUPERFAMILY ANCYLOSTOMATOIDEA Stoma thick-walled, globose, armed or unarmed anteriorly with teeth or cutting plates; without lips or corona radiata; copulatory bursae of males with greatly reduced branches; adults parasitic in intestine of mammals; LI and L, free living; commonly known as hookworms. (Genera mentioned in text: *Ancylostoma*, *Necator*.)

ORDER ASCARIDA

Oral opening usually surrounded by three lips (absent in some species); paired porelike amphids present; esophagus of some species with short, swollen region in stomatal region, followed by cylindrical to club-shaped region, often ending in terminal bulb with three-lobed valve; in a few exceptions there are appendages (ceca) extending from posterior region of esophagus; excretory system with lateral collecting tubules, in some species extending posteriorly and anteriorly (H shaped); males usually with two spicules (none or one in others); with or without gubernaculum; females usually with two ovaries (some have multiple ovaries); adults parasitic in vertebrates.

SUPERFAMILY ASCAIDOIDEA Bodies 1–40 cm long; cuticle thick in larger species, superficially annulated; terminal oral opening usually surrounded by three well-developed lips; porelike amphids on subventral lips; stoma poorly developed (collapsed); esophagus cylindrical to clavate; appendage (cecum) may extend from posterior portion of esophagus over anterior portion of intestine; second cecum sometimes present, extending forward past base of esophagus; females usually with paired ovaries; males with two spicules; small gubernaculum present in few species; adults parasitic in vertebrates; life cycle direct or indirect. (Genera mentioned in text: *Ascaris*, *Toxocara*, *Anisakis*.)

SUPERFAMILY OXYUROIDEA Lips greatly reduced or absent; cephalic sensilla in whorl of eight or four; ventrolateral sensilla absent; stoma vestibular; esophagus variable but posterior bulb always valved; intestinal ceca absent; males may have precloacal suckers; copulatory spicules may be greatly reduced; adults usually parasites of amphibians, reptiles, and mammals. (Genus mentioned in text: *Enterobius*.)

ORDER SPIRURIDA

Frequently with two lateral lips or pseudolabia (some species with four or more lips, rare species without lips); oral aperture variable in shape, encircled by teeth; amphids laterally situated on anterior extremity; stoma varying from cylindrical and elongate to rudimentary; esophagus generally divided into narrow anterior portion and expanded postcorpus enclosing multinucleate glands; hatched larvae generally provided with cephalic hook and porelike phasmids on tail; parasites of annelids, arthropods, molluscs, and terrestrial and aquatic vertebrates.

SUPERFAMILY FILARIOIDEA Oral aperture circular or oval, usually surrounded by eight sensilla of external circle (internal circle absent or consisting of two or four papillae); stoma small and rudimentary esophagus with multincleate glands; corpus and postcorpus not distinct; vulva usually in anterior portion of body; copulatory spicules of males equal or unequal; caudal alae present or absent; no gubernaculum; parasites of amphibians, reptiles, birds, and mammals. (Genera mentioned in text: *Wuchereria*, *Brugia*, *Onchocerca*, *Loa*, *Dirofilaria*.)

SUPERFAMILY DRACUNCULOIDEA Stoma commonly reduced to small vestibule; full complement of sensilla surrounding oral opening, with internal circle comprised of six well-developed sensilla and external circle of eight separate and well-developed sensilla; vulva in midbody region, atrophied in mature females; posterior intestine atrophied in females; males without well-developed caudal alae, small and postcloacal if present; adults tissue parasites of fish, reptiles, and mammals. (Genus mentioned in text: *Dracunculus*.)

Suggested Readings

Ashton, F. T., & Schad, G. A. (1996). Amphids in *Strongyloides stercoralis* and other parasitic nematodes. *Parasitology Today*, *12*, 187–194.

Bird, A. F. (1971). *The structure of nematodes*. New York: Academic Press.

Lee, D. L., & Atkinson, H. J. (1977). *Physiology of nematodes*. New York: Columbia University Press.

Roberts, M. C., & Modha, J. (1997). Probing the nematode surface. *Parasitology Today*, *13*, 52–56.

Rosenbluth, J. (1965). Ultrastructural organization of obliquely striated muscle fibers in *Ascaris lumbricoidcs*. *Journal of Cell Biology*, *25*, 495–515.

Ruiz-Tiben, E., & Hopkins, D. R. (2006). Dracunculiasis. In D. H. Molyneux (Ed.), *Advances in parasitology* (pp. 275–309). (Vol 61). London: Elsevier.

Wright, K. A. (1987). The nematode's cuticle-Its surface and the epidermis: Function, homology, analogy—A current consensus. *The Journal of Parasitology*, *73*, 1077–1083.

CHAPTER

16

Intestinal Nematodes

Chapter 16 introduces the major intestinal nematodes, the Adenophorea (*Trichuris trichiura* and *Trichinella spiralis*) and the Secernentea (*Strongyloides stercoralis*, and the two hookworm species, *Ancylostoma duodenale* and *Necator americanus*). These nematodes display varying requirements for oxygen during their life cycle. As with other parasitic infections, nematode infection also has effects on the immune response of the human host. *T. trichiura* has a worldwide distribution with an estimated 1 billion human infections. It is chiefly tropical, especially in Asia, and to a lesser degree in Africa and South America. In the US, infection is rare overall but may be common in the rural Southeast. *T. spiralis* is a small (1.5 by 0.04 mm) nematode parasite of humans. It has an unusual life cycle and is widespread and clinically important throughout the world. The Secernentea are the main class of nematodes and are characterized by having an excretory system possessing lateral canals and numerous caudal. As like all nematodes, Secernentea do not possess a circulatory or respiratory system.

Intestinal adult nematodes display two interesting phenomena. One is their varying requirements for oxygen during their life cycles. The other is the effect of the infections on the immune responses of the human host.

While discussion of the various intestinal nematodes that infect humans is structured on their evolutionary or taxonomic status, an alternate means for discussion can be based on their increasing adaptation to parasitism. Such a classification uses dependency on oxygen environments as the criterion for determining their adaptation to parasitism. Nematodes that are least dependent on such environments may be considered the most advanced. Those that demonstrate the greatest dependency on such environments (e.g., *S. stercoralis*, which demonstrates a free-living cycle as well as a parasitic cycle, are considered the most primitive. In such cycles, the egg and larva usually require a period of time in the soil. The larva also displays a mandatory lung migratory phase. The hookworms, *A. duodenale* and *N. americanus*, on the other hand, have no free-living cycle, but the egg and larva require a soil phase and the larva requires an additional lung migratory phase. The egg of *Ascaris lumbricoides* requires a period of time in soil and the larva, which hatches from the egg in the host, requires a lung migratory phase. In the life cycle of *T. trichiura*, the lung migratory phase for the larva is not present, but the egg still requires a soil phase. The intestinal nematode most adapted to parasitism according to this scheme is *Enterobius vermicularis*. No stage in its life cycle requires a period of time either in the soil or in the lungs. In fact, autoinfection is not uncommon. Intestinal nematode infections slant the host immune response towards a Th2 type profile leading to IgE production, mast cell activation, mucus secretion, and ultimately parasite clearance.

Human Parasitology. http://dx.doi.org/10.1016/B978-0-12-813712-3.00016-3

THE ADENOPHOREA

Parasitic nematodes belonging to the class Adenophorea possess neither phasmids nor excretory system. Two parasites of the human intestinal tract, *T. trichiura* and *T. spiralis*, belong to this class.

Trichuris Trichiura

In adult *T. trichiura* (Fig. 16.1), the anterior portion of the body is long and slender, whereas the posterior portion widens abruptly and thickens, giving the worm the appearance of a bullwhip; hence, its common name "whipworm." Males are slightly smaller than females, the latter measuring 30–50 mm in length. In both sexes, a capillary-like esophagus extends two-thirds of the body length and is encircled along much of its length by a series of unicellular glands, the stichocytes. The posterior extremity of males is characteristically coiled and equipped with a single spicule enclosed in a spinose, retractile, and cuticular sheath.

Life Cycle

Adult whipworms occur primarily in the human host's colon but may inhabit the appendix and rectum as well (Fig. 16.2). The female deposits up to 5000 eggs daily; these are typically barrel-shaped with two polar plugs. The eggs measure 50 by 22 μm and contain an uncleaved zygote at oviposition, after which the unembryonated eggs pass to the exterior in feces and develop slowly in warm, damp soil. An unhatched, infective, third-stage larva develops in 3–6 weeks.

New human hosts become infected when these embryonated eggs are ingested with contaminated food or water or from fingers. The larvae hatch in the upper portions of the small intestine and quickly burrow into the cells of the intestinal villi near the Crypts of Lieberkuhn, where they mature, undergoing two molts in about 3–10 days. Subsequently, they migrate to the caecal region, and develop to sexual maturity in 30–90 days from the time the eggs were ingested. Adult worms embed their long, slender bodies deeply into the colon submucosa. Their posterior ends break free into the lumen, allowing fertilization to occur with eggs subsequently voided with feces. Little is known about their nutritional requirements, but there is no evidence that they depend on host blood for any nutritional requirements. While these worms normally survive approximately 2 years in the human host, there have been reports of infections lasting 8 years or longer.

Epidemiology

Whipworm infection occurs worldwide, most frequently in tropical countries, and, while the prevalence is very high (estimated number of infections is more than a billion, making it the third most common nematode infecting humans), worm burdens, fortunately, are light in the majority of cases. In the United States, it is the second most common nematode infecting humans (after *Enterobius*) and is found predominantly in the Southeast, paralleling human Ascaris infections. Generally, the worm is found in areas of warm climate, heavy rainfall, dense shade, and sanitary conditions conducive to soil pollution. Children are more heavily infected than adults because they are more apt to come into close physical contact with contaminated soil. A degree of acquired immunity is also probable.

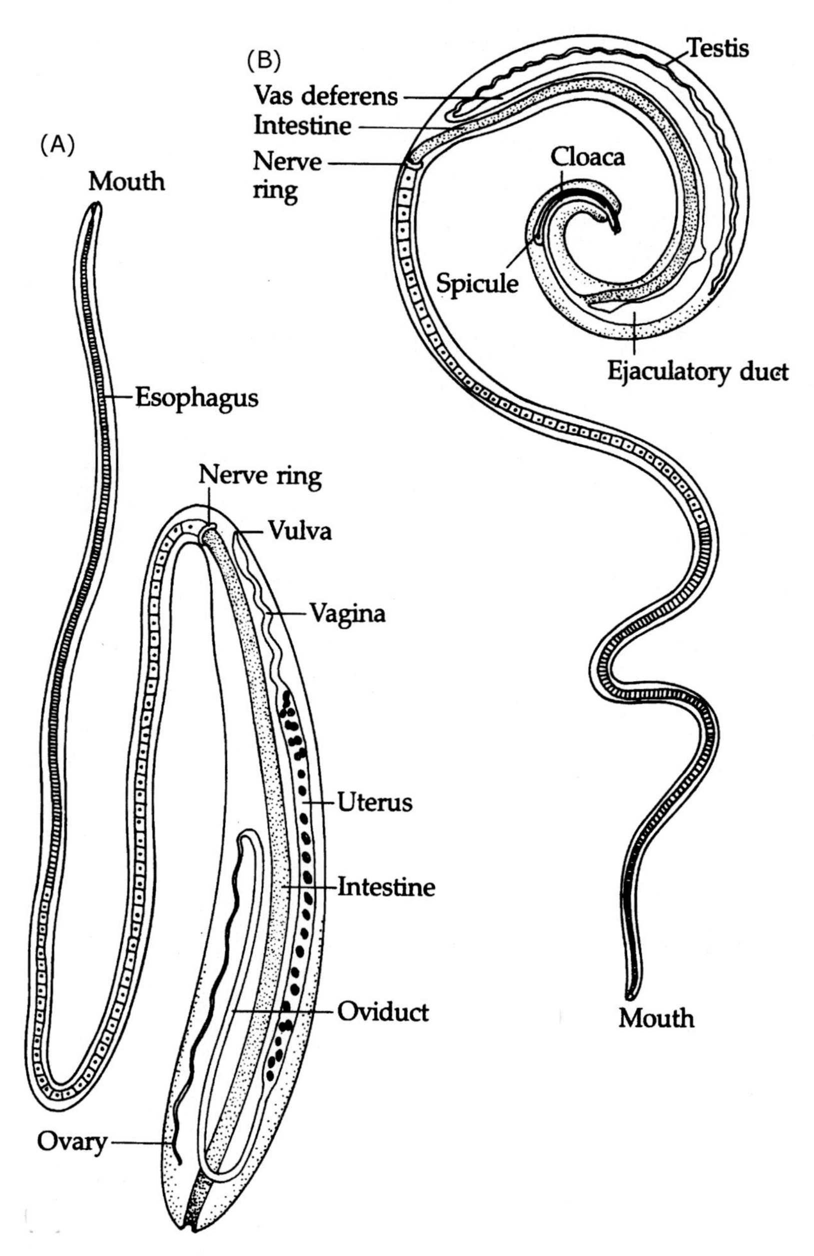

FIGURE 16.1 **Adult Trichuris trichiura.** (A) Female. (B) Male.

Symptomatology and Diagnosis

Most infections are light with no clinical symptoms. Chronic infections, however, produce such characteristic symptoms as bloody stools, pain in the lower abdomen, weight loss, rectal prolapse, nausea, and anemia. In the case of rectal prolapse, adult worms can be observed

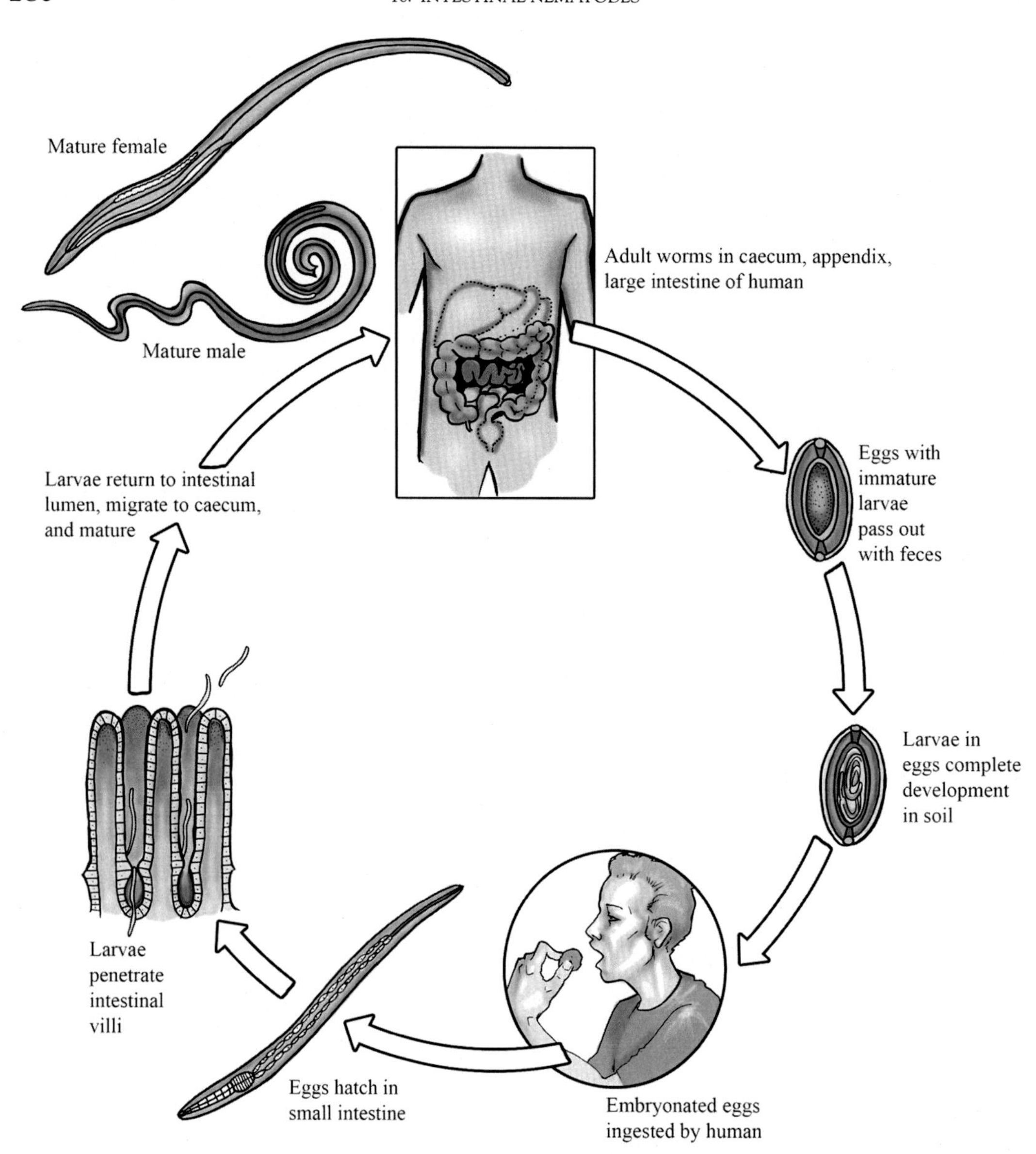

FIGURE 16.2 **Life cycle of Trichuris trichiura.** *Credit: Image courtesy of Gino Barzizza.*

externally, embedded in the rectal mucosa. Anemia results primarily from hemorrhage when the worms penetrate the intestinal wall, although some blood loss may be attributable to the worms ingesting host blood. In heavy infections, secondary bacterial infections are common, a result of the worms' penetration of the mucosal lining, providing entry for pathogenic bacteria. Mixed infections of whipworm and *E. histolytica*, hookworm, or *A. lumbricoides* are not uncommon. Identification of eggs from fecal material constitutes diagnosis.

Chemotherapy

Mebendazole or albendazole, the drugs of choice, are most effective when administered orally for 3 consecutive days. Mebendazole and albendazole are two of the benzimidazole-carbamate class of compounds that generally cause degenerative changes in the nematode intestine. They are believed to have a depolymerizing effect on cytoskeletal elements, such as microtubules. The drugs are contraindicated during pregnancy or the first year of a child's life.

Host Immune Response

Recent research indicates an IgE-associated level of protection in *T. trichuris* infections. In addition, an elevated IL-10 response is observed. The presence of IL-10, produced by monocytes and B and T cells, is known to downregulate Th1T cell responses. In *T. trichuris* infections, a negative correlation is observed between the IL-10 production and age of the infected individual, suggesting that any regulatory effects of the cytokine on Th1 pro-inflammatory responses are more likely found in younger patients. Conversely, Th1 pro-inflammatory responses with IFN-γ production increases in older patients with pathogenesis at the mucosal surfaces.

Trichinella Spiralis

Small and slender, adult trichina worms are rarely observed (Fig. 16.3). The male, measuring 1.5 by 0.04 mm, has a curved posterior end with two lobed appendages called alae. The male reproductive system, with its single testis, is located in the posterior third of the body. The female, measuring 3.5 by 0.06 mm, has a bluntly rounded posterior end and is monodelphic, with the vulva in the anterior fifth of the body.

Life Cycle

T. spiralis requires only one host in its life cycle, with larval and adult stages occurring in different organs (Fig. 16.4). Infection results from the consumption of meat, most commonly poorly cooked pork, containing encapsulated first-stage larvae. Once ingested, the larvae are released from their capsules in the duodenum by the action of the host's digestive enzymes. Shortly thereafter the freed larvae penetrate the absorptive and goblet cells of the mucosa. There, undergoing four molts within 24–30 h, they reach sexual maturity.

Soon after copulation, the male passes out of the host, while the female burrows deeper into the mucosa and submucosa, sometimes entering the lymphatic ducts to be carried to the mesenteric lymph nodes. About 5 days after initial ingestion of the infective larvae by the host, the adult ovoviviparous female begins depositing first-stage larvae. It is estimated that one female, in a period of 5–10 days, produces about 1500 larvae. Eventually, the female dies, and the first-stage larvae, about 0.1-mm long, are carried by the lymphatic and blood vessels to the right side of the heart in venous blood.

From the heart, the larvae enter the peripheral circulation and are carried to various tissues of the body. It is only in striated muscles, especially those of the diaphragm, jaws, tongue, larynx, and eyes, that larvae develop into the infective stage. Penetration of muscle cells and establishment as intracellular parasites within myofibers occurs about 6 days after initial infection. Usually a single larva occupies each muscle fiber (Fig. 16.5). After penetration,

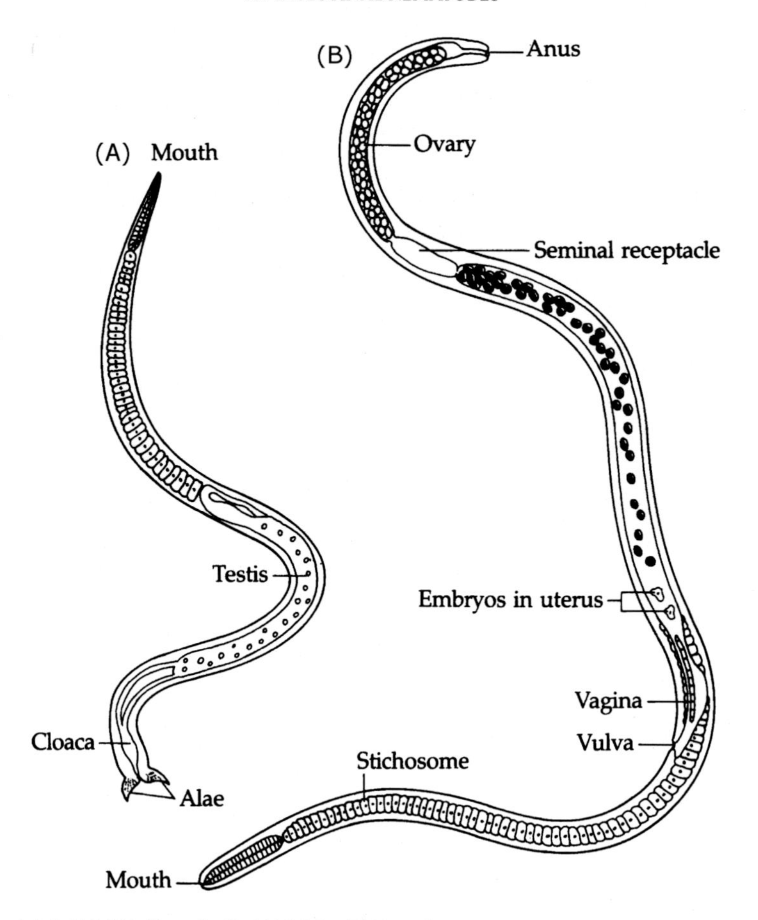

FIGURE 16.3 **Adult Trichinella spiralis** (A) Male. (B) Female.

several changes occur in the parasitized contractile fiber. The myofilaments disappear, mitochondria degenerate, the smooth endoplasmic reticulum increases, and the nuclei enlarge forming a complex surrounding the larva. Shortly thereafter, collagen is produced by neighboring fibroblasts. The cysts-like nucleated mass surrounding the larva is known as a nurse cell. On or about the 17th day, the larva begins to coil, absorbing nutrients from the host muscle sarcoplasm. A network of capillaries eventually surrounds the parasite–nurse cell complex probably affording additional nourishment to the growing larva. Growth is rapid, the larva reaching a length of about 1 mm in approximately 8 weeks, at which time it becomes infective. Encapsulation begins at about the 21st day as the larva is gradually enveloped by a double, ellipsoidal capsule (0.25–0.5-mm long) of host origin. The outer capsule membrane is derived from the sarcolemma, while the inner membrane is a combination of degenerative myofibers and other cells such as fibroblasts. Capsule formation is complete in about 3 months.

Eventually, the capsule becomes calcified, a process that may begin as early as 6 months after initial infection and requires about 18 months for completion. If calcification is delayed,

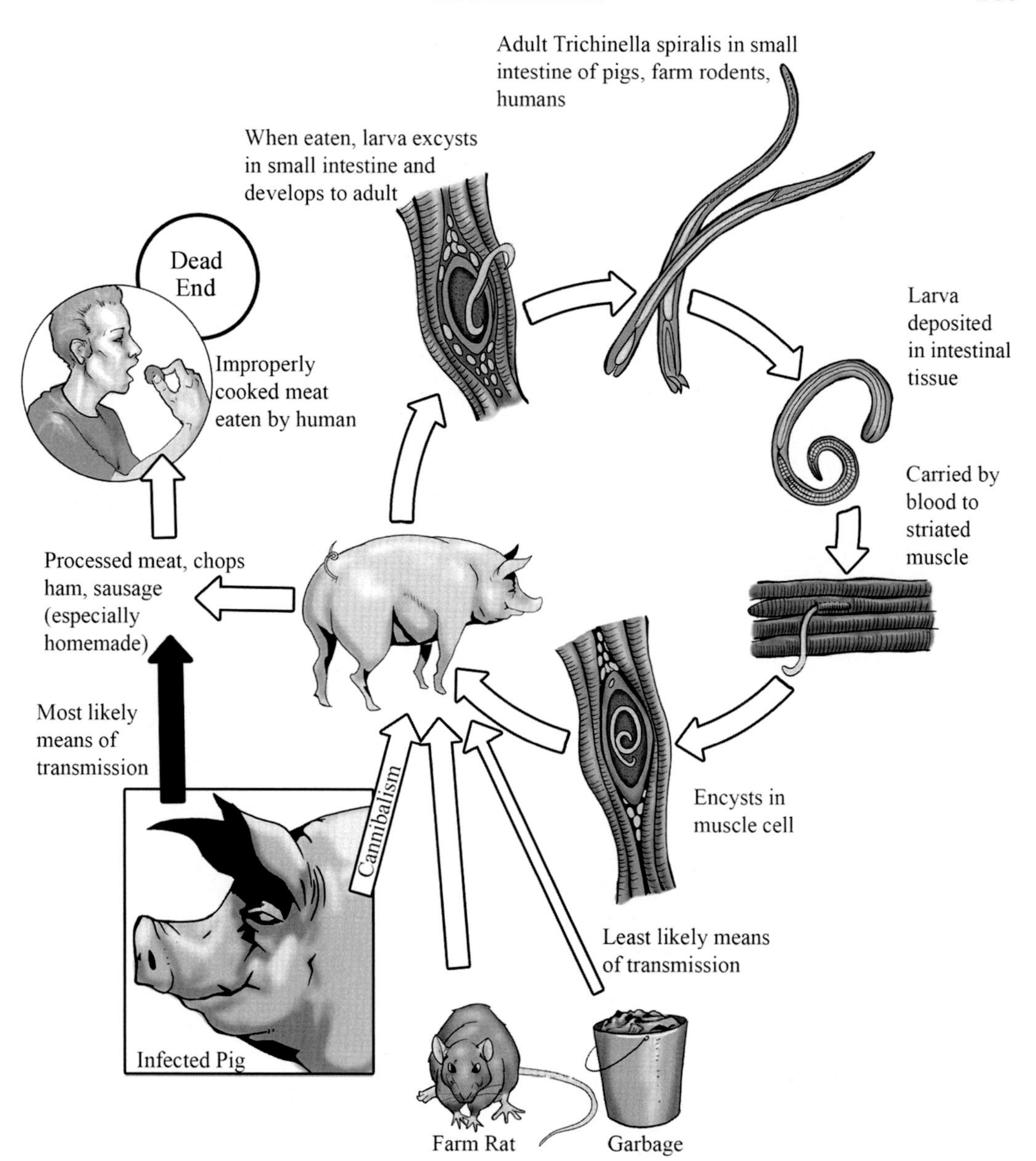

FIGURE 16.4 **Life cycle of Trichinella spiralis.** *Credit: Image courtesy of Gino Barzizza.*

a *T. spiralis* larva can remain viable for several years. During capsule formation, the enclosed larva enters developmental arrest, a state in which it can survive indefinitely. When muscle harboring the encapsulated larva is eaten by a carnivorous mammal, the larva excysts and reinitiates the life cycle.

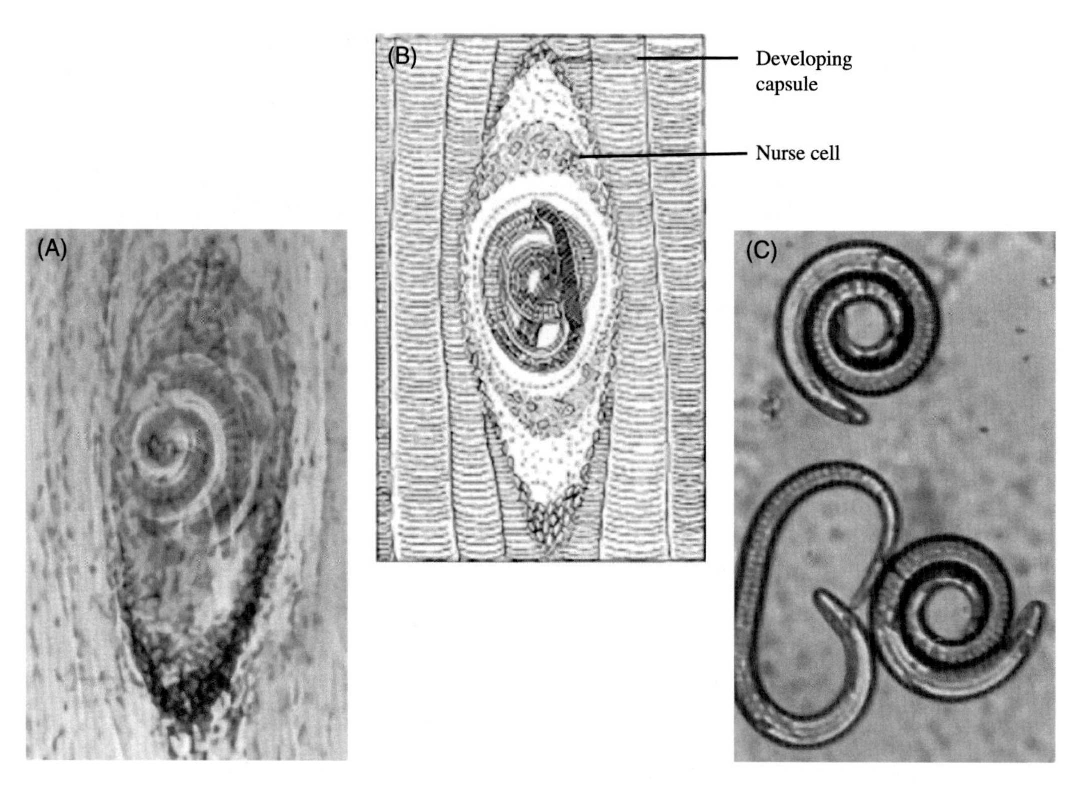

FIGURE 16.5 **Trichinella spiralis larvae encysted in skeletal muscle.** (A) Larva–nurse cell complex. (B) Schematic representation of larva–nurse cell complex. (C) Unstained, living larvae freed from nurse cells by pepsin–HCl digestion.

Epidemiology

The term sylvatic trichinellosis denotes the cycling of the disease between wild carnivores and their prey or carrion. Urban trichinellosis, on the other hand, is the term used to designate the cycling of the disease among humans, rats, and pigs. Rats and pigs feeding on garbage that includes infected pork waste, in turn, become infected. Dead or dying infected rats are themselves eaten by the pigs. Raw or poorly cooked pork, usually sausage, harboring infective larvae then becomes the vehicle for human infections. In nature, the cycle is also maintained by cannibalistic rats.

In Alaska, where polar and black bears are common sources of human infection, there is overlapping of sylvatic and urban trichinellosis. The disastrous 1897 Andre hydrogen balloon expedition to the Arctic provides dramatic evidence of trichinellosis in polar bears. A book and a movie, both titled *The Flight of the Eagle*, recount the story of these ill-fated Swedish explorers who, having lost most of their supplies, perished after they resorted to eating the meat of a polar bear they had killed. Lacking any means of kindling fires, they were forced to eat the meat raw. Unfortunately, the meat was infected, and the explorers died, not from exposure but from trichinellosis. Evidence of the cause of the tragedy was discovered 33 years later in the frozen, stored carcass of the bear.

Trichinellosis is a cosmopolitan disease that occurs most commonly in Europe and the United States, where there are estimated to be about 150,000–300,000 cases a year. However, the number of clinical cases reported is less than 150. The disease is rare in parts of the tropics and subtropics for the opposite reason that it is found in the United States: low consumption of pork, and of meat in general, by peoples whose diet consists primarily of fish. However, there are parts of the tropics where the people derive their protein from pork and meat, yet, ironically the disease is rare. Religious bans keep still other peoples, such as Jews, Hindus, and Moslems, free of the disease, and, obviously, vegetarians are not exposed to infection.

Symptomatology and Diagnosis

The primary symptoms of trichinellosis result from larval invasion of muscle and other tissues and the hyperimmune reaction of the host to the metabolic by-products and secretions of the larvae. While relatively few victims have infections heavy enough to produce clinical symptoms, those that do occur appear during three clinical phases: (1) mild, following penetration of adult females into the mucosa and submucosa; (2) severe, during migration of larvae; and (3) moderate, after penetration and encapsulation of larvae in muscle cells. Symptoms usually abate after 30 days.

Symptoms resulting from the first phase appear 12 h to 2 days following ingestion of infective larvae. The microscopic lesions formed as a result of penetration become inflamed from host reactions against concomitant bacterial invasion and the worms' excreta. Nausea, fever, profuse perspiration, and diarrhea commonly occur. Some facial edema may be present also, accompanied by a slight rash. These symptoms subside within 5–7 days following onset.

The second phase may last for 3 weeks and is characterized by symptoms resembling such diseases as rheumatism, pneumonia, encephalitis, pleurisy, meningitis, myocarditis, and peritonitis.

During the third phase, there may be intense muscle pain, difficulty in breathing, swelling of facial muscles, weakening of blood pressure and pulse, heart damage, and nervous disorders, including hallucinations. Death may result from heart failure, respiratory complications, peritonitis, or cerebral involvement.

Most cases of trichinellosis are asymptomatic and go undetected. In suspected cases, several diagnostic laboratory procedures are available. Positive results from skin and serological tests are significant. Negative tests results, especially in the early stages of the disease, are inconclusive. Intradermal tests using a suspension prepared from larvae are sensitive enough to give positive results within an hour provided the suspected infection is at least 2–3 weeks old. The appearance of a weal of about 5 mm in diameter indicates exposure to the worm. Several other serological procedures, such as flocculation and agglutination tests, are available and are similar in degree of sensitivity. Enzyme-linked immunosorbent assay (ELISA) can detect antibodies in the serum as early as 12 days postinfection.

The definitive diagnostic procedure is the demonstration of live larvae in a specimen of biopsied muscle. In such a procedure, usually performed about the 3rd or 4th week of infection, a muscle section, usually taken from the deltoid or gastrocnemius muscle, is placed either between two slides or in a muscle press and examined microscopically. Alternatively, the muscle section can be first digested with pepsin and then examined microscopically for larvae.

Chemotherapy

No satisfactory chemotherapeutic regimen for trichinellosis has been devised. The therapeutic value of albendazole and mebendazole remains inconclusive. Bed rest and supportive treatment, such as administration of analgesics to relieve symptoms, are beneficial.

In selected cases involving the heart or the central nervous system, steroid therapy has been used successfully to relieve inflammatory symptoms.

Host Immune Response

In the life cycle of *T. spiralis*, the larval migratory phase and the adult intestinal phase are most likely to elicit host immune responses. During early infection (4th–8th day), the migrating larvae are found in various tissues such as lung alveoli. At this time, an increase in lymphocyte, macrophage, and eosinophil infiltration is associated with a transient increase in the Th2 response. In later infections (14th day), a very limited activation of these cells is evident pointing to a suppressed innate immune response. The eviction of adult worms from the gut has been reported to be induced by IL-9 up-regulation of Th2 cytokines IL-4 and IL-13.

Prevention

Education of the public is the most effective way to control the disease in the human population. As in California, laws governing pork production must be strengthened to require that garbage containing raw scraps intended for use as hog feed first be sterilized. Although costly, microscopic examination of pork should be reinstituted and updated. Finally, the public should be informed of the need to cook pork products thoroughly (at temperatures higher than 70°C), to the point at which none of the meat shows pink, in order to kill the infective larvae. Microwave cooking of pork, especially roasts, should be monitored with a meat thermometer, and the temperature should reach at least 77°C.

THE SECERNENTEA

The second group of nematodes infective to humans belongs to the class Secernentea Although species belonging to this class exhibit morphological and life cycle differences, all possess phasmids, the minute, usually paired, chemoreceptors located posteriorly to the anus. The adult forms of five nematodes that infect the human intestine are discussed: the thread worm; *S. stercoralis*, the two hookworms, *A. duodenale* and *N. americanus*; the large intestinal roundworm, *A. lumbricoides*; and the pinworm, *E. vermicularis*.

Strongyloides Stercoralis

S. stercoralis and other members of this genus are unique in that they may exhibit either a direct or homogonic, exclusively parasitic life cycle or an indirect, or heterogonic, life cycle in which free-living generations may be interrupted by parasitic generations, depending upon environmental conditions. Both the homogonic life cycle and the parasitic phase of the heterogonic life cycle involve only **protandrogonous** females, while the free-living life cycle involves both adult males and females. In protandrogonous forms, the male reproductive organs develop first, and then disappear. Subsequently the female reproductive organs develop giving the impression that the worms reproduce parthenogenetically. Protandrogonous, parasitic females are approximately 2.0-mm long and 0.04-mm wide (Fig. 16.6A). The esophagus, lacking a posterior bulb, extends one-third the body length and contains a shallow buccal capsule. The vulva lies in the posterior third of the body and the didelphic uteri

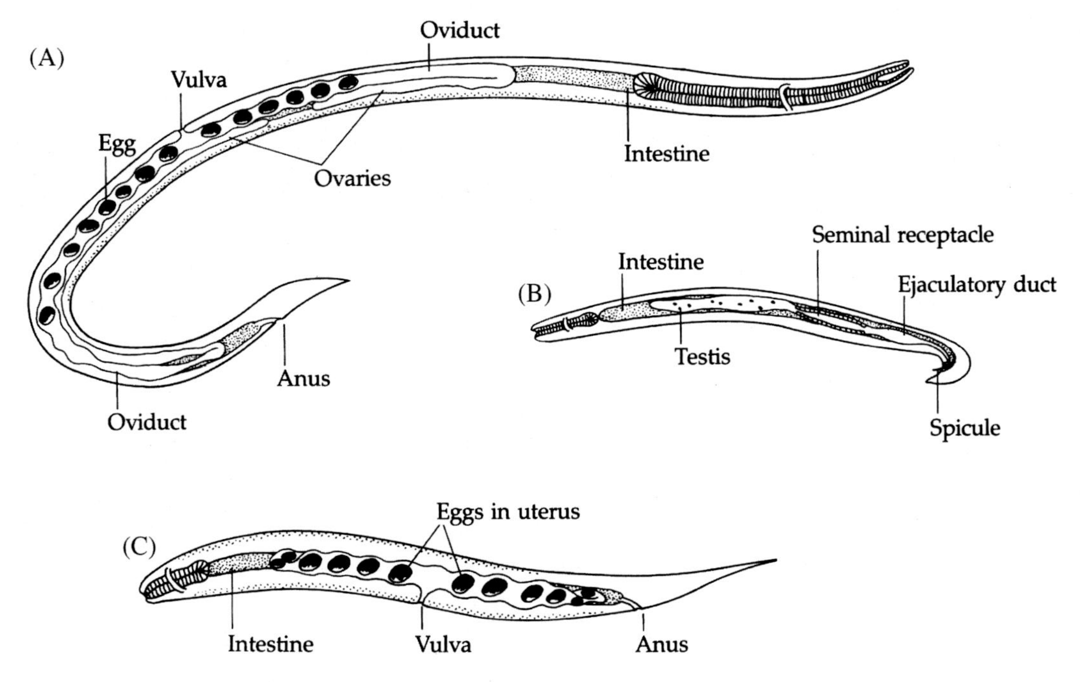

FIGURE 16.6 **Morphology of Strongyloides stercoralis.** (A) Parasitic female. (B) Free-living male. (C) Free-living female.

contain few eggs at any one time. Free-living males (Fig. 16.6B) are about 1.0-mm long; free-living females (Fig. 16.6C) are about 2.0–2.5-mm long. The uteri of a free-living female contain far more eggs than do those of its parasitic counterpart. The free-living female is also somewhat larger with the vulva situated in the midsection of the body.

Life Cycle

The life cycle of *S. stercoralis* can be divided into three phases: free-living, parasitic, and autoinfectious (Fig. 16.7).

Free-living Phase: Free-living *S. stercoralis* dwell in moist soil in warm climates. Copulation occurs in the soil. When the sperm penetrates an oocyte, the sperm nucleus disintegrates; sperm penetration merely activates the oocyte to develop parthenogenetically with no contribution to the genetic material of the developing embryo. Following oviposition, the first-stage, rhabditiform larvae (Fig. 16.8) are well developed and require only a few hours for complete development. The eggs hatch in the soil where the liberated larvae feed actively on organic debris, pass through four molts, and develop into sexually mature adults.

This free-living, or heterogonic, cycle may continue without interruption. However, if the environment becomes inhospitable, the rhabditiform larva molts twice to become a nonfeeding, filariform larva, the form infective to humans.

Parasitic Phase: When filariform larvae encounter a human or other suitable host, they readily penetrate the skin (Fig. 16.9A) and are carried by the lymphatics or the small cutaneous veins to the postcaval vein, whence they enter the right side of the heart and are carried

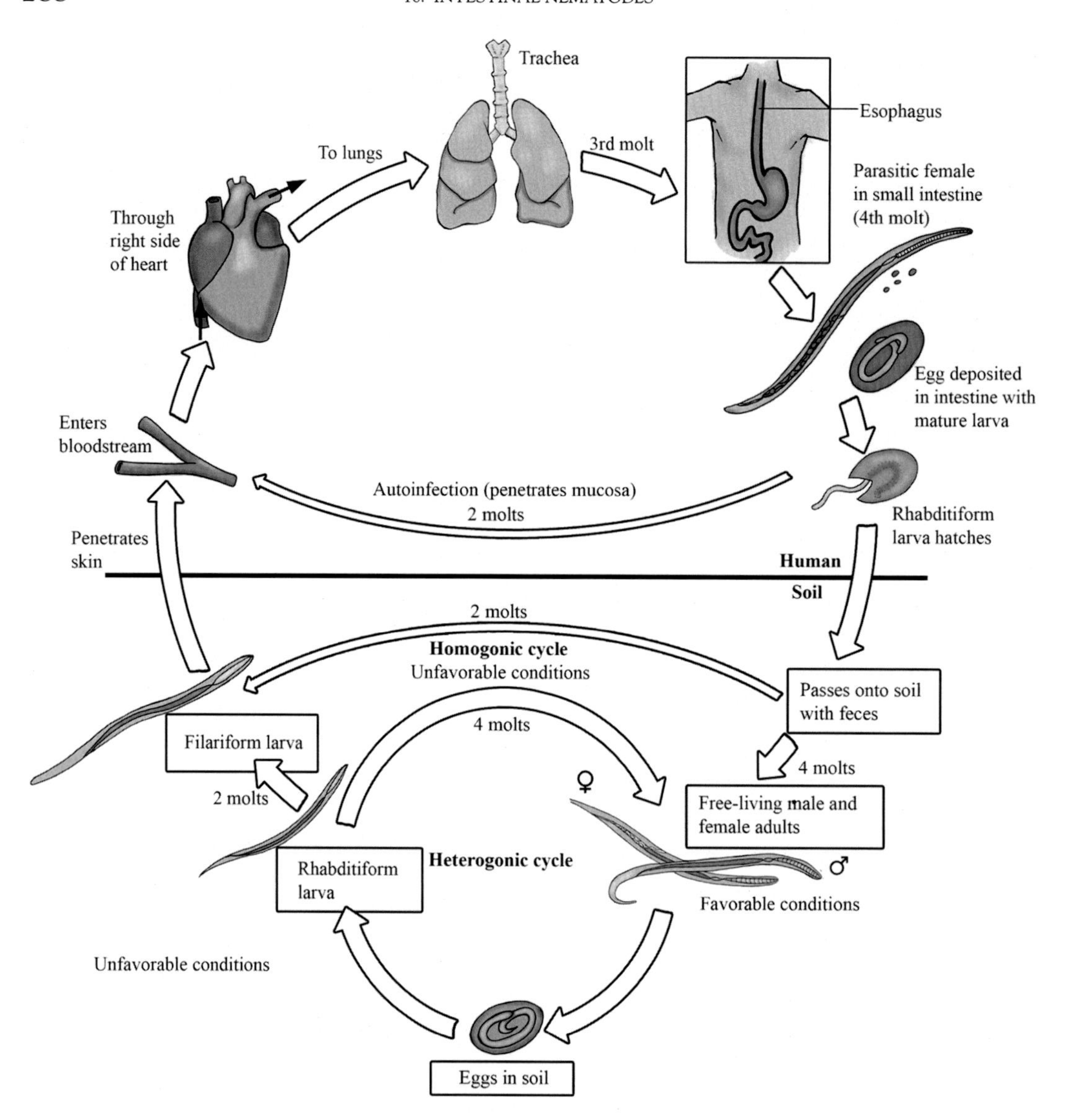

FIGURE 16.7 **Life cycle of Strongyloides stercoralis.** *Credit: Image courtesy of Gino Barzizza.*

to the lungs via the pulmonary artery (Fig. 16.9B). In the lungs, following a third molt, the larvae rupture from the pulmonary capillaries and enter the alveoli. There is some laboratory evidence, using experimentally infected animals, that not all larvae follow the lung route to reach the intestinal tract. However, since symptoms in most infected patients involve the lungs, it appears that in human cases a majority of the larvae follow the lung route.

From the alveoli, the larvae move up the respiratory tree to the epiglottis. Abetted by coughing and subsequent swallowing by the host, they migrate over the epiglottis to the esophagus and down to the small intestine where they undergo a final molt and become

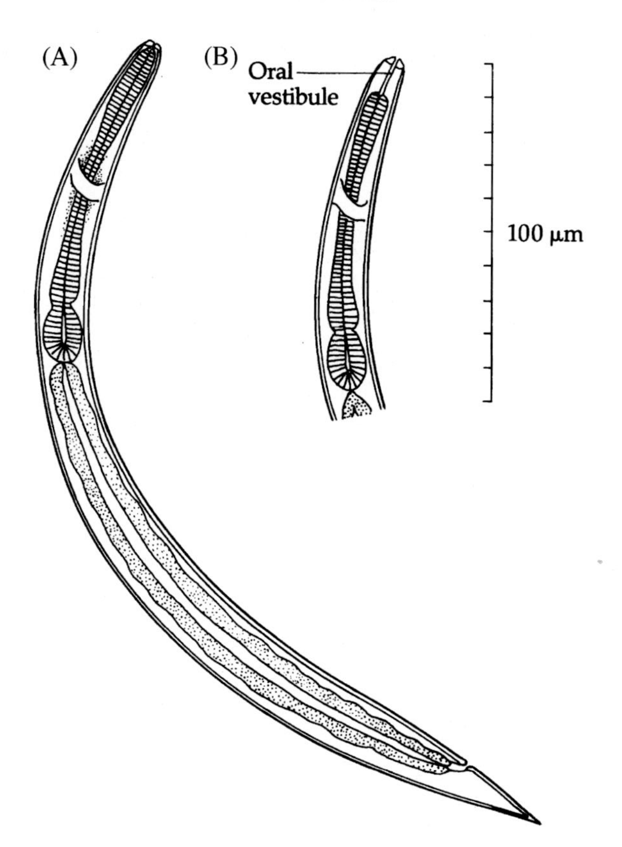

FIGURE 16.8 **Rhabditiform larvae.** (A) *Strongyloides stercoralis*. (B) Anterior portion of hookworm. Note the elongatedoral vestibule.

protandrogonous worms. The worms burrow into the mucosa of the small intestine and produce embryonated eggs within 25–30 days postinfection (Fig. 16.9C and D). Some investigators report that embryos are produced parthenogenetically, while others report that fertilization occurs. The eggs, averaging 54 by 32 μm and covered by a thin, transparent shell, hatch in the mucosa into first-stage, rhabditiform larvae, which feed during passage through the lumen of the intestine and exit the host body with feces. Eggs are seldom found in feces. Under conditions favorable for development, the larvae become established in the soil, undergo four molts, and become free-living adults. However, under adverse conditions, the rhabditiform larvae metamorphose into infective filariform larvae after two molts.

Autoinfectious Phase: During passage through the host digestive tract, rhabditiform larvae may rapidly undergo two molts into filariform larvae and, by penetrating the intestinal mucosa or perianal skin, enter the circulatory system and continue their parasitic lives without ever leaving the host. Such a cycle is not uncommon and accounts for some World War II veterans having harbored infections for more than 50 years, as well as for the development of increasingly heavy, even lethal, infections.

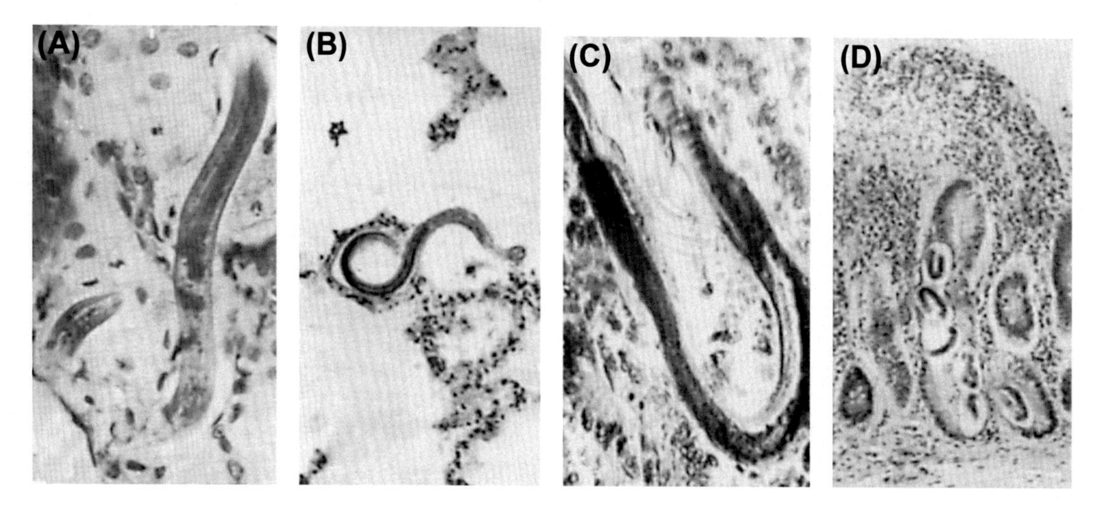

FIGURE 16.9 **Strongyloides stercoralis.** (A) Filariform larva in skin. (B) Larva in lung. (C) Adult and (D) Eggs in mucosa of duodenum.

Epidemiology

Humans usually contract infection through contact with infective larvae in the soil and less frequently from larvae in contaminated water. It has been estimated that human cases of strongyloidiasis currently number 100–200 million worldwide. Estimated cases number 21 million in Asia, 900,000 in the former USSR, 8.6 million in Africa, 4 million in tropical America, 400,000 in North America, and 100,000 in the Pacific Islands. The free-living forms thrive best in warm, moist climates where sanitation is substandard. In the United States, among residents of mental institutions, the prevalence in feces of infective larvae or larvae capable of rapidly becoming infective combined with poor sanitation and/or personal hygiene, results in a high incidence of infection. A study of 1437 mental patients in New York City institutions revealed an 18% rate of infection. Since dogs and cats also serve as sources of human infection, the disease can be considered zoonotic.

A second species causing human strongyloidiasis, *Strongyloides fuelleborni*, has been described from Papua New Guinea and sub-Saharan Africa. This species has a predilection for young children under 4 years of age causing a frequently fatal condition known as swollen belly syndrome. The source of the infective larva is believed to be mothers' milk.

Symptomatology and Diagnosis

Symptoms of human strongyloidiasis appear in three phases: cutaneous, pulmonary, and intestinal.

The cutaneous phase is characterized by slight hemorrhaging, swelling, and intense itching ("ground itch") at sites that have been invaded by infective filariform larvae. Occasionally, the invasion sites are secondarily invaded by infectious microbial agents, which results in severe inflammation.

Larval migration through the lungs produces the pulmonary phase. Lung damage due to massive, cellular reactions to the migrating larvae may delay or prevent further migration.

When this happens, the larvae may develop in the lungs and commence reproducing as they would in the intestine. In this case the patient develops burning sensations in the chest, a cough, and other symptoms of bronchial pneumonia.

Intestinal symptoms appear when female worms become embedded in the mucosa and, rarely, beyond the muscularis. Moderate to heavy infections produce pain and intense burning in the abdominal region, accompanied by nausea, vomiting, and intermittent diarrhea. Long-standing infections result in chronic dysentery and weight loss. Very heavy infections may be fatal. In most instances, this is attributable to massive invasion of tissues by filariform larvae, to secondary bacterial infections from ulceration of intestinal mucosa, or to immunosuppression (as in AIDS patients).

The surest means of diagnosis is microscopical identification of rhabditiform or filariform larvae in feces. However, children infected with *S. fuelleborni* shed eggs in their feces. Accurate diagnosis requires that larvae of *S. stercoralis* be distinguished from those of hookworms, which they resemble (Fig. 16.8). Occasionally, eggs of *S. stercoralis* are passed in the feces, in which case these, as well as those of *S. fuelleborni*, must be distinguished from hookworm eggs (Fig. 15.13). The same is true when duodenal fluid is aspirated for examination. Sputum should also be examined for larvae. Serological tests have proven useful. ELISAs, which produce few cross-reactions, have been used successfully.

Chemotherapy

Oral administration of 400 mg albendazole daily over a period of 3 consecutive days is the therapy of choice. In trials, a single oral dose of Ivermectin (200 mg/kg) has proven very successful with a cure rate exceeding 90%. Relapses of the intestinal phase have been reported, especially in patients who are immunologically compromised (irradiated cancer patients, tissue transplant patients, AIDS patients, etc.).

Host Immune Response

S. stercoralis is considered an opportunistic parasite (see p. 32).

Prevention

Prevention requires the sanitary disposal of human excrement, protection of skin from contact with contaminated soil, and appropriate treatment in cases of autoinfection. The screening of pregnant women and patients who are candidates for immunosuppressive therapy for *S. fuelleborni* and *S. stercoralis infections*, respectively, should be followed by treatment where infection is present.

HUMAN HOOKWORM DISEASE

Hookworm disease has been, and remains, among the most prevalent and important of human parasitic diseases. Unlike malaria, amoebiasis, or schistosomiasis, hookworm disease may not be clinically spectacular; yet, it can profoundly affect entire populations by gradually sapping the victims' strength, vitality, and overall well-being. As commonly seen in parts of the Middle East and Far East, and until several decades ago, in the Southeastern United

States, victims become lethargic and nonproductive, resulting in economic losses beyond calculation. While great progress has been made in combating this scourge, hookworm disease has by no means been controlled or eradicated and remains a major public health problem in many parts of the world, especially in developing countries.

Adults of two species of hookworms, *A. duodenale* and *N. americanus*, cause infection among humans. Since these worms are similar in morphology and life cycle, they will be described together with notations on dissimilarities.

Necator Americanus and Ancylostoma Duodenale

A. duodenale is considered the more pathogenic of the two species, and its adults are somewhat larger than those of *N. americanus* (Fig. 16.10). Female adults measure about 9–13 mm

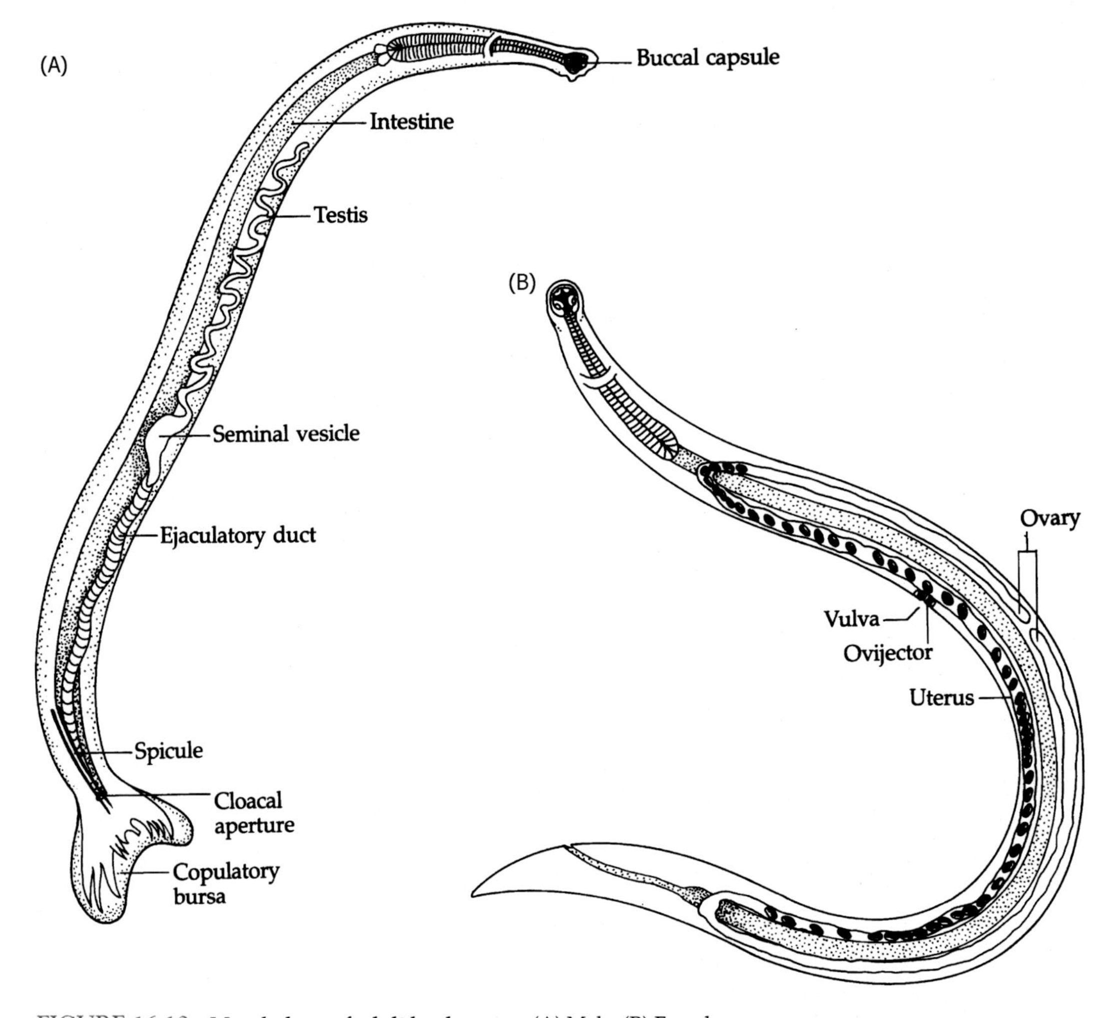

FIGURE 16.10 **Morphology of adult hookworms.** (A) Male. (B) Female.

in length, while males are 5–11-mm long. The female reproductive system is didelphic; males have a single testis. The posterior end of the male has an umbrella-shaped bursa, with riblike rays, that expands over and envelops the vulva of the female to anchor the male during copulation (Fig. 16.11). The vulva is located in the anterior half of the body in female *N. americanus* and in the posterior half in *A. duodenale*. There are chitinous specializations in the buccal capsules of both species (Fig. 16.12). *N. americanus* has a conspicuous pair of semilunar cutting plates on the dorsal wall, a concave tooth on the dorsal medial wall, and a pair of triangular lancets deeper on the ventral wall of the buccal capsule; *A. duodenale*, on the other hand, has two pairs of teeth on the ventral wall of its buccal capsule.

Eggs of the two species are indistinguishable other than that those of *N. americanus* are slightly larger, measuring 64–76-μm long and 36–40-μm wide. Eggs from both species have thin, transparent shells and bluntly rounded ends and upon oviposition enclose uncleaved embryos.

Life Cycle

Humans, almost exclusively, are hosts for *A. duodenale*, while dogs also are common hosts for *N. americanus* (Fig. 16.13). Eggs are expelled in feces; under optimal conditions (temperature of 23–33°C, shade, and a sandy soil rich in organic materials), a rhabditiform larva matures in 1–2 days and hatches from the thin-shelled egg. The newly emerged larva, about 275-μm long, feeds on bacteria and organic materials in the soil and doubles its size in 5 days. After two molts, the rhabditiform becomes a nonfeeding, infective, filariform larva. The cuticle of the last molt is retained and encloses the larva as a sheath.

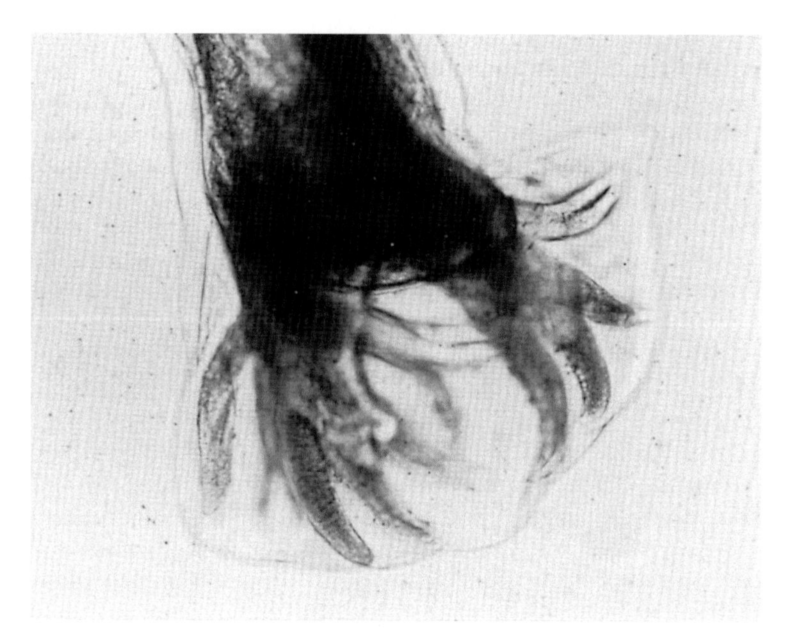

FIGURE 16.11 **Copulatory bursa of male hookworm.**

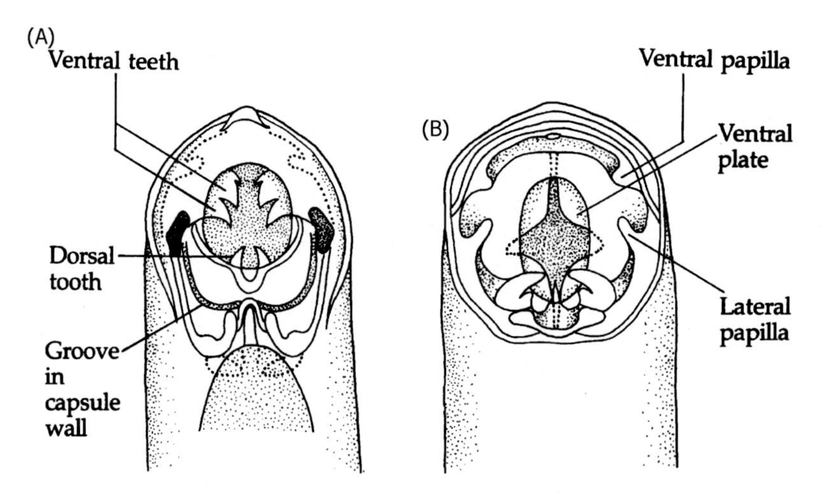

FIGURE 16.12 **Hookworm buccal capsules.** (A) Ancylostoma. (B) Necator.

The active, ensheathed, filariform larvae inhabit the upper 10 cm of soil, usually remaining within 50 cm of the initial site of oviposition, where they can live up to 6 weeks. Human infection occurs when these larvae penetrate the skin, usually of the feet and legs. Entry is most often gained through hair follicles, pores, and skin abrasions. Upon penetration, the larvae enter the host's lymphatic system, migrate to the right side of the heart, and enter the lungs via the pulmonary artery. Rupturing from lung capillaries, they enter the alveoli and migrate up the respiratory tree, molting en route, and then are coughed up and swallowed. The migratory period lasts about 1 week. At the third molt, each larva develops a temporary buccal capsule enabling it to develop into a feeding worm. Once the larvae reach the small intestine, they actively burrow into the intervillous spaces where, at about the 13th day, they undergo their fourth molt. They become sexually mature adults 5–6 weeks postpenetration.

A. duodenale infection can also be acquired by humans orally, and, in some endemic regions, this is the primary means of transmission. Following ingestion, the filariform larva is swallowed and, molting twice en route, develops to sexual maturity in the small intestine.

Epidemiology

An estimated 72.5 million humans harbor *A. duodenale*, the majority (59 million) in Asia. Some 384.3 million are infected with *N. americanus* worldwide, of which 1 million live in the United States.

A. duodenale, commonly known as the Old World hookworm, occurs in southern Europe, North Africa, India, China, Japan, and Southeast Asia. It also has been reported in the New World among Paraguayan Indians, in isolated areas of the United States, and in the Caribbean. Among coal miners in Belgium and Great Britain, the infection produces a classic form of anemia. Infected tunnel construction workers in Switzerland, Germany, and Italy are similarly affected.

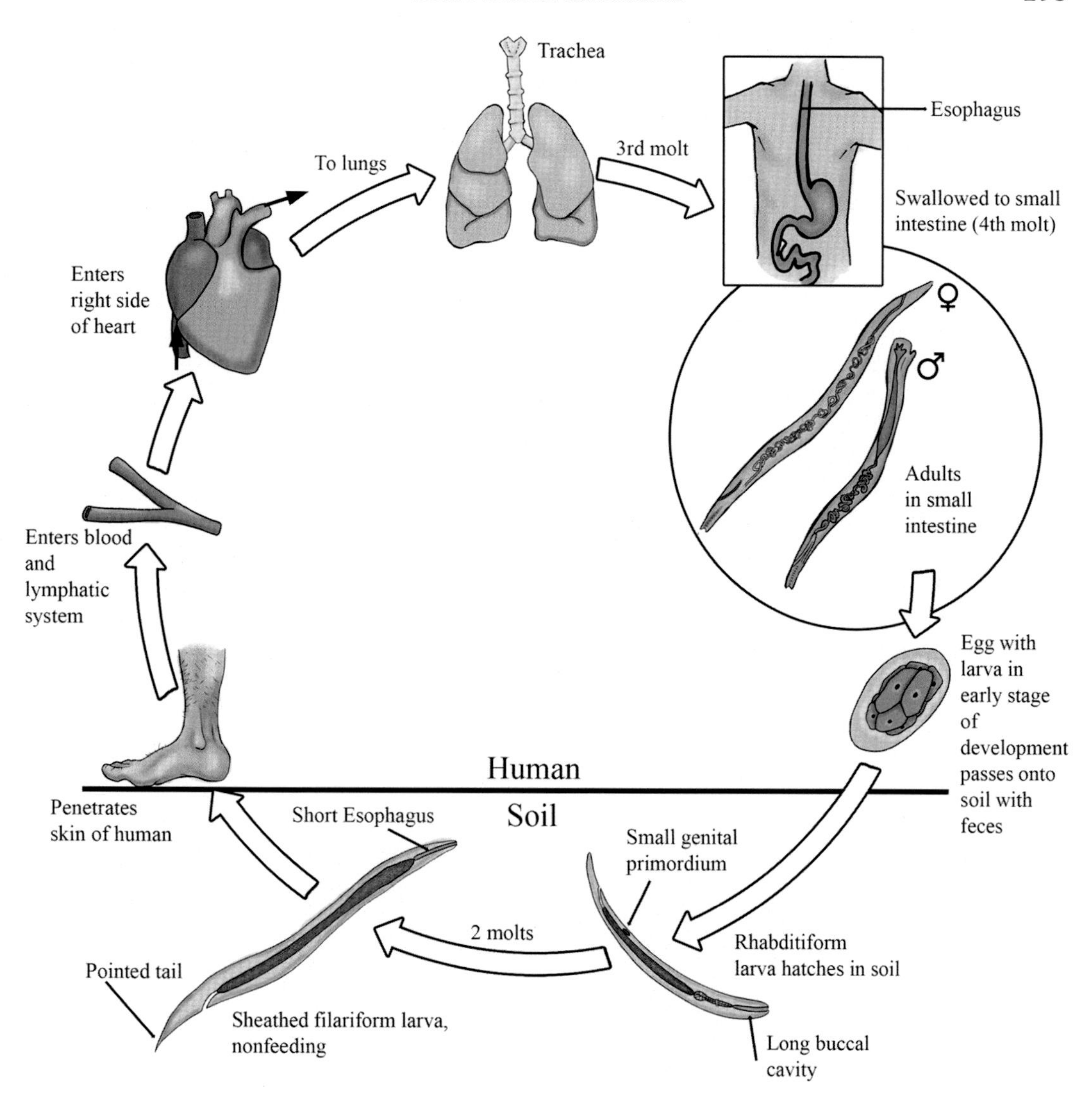

FIGURE 16.13 **Life cycle of hookworms.** *Credit: Image courtesy of Gino Barzizza.*

N. americanus, the New World or American hookworm, is found in the southern United States, Central and South America, and the Caribbean. This species is also indigenous to Africa, India, Southeast Asia, China, and the southwestern Pacific Islands. It is believed to have been introduced into the Americas during the slave trade era or even earlier.

Four essential factors in the spread of hookworm are (1) shaded sandy or loamy soil; (2) sufficient moisture to assure development of eggs and larvae (i.e., rainfall of 75–125 cm during the warm months of the year); (3) contamination of the soil by egg-containing feces, introduced as a result of poor sanitation and/or use of human excrement for fertilizer; and (4) a population that, by choice or necessity, comes in contact with contaminated soil.

Symptomatology and Diagnosis

The course of human hookworm disease can be divided into three phases: invasion, migration, and establishment in the intestine.

Invasion commences when infective larvae penetrate human skin. Although little damage is inflicted upon superficial skin layers, host cellular reaction stimulated during blood vessel penetration may isolate and kill the larvae. Local irritation from invading larvae, combined with inflammatory reaction to accompanying bacteria, evokes an urticarial condition commonly known as ground itch.

The migration phase is the period during which larvae escape from capillary beds in the lung, enter the alveoli, and progress up the bronchi to the throat. This migration can produce severe hemorrhaging when large numbers of worms are involved; otherwise, a dry cough and sore throat may be the only symptoms. In areas where reinfection is continual, some filariform larvae of *A. duodenale*, following penetration, may invade host skeletal musculature where they remain dormant, resuming development at a later time. While there has been no explanation for this phenomenon, it is known that dormancy can be caused by pregnancy with development resuming at the onset of parturition. These larvae subsequently may appear in breast milk, which then becomes a vehicle for transmission to breast-feeding infants.

The most serious stage of hookworm infection arises when the parasites become established in the host's intestine. Upon reaching the small intestine, young worms use their buccal capsule and "teeth" to burrow through the mucosa where they vigorously begin feeding upon blood (Fig. 16.14). Salivary secretions of the worms contain anticoagulants to facilitate blood-feeding. Blood loss caused by *A. duodenale* adults is estimated to be ten times that caused by a comparable number of *N. americanus* adults. When more than 75 *N. americanus* or

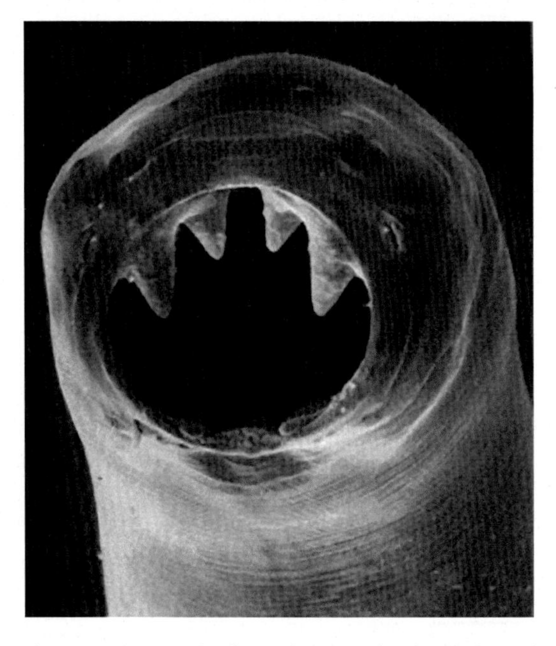

FIGURE 16.14 **Scanning electron micrograph of Ancylostoma duodenale buccal area.**

10 *A. duodenale* are present, even if 40% of the iron removed by the worms is reabsorbed by the host, an iron-deficiency anemia develops, accompanied by intermittent abdominal pain, loss of appetite, and a craving to eat soil (geophagy). Heavy infections often produce severe anemia, protein deficiency, dry skin and hair, edema, distended abdomen (especially in children), stunted growth, delayed puberty, mental dullness, cardiac failure, even death.

Diagnosis based on clinical symptoms can be misleading because the same symptoms may result from nutritional deficiencies or from a combination of infection and such deficiencies. Positive diagnosis requires identification of eggs in the feces. For light infections, concentration-type diagnostic techniques, such as zinc sulfate flotation or several modifications of the formalin-ether method, are employed.

Meticulous care in identification of larvae is essential, especially from stools that are several days old, since the rhabditiform larvae of hookworms strongly resembles that of *Strongyloides* (Fig. 16.8) and even those of ruminant parasites (e.g., *Trichostrongylus* spp.), which occasionally infect humans.

Chemotherapy

Several drugs provide effective treatment for both human hookworm species. Mebendazole or albendazole administered orally for 3 consecutive days results in a very high rate of cure. The benzimidazoles, because of possible side effects, are usually contraindicated for treatment of children. Pyrantel pamoate is prescribed as an alternative drug for infantile hookworm infection. The dormant state of *A. duodenale* is not treated until the larvae leave the musculature and become established in the intestine. When severe anemia has developed due to hookworm infection, the anemia should be treated first. While oral administration of iron prior to treatment for hookworm quickly restores hemoglobin levels, reversing the course of treatment delays for months restoration of hemoglobin to normal levels.

Host Immune Response

Hookworm infections induce strong immune responses, but there is little evidence that these responses are protective. Information on T-cell activity in hookworm infections is sparse. Available data indicate that a Th2 response predominates, generating IgE and eosinophils. In vivo experiments suggest that eosinophils can kill infective L_3 larval stages but not adult worms. The dominant Th2 cytokines in response to adult worm infections are IL-4, IL-5, and IL-13, with IL-4 promoting IgE synthesis.

Prevention

Obvious precautions to prevent the spread of hookworm infection include: improved sanitation, including sanitary disposal of human excrement; treatment of infected individuals; protective measures to prevent contact with infective larvae; and correction of nutritional deficiencies to reduce susceptibility. Proper disposal of dog feces is important in programs to control *N. americanus*. Finally, education is always an important aspect of any control program.

Cutaneous Larval Migrans

Similar to the manner in which animal schistosome larvae attack humans, infective filariform hookworm larvae of animals, for which humans are incompatible hosts, often penetrate human skin. Such larvae normally fail to pass beyond the stratum germinativum, instead persisting and migrating for some time at that level causing a skin condition, not unlike schistosome dermatitis, known as cutaneous larval migrans, or creeping eruption (Fig. 16.15). The most common agents of this condition are the dog and cat hookworms, *Ancylostoma braziliense* and *Ancylostoma caninum*.

Epidemiology

Creeping eruption is prevalent in many parts of the world, particularly in tropical and subtropical regions. In the United States, incidence is high along the Gulf Coast and in the southern Atlantic states. Humans become infected with animal hookworms by contact with soil upon which infected cats and dogs have defecated. A frequent source is children's sandboxes, which, during summer days, afford optimum conditions of shade, sandy soil, and warmth. Infective larvae also thrive in the soil under houses. It is not surprising, therefore, that infection rates are highest among children, plumbers, electricians, etc.

Symptomatology and Diagnosis

The feet, arms, and face are the most common sites of infection; however, any part of the body that comes in contact with contaminated soil is susceptible. Red, itchy papules develop

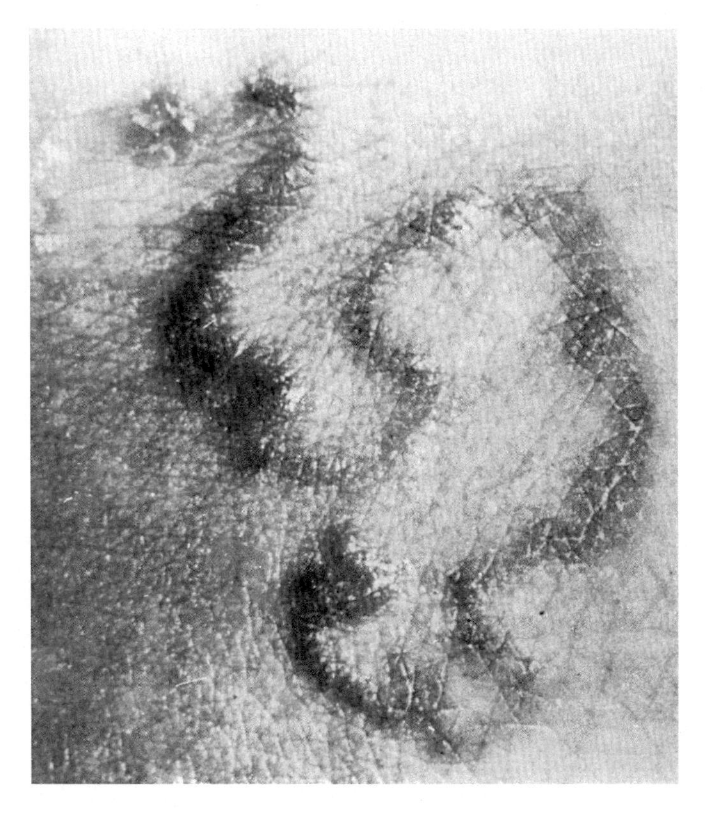

FIGURE 16.15 **Cutaneous larval migrans.**

at the invasion site, and the migratory paths of the larvae appear as slightly elevated ridges. These ridges represent an inflammatory response to the burrowing larvae as they make cutaneous tunnels. This tunneling phenomenon, probably an attempt by the larvae to find a point of entry into the circulatory system, produces intense itching along the migratory pathways. The larval infection may persist for weeks or even months, and secondary bacterial infection is common.

Chemotherapy

Treatment generally targets alleviation of symptoms, such as the intense itching, rather than destroying the larvae. A topical ointment consisting of a 10% suspension of thiabendazole has proven effective, and light infections often respond to chilling of the active portion of the lesion with ethyl chloride. The latter treatment must be administered with extreme caution since prolonged exposure to ethyl chloride can produce second-degree burns. Any accompanying microbial infection should be treated with antibiotics and/or fungicides.

Prevention

Obviously, prevention of cutaneous larval migrans caused by hookworms depends upon avoiding contact of bare skin with soil contaminated by feces from infected cats and dogs. Toward that end, animals should be denied access to underhouse crawl spaces, sandboxes should be kept covered when not in use, and pets should be treated with appropriate anthelmintics. In addition, service personnel should always keep their extremities covered when working in areas that are suspect.

Ascaris Lumbricoides

In *A. lumbricoides*, known as the large intestinal roundworm of humans, females may attain a length of 40 cm while male worms may reach 30 cm (Fig. 16.16). In both sexes, the mouth is surrounded by one dorsal and two ventrolateral lips. The posterior end of the female is straight while that of the male curves ventrally. The didelphic female reproductive system is located in the posterior two-thirds of the body with the vulva situated about one-third of the body length from the anterior end. The female is a prodigious egg producer, depositing about 200,000 eggs daily; the uterus may contain up to 27 million eggs at a time. The fertilized egg measures 45–75-μm-long and 35–50-μm-wide.

A unique phenomenon termed chromosome diminution is observed in *Ascaris spp.* and *Strongyloides spp.* Chromosome diminution is a programmed process that eliminates specific DNA sequences from the genome of these species. Approximately 13% of the somatic cell genome of Ascaris *spp.* is eliminated during cleavage of the embryo. The germ line is apparently unaffected. Many functional aspects of this phenomenon still remain a mystery but recent data have indicated that diminution may serve as a mechanism for gene regulation and gene silencing.

Life Cycle

Adult worms inhabit the lumen of the small intestine and draw nourishment from semi-digested food of the host (Fig. 16.17). Copulation occurs at this site, and eggs are passed with host feces. The outer, albuminous coat of the fertilized egg is golden brown due to bile

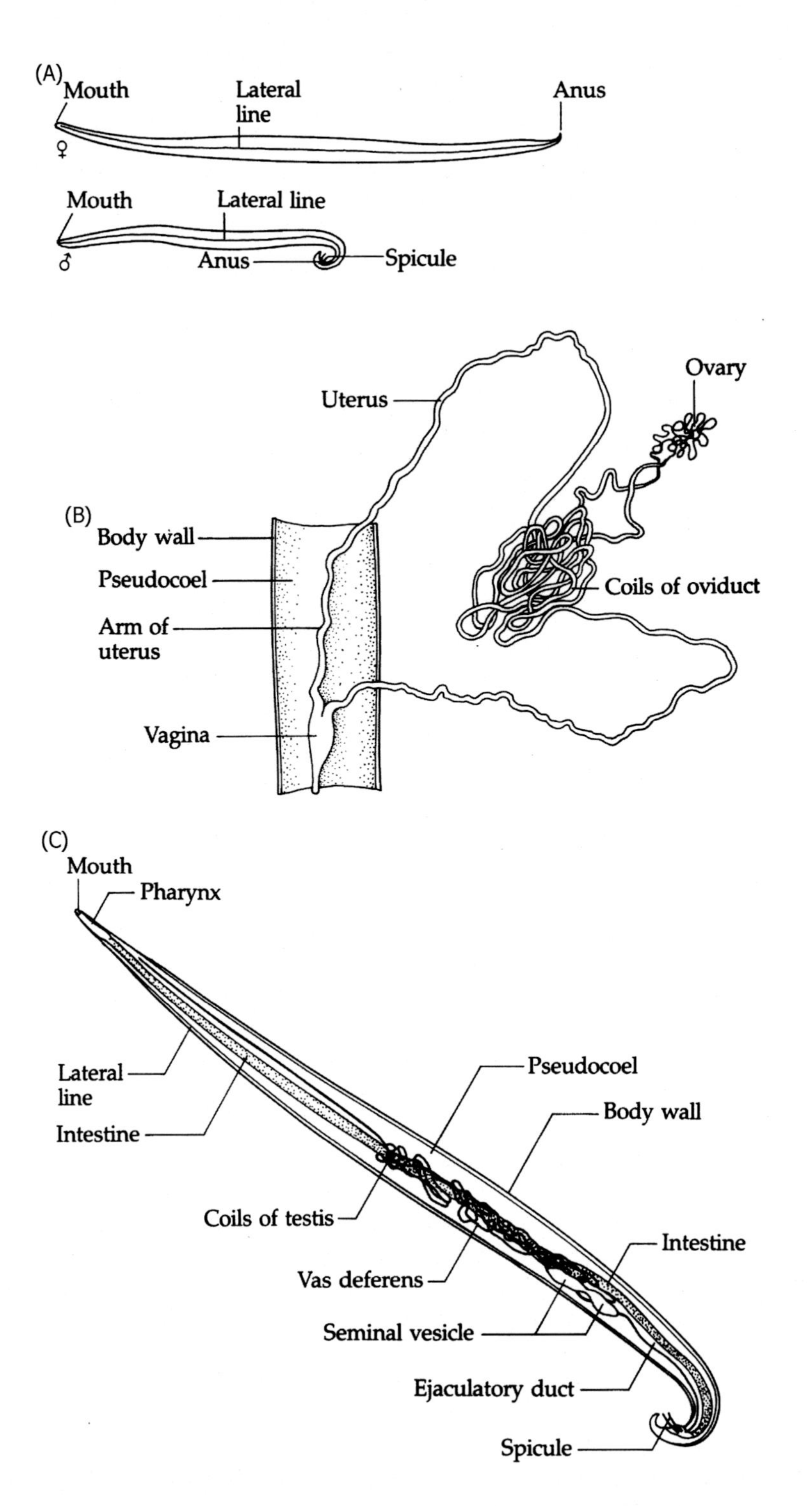

FIGURE 16.16 **Ascaris lumbricoides.** (A) Male and female adults. (B) Female reproductive system teased out of body. (C) Cutaway view of adult male.

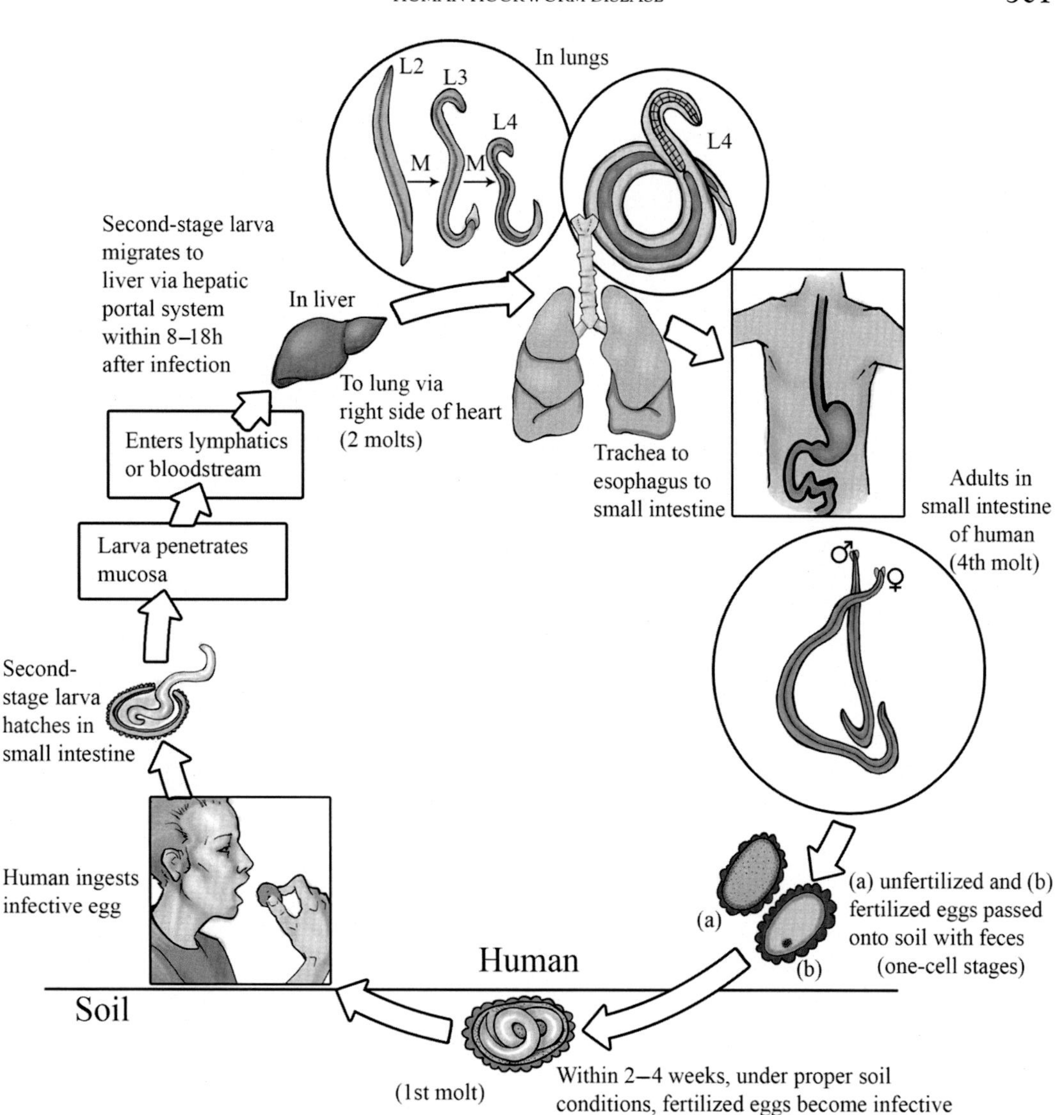

FIGURE 16.17 **Life cycle of Ascaris lumbricoides.** *Credit: Image courtesy of Gino Barzizza.*

pigment adsorbed from feces (Fig. 15.13C). Among the oval, fertilized eggs are found numerous unfertilized eggs, identifiable by their elongated shape and the absence of the albuminous coat. When fertilized eggs are deposited, the zygote is uncleaved, and it remains in this state until the eggs reach soil. Eggs deposited in soil are resistant to desiccation but are, at this stage of development, very sensitive to environmental temperatures. The zygote within the eggshell develops at an environmental (soil) temperature (about 25°C) lower than the body temperature of the host (i.e., 37°C for humans). However, development ceases at temperatures below 15.5°C, and eggs cannot survive temperatures more than slightly above 38°C.

After 2–4 weeks in moist soil at optimal temperature and oxygen levels, the embryo molts at least once in the shell and develops to an infective second-stage larva. Eggs containing infective larvae may remain viable in the soil for 2 years or longer.

After being ingested by a human, eggs containing infective larvae hatch in the duodenum. The larvae actively burrow into the mucosal lining, enter the circulatory system, and are carried via the venous system to the liver, through the right side of the heart, and to the lungs propelled by the pulmonary arterial flow. This migration requires approximately 1 week. The larvae remain in the lungs for several days, molting twice, and eventually rupture from the pulmonary capillaries to enter the alveoli. From there, they move up the respiratory tree and trachea to the epiglottis to be coughed up, swallowed, and passed again to the small intestine. During this complex migratory process, individual worms grow from 200–300 μm in length to approximately ten times that length. A fourth molt in the small intestine is essential to the worms' survival, and only those that undergo this final molt develop to sexual maturity. The interval from ingestion of infective eggs to the appearance of sexually mature worms in the small intestine is about 3 months.

Epidemiology

Distribution of *A. lumbricoides* is worldwide, but it is most prevalent in warmer climates. Dependent upon poor sanitation for its spread, human ascariasis has been described as a household and backyard infection. An estimated 1008 million people are infected, making it the most common nematode parasitizing humans. It is most prevalent in children, particularly between the ages of 5 and 9 years, the group most frequently exposed to contaminated soil and least likely to observe basic sanitary practices such as washing hands before eating and keeping hands out of the mouth. Hand-to-mouth transmission is most common; however, in countries where human excrement is used as fertilizer, contaminated vegetables are also a common source of infection. Water is rarely implicated in the transmission of *A. lumbricoides*.

Since ascariasis is more prevalent in humid climates than arid ones, its occurrence is often patchy within individual countries. For instance, prevalence in Nigeria ranges from 0.9% to 98.2%; and, in Ghanian villages, from 0% to 76%. Generally, prevalence is highest in Africa and Asia, with 40% of the population infected. Latin America follows closely with 32%.

Symptomatology and Diagnosis

About 85% of ascariasis cases are symptomless; however, the most frequent symptom is upper abdominal discomfort of varying intensity. Symptoms such as asthma, insomnia, eye pain, and rashes represent allergic responses of the host to metabolic excretions and secretions of adult worms, as well as to dead and dying worms. Little damage results from larval penetration of the host's intestinal mucosa. However, aberrant larvae migrating to such organs as the spleen, liver, lymph nodes, and brain usually elicit an inflammatory response. Also, larvae escaping from capillaries in the lungs and entering the respiratory tree cause small, hemorrhagic foci accompanied by coughing, fever, and difficulty in breathing. Larvae in large numbers can produce numerous small blood clots, which can lead to potentially fatal pneumonitis if large areas of the lungs are affected.

Large numbers of adult worms sometimes cause mechanical blockage of the intestinal tract, and adult worms penetrating the intestinal wall or appendix may cause local hemorrhage, peritonitis, and/or appendicitis. Adult female worms may even wander up the bile duct to the liver, causing abscesses, or down the pancreatic duct, causing fatal, hemorrhagic pancreatitis. Loss of appetite and insufficient absorption of digested food also occur as a result of heavy infections. Migration of worms is sometimes abetted by high fever, chemotherapy, administration of anesthesia, etc.

Diagnosis is made by identification of eggs in feces (Fig. 15.13). Since egg production is fairly constant (about 200,000 eggs per female daily), egg counts can provide reasonably accurate estimates of the number of adult worms present provided uniform samples are used.

Chemotherapy

For treatment of individuals in whom adult worms have been verified in the intestine but who do not require hospitalization, a single dose of pyrantel pamoate is highly effective. One-time treatments with mebendazole or albendazole are acceptable alternatives. Piperazine citrate is highly effective in cases of intestinal obstruction. The drug paralyzes the worm, nullifying its ability to counter host intestinal peristalsis, and causes it to be passed. If the obstruction persists, surgery may be necessary.

Host Immune Response

Chronically infected individuals display reduced Th1 responses with dramatically lower TNF-α, and IL-12 activities. Th2 and IgE (induced by IL-4 and IL-5) immune responses are elevated. There appears to be little difference in host immune responses between the various stages of the life cycle. The lung stage parasites (L_3/L_4) cause significant and highly polarized Th2 responses, which persist throughout the life cycle. It has been hypothesized that a substantial Th2 cytokine release promotes protective immunity, suggesting a host adaptation to control the parasite burden while keeping immune-mediated host self-damage at a minimum.

Prevention

The most reliable preventive measure is a multipronged attack emphasizing scrupulous personal hygiene, public sanitation, health education, and environmental sanitation, especially the processing of night soil. For optimal effectiveness, such a program should be combined with treatment of the population with broad spectrum anthelmintics two or three times annually.

Visceral Larval Migrans

Visceral larval migrans usually results from migration of second-stage larvae of ascaroids, the adults of which normally are found in dogs, cats, and within the internal organs of accidental hosts, primarily young children. Most commonly, human visceral larval migrans develops following accidental ingestion of infective eggs of *Toxocara canis*, although several other nematode species, such *as Bayliascaris procyonis, Angiostrongylus cantonensis, Angiostrongylus costaricensis, Gnathostoma spinigerum*, and *Toxocara cati*, can also cause the condition. In humans, following such accidental ingestion of eggs, second-stage larvae hatch, penetrate the intestinal wall, and quickly invade the liver. Although the majority of these larvae remain in the liver some travel to the lungs and, sometimes, the central nervous system and eyes.

Although most of the larvae eventually gravitate to a single location and become encapsulated by host tissues, for a period of at least several weeks they actively migrate through tissues, leaving long trails of inflammatory and granulomatous reactive cells.

It should be emphasized that, while hookworms of normally nonhuman hosts are the usual suspects in cases of cutaneous larval migrans, and *T. canis* is most often implicated in visceral larval migrans in humans, location of lesions and even presence of characteristic symptoms are not invariably reliable criteria for specific identification of the etiologic agent.

Epidemiology

As symptoms of visceral larval migrans are imprecise and inconsistent and as dog parasites have long been considered noninjurious to humans, confirmed cases of this disease have been rare. However, available reports indicate that the disease occurs worldwide and probably involves several nematode species. In the United States, a high percentage of puppies and kittens are infected with *Toxocara*, perhaps as many as 98% according to some reports. The life cycle of *T. canis* appears to be completed only in puppies. In adult dogs, the second-stage larva encysts in various tissues. In pregnant bitches, these larvae can become active and migrate across the placenta, infecting the fetal pup where the life cycle is completed. The close association of young children with their pets has been cited frequently as a factor in the transmission of parasitic disease; hence, it is not surprising that this segment of the population is the most vulnerable to this disease. The ubiquitous sandbox provides an ideal medium for survival of eggs, as do park areas and beaches where owners walk their dogs.

A. cantonensis infects humans in Southeast Asia, Hawaii, the Pacific Islands (Tahiti, Samoa, and Cook Islands), the Philippines, Taiwan, parts of China, the Caribbean, and Madagascar, while *A. costaricensis* is prevalent in Central and South America, notably Costa Rica, where approximately 300 cases are reported annually. The normal definitive hosts for both species are wild rats. Humans contract the disease by ingestion of third-stage larvae in insufficiently cooked intermediate hosts: mollusks and crustaceans for *A. cantonensis*; the slug, *Vaginulus plebius*, for *A. costaricensis*.

Symptomatology and Diagnosis

In visceral larval migrans attributable to *Toxocara*, the degree of pathology is related to the number of infective eggs ingested and the site at which the larvae settle. Most infections are light with symptoms including fever, pulmonary congestion, and eosinophilia. Characteristic lesions most often occur in the liver and are accompanied by increased levels of various leukocytes, especially eosinophils. The lesion is a protective response of the host, but it also protects the larva since it isolates the parasite from further contact with host defense mechanisms.

In heavy infections, some children develop anemia from the excessive leukocyte buildup. Ocular disease may develop when larvae become entrapped in the eye. The severest consequences of infection are usually allergic reactions, especially if the patient is hypersensitive to metabolites produced by the larval nematodes. In rare instances, fatalities due to toxicariasis have been reported.

Diagnosis is complicated by the lack of a specific body of symptoms. Eosinophilia and hepatomegaly occurring in conjunction with a history of proximity to pets are clinically significant. While the only positive diagnosis is identification of the larvae, this is exceedingly difficult since there are usually too few of them to be retrieved by needle biopsy. An effective

ELISA test is currently being used to detect antibodies against the excretory–secretory antigens of *Toxocara* larvae.

Third-stage larvae of *A. cantonensis* usually migrate to the capillaries of the meninges in human patients causing a type of eosinophilic meningoencephalitis in humans. The condition is characterized by severe headache, fever, and some central nervous system involvement. Although the disease is normally self-limiting after approximately 3 weeks, there have been instances of fatalities in humans. The larvae of *A. costaricensis*, on the other hand, when ingested by humans, lodge in mesenteric venules causing thromboses and infarcts, which lead to ulceration and even peritonitis. Most damage occurs in the large intestine where eosinophilic granulomas sometimes mimic appendicitis symptoms.

Chemotherapy

Most toxocariasis infections are self-limiting, and only severe cases warrant treatment. Albendazole or mebendazole administered for 5 days, augmented with corticosteroids when allergic symptoms are also present, are the drugs of choice. There is no reliable documentation of beneficial effects from treatment of *Angystrongylus* infections.

Prevention

Generally, the best protection for children from exposure to Toxocara is routine treatment of pets for worms. Puppies and kittens should be treated every 6 months and adult pets every 2 months. Sandboxes should be covered when not in use.

Education in the proper cooking of crustaceans, mollusks, and other organisms, which serve as intermediate and paratenic hosts for *Angiostrongylus* is, perhaps, the most effective means of limiting human infection.

Anasakis Spp

Anisakid nematodes include a number of ascaroid species that normally infect the stomach and intestines of various marine fishes, birds, and such fish-eating mammals as dolphins, whales, seals, and porpoises. Third-stage larvae, usually measuring about 2–3-cm long and 0.5–1.0-mm wide, are found in the body cavities, liver, and/or musculature of a number of marine fishes that serve as intermediate or paratenic hosts in the life cycle of the worms (Fig. 16.18).

Anisakis and certain other anisakid nematodes, especially *Pseudoterranova* and *Phocanema*, represent a public health concern in many parts of the United States among people who eat raw or inadequately cooked fish harboring infective larvae in their flesh. These parasites pose a major problem in Japan and parts of Scandinavia and, at one time, in the Netherlands as well. Human anisakiasis has essentially disappeared from the Netherlands due to recent laws regulating fish processing that prohibit fish being held on boats without immediate refrigeration. At ambient temperature, anisakid larvae migrate from the intestinal tract of fishes into the flesh where they present a greater threat of being ingested by humans.

When ingested by a human, larvae burrow into the stomach or intestinal walls and cause inflammatory responses ranging from localized granulomata to massive, eosinophilic, hemorrhagic, tumorlike growths (neoplasms with larvae at the center of the lesion) (Fig. 16.19). Consequent swelling of the intestinal wall may cause intestinal obstruction, peritonitis, and

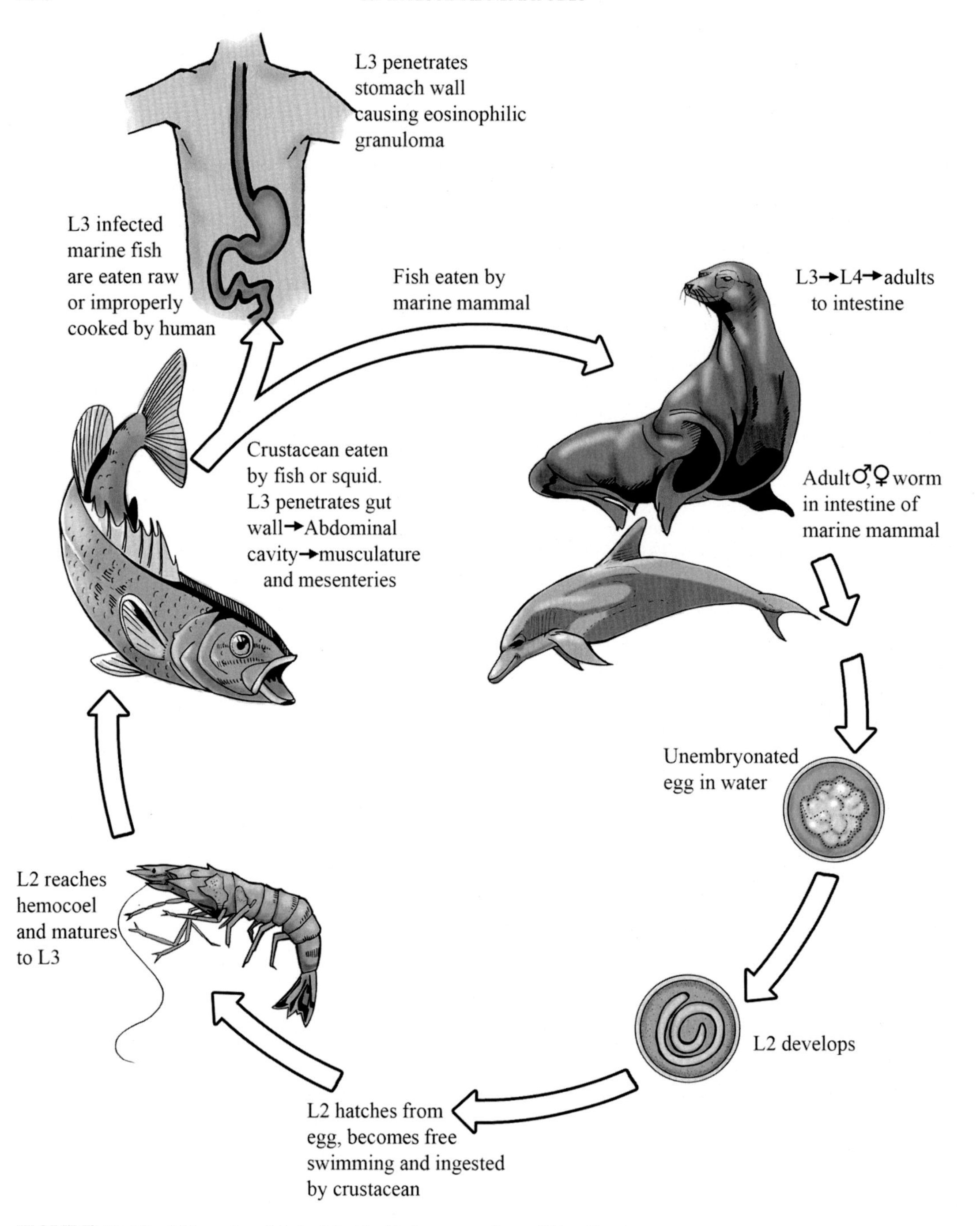

FIGURE 16.18 **Life cycle of Anisakis.** *Credit: Image courtesy of Gino Barzizza.*

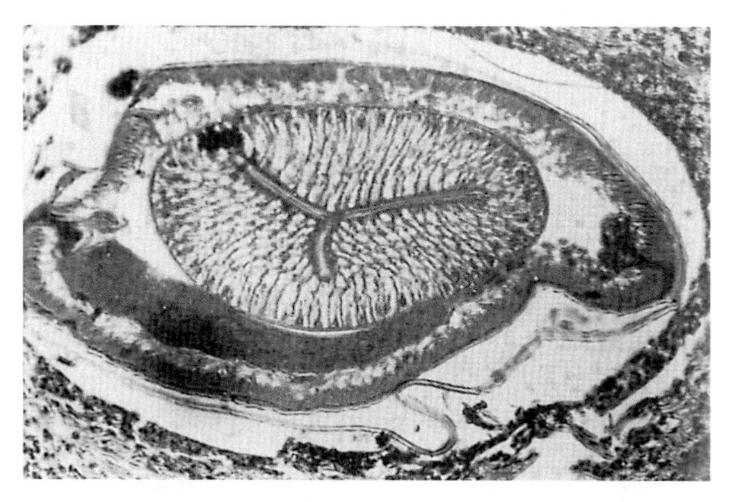

FIGURE 16.19 **Light micrograph of a section through the esophagus of an Anisakis larva in the human intestine.**

the development of abscesses. Most cases have been reported from countries where fish, such as herring in Scandinavia and sashimi in Japan, are eaten raw. With the growing popularity of sushi restaurants in the United States (especially California and Hawaii) there has been a marked increase in anisakiasis, and a number of cases have been fatal. Definitive diagnosis is made by endoscopy or biopsy. The only treatment is surgical removal of the larva. Prevention consists of thorough cooking or proper freezing of fish.

Enterobius Vermicularis

This nematode, commonly known as pinworm or seatworm, is parasitic only to humans (Fig. 16.20). It is familiar to parents of young children worldwide. Female *E. vermicularis,* measuring 8–13-mm long by 0.4-mm wide, are characterized by the presence of winglike expansions (alae) of the body wall at the anterior end, distension of the body due to the large number of eggs in the uteri, and a pointed tail. Males, smaller in size, are 2–5-mm long and possess a curved tail.

Life Cycle

Sexually mature worms usually inhabit the ileocaecal area of the human intestinal tract, but they can spread to adjacent regions of the small and large intestines (Fig. 16.17). Adhering to the mucosa, the worms feed on bacteria and epithelial cells. Males die following copulation, while egg-bearing females, with up to 15,000 eggs in their uteri, migrate to the perianal and perineal regions. There, stimulated by lower temperature and aerobic environment, they deposit their eggs and then also die. More eggs are released when the female's body ruptures. The elongate eggs, each measuring approximately 50–60 by 20–30 μm, are characteristically flattened on one side and each contains, upon deposition, an immature larva. The infective, third-stage larva completes development within the egg several hours after leaving the body of the female worm. Infection and reinfection occur when eggs containing the infective larvae are ingested by the host. This may happen when eggs are picked up on the hands from bedclothes

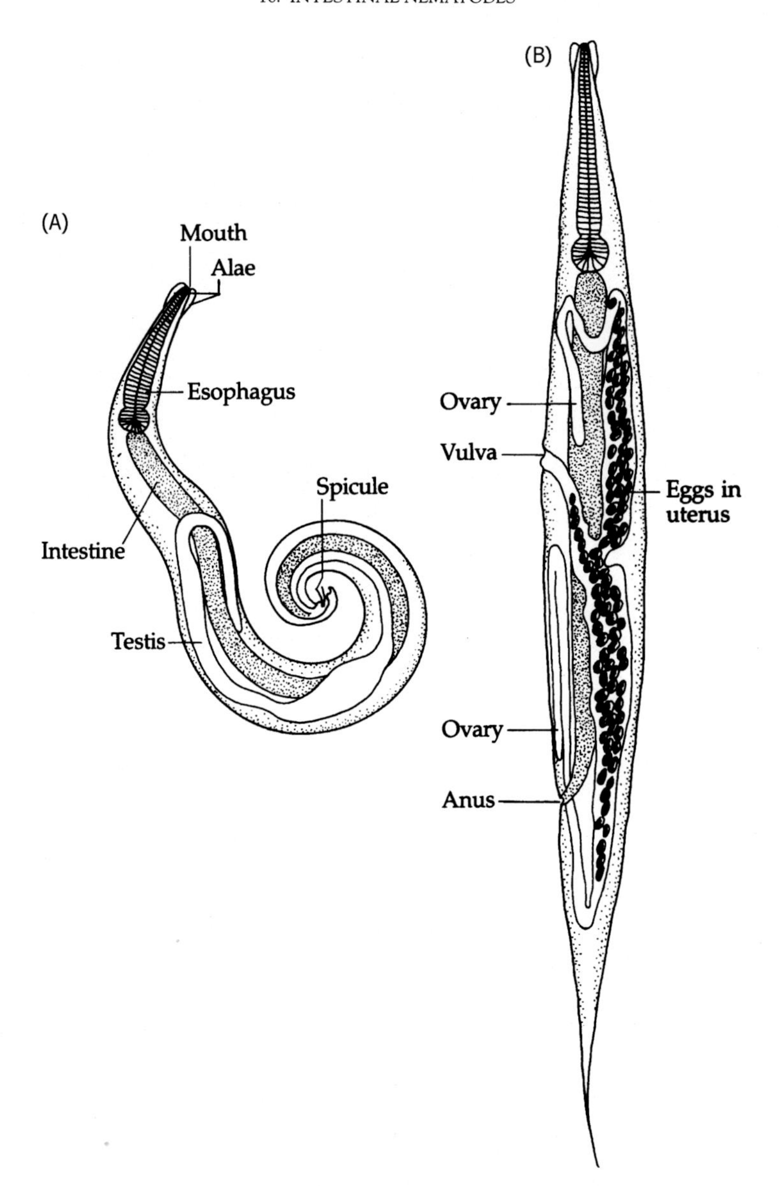

FIGURE 16.20 **Morphology of adult Enterobius vermicularis.** (A) Male. (B) Female.

or beneath fingernails contaminated when the host scratches the perianal zone to relieve itching caused by nocturnal migration of the female worms. However, the lightweight eggs are sometimes airborne and, therefore, can also be inhaled. Retroinfections occur when third-stage larvae hatch from perianally located eggs and migrate back up the host's intestinal tract.

Ingested eggs usually hatch shortly after reaching the duodenum. The escaping larvae undergo molts and development as they migrate posteriorly, reaching sexual maturity by the time they arrive at the colon. The life cycle of *E. vermicularis* spans about 2 months. While

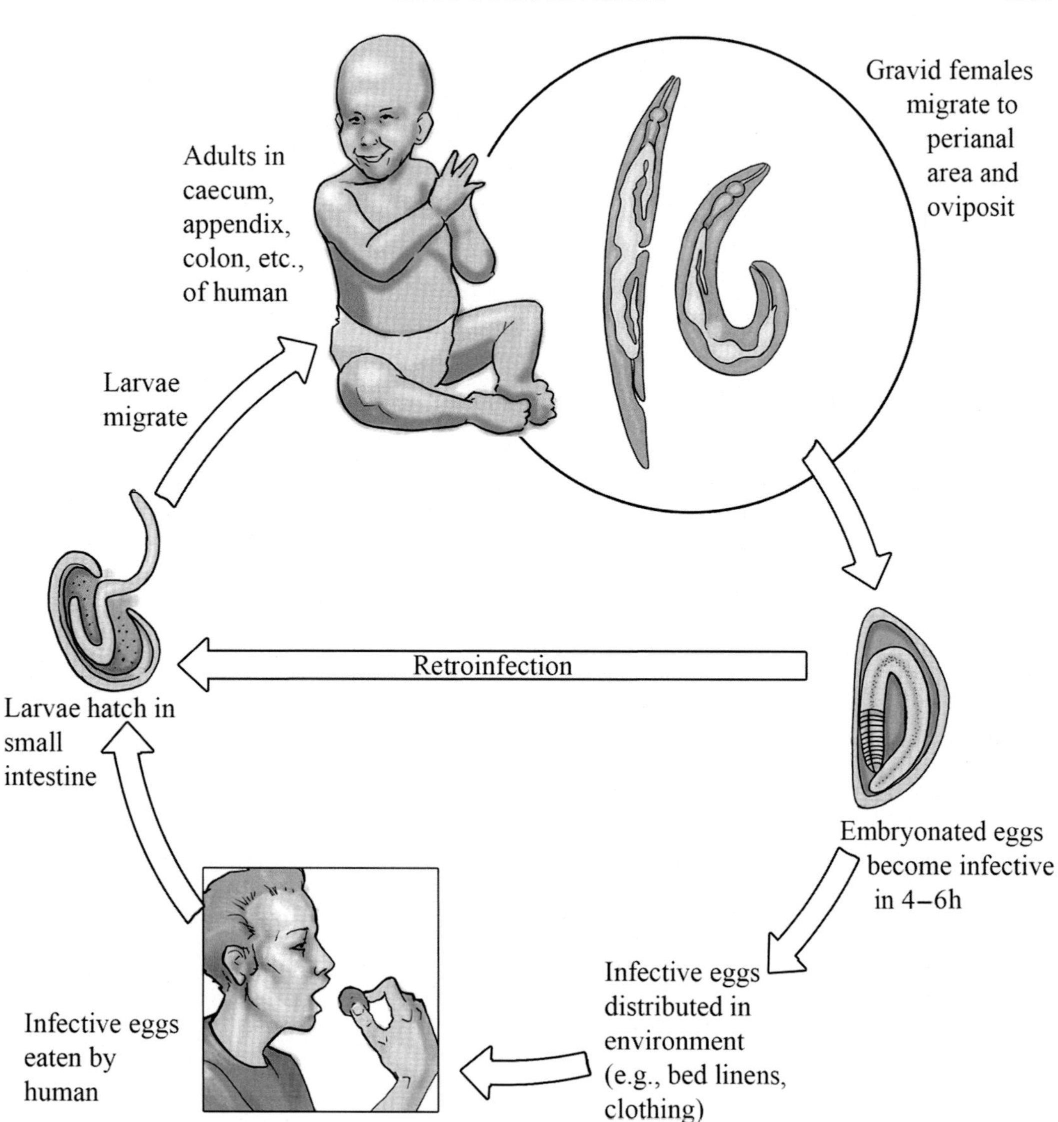

FIGURE 16.21 **Life cycle of Enterobius vermicularis.** *Credit: Image courtesy of Gino Barzizza.*

vulnerable to even moderately high temperatures, eggs are highly resistant to drying and remain viable for a week or more under cool, humid conditions. Some evidence suggests that eggs can remain viable for years under favorable conditions.

Epidemiology

Children, especially of early school-age, are most vulnerable to *E. vermicularis* infection. Geographic distribution of the worm is global. It is especially prevalent in temperate zones, where an estimated 500 million persons are infected. Prevalence, however, varies with each locale. For instance, Alaskan Inuits display a 51% prevalence; elementary students in and around Tallahassee, Florida, 27%; preschoolers in San Francisco, 58%; Sicilian children, 77%; children overall in the United States, 33%. It is the most common nematode parasitizing humans in the United States.

Infections occur in one of four ways: (1) retroinfection when hatched larvae migrate back into the large intestine; (2) self-infection when the patient is reinfected by hand-to-mouth transmission; (3) cross-infection when infective eggs are ingested, either with contaminated food or from fingers that have been in contact with a contaminated surface or body parts from infected humans; or (4) inhalation of airborne eggs. In households with heavily infected individuals, infective eggs have been found in samples of dust taken from chairs, tabletops, dresser tops, floors, baseboards, etc. In a survey to determine the distribution of airborne pollen in public places, such as theaters, pollen and pinworm eggs were found on sample plates not only from arm rests and baseboards but also from chandeliers high above the seats; most of these eggs, however, were no longer viable. Experiments show that at room temperature or above, fewer than 10% of such eggs survive more than 2 days, probably accounting for the less than universal infection in such public places.

Symptomatology and Diagnosis

Pinworms are not highly pathogenic. Clinical symptoms such as itching and irritation are caused by the migration of gravid females around the perianal, perineal, and vaginal areas. Heavy infections in children may also produce such symptoms as sleeplessness, weight loss, hyperactivity, grinding of teeth, abdominal pain, and vomiting. Gravid females may also migrate up the female reproductive tract, become trapped in the tissues, and cause granulomata in the uterus and fallopian tubes. They may also migrate to the appendix, the peritoneal cavity, or even the urinary bladder.

Diagnosis is verified when adult worms and/or eggs are detected. Female worms emerge at night and are frequently visible in the perianal and perineal regions. Adult worms can often be observed on feces as well; however, eggs are found in feces in only about 5% of cases. The most reliable procedure for finding eggs is to press a strip of scotch tape on the perianal skin, remove it, and place it on a clean microscope slide for examination. Negative results from this protocol for 7 consecutive days constitute confirmation that the patient is free of infection.

Chemotherapy

Following positive diagnosis in any individual, treatment should be administered to all members of the household. Several relatively inexpensive and essentially nontoxic drugs are available. Either pyrantel pamoate, albendazole, or mebendazole, usually administered in a single dose and repeated once after 2 weeks, are the treatments of choice. Mebendazole and albendazole are contraindicated for pregnant women since they are teratogenic in experimental animals.

Prevention

Complete eradication of pinworm infection from a population is highly unlikely. Scrupulous personal hygiene is the most effective deterrent. Fingernails should be cut short, and hands should be washed thoroughly after toilet use and before food is prepared or eaten. Since infection is most prevalent in urban areas where relatively large populations intermingle, education of parents has proven most effective. Parents should be informed that it is a self-limiting, nonfatal infection, widespread among children and that no social stigma should be attached to it. There is no evidence that dogs can transmit the infection. Infected children as well as other members of the household should be treated promptly. Bedclothes, towels, and washcloths from infected homes should be carefully laundered in hot water and aired in sunlight.

Suggested Readings

Audicana, M. T., Ansotegui, I. J., Fernandez de Corres, L., & Kennedy, M. W. (2002). *Anisakis simplex:* dangerous-dead or alive? *Trends in Parasitology, 18*, 20–25.

Bundy, D. A. P., & Cooper, E. S. (1989). *Trichuris* and trichuriasis in humans. *Advances in Parasitology, 28*, 108–173.

Cheng, T. C. (1982). Anisakiasis. Sect. C. "Parasitic Zoonoses". In J. H. Steele, & M. G. Schultz (Eds.), *Handbook series in zoonoses* (pp. 37–54). (II). Boca Raton, Florida: CRC Press.

Cooper, P. J., Chico, M. E., Sandoval, C., Espinel, I., Guevara, A., & Kennedy, M. W. (2000). Human infection with *Ascaris lumbricoides* is associated with a polarized cytokine response. *The Journal of Infectious Diseases, 182*, 1207–1213.

Crompton, D. W. T. (1988). The prevalence of ascariasis. *Parasitology Today, 4*, 162–169.

Grove, D. I. (1996). Human strongyloidiasis. *Advances in Parasitology, 38*, 252–309.

Liu, M., & Boireau, P. (2002). Trichinellosis in China: epidemiology and control. *Trends in Parasitology, 18*, 553–556.

Loukas, A., & Prociv, P. (2001). Immune responses in hookworm infections. *Clinical Microbiology Reviews, 14*, 689–703.

Miller, T. A. (1979). Hookworm infection in man. *Advances in Parasitology, 17*, 315–384.

Oshima, T. (1987). Anisakiasis—Is the sushi bar guilty? *Parasitology Today, 3*, 44–48.

Pritchard, D. I., & Brown, A. (2001). Is *Necator americanus* approaching a mutualistic symbiotic relationship with humans? *Trends in Parasitology, 17*, 169–172.

Smith, J. H., & Wootten, R. (1978). *Anisakis* and anisakiasis. *Advances in Parasitology, 16*, 93–163.

CHAPTER

17

Blood and Tissue Nematodes

Chapter 17 is the chapter in which those nematodes that inhabit the blood and tissues of their human hosts are introduced. Seven filarial nematodes belonging to five genera are considered. Because the life cycles of all these organisms with the exception of one form, *Dracunculus medinensis*, are similar, a composite life cycle illustration of the group is presented in this chapter. Since *D. medinensis* parasitizes the connective tissue of its host while the other forms are found primarily in the circulatory systems of their hosts, its life cycle is presented separately. Also considered in this chapter is the periodicity of surges displayed by the microfilariae of the filarial worms and its influence on control of the diseases imposed by these organisms. The impact of microfilarial surges and the feeding habits of the mosquito vectors is also discussed. Within the consideration of each genus, the epidemiology, symptomatology and diagnosis, chemotherapy, host immune response, and prevention are considered.

Seven nematodes of the superfamily Filariodea are parasitic to humans. Generally referred to as filarial worms, these are *Wuchereria bancrofti*, *Brugia malayi*, *Onchocerca volvulus*, *Loa loa*, *Mansonella perstans*, *Mansonella ozzardi*, and *Mansonella streptocerca*. Since the life cycles of all seven are similar, only significant variations will be noted in the discussion of each species.

One nematode of the superfamily Dracunculoidea, *D. medinensis*, is parasitic in the cutaneous tissues of humans. Its life cycle differs significantly from that of the filarial worms and is discussed separately (see p. 325).

LIFE CYCLE

The long, threadlike, adult filarial worms inhabit the lymphatic glands, tissues, and body cavities of the definitive host (Fig. 17.1). Females are ovoviviparous, the larvae hatching in the uterus. At the time of larviposition, the larvae, known as microfilariae, are less well developed than typical first-stage (L1) larvae and are considered prelarvae or advanced embryos (Figs. 15.15A–D). Once deposited, microfilariae migrate into the blood vessels via the thoracic lymph duct or by penetrating the walls of the lymph vessels to invade neighboring small blood vessels. Larvae can survive in blood for several years until ingestion by a suitable insect vector.

Blood-dwelling microfilariae are usually sheathed, retaining the flexible eggshell as a covering membrane. In tissue-dwelling species, however, the sheath is usually sloughed, and the larva is said to be unsheathed.

Human Parasitology. http://dx.doi.org/10.1016/B978-0-12-813712-3.00017-5

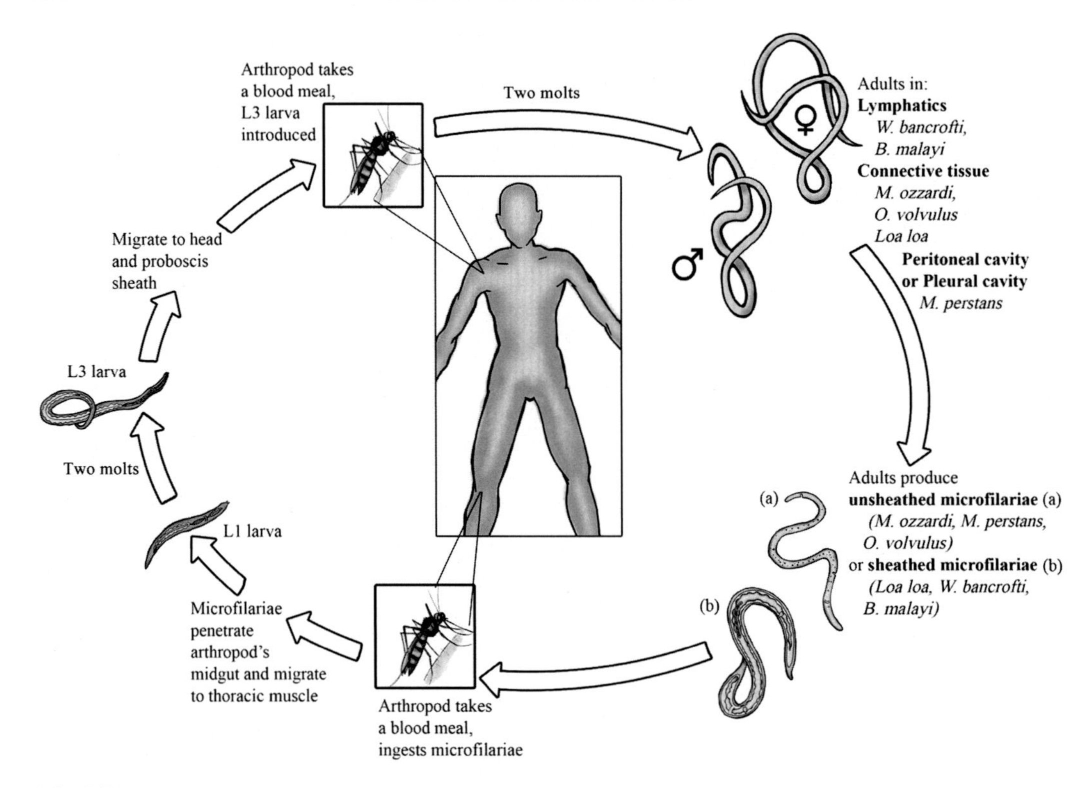

FIGURE 17.1 **Life cycles of filarial worms.** *Credit: Image courtesy of Gino Barzizza.*

After being ingested with a blood meal by a suitable insect vector, microfilariae develop in the digestive tract of the insect into L1, rhabditiform larvae. The latter penetrate the midgut wall into the hemocoel and migrate to the thoracic musculature where they undergo two molts and metamorphose within 3 weeks into the infective, L3, filariform larvae. These then migrate to the proboscis sheath and gain access to the human circulatory system through the puncture wound made by the feeding insect. In the human host, larvae undergo two molts, metamorphosing into adult worms during migration to the definitive site of infection. Approximately 6 months later, microfilariae appear in the bloodstream.

PERIODICITY

In 1879, Patrick Manson, while studying *W. bancrofti* infections among inhabitants of southern China, noted a nocturnal surge in the microfilarial population in the peripheral circulation. It was later observed in a number of species of filarial worms that this periodicity varies with species (or strain) as well as endemic area. For example, microfilariae of *W. bancrofti* in the Caribbean, parts of North and South America, Africa, and Asia are nocturnal. Those of the strain endemic to the South Pacific islands are diurnally subperiodic; that is, while present in the peripheral circulation throughout the entire 24-hour period, their numbers increase during the daytime. In the Philippines, on the other hand, this same species shows a modified nocturnal periodicity in that twice as many larvae are present at night as during the day. When microfilariae vacate the peripheral circulation, they accumulate in the small vessels of the lungs and liver.

Evolution of the phenomenon of microfilarial periodicity is of obvious survival value since it enhances opportunity for the microfilariae to be ingested by the insect vectors at certain times. Not surprisingly, the surge of microfilariae coincides with the active feeding periods of the various insect vectors. For example, in the parasites discussed above, the insect vector for the nocturnal *W. bancrofti* strain is the nocturnal feeding mosquito *Culex fatigans*, while the vector for the subperiodic strain is the diurnal feeder, *Aedes polynesiensis*.

Although numerous studies have been undertaken in an effort to identify the mechanism(s) responsible for microfilarial periodicity, the explanation for this phenomenon of helminth physiology remains elusive. It has been determined that periodicity is neither dependent on light and dark conditions nor upon the circadian rhythm of the definitive host. For example, if the routine is altered so that the host sleeps by day and is active at night, the periodicity of microfilariae is reversed. The sleeping period of the host is characterized by physiological changes such as decreases in body temperature and oxygen tension, increases in carbon dioxide tension and body acidity, lower excretion of water and chlorides by the kidneys, less adrenal activity, etc., some or all of which may trigger the rhythmic behavior of microfilariae. It should be reemphasized, however, that different species and strains of microfilariae respond differently to similar stimuli.

GENOMIC STUDIES

Filarial worms, and nematodes in general, display a great deal of diversity in their genetic expressions. According to current data, in excess of 93,000 genes have been identified on nematodes and more than half of those number are unique with no known homologs indicating that no single nematode genome can serve as a general model.

Of the filarial worms that cause human disease, with the exception of *L. loa* and *Mansonella* spp., a relationship exists between the mutualistic endosymbiont bacterium *Wolbachia* and the parasite. It has been reported that the bacterium greatly influences the parasite's development and viability, while the parasite provides the necessary growth environment as well as certain essential metabolites (e.g., amino acids, vitamins, coenzymes, etc.) for the symbiont. Since it is apparent that the genomes of both partners are metabolically entwined, they should be considered together when examining their respective genomes for potential targets for drug and vaccine interventions.

Wolbachia provides a number of biochemical pathways whose products are utilized by *B. malayi*, one of which is heme. Heme is required for the synthesis of a number of essential enzymes such as cytochromes, peroxidases, catalases, etc. Heme is also required for the synthesis of ecdysteroid-like hormones which are involved with ecdysis.

Almost all the genes involved in heme biosynthesis have been identified in the *Wolbachia* genome while *B. malayi* has an incomplete gene sequence contributing only a single gene for the pathway. This imbalance strongly suggests that the parasite must either rely on *Wolbachia* for its heme requirement or depend on some other means of extracting heme from its environment. Further research has revealed that *B. malayi* lacks the latter property, lending further credence that this pathway is promising as a potential drug target.

RNA interference technology (see page 206) provides another means to control filariasis caused by *B. malayi*. As noted previously, RNAi technology can be used in at least two ways. For instance, it can have a lethal effect by blocking the translation of certain essential proteins such as microfilaria sheath enzyme. Inhibition of this protein caused abnormalities in approximately 50% of released microfilariae by *B. malayi*. Also, the technology can be used as an aid in determining gene functions that are important for various processes in the life cycle of the parasite, for example, such studies have revealed a gene that expresses a cathepsin-like cysteine protease that is important for microfilariae development and their subsequent release. These examples are just of a number of approaches whereby genomics have the potential of opening new avenues for treatment.

FILARIAL WORMS

Wuchereria Bancrofti

This filarial worm, parasitic only in humans, causes a lymphatic disease known as Bancroft's filariasis, which is characterized by extensive enlargement of extremities (Fig. 17.2). Ancients likened the thickened skin to that of elephants; hence, the misnomer elephantiasis (which literally means "caused by elephants" rather than "like elephants"). The adult female worm is 8–10 cm long, with the vulva situated anteriorly near the middle of the esophagus. Male worms are smaller, attaining a length of only 40 mm, and are further distinguishable by their curved posterior ends and genital spicule apparatus.

Life Cycle (Fig. 17.1)

Adults live intertwined with each other in the major lymphatic ducts. Sheathed microfilariae, following deposition, utilize as vectors mosquitoes of several genera, including *Culex*, *Aedes*, *Mansonia*, *Anopheles*, and *Psorophora*. Development in the mosquito requires 1–3 weeks. Once introduced into the definitive host, larvae molt twice and migrate to the varices of lymphatic glands of the groin and epididymis of males and labial and mammary glands of females, where they require 6 months or more to develop to maturity.

Epidemiology

More than 120 million people are reported to be afflicted with lymphatic filariasis, with *W. bancrofti* the most common source of infection. The parasite is estimated to affect more than 100 million inhabitants of the Nile delta, central Africa, Turkey, India, Southeast Asia, the Philippines, Pacific islands, Indonesia, Australia, Caribbean, and parts of South and Central America (Fig. 17.3), with India having by far the largest number of cases. It was probably introduced into the New World during the slave trade. It is certain that infection was introduced into the United States via slaves brought to Charleston, SC, and that it persisted in the Southeastern United States until the 1920s. The Pacific strain occurs throughout the Pacific islands except Hawaii.

Human infection is closely related to the ecology of the mosquito vectors as well as to human habits. The occurrence of periodic filariasis, found in areas of dense population and poor sanitation, parallels the distribution of its principal vector, *C. fatigans*, which breeds in sewage-contaminated water. On the other hand, subperiodic filariasis in the Pacific islands often occurs in rural areas, and this correlates ecologically with its principal vector, *A. polynesiensis*, a mosquito that breeds in the brush.

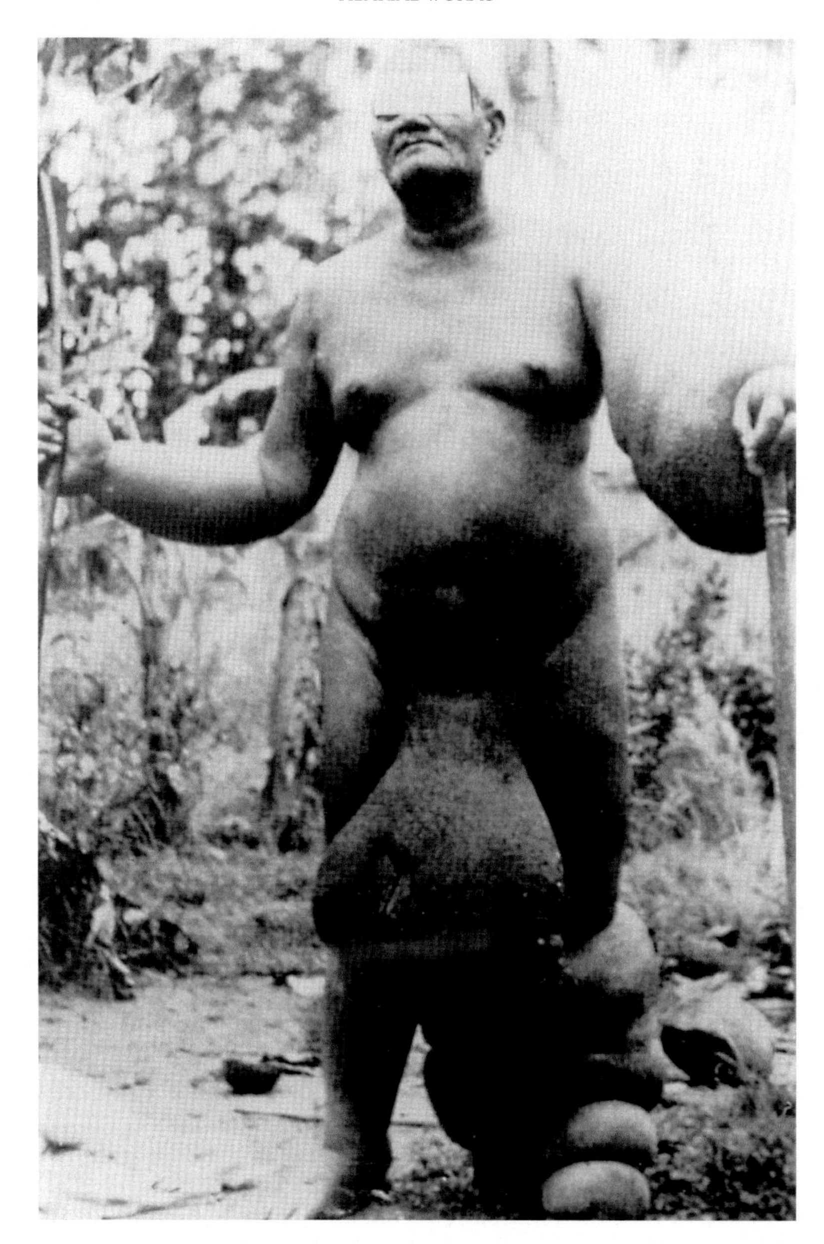

FIGURE 17.2 **South Pacific native severely affected by Bancroft's filariasis.**

Symptomatology and Diagnosis

Pathology in *W. bancrofti* infection is due largely to living, dead, and degenerating adult worms. As lymph vessels and glands become blocked by such worms, edema develops; in time, the accumulation of connective tissue cells and fibers contributes to the enlargement of limbs, scrotum, and other extremities.

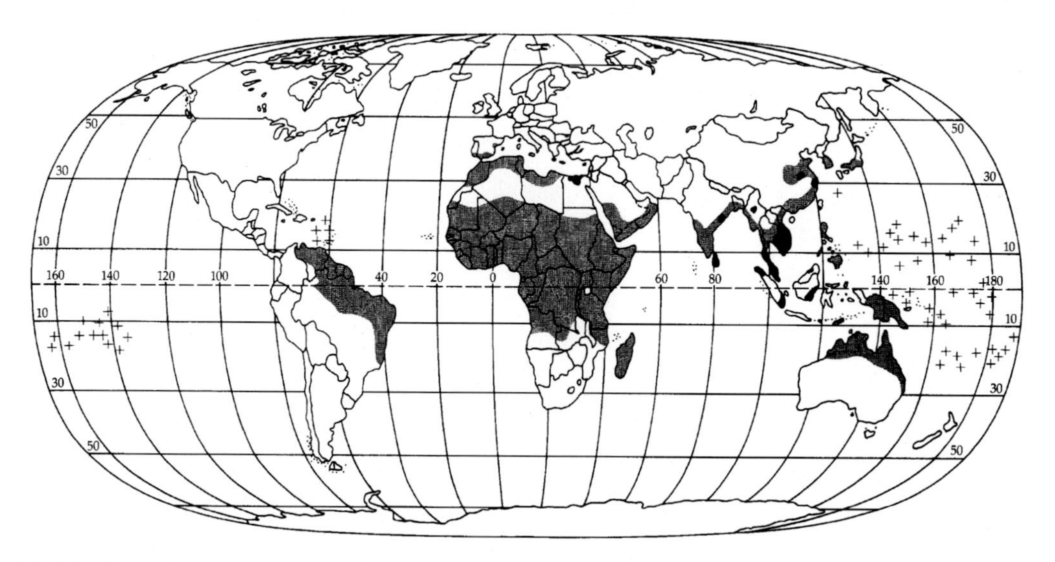

FIGURE 17.3 **Distribution of *Wuchereria bancrofti*.** (+, Islands; *black areas*, concurrent *Brugia malayi*).

Clinically, the disease can be divided into incubation, acute (inflammatory), and obstructive (chronic) phases. The incubation phase is largely asymptomatic and may last for a year or more. Symptoms that do appear are usually mild and may include low-grade fever caused by lymphatic inflammation. In time, the adult worms die, microfilariae disappear, and the patient is often unaware of having been infected. In endemic areas, children frequently harbor microfilariae in their blood while exhibiting few symptoms. During World War II, US armed forces in the south Pacific were constantly exposed to infection, but surprisingly few showed any microfilariae in the peripheral circulation.

The acute phase commences when the female parasites reach maturity and begin releasing microfilariae. This phase is actually an allergic response to the products of dying and degenerating adult worms and is characterized by intense inflammation of the lymph areas of the lower body of the patient. In males, the scrotum is frequently affected. Chills, fever, and toxemia, accompanied by localized swellings in the arms and legs, may persist for days and/or recur at frequent intervals.

The obstructive phase is characterized by blockage of lymph flow resulting from acute granulomatous response in the lymphatic system to dead and degenerating adult worms. This phase may eventually lead to the condition known as elephantiasis; however, only about 10% of the infected population manifests this chronic condition, and, even then, it develops only after many years of continual filarial reinfection. Elephantiasis is rarely seen in people less than 25 years old and is most prevalent in humans older than 40.

Diagnosis requires demonstration and accurate identification of microfilariae in the blood. Stained thick and thin blood smears are used for species identification. Such criteria as presence or absence of sheath and distribution of internal nuclei and organ primordia are discernible after staining. In early infections, before microfilariae are present in the blood, intradermal tests using antigen prepared from *Dirofilaria immitis* (the dog heartworm) yield almost 100% accuracy. Diagnostic procedures should be supplemented with a complete history of patient exposure in endemic areas.

Chemotherapy

A single annual or semiannual treatment with diethylcarbamazine typically reduces microfilaremia by 90%. Administered with care, it may kill adult worms as well. The drug is commonly administered in combination with diethylcarbamazine-fortified salt. Daily consumption of fortified salt for a period of 1 year has resulted in the total elimination of lymphatic filariasis in a number of tested populations in tropical Africa. This regimen is contraindicated in regions where onchocerciasis or loiasis are endemic. A single-dose treatment of ivermectin plus diethylcarbamazine has also proven highly effective, producing a 99% reduction in the number of microfilariae.

Edema may be alleviated by pressure bandaging of the affected region to force excess lymph from the area. Granulomatous tissue can sometimes be removed surgically, but such treatment is rarely attempted in advanced cases of elephantiasis.

As observed, antifilaricidal drugs are effective in killing microfilariae but are much less effective on adult worms. Since drug treatment for filariasis has not been changed for the last two decades, the need for finding either new drugs or new targets is imperative. It is hopeful that the introduction of gene techniques will present new avenues for treatment and, possibly, prevention.

Host Immune Response

The extent of pathology from lymphatic filariasis depends on a variety of host immune responses peculiar to the various phases of infection. While these phases may be categorized as incubation, acute, and obstructive, a wide spectrum of responses from asymptomatic to symptomatic are observed. These in turn may vary from a mild to an intense nongranulomatous response or to a variety of granulomatous obstructive reactions. Some investigators believe these phases may be merely a continuum reflecting: (1) the initial infection (incubation or asymptomatic phase), characterized by dominant Th2 cytokine responses (e.g., IL-4, IL-5, IL-10); (2) the inflammatory (acute) phase, characterized by a shift to Th1 cytokine responses (e.g., IFN-γ); and (3) the culminating obstructive (chronic) phase, associated with an elevated level of both Th1 and Th2 cytokine responses to parasite antigens. The intensity of exposure to the parasite has been shown to affect lymphocyte responsiveness and cytokine bias.

Prevention

Control of and protection from mosquitoes in endemic areas should accompany mass chemotherapy of the indigenous population. The mosquito population, however, has proven difficult to control because of increased resistance to insecticides. Also, hindering control efforts is the spread of urbanization in tropical regions resulting in an increase in the prevalence of the disease and a concomitant increase in nonsylvatic breeding sites. However, the use of screens, insect repellents, and insecticides, coupled with mass treatment, has proven effective in controlling Bancroftian filariasis in the US Virgin Islands, Puerto Rico, and Tahiti.

Brugia Malayi

Until 1960, several species of filarial worms with similar microfilariae were assigned to the genus *Wuchereria.* In that year, based on the study of adult worms, the genus *Brugia* was established to designate the "malayi" group, including *B. malayi*, which is a parasite of primates, including humans and cats.

Although only about half the size, adult worms closely resemble those of *W. bancrofti*. Females are about 55 mm long; males, about 23 mm. The sheathed microfilariae usually appear nocturnally in the peripheral circulation, but there is also a subperiodic strain. The nocturnal strain is host specific, infecting humans exclusively, while the subperiodic strain infects not only humans but also cats, macaque monkeys, and leaf monkeys.

Life Cycle

The life cycle of *B. malayi* closely parallels that of *W. bancrofti*. The principal mosquito vectors are members of the genus *Mansonia*; however, members of the genera *Aedes*, *Culex*, and, occasionally, *Anopheles* also serve in this capacity.

Epidemiology

The Pacific geographic range of *B. malayi* overlaps that of *W. bancrofti*, extending from India to China, Japan, Taiwan, Malaysia, and Indonesia (Fig. 17.3). It is found most often in low-lying regions, which provide optimal breeding conditions for the vectors.

Symptomatology and Diagnosis

While *B. malayi* rarely affects the genitalia, adult worms live in the lymphatics and cause the same symptoms as *W. bancrofti*, culminating in elephantiasis. Diagnosis is identical to that for *W. bancrofti*.

Chemotherapy and Prevention

Treatment and prevention are also the same as for *W. bancrofti*.

B. malayi has been virtually eliminated in Japan, Taiwan, and South Korea due to strict compliance with control programs.

Host Immune Response

Human lymphatic filariasis is generally considered to be a result of infections by either *W. bancrofti* or *B. malayi*. The immune responses are very similar and are discussed in the previous section.

Onchocerca Volvulus

Female worms measure about 50 cm long by 300 μm in width, while males are about 42 cm long and 105 μm wide. The male has a transversely striated cuticle reinforced externally with spiral thickenings.

Life Cycle

Adult worms are found in fibrous nodules, onchocercomas, in the subcutaneous connective tissues and viscera of humans (Fig. 17.4), who are the only known hosts for this species. The nodules are produced by a host inflammatory response to the protein of the worms. Usually a male and a female are coiled in each nodule but, occasionally, several worms are coiled within a single onchocercoma. Immediately upon deposition within the nodules, eggshells

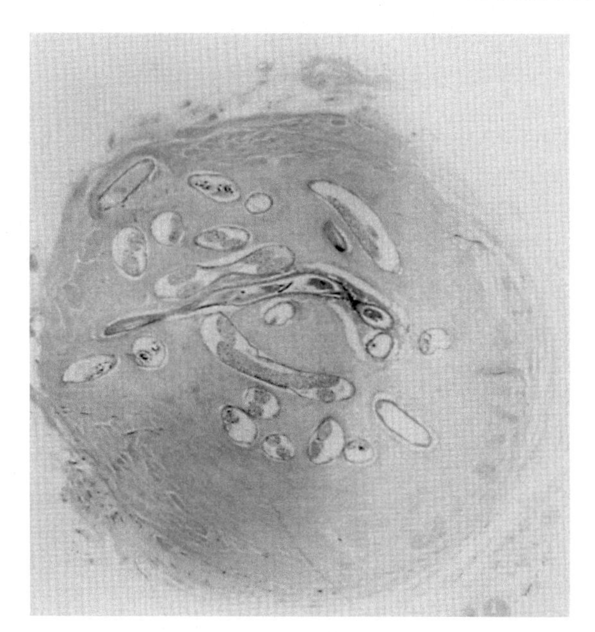

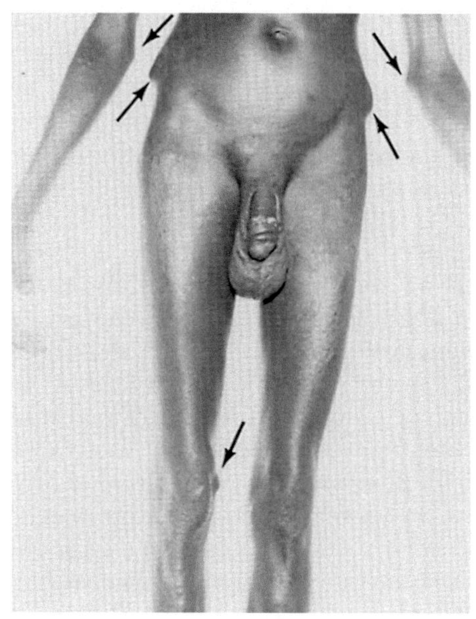

FIGURE 17.4 ***Onchocerca volvulus.*** (*left*) Section through a nodule showing adult worms with microfilariae. (*right*) Onchocercomas on a native in the Congo. Arrows point to onchocercomas on elbows, hips, and knee.

rupture, and the unsheathed microfilariae migrate to the lymphatics or, more often, to the connective tissues of the skin and eyes. An adult female worm can produce millions of microfilariae. Since they do not live in the blood, they do not display periodicity, but are positively phototactic. Blackflies, *Simulium* spp., become infected with the microfilariae while taking a blood meal. The average developmental time for infective larvae in the vector is 2 weeks during which they molt twice, migrate from the hemocoel and wing musculature to the proboscis, and are deposited on human skin when the fly feeds. Infective L3 larvae migrate into the subcutaneous tissue through the puncture site. Development to sexual maturity takes approximately a year; life span of the worms can be as long as 14 years.

Epidemiology

The World Health Organization (WHO) considers onchocerciasis the world's second leading infectious cause of human blindness. Worldwide, there are estimated to be 20 million cases of onchocerciasis, with another 120 million people at risk for the disease. Approximately 96% of the cases are in tropical Africa, although there are significant numbers in the highlands of western Guatemala, Colombia, and northeastern Venezuela as well. The disease also occurs in Mexico and the Near East. In endemic areas in Central America, infected flies abound and breed in the high mountain streams, usually 1000–4000 feet above sea level. Infection is common among workers on highland coffee plantations. Recent estimates show that, while it is being controlled to some extent in Central and South America, prevalence is increasing in Africa where the introduction and increasing use of irrigation techniques and construction of hydroelectric dams and attendant lakes has led to the spread of *Simulium* breeding sites.

Symptomatology and Diagnosis

In general, onchocerciasis is a disease that manifests itself in a cell-mediated, inflammatory, host response to foreign proteins from live, dead, and/or dying worms. One of the diseases caused by *O. volvulus* is commonly known as "river blindness." It is estimated that more than one-quarter of a million people suffer from this affliction. *O. volvulus* microfilariae invade the cornea, causing inflammation of the sclera, cornea, iris, and retina. Formation of fibrous tissue usually follows, leading to impaired vision or total blindness. Such ophthalmic changes are gradual and require 7–9 years to develop fully.

In addition to the dramatic condition described above, the presence of microfilaria in the connective tissues of the skin often produces severe dermatitis resulting from either allergic responses or toxicity. Affected areas of the skin become thickened, depigmented, wrinkled, and cracked. Since the symptoms resemble those accompanying vitamin A deficiency, it has been suggested that they reflect the parasite's competition for vitamin A or interference with its metabolism in the host.

Adult worms may also cause minor pathological alterations such as subcutaneous nodules, especially over bony prominences (Fig. 17.4). Onchocercomas caused by the Venezuelan and African strains of *O. volvulus* not only appear in the pelvic area but also occur less frequently on the chest, spine, and knees. On the other hand, infection with the Central American strain more commonly produces nodules above the waist, especially on the head and neck. Although subcutaneous onchocercomas are readily excised, adult worms in deep-seated nodules continue to produce microfilariae that can migrate to the surface for transmission and can continue to cause damage to the eyes.

Superficial nodules, cutaneous reactions, eosinophilia, or ocular symptoms in patients from endemic areas are strong indicators of onchocerciasis. Microscopical demonstration of microfilariae from dermal lymph or skin biopsy is proof positive, as is the identification of adults in skin nodules.

Chemotherapy

Ivermectin, administered in a single dose, has replaced diethylcarbamazine and suramin as the most effective treatment for onchocerciasis. The drug paralyzes the microfilariae, allowing host macrophages to remove them before they can degenerate and release allergenic materials into the circulation. Ivermectin treatment also improves the adverse skin conditions that result from the infection. Ivermectin, however, does not affect the adult worms enclosed in the onchocercomas or the release of microfilariae. Ivermectin treatment is contraindicated in patients with concurrent infection with the eye worm, *L. loa*, since it precipitates severe reactions.

Host Immune Response

In a community where onchocerciasis is endemic, three categories of individuals can be identified. One category consists of individuals continuously exposed to *O. volvulus* but apparently resistant. These individuals display a dominant Th1 cellular response. A second category of individuals displays microfilariae in the skin but are clinically asymptomatic and are characterized by their inability to produce Th1-type cytokines (e.g., interferon-γ [IFN-γ] and interleukin-2 [IL-2]). The third category of individuals display onchocercal skin disease and develop pathogenic immune responses with vigorous reactions to filarial antigens. The latter two categories of individuals display an increase in IL-10 production. IL-10 is thought to downregulate Th1-type cytokines and promotes cellular unresponsiveness with chronic filarial infections.

Prevention

Preventive measures are three-fold: surgical and chemical treatment of patients to prevent further spread of the disease, control of the insect vector population, and protection and education of potential victims. Patient treatment has already been discussed. Control of *Simulium* requires judicious use of insecticides on aquatic larvae, especially during the dry seasons, and on vegetation along the banks of swift-moving streams and rivers. Protective netting and screening and use of insect repellents effectively shield individuals from biting by infected flies.

Due to a combination of health education and mass drug administration with ivermectin, the Onchocerciasis Elimination Program for the Americas reports that the disease has been eliminated from Colombia (2013), Ecuador (2014), Mexico (2015), and most recently from Guatemala (2016). In addition, the WHO has reported that 12 countries in Africa are expected to eliminate the disease by 2020.

Loa Loa

Adult females *L. loa* are 50–70 mm long by 0.5 mm wide, with the vulva located at the extreme anterior end; adult males measure 30–35 mm long by 0.4 mm wide. The adults live in the subcutaneous tissues of the body and migrate freely. Because they are often seen moving beneath the conjunctiva, these parasites are known as African eye worms.

Life Cycle

Humans and baboons are the only definitive hosts of *L. loa*. The sheathed microfilariae display diurnal periodicity, retreating to the pulmonary capillaries at night. Various members of the mango fly genus *Chrysops* serve as vectors in which the developmental period to infective third-stage larvae is approximately 10–12 days. Within an hour after introduction into the definitive host as the vector takes a blood meal, the L3 larvae penetrate to the subcutaneous and muscle tissues. There, over the next 12 months, they molt twice and metamorphose into adult worms. The life span of adult worms is estimated to be 4–17 years.

Epidemiology

An estimated 20 million patients suffer from loaiasis. Although the parasite was introduced into the Caribbean during the slave trade era, it did not persist there, and the disease is now limited to the African equatorial rain forest and southern Sudan. The vector breeds in muddy ponds and swamps; not surprisingly, infection rates are highest in these regions.

Symptomatology and Diagnosis

L. loa is only mildly pathogenic. Adult worms wander throughout the body, moving through the tissues at a maximum rate of about 1.5 cm/min. The most troublesome infection sites are the conjunctiva (Fig. 17.5) and the bridge of the nose, where impaired vision, irritation, and pain may result. Most symptoms are general inflammatory reactions to adult worms and microfilariae and are often transient, appearing and disappearing at irregular intervals. A typical manifestation takes the form of transient, painful, subcutaneous swellings,

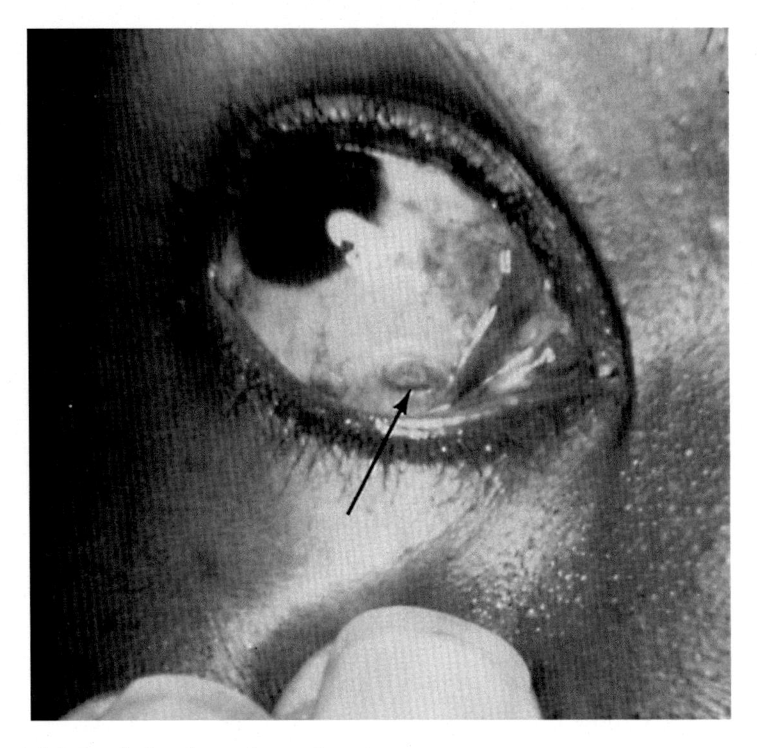

FIGURE 17.5 **Adult female *Loa loa* under conjunctiva.**

commonly termed fugitive or Calabar swellings. These are most often seen on the hands and forearms or near the eyes and may grow to the size of a hen's egg.

Sheathed microfilariae in the spleen can cause eosinophilia and fibrosis. Victims of the infection commonly exhibit a wide range of symptoms attributable to the wandering worms, such as low-grade fever, dermatitis, pain in the limbs, edema, and eosinophilia.

Diagnosis is usually based on sightings of the wandering worm in the conjunctiva, the presence of Calabar swellings, eosinophilia, and/or diurnal demonstration of microfilariae in blood. Antigens prepared from the dog heartworm, *D. immitis*, are useful diagnostic tools when other techniques are inconclusive.

Chemotherapy

Surgical removal of wandering adult worms from the conjunctiva is advisable. Diethylcarbamazine is the drug of choice to kill microfilariae; however, it can produce serious side effects, ranging from encephalitis to death. Ivermectin has been used effectively, but in mixed infection with *O. volvulus*, this drug too can have severe side effects, a fact that takes on special significance in view of the overlap in the distribution of onchocerciasis and loaiasis in West and Central Africa. Neither drug affects adult worm.

Prevention

The protocol recommended for control and prevention of infection with *O. volvulus* and other filarial worms applies for *L. loa* infection as well.

Mansonella Ozzardi, Mansonella Perstans, and Mansonella Streptocerca

In the genus *Mansonella*, the precise number of species infective to humans has not been firmly established. There are striking similarities among all these species, and the validity of assigning them to separate species may warrant challenge. In all instances, members of the midge genus *Culicoides* serve as vectors and adult worms reside in the body cavities and neighboring associated tissues of the definitive hosts. *M. ozzardi* favors visceral adipose tissue, *M. perstans* prefers the peritoneal cavity and occasionally the pericardial cavity, and *M. streptocerca* adults and microfilariae favor the subcutaneous tissue. The unsheathed microfilariae of the three species display no periodicity and are readily visible in blood and other host tissues. *M. ozzardi* is endemic in northern Argentina, the northern coast of South America, and throughout Central America. *M. perstans* is found primarily not only in tropical Africa but also to a lesser degree in South America and the Caribbean. *M. streptocerca* is common in East Africa, including Uganda, Kenya, and southern Sudan.

Other than local tissue reaction in the form of hydrocoels, no dramatic symptoms are associated with infection by these parasites.

THE GUINEA WORM

Dracunculus Medinensis

Awareness of *D. medinensis* dates back to antiquity. The "fiery serpent" of the biblical Israelites (numbers 21:6), *D. medinensis* is today commonly called guinea worm or Medina worm. Long and thin, the adult female measures 500–1200 mm by 0.9–1.7 mm and the adult male 12–29 mm by 0.4 mm.

Life Cycle

Adult worms inhabit the body cavity, its surrounding membranes, and the connective tissue of the human host. Male worms are rarely observed. The vulva, positioned equatorially in young females, is atrophied and nonfunctional in adults. The branched, gravid uterus, filled with thousands of larvae, compresses the intestine of the female and renders it nonfunctional. Gravid females migrate to the subcutaneous tissues of infected humans.

In the subcutaneous tissues, gravid worms direct their heads toward the skin and secrete an irritant that causes papules to form in the host dermis, most frequently on the ankles and wrists (Fig. 17.6). As each papule grows, it assumes the external appearance of a blister, eventually rupturing and leaving a cup-shaped ulcer in the skin. When the open ulcer comes in contact with water, a loop of the worm's uterus prolapses, either through the broken anterior end of the body or through the mouth, and ruptures releasing numerous first-stage rhabditiform larvae into the water. The larvae can survive in the aquatic environment for several days. Cold water stimulates contraction of the female body wall, causing larvae to be ejected in spurts. As the larvae are ejected, the body wall continually eases out of the ulcer and atrophies, and the remaining portion of the worm shortens proportionately. Larvae ingested by a suitable species of the copepod genera *Cyclops*, *Mesocyclops*, and *Thermocyclops* burrow through the midgut and enter the hemocoel. The presence of more than five or six larvae is fatal to the arthropod.

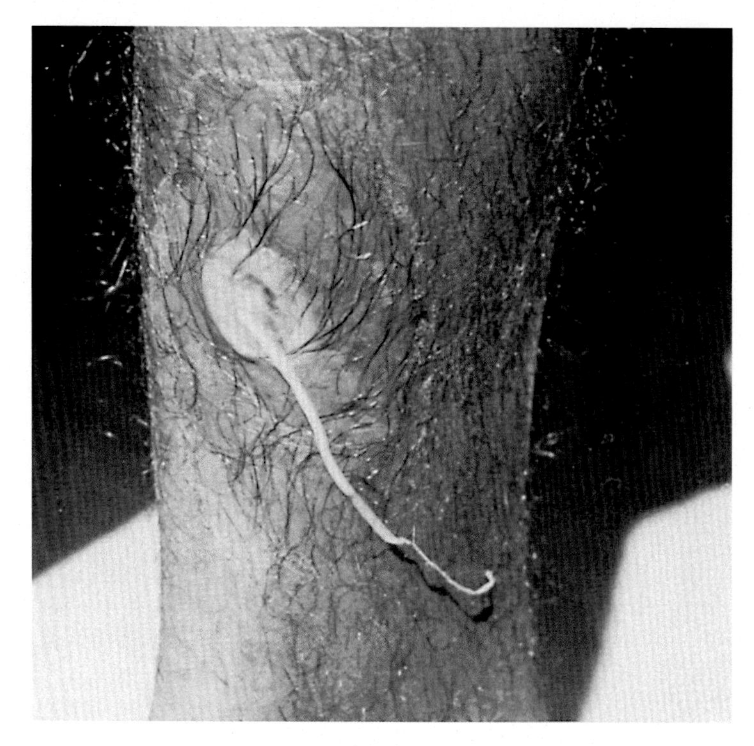

FIGURE 17.6 **Female *Dracunculus medinensis* partially protruding from blister on leg.**

Within the hemocoel of the copepod, the rhabditiform larvae undergo two molts, metamorphosing into infective, sheathed, L3 larvae in approximately 20 days. When drinking water contaminated with infected copepods is consumed by a human, the larvae, freed from the crustacean during digestion, exsheath in the duodenum of the human host. The larvae then burrow through the mucosa, undergo two additional molts, and lodge in the liver, body cavity, or subcutaneous tissues, where they mature in 8–12 months. The adult female is fertilized about 3-months postinfection; males usually die and degenerate 3–7 months after infection. A period of 10–14 months elapses between time of initial infection of the human host and the eruption of skin blisters.

Epidemiology

Human infection occurs throughout Africa except for the southern regions, in southwestern Asia (including southern India and Nepal), and to a lesser extent in northeastern South America and the West Indies. It is estimated that until recently the number of guinea worm cases worldwide was 3.5 million. Recent estimates show only 160,000 cases persist with approximately one-third of them in the Sudan. Two criteria are requisite for the completion of the life cycle: ingestion of infected copepods and contact of the infected human host with water. In drought-stricken areas of Africa, pools of stagnant water abundant with copepods provide ready sources of infection for individuals in search of drinking water. In southern India, the step-well is a prime source of infection. The practice of standing ankle or knee deep in the well to fill water containers allows gravid female worms in open ulcers of infected persons to release their larvae. Copepods, previously infected by larvae released in similar fashion are drawn with the drinking water, providing a source of new infections.

A number of animals, including canines, felines, horses, and raccoons, have been implicated as possible reservoir hosts for the infection. However, since there are a number of other species of *Dracunculus* with which *D. medinensis* can be confused, the importance of these animals as reservoirs is as yet unknown.

Nigeria was one of the "hot beds" for *D. medinensis* with 3 million cases reported in 1986. In 2006, only 12,000 cases were reported. It is now estimated that the Guinea worm is poised to be the first disease since smallpox to be "pushed into oblivion." This has been largely accomplished by the sponsorship of the Carter Center and the use of the insecticide Abate.

Symptomatology and Diagnosis

D. medinensis infection causes a broad spectrum of nonspecific symptoms, such as eosinophilia, nausea, diarrhea, asthma, and fainting. These symptoms are believed to result from absorption of metabolic wastes produced by female worms during papule formation. In addition, cutaneous ulcers caused by female worms are common sites for secondary bacterial and fungal infections that often produce permanent scars and muscle damage.

Female worms failing to reach host skin sometimes cause reactions in deeper tissues of the body. Commonly, they degenerate or become calcified. Degeneration of the worms stimulates the release of strongly antigenic molecules that can cause fluid-filled abscesses, while calcification near a joint may produce a type of chronic arthritis.

Appearance of localized blisters or ulcers or microscopical identification of the female worm, especially the protruded head, or larvae constitutes diagnosis. X-ray examination may reveal calcified worms. An ELISA test is available, but its use to date has been limited.

Chemotherapy

Both chemotherapeutic and mechanical techniques, including surgery, are employed for removal of adult worms. The drug of choice is metronidazole administered over a 7–10 day period. Metronidazole acts not only on the worm itself but also as an anti-inflammatory agent. Mechanical withdrawal of the gravid female from the ulcerated area requires painstaking care, and the worm must be extracted slowly. If the worm breaks during the extraction process, larvae escape into the subcutaneous tissues, causing severe and painful inflammatory reactions. The ancient method, dating back to biblical days, of winding the worm on a stick is still used in parts of Africa and Asia (Fig. 17.7). During the procedure, the worm is wound around a stick, which is slowly turned, withdrawing the worm a few centimeters a day. Applying cold water to the area hastens the process by causing expulsion of more larvae to the exterior and allowing exposure of an additional 5 cm or so of the worm. The caduceus, the official emblem of the medical profession, includes a pair of serpents wrapped around a staff. It is not inconceivable that it may have been derived from this ancient method of *D. medinensis* removal.

Prevention

As with other parasitic diseases, prevention and control require interruption of the life cycle. It is essential that the practice of bathing and washing in sources of drinking water be discontinued. Water suspected of being contaminated should be boiled or filtered before use

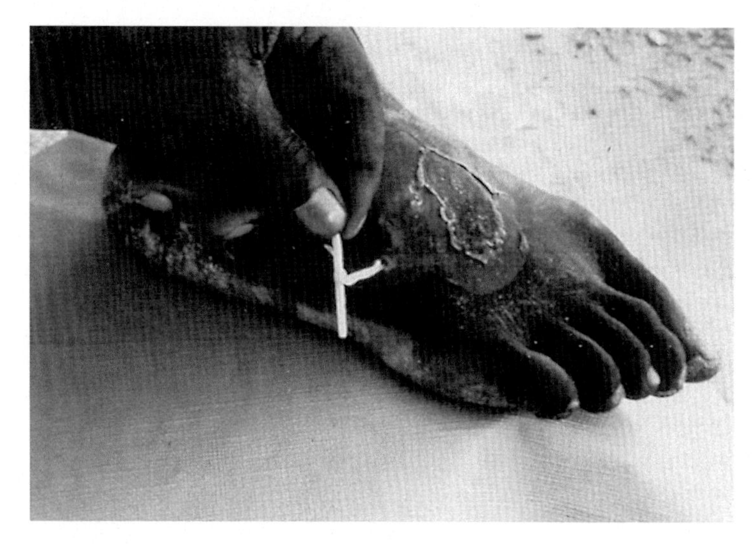

FIGURE 17.7 **Female *Dracunculus medinensis* slowly being withdrawn from ulcer on foot with twig.**

and, whenever possible, drinking water should be obtained from swift-running waterways, which are usually free of copepods. Chemical treatment of water with chlorine or copper sulfate to destroy copepods is an alternative. The latest treatment of infested waters has been the use of the insecticide Abate. The latter treatment has already proven so successful that an end to guinea worm disease is predicted for the very near future.

Suggested Readings

Bockarie, M. J., Tavul, L., Ibam, I., et al. (2007). Efficacy of single-dose diethylcarbamazine compared with diethylcarbamazine combined with albendazole against Wuchereria bancrofti infections in Papua, New Guinea. *American Journal of Tropical Medicine and Hygiene*, *76*, 62–66.

Cheng, T. C. (1983). Cutaneous lesions due to nonarthropod parasites. In L. C. Parish, W. B. Hutting, & R. M. Schwartzman (Eds.), *Cutaneous infestations of man and animal* (pp. 237–254). New York: Praeger Publishers.

Chernin, E. (1983). Sir Patrick Manson's studies on the transmission and biology of filariasis. *Reviews of Infectious Diseases*, *5*, 148–166.

Denham, D. A., & McGreevy, P. A. (1977). Brugian filariasis: epidemiological and experimental studies. *Advances in Parasitology*, *15*, 244–309.

Duke, B. O. L. (1984). Filtering out the guinea worm. *World Health*, *3*, 29.

Duke, B. O. L. (1990). Onchocerciasis (river blindness)—can it be eradicated? *Parasitology Today*, *6*, 82–84.

Hopkins, D. R., Ruiz-Tiben, E., Diallo, N., Withers, P. C., & Maguire, J. H. (2002a). Dracunculiasis eradication: and now Sudan. *American Journal of Tropical Medicine and Hygiene*, *67*, 415–422.

Hopkins, D. R., Eigege, A., Miri, E. S., Gontor, I., Ogah, G., Umaru, J., et al. (2002b). Lymphatic filariasis elimination and schistosomiasis control in combination with onchocerciasis control in Nigeria. *American Journal of Tropical Medicine and Hygiene*, *67*, 266–272.

O'Connor, R. A., Jenson, J. S., Osborne, J., & Devaney, E. (2003). An enduring association? Microfilariae and immunosupression in lymphatic filariasis. *Trends in Parasitology*, *19*, 565–570.

Pfaff, A. W., Kirch, A. K., Hoffman, W. H., Banla, M., Schulz-Key, H., Geiger, S. M., et al. (2003). Regulatory effects of IL-12 and IL-18 on *Onchocerca volvulus* and *Entamoeba histolytica* specific cellular reactivity and cytokine profiles. *Parasite Immunology*, *25*, 325–332.

Pinder, M. (1988). Loa loa—a neglected filaria. *Parasitology Today*, *4*, 279–284.

Richards, F. D., Boakye, B., Sauerbrey, M., & Seketeli, A. (2001). Control of onchocerciasis today: status and challenges. *Trends in Parasitology*, *17*, 558–563.
Sobosly, P. T., Luder, C. G., Riesch, M., Geiger, S. M., Banla, M., Batchassi, E., et al. (1999). Regulatory effects of Th1-type (IFN-γ, IL-12) and Th2-type cytokines (IL-10, IL-13) on parasite-specific cellular responsiveness in *Onchocerca volvulus*-infected humans and exposed endemic controls. *Immunology*, *97*, 219–229.
Van de Waa, E. A. (1991). Chemotherapy of filariases. *Parasitology Today*, *7*, 194–199.

CHAPTER

18

Arthropods as Vectors

Chapter 18 is aptly entitled "Arthropods as Vectors." The opening sections of the chapter deal with the significance of arthropods as vectors of a number of diseases of humans. In addition to referring to those diseases caused by parasites as outlined in previous chapters, this chapter also takes up those diseases with which the arthropod plays a significant role in its spread to humans. In this context, the arthropods are grouped according to their classification which is seen at the end of the chapter. The biting dipterans include mosquitoes, blackflies, sandflies, tsetse flies, and tabanid flies with their associations with diseases such as anthrax, tularemia, bartonellosis, and hemorrhagic fever, to name a few. Other arthropods such as fleas, lice, and reduviid bugs are discussed relative to such diseases as the plague and relapsing fever. A section related to the acarines (ticks and mites) is at the end of the chapter. Diseases such as Lyme disease, typhus, ehrlichiosis, and Rocky Mountain spotted fever are discussed.

Members of the phylum Arthropoda probably constitute the largest number of individuals and species of any phylum in the animal kingdom. There are more than 760,000 known species of arthropods. According to current classification systems, the phylum is divided into four subphyla: Trilobitomorpha, Chelicerata, Crustacea, and Uniramia. This chapter deals with the role of Chelicerata (ticks and mites) and Uniramia (insects) as vectors of human disease-producing organisms (Fig. 18.1).

Although most arthropods generally are of little if any medical importance, they are of considerable biological interest, particularly to parasitologists. Furthermore, parasitic arthropods, such as ticks, mites, and certain insects, are of considerable medical and veterinary importance not only because they inflict direct injury upon their hosts but also because many serve as vectors for various pathogenic microorganisms and viruses.

SIGNIFICANCE OF ARTHROPODS AS VECTORS

Disease-producing organisms transmitted to humans by arthropods have significantly influenced the history and demography of the human race. The plague, or Black Death, that so tragically decimated the population of Europe in the 14th century is vivid documentation of the toll such a disease can exact. Trench fever, the scourge of World War I, also attests the gravity of vector-borne diseases. Other such diseases that have had dramatic impact upon historic events and eras include trachoma during Napoleon's invasion of

Human Parasitology. http://dx.doi.org/10.1016/B978-0-12-813712-3.00018-7

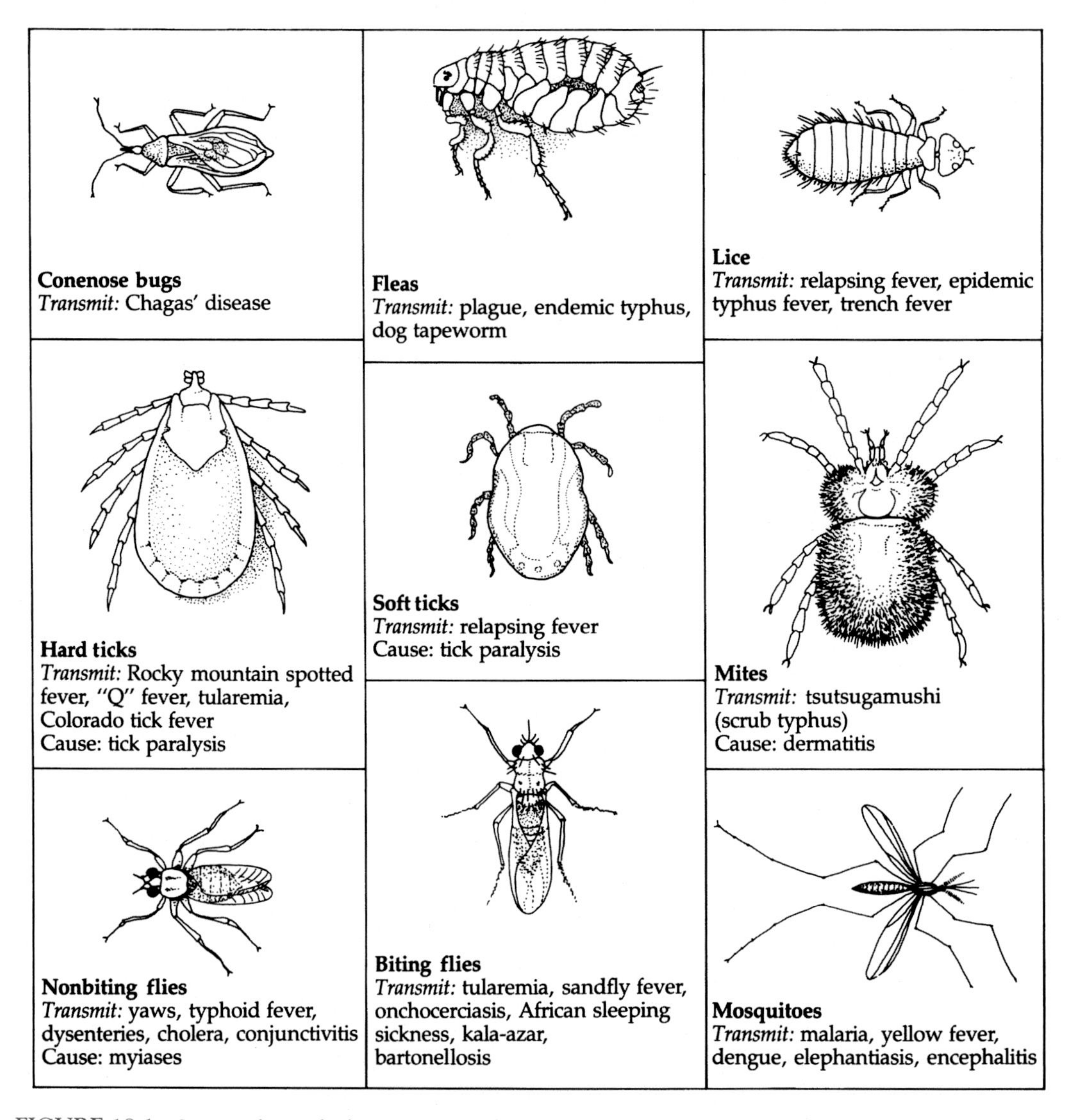

FIGURE 18.1 **Some arthropods that transmit pathogens of human disease.**

Egypt, malaria and yellow fever during the construction of the Suez and Panama Canals, and African trypanosomiasis and malaria during the European explorations of Africa in the 19th century.

The disease agents transmitted by various arthropods vary widely in size and degree of pathogenicity. The smallest are submicroscopic viruses no larger than a large protein molecule and are obligate parasites of cells. Rickettsiae, currently considered a highly specialized form of bacteria, are somewhat larger than most viruses although their size range overlaps the larger viruses and smaller bacteria. Like viruses, rickettsiae are obligate parasites, and they are at some point dependent upon arthropods for transmission. Unlike viruses, but similar to many bacteria, rickettsiae are vulnerable to antibiotics. Next in order of size, bacteria thrive in a wide range of environments and display equally diverse metabolic patterns and host relationships. While many bacteria are free-living, many others form symbiotic relationships of commensalistic, mutualistic, or parasitic nature. The role of insects in the transmission of protozoans such as *Plasmodium*, *Leishmania*, and *Trypanosoma* and nematodes such as *Wuchereria*, *Onchocerca*, and *Brugia* has already been discussed. A number of tapeworm infections are also arthropod transmitted. For instance, *Dipylidium caninum* is transmitted by fleas, *Diphyllobothrium latum* by copepod crustaceans, and *Hymenolepis diminuta* by beetles. The lung fluke *Paragonimus westermani* uses a crustacean as the second intermediate host in its life cycle.

The manner in which the various organisms parasitic to humans are transmitted dictates the type of association the arthropods establish with the parasites. The simplest relationship is one in which the arthropod is a **mechanical vector**, functioning merely as a passive carrier of the etiologic agent. Examples of this type include typhoid organisms and *Entamoeba histolytica* cysts from contaminated excreta, which adhere to body parts or pass through the digestive tracts of the common housefly and roach and are subsequently transferred to food and drink touched by the insects. As a **biological vector**, the arthropod is used by the disease-producing organism not only as a vehicle of transmission but also as an environment for development and/or reproduction prior to its infective stage.

Biological transmission is of four types. In **propagative biological transmission**, the disease-producing organism reproduces in the arthropod but undergoes no further development; examples are the plague bacillus in the flea and the yellow fever virus in the mosquito. In **cyclopropagative biological transmission**, the disease-producing organism not only reproduces but undergoes cyclical changes in the arthropod as well. *Plasmodium* spp. and trypanosomes transmitted by mosquitoes and by tsetse flies, respectively, are examples of this type. In **cyclodevelopmental biological transmission**, the disease-producing organism undergoes vital cyclical changes in the arthropod vector but does not multiply there. For example, filarial worms must spend a portion of their life cycle in their mosquito vectors although they reproduce elsewhere. Finally, in **transovarial transmission**, certain disease-producing organisms, such as rickettsiae that cause Rocky Mountain spotted fever and scrub typhus, are transmitted from infected parent arthropods (i.e., ticks and mites) to their offspring.

Once the etiologic agent has reached the stage infective to humans, there are several means by which it goes from arthropod to human host. Infective forms of parasites carried by certain blood-sucking flies exit the insect's mouthparts during the blood meal and enter human skin through the puncture. The malarial parasite uses a more efficient mechanism to reach its human victim: the infective form reaches the salivary gland of the mosquito and enters the human blood stream via saliva secreted by the insect into the human skin while feeding. Other insect-transmitted infections, such as viral encephalitis, yellow fever, dengue, and sleeping sickness, are transmitted in similar fashion. In some nonblood-feeding arthropods, the feeding larva (e.g., maggot) ingests the agent, retaining it in the digestive tract during metamorphosis; later, the resulting adult arthropod transmits it to the human via vomit or excreta on food and drink.

GENERAL STRUCTURAL FEATURES

In the early phases of their evolution, primitive arthropods were conspicuously multisegmented, each segment equipped with a pair of appendages; in present forms, some segments are fused, the number of appendages is reduced, and most have evolved wings, all of which markedly enhances their mobility. In insects, the most successful of terrestrial animals, the body is divided into three regions (Fig. 18.2): the **head**, with a variety of mouthparts and sense organs; the more or less rigid **thorax**, bearing three pairs of walking appendages and usually wings; and the segmented **abdomen**, lacking appendages. Acarines, the ticks and mites, represent the second most successful terrestrial arthropod group. They also have three body regions but are wingless and possess four pairs of walking appendages (Fig. 18.3). Segmentation in ticks and mites is so indistinct that they appear to be unsegmented.

Arthropods have a chitinous exoskeleton, the **cuticle**, which extends to all external openings. The rigid cuticle limits growth; therefore, periodic molting is required. The

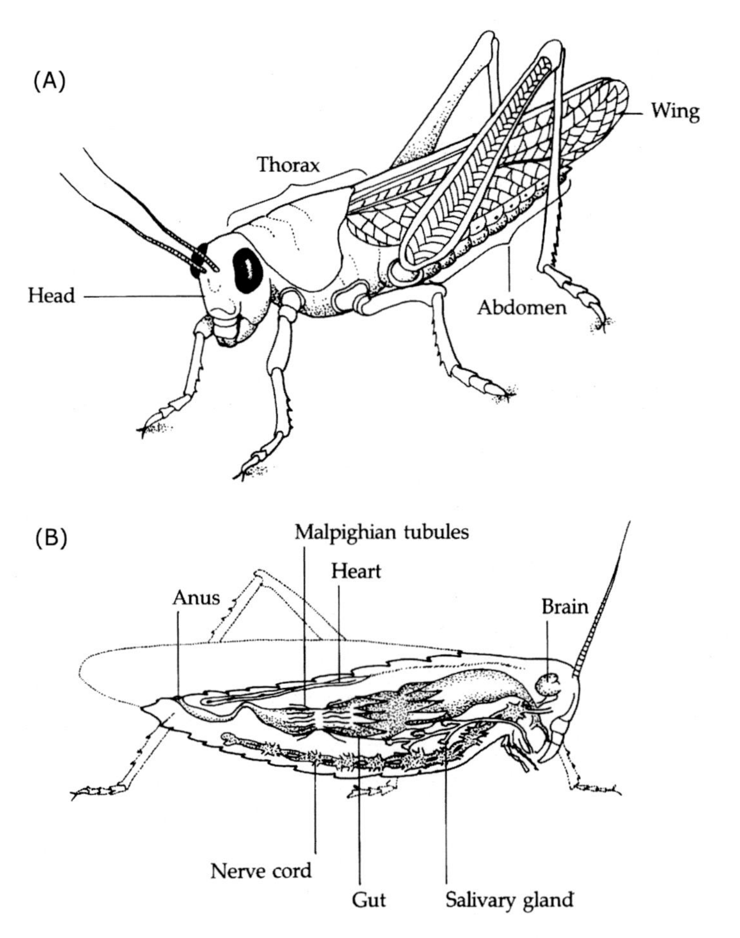

FIGURE 18.2 **Anatomy of a typical insect.** (A) External. (B) Internal.

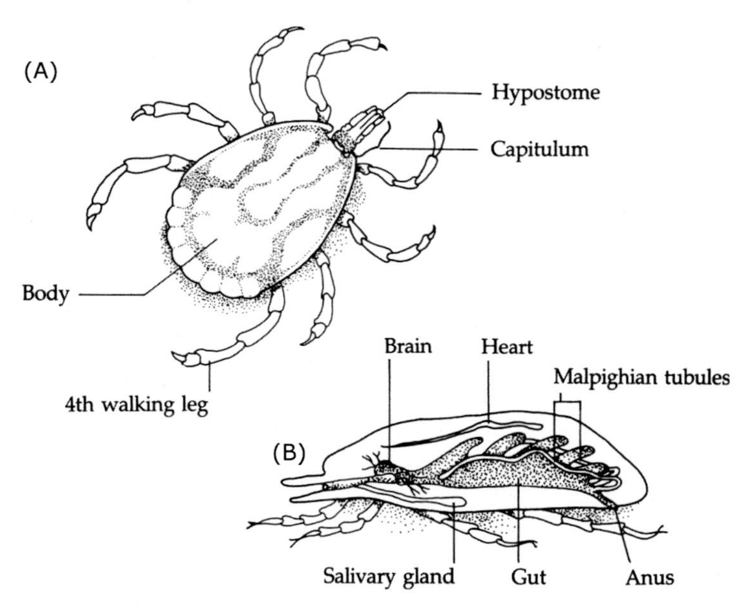

FIGURE 18.3 **Anatomy of a typical acarine.** (A) External. (B) Internal.

open circulatory system consists of a dorsal tubular heart that slowly pumps hemolymph through large sinuses that make up the major body cavity or **hemocoel**. Hemolymph is not always involved in oxygen transport; in many terrestrial forms, gaseous exchange between cells and environment is accomplished through a system of branched tubules, or **tracheae**. The nervous system consists of a pair of ventral nerve cords with segmentally arranged ganglia; the typical "brain" is a major ganglion located dorsally in the head region and linked to the nerve cords by circumesophageal connectives from the anterior most ventral ganglion. Nitrogenous wastes are excreted by terrestrial insects and most acarines in the form of uric acid produced in **Malpighian tubules** that empty into the hindgut.

Arthropods acquire disease-producing organisms primarily during feeding. Because their food sources are so varied, the mouthparts of arthropods likewise vary widely. The mouthparts of primitive insects were adapted for chewing and many extant species retain that adaptation. Grasshoppers, bees, ants, wasps, cockroaches, and termites retain the basic mouthparts of chewing insects. Although chewing insects are ineffective as biological transmitters of pathogenic organisms, knowledge of their basic mouthparts is essential for understanding the evolutionary modifications that have produced the more specialized feeding apparatuses.

Lying directly behind the "upper lip," or **labrum**, are the first of the true mouthparts, the paired **mandibles**, which are heavily sclerotinized and bear jagged teeth along their medial margins. Behind them are the paired, jointed **maxillae**. The **hypopharynx** is not a true mouthpart but an unsegmented, tubular outgrowth of the body wall arising from the ventral, membranous floor of the head. The "lower lip" of insects is the heavily sclerotinized, segmented **labium**. The mandibles masticate the food, and the maxillae and labium push the pulverized food into the mouth. In more advanced insects, the major modification of the feeding apparatus is the evolution from the chewing type to one of several cutting and/or piercing types (Fig. 18.4).

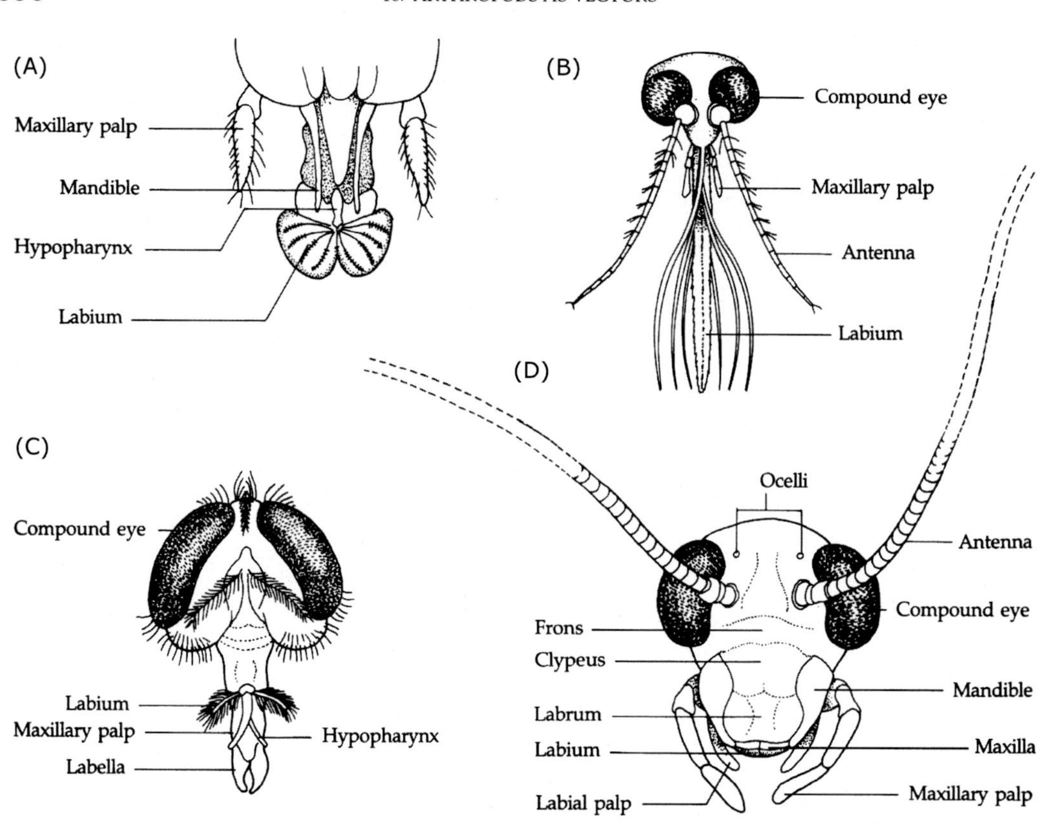

FIGURE 18.4 **Types of some insect mouthparts.** (A) Cutting-sponging. (B) Piercing-sucking. (C) Sponging. (D) Chewing.

The **cutting-sponging type**, characteristic of horseflies, features sharp-bladed mandibles and long, stylet-like maxillae. The mandibles cut and tear the skin of the host, and the sponge-like labium collects the blood and conveys it to the esophagus by means of a tube formed partially by the hypopharynx.

The mouthparts of most nonbiting dipterans, such as the common housefly, are of the **sponging type**, similar to the cutting-sponging type except that the mandibles and maxillae are nonfunctional. The remaining parts form a proboscis with a spongelike apex called the **labella**. Liquid food is conducted to the mouth through minute capillary channels on the labella. Solid food is ingested only after being dissolved or suspended in deposited saliva.

The **piercing-sucking type**, characteristic of mosquitoes (Fig. 18.5), flies, lice, and bedbugs, features mandibles, maxillae, and hypopharynx modified into a long, thin, tubular, sharp-tipped stylet for piercing skin. This narrow tube is enclosed by the labrum to form a **stylet bundle** that is held in a groove on the labium. Together, the stylet bundle and labium make up the **proboscis**. While not itself penetrating the skin, the labium guides the bundle

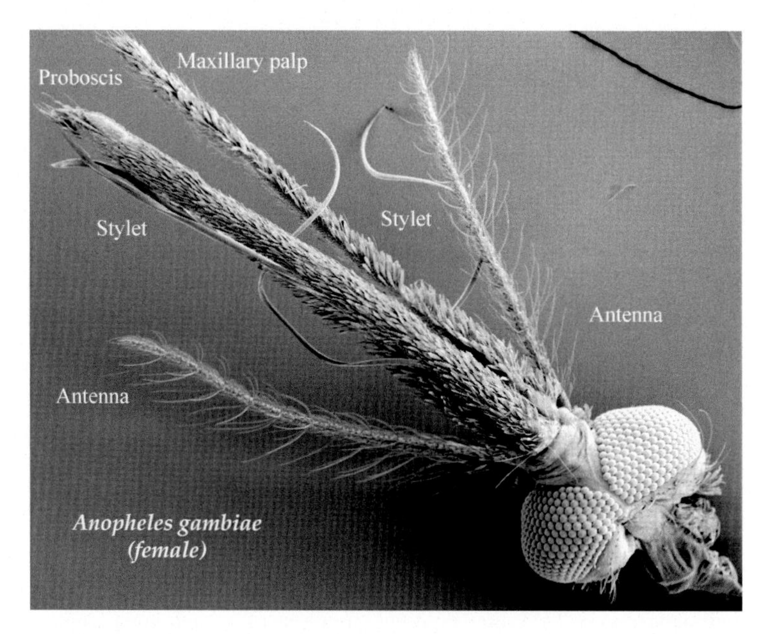

FIGURE 18.5 **Scanning electron micrograph of head of female *Anopheles gambiae*.** Ant., antenna; m.palp, maxillary palp; prob., proboscis; sty., stylet. *Source:* From Dr. Larry Zwiebel, Department of Biological Sciences, Vanderbilt University, Nashville, Tennessee.

into the wound site. During insect feeding, the stylet bundle pierces the skin of the host like a hypodermic needle, and blood is withdrawn through it. The hypopharynx usually contains the salivary gland duct.

Among acarines, mites are more versatile in their feeding habits than ticks. While mites may feed on decaying animal matter, feces, plants, animal secretions, and blood, ticks feed only on the blood of reptiles, birds, and mammals. Since the mouthparts of ticks and mites are similar, those of the former will serve to illustrate both groups (Fig. 18.6).

The **capitulum** is a small anterior projection bearing three structures that constitute the acarine mouthparts (i.e., the elongate **hypostome**, a pair of segmented **chelicerae**, and a pair of segmented **pedipalps**). The hypostome, usually toothed, is medially located, ventral to the mouth with its free end projecting anteriorly. Bilateral chelicerae are located on the dorsolateral surfaces of the hypostome, flanking the mouth. The free end of each chelicera is forked, one branch forming a fixed, dorsal, toothed digit, the **digitus externus**, the other a lateral, movable **digitus internus**. With these appendages, the host's skin is pierced and/or torn, and either the entire capitulum or, at least, the toothed hypostome is inserted into the opening. The appendages also serve as anchors when the parasite is attached. Paired pedipalps arise from the base of the capitulum at its anterolateral margin. During feeding, the pedipalps either bend outward ("soft ticks") as the chelicerae and hypostome penetrate the flesh or remain rigidly and intimately associated with the hypostome ("hard ticks") during skin penetration. In either instance, the pedipalps serve as counteranchors while the tick is attached to the host.

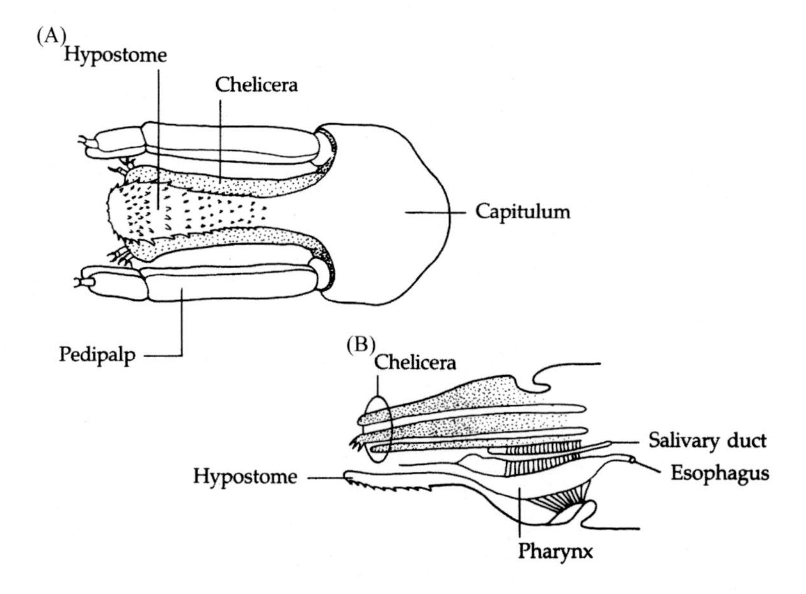

FIGURE 18.6 **Mouthparts of ticks.** (A) Ventral aspect of capitulum. (B) Longitudinal section through tick head.

THE DIPTERANS

While the previous sections dealt with structural modifications in arthropod vectors and some of the means by which they transmit pathogens, the following sections focus upon specific arthropods and some of the human pathogens they transmit. Numerous arthropod-transmitted pathogens have been discussed in preceding chapters devoted to protozoans and helminths, and appropriate references are made to those earlier chapters.

Because of their overwhelming numerical preponderance and the medical importance of the pathogens they transmit, dipterans are treated separately from other insects. Alone they may transmit as many disease organisms as all other insects combined.

The development of Dipterans is typical of higher insects (i.e., **holometabolous** or **complete, metamorphosis**). The phases of the life cycle include the **egg**, a fixed number of **larval** stages, a single **pupal** stage, and the **imago** or **adult** stage. More primitive insects, in contrast, are characterized by **hemimetabolous**, or **incomplete**, **metamorphosis**. Life cycle phases include the egg, a fixed number of **nymphal** stages, which gradually metamorphose serially to the adult stage. The body form of the nymphs resembles the adult. All the insects discussed in this chapter, with the exception of reduviid bugs and lice, display holometabolous life cycles.

Biting Dipterans

Mosquitoes

A number of species of the mosquito family Culicidae are active transmitters of organisms responsible for human disease. Two groups in this family, the anophelines and the culicines, are quite distinct biologically. The cigar-shaped anopheline eggs, for example, are equipped with side floats and are normally deposited diffusely over water. Culicine eggs, on the other hand, possess no side floats and are often deposited in raftlike arrays on the surface of water (*Culex*), in cushionlike arrangements under water plants (*Mansonia*), or on damp surfaces to await heavy rains (*Aedes*) (Fig. 18.7).

While anopheline larvae appear immediately under and parallel to the surface of the water, culicine larvae hang from breathing tubes anchored either to the water surface (*Culex* and *Aedes*) (Fig. 18.8) or to the stems of aquatic vegetation (*Mansonia*). Adults are distinguishable by the configuration and angle of their body parts (Fig. 18.9). Anopheline body parts are arranged in a straight line, inclined at an angle to the surface upon which they alight, while those of culicine adults are bent into a hump-backed posture.

Three groups of pathogenic organisms are transmitted to humans by mosquitoes. One group consists of the causative agents of malaria, *Plasmodium* spp. The role of the anopheline mosquito in the transmission of human malaria has already been discussed (see p. 117). Although there are approximately 350 known species of *Anopheles* throughout the world, only about two dozen are major transmitters of human malaria. Many species are genetically or ecologically incompatible with the malarial organism. For example, differences in susceptibility of mosquito populations to *Plasmodium* sometimes reflect variations in gene frequency in different geographic areas. Also, ecological hazards such as storms and droughts influence the mosquitoes' feeding habits and oviposition and, therefore, affect the life span of the vector and its ability to harbor the parasite effectively.

Filarial worms, the causative agents of human filariasis, constitute a second group of these mosquito-borne pathogens. Numerous anopheline and culicine species serve as vectors for

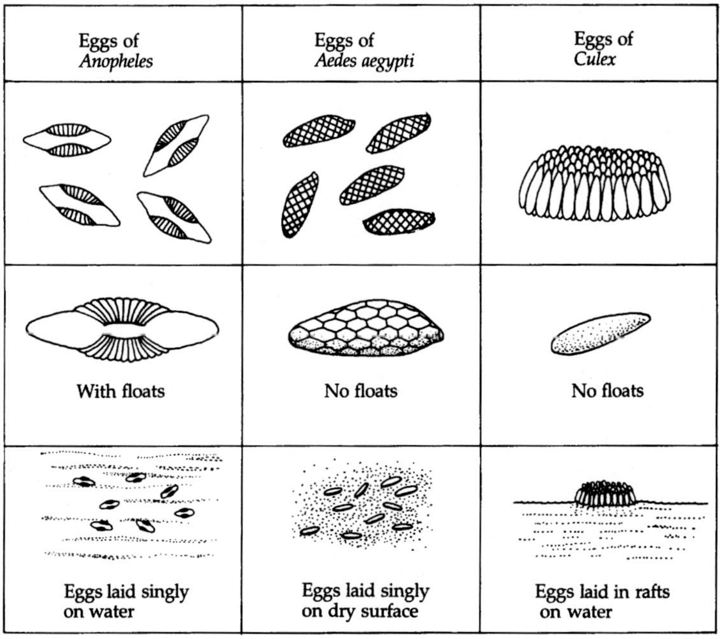

Eggs of *Anopheles*	Eggs of *Aedes aegypti*	Eggs of *Culex*
With floats	No floats	No floats
Eggs laid singly on water	Eggs laid singly on dry surface	Eggs laid in rafts on water

FIGURE 18.7 **Mosquito eggs.**

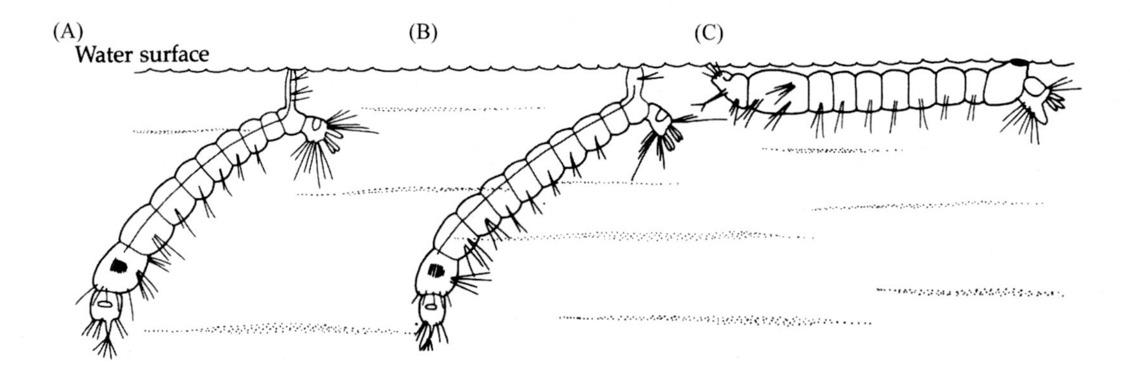

FIGURE 18.8 **Mosquito larvae.** (A) Culex adhering to water by its breathing tube. (B) *Aedes*. (C) *Anopheles*.

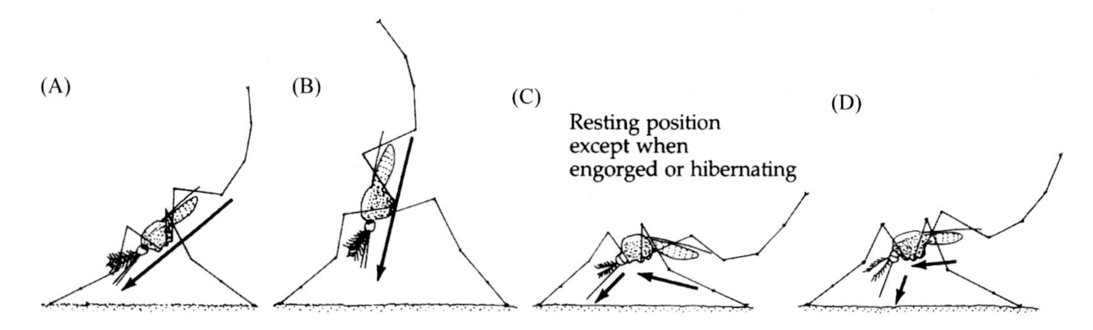

FIGURE 18.9 **Resting position of common mosquitoes.** (A, B) *Anopheles*. (C) *Aedes*. (D) *Culex*.

these organisms. It has been previously noted (see p. 333) that the ecology of the mosquito, especially its feeding cycle, is inextricably linked to the periodicity of microfilarial surge in the human host. For instance, the principal vector of the nocturnally surging *Wuchereria bancrofti* is the nocturnally feeding mosquito *Culex fatigans*, while the subperiodic variety employs the diurnal feeder *Aedes polynesiensis*.

The third group of pathogens is the arboviruses. Arboviruses (arthropod-borne viruses) are a large group of viruses that are spread mainly by blood-sucking insects or acarines. More than 200 diseases are transmitted in this manner, approximately three quarters of them by mosquitoes, a few by biting flies, and the rest by ticks. The most serious of the arboviral diseases are the hemorrhagic fevers and the encephalitides. In hemorrhagic fever, increased permeability of capillary walls caused by the viruses precipitates bleeding from kidneys, lungs, gums, nose, etc. In viral encephalitis, viral attacks upon the central nervous system cause symptoms ranging from minor back pain to temporary paralysis, with sometimes lingering spastic effects.

Yellow fever is probably one of the more prevalent of the hemorrhagic diseases. During construction of the Panama Canal, yellow fever caused such widespread illness among the workers that the project came to a halt. It was then that the Yellow Fever Commission, under the leadership of Walter Reed and his associates James Carroll, Jesse W. Lazear, and A. Agramonte, won acclaim by proving Carlos Finlay's hypothesis that the pathogen is transmitted by *Aedes aegypti*.

Female mosquitoes acquire the virus while feeding on blood of yellow fever victims. A single female ingests thousands of viruses at one feeding and usually remains infective for the rest of her normal life, 200–240 days. Under field conditions, the normal incubation in the mosquito is 12 days; however, fluctuations in temperature can alter incubation times. For instance, in mosquitoes exposed to temperatures of 36.8°C, the incubation period is reduced to 4 days; at 21°C, it is lengthened to 18 days. In addition to *A. aegypti*, other members of this mosquito genus serve as natural vectors for yellow fever; among these are *Aedes vittatus* in Egypt and *Aedes simpsoni* and *Aedes africanus* in eastern Africa. Jungle animals, especially monkeys, serve as natural reservoirs for the yellow fever virus.

Currently, world incidence of this disease, as reported by the World Health Organization, fluctuates from one hundred to several thousand cases annually. Vector control and mass inoculation have virtually eliminated endemicity in the Americas and have greatly reduced the incidence in Africa.

Three hemorrhagic fevers, **hemorrhagic dengue**, **breakbone fever**, and **Zika**, are transmitted by several species of *Aedes*. Hemorrhagic fever is endemic in Southeast Asia, afflicting an estimated 50 million people, with native children the principal victims. It has a mortality rate of about 7%. Breakbone fever is nonlethal and is characterized by a rash, high fever, and pain in the joints. It has been reported in Japan, New Guinea, northern Australia, the Philippines, Hawaii, and most recently Mexico and the southwestern United States. Zika virus was first reported in humans in Uganda in 1952. Subsequently, the virus has been reported in human populations in more than 60 areas including French Polynesia, Southeast Asia, Europe, South America, the Caribbean, and the United States. *A. aegypti* (Europe, Africa, Western hemisphere) and *Aedes albopictus* (Africa, Europe, Southeast Asia) are the principle vectors. Symptoms are usually mild ranging from low grade fever, skin rash, muscle or joint pain and headache. Zika infection during pregnancy is strongly linked to newborns with microencephalitis.

In the United States and parts of Latin America, viral encephalitides, such as **Western** and **Eastern equine encephalitis**, **St. Louis encephalitis**, and **Venezuelan equine encephalitis**, are usually nonlethal. In 1996–97, an outbreak of **West Nile** fever in Romania with more than 500 cases and a fatality rate of 10% was reported. Increases in human infections have now been reported in various parts of the world and West Nile virus is recognized as the most widespread of the flaviviruses, with geographic distribution involving Africa, Eurasia, and North and South America. The virus is apparently very "nimble." West Nile virus can spread from organ donor to organ recipient, from a pregnant mother to her fetus, by blood transfusions, and possibly through breast milk.

Culicine mosquitoes are the principal vectors of the arboviruses, and transmission is usually from birds to horses or humans, both of which are dead ends in the transmission sequence. A similar, but equally serious disease, **Japanese B encephalitis**, occurs in Japan.

Blackflies

Onchocerca volvulus, the causative agent of human onchocerciasis (see p. 320), is transmitted by the blackflies *Simulium damnosum* (Fig. 18.10) and *Simulium neavei* in Africa and *Simulium ochraceum*, *Simulium callidum*, and *Simulium metallicum* in Mexico, Central America, and

FIGURE 18.10 **Blackfly, *Simulium damnosum*, taking a blood meal from a human.**

South America. Females of *Simulium* spp. deposit eggs in fast-flowing water ranging from streams to large rivers; in the latter, eggs are deposited most abundantly where there are rapids and in areas below dams. The hatched larvae and resulting pupae remain in the aquatic habitat. Adult female flies feed on the blood of a variety of mammals, including humans. In addition to transmitting onchocerciasis to humans, the bites of the blackflies can themselves be troublesome, frequently producing severe reactions.

Sandflies

Phlebotomine sandflies (Fig. 18.11) transmit three different types of organisms pathogenic to humans. The viral disease **sandfly fever** occurs in the Mediterranean region, central Asia, south China, parts of India, Sri Lanka, and parts of South America. Of short duration, it is not considered serious. In the northwestern regions of South America, sandflies transmit the bacterium *Bartonella bacilliformis*, the causative agent of **Corrion's Disease** (=**bartonellosis**), a severe, often fatal illness. The third group of pathogens transmitted by sandflies consists of protozoans that cause the various types of **leishmaniasis** (see p. 90).

The phlebotomine sandflies that serve as vectors for these pathogens generally belong to two genera: *Phlebotomus* in the Eastern Hemisphere and *Lutzomyia* in the Western Hemisphere. A small insect, measuring 1.25–2.5 mm long, the sandfly has long, slender legs, and short setae covering most of its body parts and wings. The females use piercing/sucking mouthparts to extract juices from plants and blood from various vertebrates, including humans. Sandflies are not strong fliers, usually remaining close to their breeding sites in

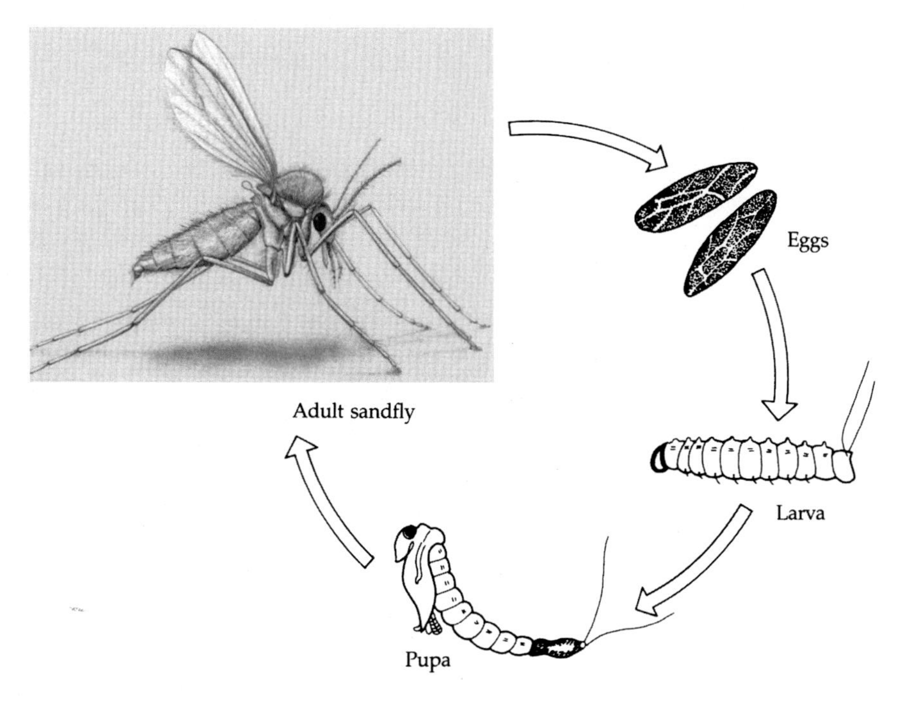

FIGURE 18.11 **Stages in the life cycle of the sandfly *Phlebotomus* spp. Tsetse flies, Glossina.**

damp areas rich in organic debris (e.g., under logs and dead leaves, inside hollow trees, and in animal burrows).

Tsetse Flies

Members of the genus *Glossina* (Fig. 18.12) are a menace not only because they are vectors for **African trypanosomiasis** (see p. 102) but also because both male and female flies feed on blood and their bites inflict large, painful welts on their victims (Fig. 18.13). Although *Glossina* was once widespread, it is now limited to continental Africa south of the Tropic of Cancer. During the life cycle of this important vector, the female harbors in her body a supply of fertilized eggs that hatch at intervals into larvae that feed from specialized "milk glands." The developing fourth-stage larva is deposited in a shady spot, usually at the base of a tree or shrub, and immediately burrows into the soil and pupates. A single female tsetse fly deposits 8–10 larvae, one at a time, at intervals of from 10 to 12 days each. The adult fly emerges in 3–4 weeks.

While several species will feed on human blood, they appear to prefer that of other animals. The various species of *Glossina* are ecologically adapted to different locales according to feeding preferences and site choices for larviposition. For instance, the two species that transmit trypanosomiasis breed either under trees near rivers and lakes, taking their blood meals from people at drifts or watering places (West African sleeping sickness), or in open woodland, feeding mainly on large animals (East African sleeping sickness).

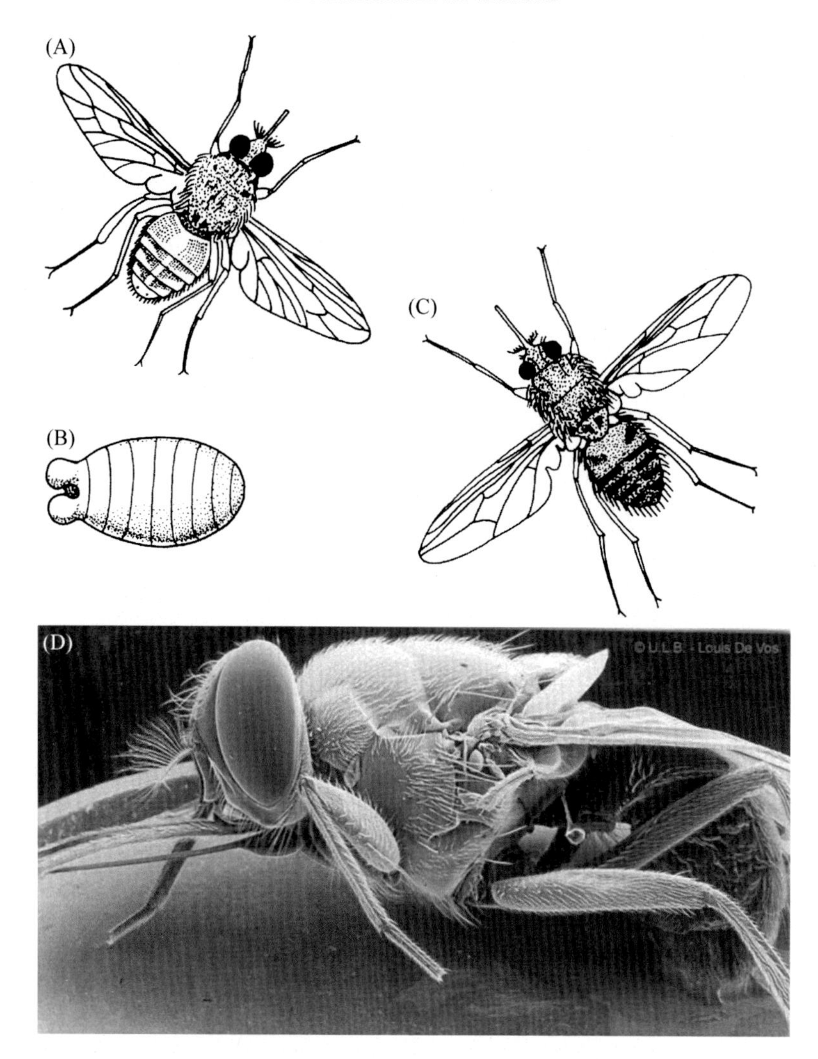

FIGURE 18.12 (A) *G. palpalis*, (B) Pupa, (C) *G. morsitans*, and (D) Scanning electron micrograph of *Glossina* spp.

Tabanid Flies

Tabanid flies (Fig. 18.14) are large, stoutly built, and often brightly colored. Strong fliers, they viciously attack a variety of mammals, including humans. Only adult females are blood feeders and have mouthparts adapted for cutting and piercing. Members of two genera, *Chrysops* and *Tabanus*, commonly known as deerflies and horseflies respectively serve as major transmitters of human pathogens.

Tabanid flies are prominent among the arthropod vectors of such bacterial diseases of humans and/or animals as **tularemia** and **anthrax**. Tularemia, a disease of humans in the United States, Canada, northern Europe, the former USSR, Turkey, and Japan, is caused by *Francisella tularensis*, for which wild rabbits often serve as sylvatic reservoir hosts. One of the vectors for *F. tularensis* is *Chrysops discalis*. Anthrax, a much dreaded disease of cattle caused by *Bacillus anthracis*,

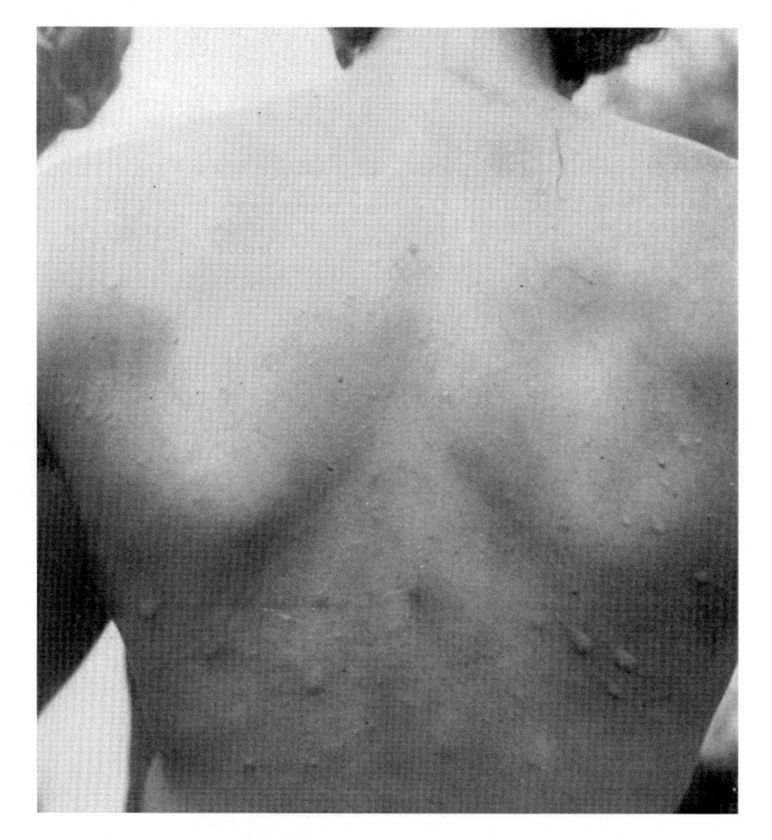

FIGURE 18.13 **Tsetse fly welts on human skin.**

is also a serious disease among humans. *Tabanus striatus* and other tabanid flies are common vectors. **Loaiasis**, caused by the African eye worm *Loa loa* (see p. 323) is transmitted to humans by several diurnally feeding species of *Chrysops*, including *C. dimidiata* and *C. silacea*.

Nonbiting Dipterans

The common housefly (*Musca*), the bluebottle fly, and a number of blowflies (*Calliphora*) are examples of nonbiting dipterans. While adults are nonparasitic and are primarily considered household and farm pests, they are, nonetheless, capable of the mechanical transmission via contaminated appendages of a variety of human pathogens, including the bacteria that cause **typhoid fever** (etiologic agent, *Salmonella typhi*) and **bacillary dysentery** (etiologic agents, *Shigella dysenteriae* and certain strains of *Escherichia coli*). In addition, the causative agents of two eye diseases, **trachoma** and **conjunctivitis**, are transmitted by this group of dipterans. The etiology of the latter two diseases has not been established conclusively. Trachoma is probably caused by a virus of the psittacosis-lymphogranuloma group, while conjunctivitis is attributable to several different bacteria. These flies are also suspected of transmitting other agents of human diseases, such as *Vibrio cholerae*, the bacterium responsible for **cholera**, and *Treponema pertenue*, the spirochete that causes **yaws**.

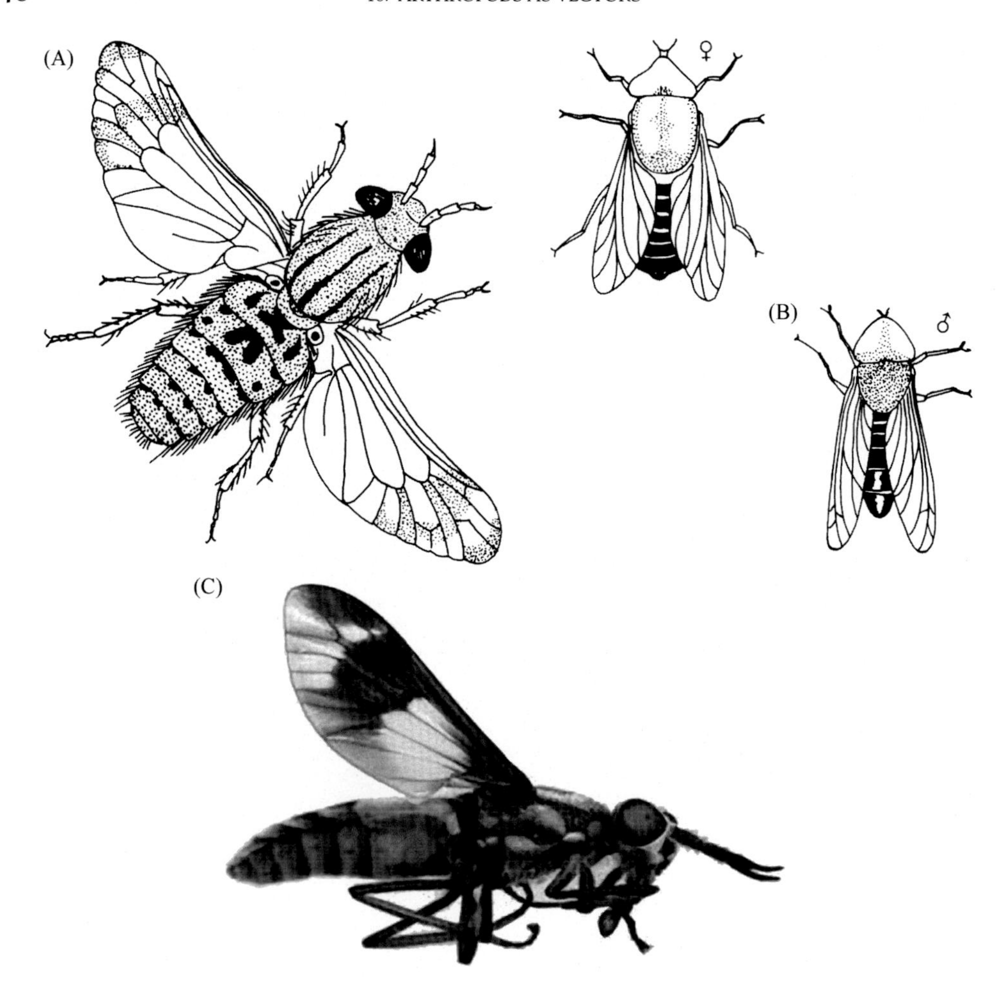

FIGURE 18.14 **Tabanid flies.** (A) *Chrysops*. (B) *Tabanus* (male and female). (C) Lateral view of *Chrysops*.

A number of other organisms that produce human disease are commonly associated with these flies although there is no direct evidence that they act as major transmitters. In fact, one body of opinion holds that only two categories of disease, namely, certain intestinal and ocular disorders, are transmitted by flies and that, like the bedbug, flies are unjustly maligned as transmitters of other pathogens.

OTHER INSECTS

A significant factor in the epidemiology of diseases transmitted by nondipteran insects is the vastly limited mobility of the three groups in this category (reduviid bugs, fleas, and lice) compared to dipterans. Although reduviid bugs possess functional wings, they are poor fliers. Fleas and lice are wingles.

Reduviid Bugs

There are about 2500 known species of reduviid, or assassin, bugs (Fig. 18.15). Characteristically, a short, three-jointed proboscis protrudes from the tip of the head, and the insects feed primarily on body fluids of other insects although some attack humans and other animals. They are comparatively large insects, measuring 1.5–2 cm long, and some are brightly colored. Members of three genera, *Triatoma, Panstrongylus, and Rhodnius*, are the major vectors for **Chagas' disease** (see p. 108). While several dozen species belonging to these genera occur in various countries of North and South America (Fig. 18.16), like the *Plasmodium*-transmitting mosquitoes, relatively few actually serve as significant vectors for Chagas' disease. Most species are arboreal and feed on the blood of wild animals, and their importance lies in their role in maintaining the infection in sylvatic reservoirs. Species that commonly inhabit or occasionally intrude into human dwellings are the major transmitters to humans. Such insects, unlike transient mosquitoes, actually invade the homes and establish stable colonies in cracks and crevices of walls and in thatched roofs. The adults emerge at night for a blood meal once or twice a week. Their bites are painful, often resulting in itchy swellings from toxins injected during feedings.

The life cycle is of the hemimetabolous type. The female deposits eggs in wall crevices, furniture, and roofs, and wingless nymphs hatch in 8–30 days. Development to adulthood generally progresses through five nymphal instars to sexual maturity. The cycle is temperature dependent and may require from 6 months to 2 years for completion.

Fleas

While the parasitologist's interest in fleas (Fig. 18.17) usually focuses on their blood-sucking habits and their role as intermediate hosts for helminth parasites, the primary concern here is with their role as vectors for pathogenic organisms.

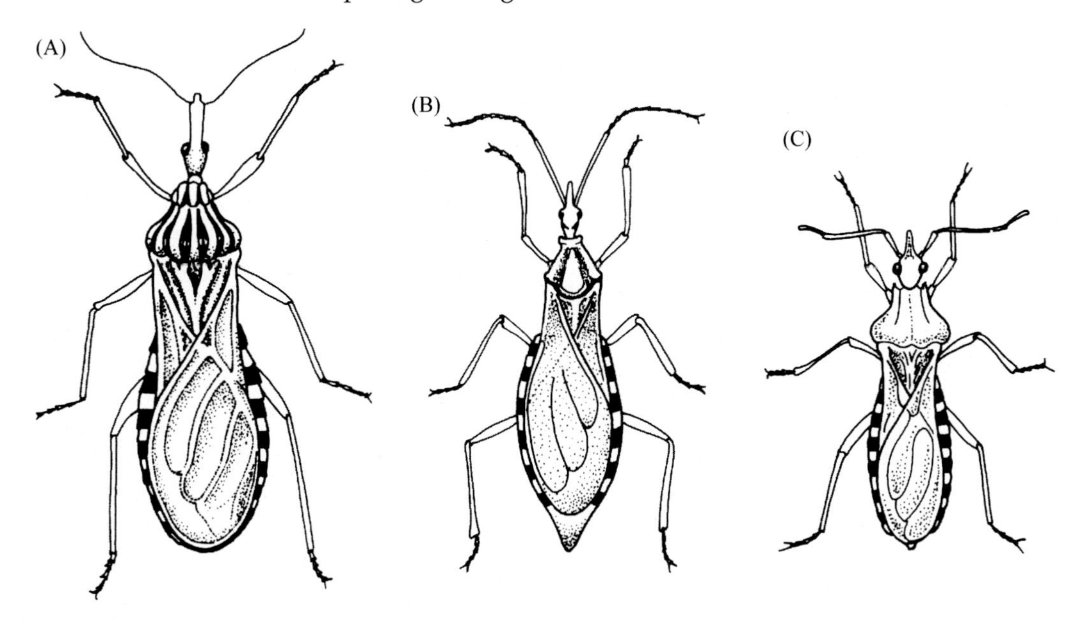

FIGURE 18.15 **Some bloodsucking reduviids.** (A) *Panstrongylus*. (B) *Triatoma*. (C) *Rhodnius*.

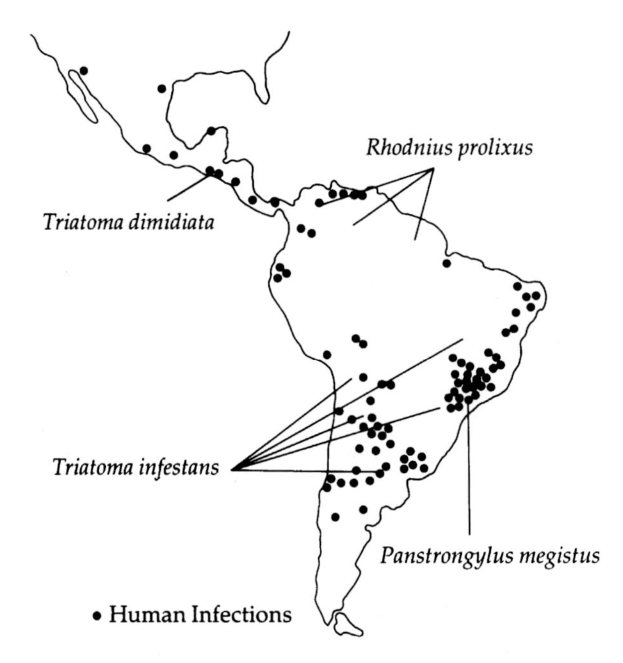

FIGURE 18.16 **Distribution of reduviids and Chagas' disease.**

Fleas constitute a small, highly specialized order of insects of obscure origin and evolution. Eggs are deposited by females a few at a time, usually 3–20, either on or off the host. When deposited on the host, they soon fall off since they are not adhesive; therefore, they are commonly found in the host's abode. A cephalic spine expedites the hatching of a whitish, legless, vermiform larva with a distinct head that, at this stage, resembles larvae of certain dipterans. The mouthparts of the larva are of the chewing type, and nourishment is derived from decaying vegetable and animal matter. The larval growth period varies from 9 to 200 days, depending on such environmental conditions as humidity, temperature, and oxygen tension. Flea larvae usually undergo two molts, which alternate with growth periods prior to pupation. Duration of the pupation period also depends upon environmental conditions, varying from 7 days to a year.

The adult flea is flattened laterally, wingless, and equipped with muscular hind legs for jumping. The body is covered with spurs and bristles, directed backward, that facilitate its movement through the fur or feathers of the host. While many fleas are host specific in their blood-feeding habits, others display little specificity.

A notorious organism transmitted by fleas is the bacillus *Yersinia pestis*, which causes **bubonic plague**. During the 14th century AD, the Black Death, as it was commonly called, killed a quarter of the population of Europe. Epidemics in the 6th century AD and the **Great Plague of London** in 1665 likewise took heavy tolls in human lives. These epidemics probably originated in central Asia and spread via caravans and, later, international shipping. Epidemics recurred as late as the 19th century. According to WHO statistics, there are now about 1000–6000 cases of plague annually. The use of antibiotics and other medications has reduced the annual mortality rate to 100–200. Presently, in the sparsely populated regions of the western and southwestern United States, large numbers of wildlife serve as sylvatic reservoirs, harboring the disease-producing bacterium and representing a continual threat of infection to humans who encroach on these environs (Fig. 18.18).

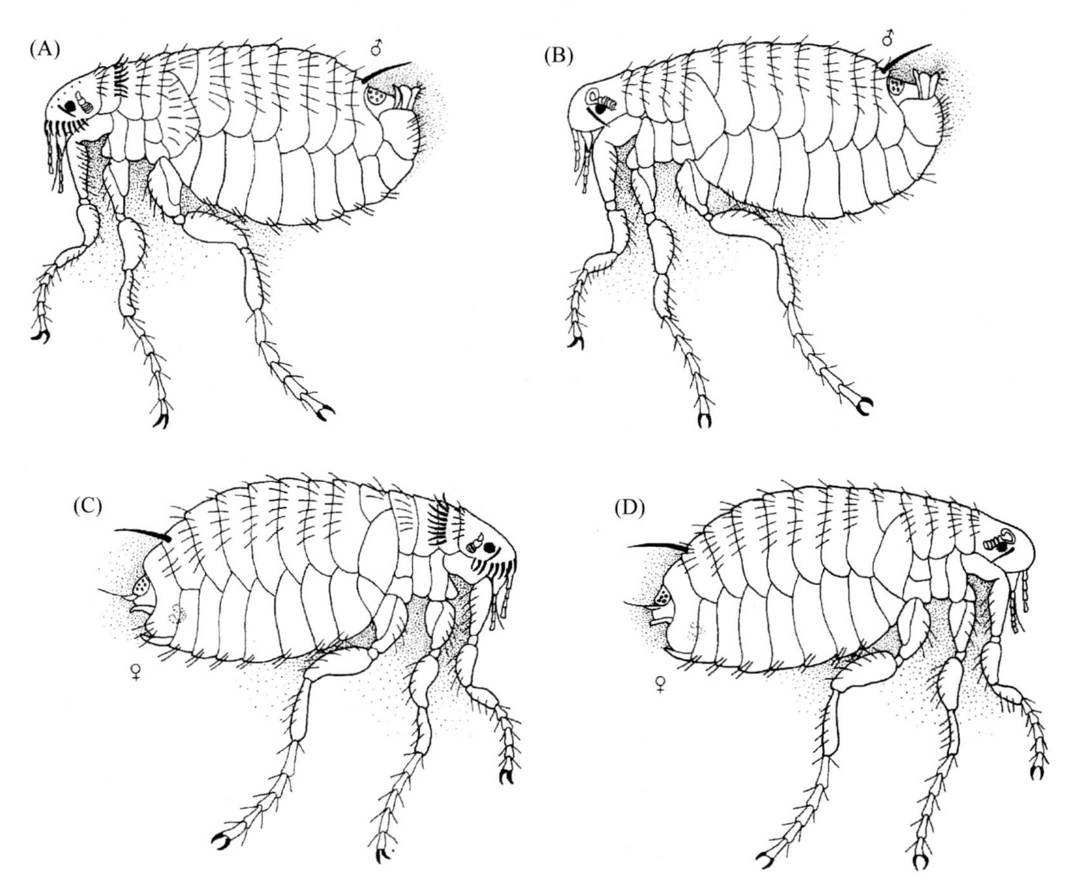

FIGURE 18.17 **Fleas.** (A) Male of the cat flea *Ctenocephalides felis*. (B) Female *C. felis*. (C) Male of the oriental rat flea *Xenopsylla cheopis*. (D) Female X. *cheopis*.

In the usual course of the infection, there is an interchange of fleas between wild, infected rodents and domestic rats. The disease spreads rapidly among the domestic rat population, killing a high percentage. Transfer of the infection from rats to humans requires that the flea feed upon both. The common human flea *Pulex irritans*, while capable of parasitizing both humans and domestic animals, rarely feeds on rats and, therefore, is not an important vector of the plague bacillus. However, several species of the genus *Xenopsylla* do feed on both humans and other animal hosts, including rats. *Xenopsylla cheopis*, the Asiatic rat flea, is the most commonly encountered species in this genus and is the major vector for plague bacilli. Presumed to have originated in the Nile valley, it has spread throughout the world on rat hosts. Male and female fleas acquire the bacilli during feeding on infected rats. The ingested bacilli proliferate in the flea's digestive tract and form a semisolid plug, blocking the gut and rendering the flea unable to completely ingest food (Fig. 18.19). In spite of increasing hunger, the flea's repeated attempts to take in blood are futile, and during each attempt, extracted blood is regurgitated into the host bloodstream along with portions of the bacterial plug, resulting in the introduction of large numbers of bacilli into the host.

In humans, plague bacilli proliferate in the blood, causing highly lethal **septicemic plague**. Bacilli localize in swellings, called **buboes** (Fig. 18.20), in the groin and armpits, and the

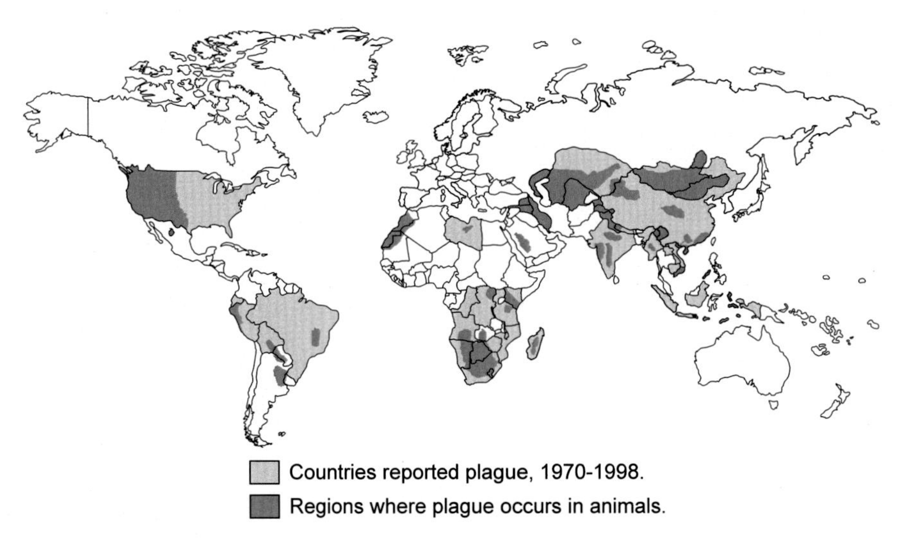

FIGURE 18.18 **Distribution of plague throughout the world as of 1998.**

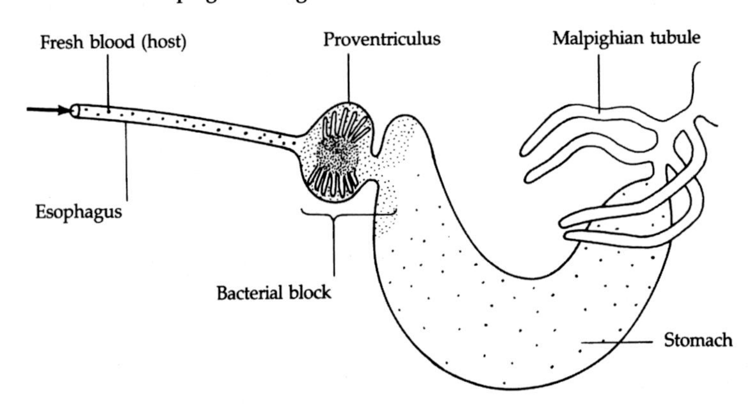

FIGURE 18.19 **Foregut of flea showing bacterial blockage.** Esophagus is distended due to accumulation of host blood.

bacterial count in the blood is low, rendering bubonic patients virtually noninfectious to fleas or other humans. This is not the case, however, with the highly contagious and lethal **pneumonic plague**, the form of the disease that affects the lungs. Interhuman transmission of bacilli can result from airborne, contaminated sputum expelled during patients' coughing attacks. In rare instances, transmission to humans results from direct contact with the pelts of infected animals.

In addition to carrying plague, fleas serve as vectors for the **murine** or **endemic typhus** organism *Rickettsia typhi* (Fig. 18.21), transmitted when contaminated flea feces is rubbed into the bite wound. Fleas acquire the rickettsia during feeding from a wide variety of infected animals, including humans, rats, and mice. *Xenopsylla cheopis* is the principal vector for this

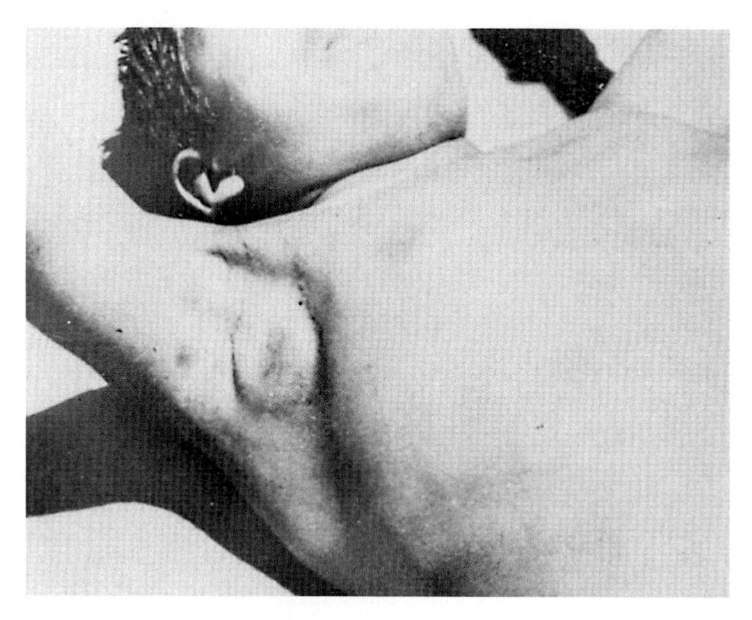

FIGURE 18.20 **Plague buboes on human patient.**

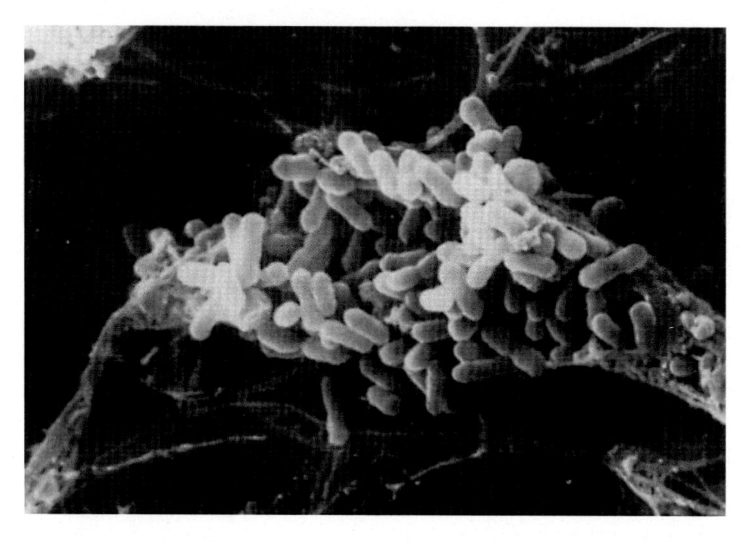

FIGURE 18.21 **Scanning electron micrograph of *Rickettsia prowazecki*, the causative agent of epidemic typhus.**

rickettsia, which causes a mild disease found throughout much of the temperate region of the world.

Pasteurella tularensis, the tularemia-causing organism, and *Salmonella enteritidis*, the salmonellosis-causing bacterium, are also transmitted to humans by fleas. These insects also serve as intermediate hosts for two tapeworms that infect humans: *D. caninum* (see p. 240) and *Hymenolepis nana* (see p. 236).

Lice

There are two orders of lice (Fig. 18.22): Mallophaga, the biting lice, and Anoplura, the sucking lice. Mouthparts of the biting louse are of the chewing type but are greatly reduced in size and number and are difficult to analyze without intensive study. Mouthparts of the sucking louse consist of an eversible set of five stylets through which it can suck host blood.

One of the mallophagans, *Trichodectes latus*, merits attention as a transmitter of a potential human pathogen. It serves as an intermediate host for the dog tapeworm *D. caninum*, an occasional parasite of humans (see p. 240).

The anopluran, *Pediculus humanus*, or body louse, is an ectoparasite of several animals, including humans. It is the major vector for three important human diseases: **relapsing fever**, **louse-borne** or **epidemic typhus**, and **trench fever**. A second anopluran species of medical importance is *Phthirus pubis*, the crab louse, which has been induced to transmit typhus-producing rickettsiae to laboratory animals. However, the body louse is the chief vehicle for transmission of the disease in nature.

Lice live in intimate contact with their human hosts. Because lice are wingless and sluggish in movement, new infestations usually occur only through direct physical contact between humans. Low standards of personal hygiene, especially when exacerbated by warfare or disaster, create a favorable climate for louse infestation and, therefore, the spread of louse-borne diseases.

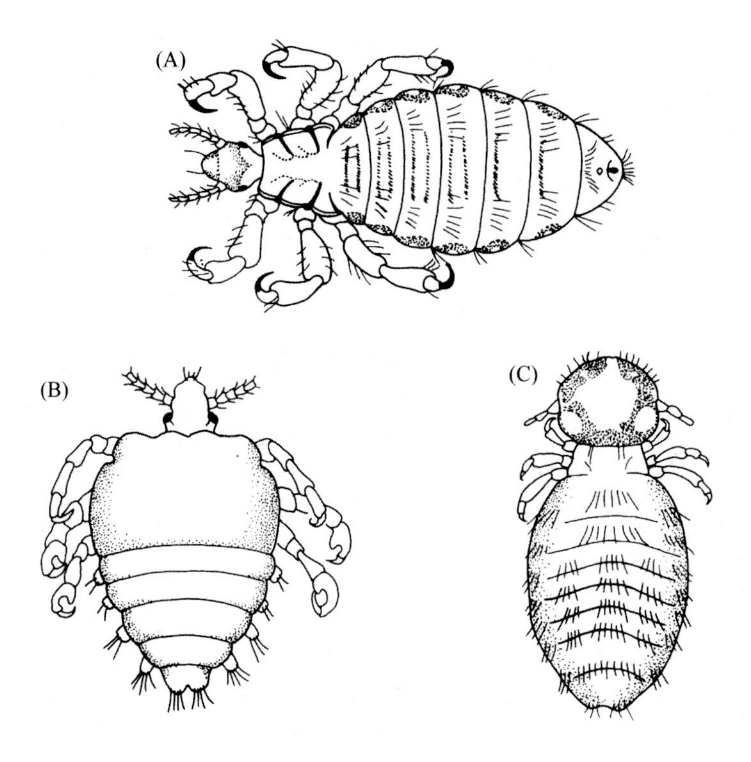

FIGURE 18.22 **Lice.** (A) *Pediculus humanus.* (B) *Phthirus pubis.* (C) *Trichodectes canis* of dogs.

Relapsing fever is a cosmopolitan disease caused by the spirochaete *Borrelia recurrentis*. The body louse ingests the spirochaete while feeding upon an infected host. *B. recurrentis* lives and reproduces in the hemocoel of the louse, where it apparently can survive for the life of the vector with no adverse effect upon the insect. Since the spirochaete has no available egress from the louse, transmission occurs when the louse's body is crushed and its contaminated body fluids enter the human host through mucus membranes or breaks in the skin. Louse-borne relapsing fever, an exclusively human disease, is characterized by intermittent high fever and rash and has a mortality rate of under 10%. The symptoms resemble those of typhus and the two were not recognized as distinct diseases until 1840.

Louse-borne, or epidemic, typhus is an ancient disease; once much dreaded, it is now limited to Asia, North Africa, and Central and South America. Caused by the rickettsia *Rickettsia prowazeki*, the last great typhus epidemic occurred in Europe immediately after World War I, with an estimated death toll of more than 3 million people. Unlike the spirochaete responsible for relapsing fever, the typhus organism is transmitted from the feces of the louse. When ingested by the louse in a blood meal from an infected human, rickettsiae invade the epithelial cells of the insect's stomach, where they multiply. The cells eventually burst, releasing large numbers of the infective organism into the louse's gut, whence they exit with feces. Damage to the louse's intestinal tract is so extensive that ingested blood diffuses into the hemocoel in such quantities that the insect turns red and usually dies. The contaminated feces quickly dries into a fine powder that remains infective for several months. Rickettsiae are inhaled by human hosts, penetrate the mucus membranes of the eyes, or are rubbed into breaks in the skin along with contaminated louse feces. The disease is characterized by high fever accompanied by headache, nausea, delirium, and stupor, usually followed by the appearance of a dull, mottled rash on the body. Untreated victims either recover spontaneously or die in about two weeks, with a higher mortality rate among elderly victims than among children. Those who survive the disease may harbor infective organisms for many years.

Similar to but milder than typhus, trench fever is caused by a rickettsia, *Rochalimaea quintana*, and transmitted by contaminated louse feces. Unlike *R. prowazeki*, *R. quintana* is an extracellular pathogen in humans, causing typhus-like symptoms that culminate in a rash that disappears within 24 h. The disease is debilitating but rarely fatal and, except for a few isolated outbreaks, rarely occurs today. However, it was one of the most common diseases during World Wars I and II.

THE ACARINES

Ticks are the major acarine vectors of human disease-producing organisms, with mites playing a far lesser role. The acarine life cycle is similar to that of a number of other blood-feeding arthropods. Females lay clusters of eggs, from which hexapod larvae emerge in 1–4 weeks. The larvae and all subsequent stages are blood feeders. Larvae metamorphose through four or five eight-legged nymphal stages, the final molt culminating in the adult stage. In some tick species, such as *Ixodes dammini*, the life cycle may extend over a period of up to 2 years.

Ticks

These acarines (Fig. 18.23) are classified into two groups: soft ticks and hard ticks. One distinguishing characteristic is the relationship of the mouthparts to the rest of the body. In soft ticks, the mouthparts are completely concealed by the soft body; in hard ticks, they project from the body. Hard ticks also possess hard, shiny shields that, in the male, cover the back completely but, in the female, do so only partially.

The two tick groups are also ecologically distinct. Soft ticks hide in cracks and crevices of houses, animal burrows, and similar areas during daylight hours, emerging at night to feed on host blood and to lay eggs. Hard ticks, on the other hand, spend most of their lives on host animals, the major portion of that time gorging on blood. Female hard ticks produce far more eggs than their soft-bodied counterparts and usually die following oviposition. Some species of hard ticks spend their entire lives on one host; other species utilize two, some even three,

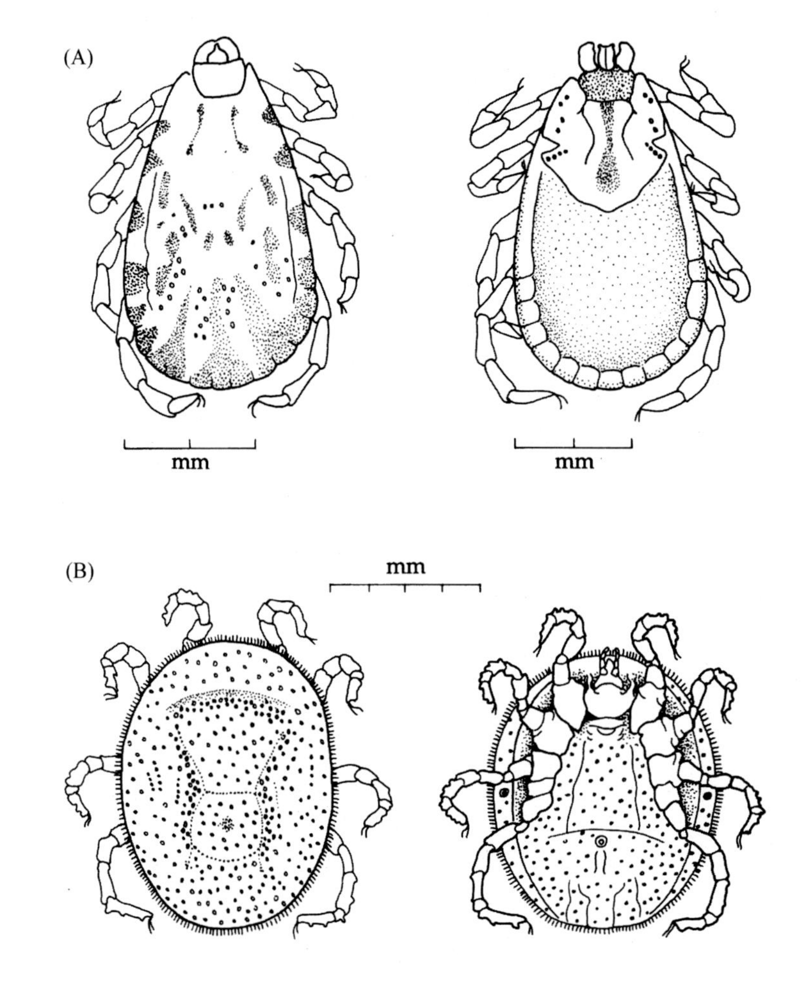

FIGURE 18.23 **Ticks.** (A) Hard tick, *Dermacentor andersoni* (male left, female right). (B) Soft tick; *Ornithodoros moubata* (dorsal aspect left, ventral aspect right).

hosts. In multihost species, the ticks drop off the host animals indiscriminately, thereby infesting large areas of land.

Ornithodoros spp., soft-bodied ticks, transmit to humans several species of the spirochaete genera *Spirochaeta* and *Borrelia.* These cause **tick-borne relapsing fever**, a disease often endemic in tropical and subtropical regions of the world. Because of this endemicity, visitors, such as campers and hunters, are apt to suffer more acutely than natives, whose symptoms tend to be mild as a result of an acquired tolerance.

Like most tick-borne diseases, relapsing fever is zoonotic, being transmitted from wild rodents and other small mammals to humans. Spirochaetes, ingested by a tick during a blood meal, gradually spread throughout the body of the arthropod. Those invading the salivary glands are most likely to be transmitted to a vertebrate host during subsequent feeding.

Symptoms of the disease are similar to those of the louse-borne variety. The primary attack, characterized by severe headache and high fever, lasts a few days and, if left untreated, is usually followed by several relapses occurring at short intervals. The patient usually recovers after 3–6 relapses although in the African variety there may be as many as 11 relapses.

A group of closely related rickettsial diseases are transmitted by hard ticks. **Rocky Mountain spotted fever**, caused by *Rickettsia rickettsia* is indigenous to North America, and **tick-borne typhus**, caused by *Rickettsia conorii*, is found throughout the world. While Rocky Mountain spotted fever is considered a dangerous disease if left untreated, most tick-borne rickettsial diseases are not particularly serious. They are usually zoonotic, with humans rarely being sources of the organisms.

Dermacentor andersoni is the principal vector for *R. rickettsia* in the Rocky Mountain states, and *Dermacentor variabilis*, the American dog tick, in the central and eastern states. Several species of *Rhipicephalus* serve as vectors for *R. conorii* in Africa and the Mediterranean area. The rickettsiae, acquired when the tick feeds on an infected vertebrate, spread throughout the body of the arthropod, causing no apparent harm. The microorganisms penetrate both the egg cells in the ovaries and the cells of the salivary glands, allowing transmission of the rickettsiae to progeny as well as to other animals. Numerous wild animals serve as reservoirs for R. *rickettsia* while *R. conorii* is harbored only by canines.

Symptoms of Rocky Mountain spotted fever are high fever accompanied by chills and headache and, at times, a typhus-like stupor. These are followed, in 4–5 days, by the appearance of a rash over the entire body. The disease, sometimes fatal to older patients, is far more pernicious than tick-borne typhus, which is rarely fatal.

Another hard tick, *Ixodes scapularis* (Fig. 18.24), is the principal vector for the spirochaete *Borrelia burgdorferi*, the causative agent of **Lyme disease** or **Lyme borreleosis** in the United States. Lyme disease is currently considered the most frequently diagnosed tick-transmitted disease of humans in the United States and, possibly, the world. It was estimated in 2002 that there were more than 23,000 cases in the United States.

First reported in 1975 in Lyme, Connecticut, Lyme disease has now been reported in Europe, Australia, the former Soviet Union, China, Japan, and Africa. The disease, which produces symptoms similar to rheumatoid arthritis, begins as a rash at the site where the tick takes a blood meal. Fatigue, fever, chills, and headache may also occur during this stage, which may last up to 30 days. Neurological complications with muscular pains may then become evident. A small percentage of patients (5%) may also show transitory cardiac malfunctions lasting from 3 days to 6 weeks. Several months after the appearance of the rash, rheumatoid arthritis-like symptoms affecting the knees and other large joints ensues.

FIGURE 18.24 **Light microscope view of Ixodes spp.**

These symptoms persist indefinitely unless treated. Treatment with any of several broad-spectrum antibiotics has proven effective.

Several hard tick species serve as transmitters of the bacterial genus *Ehrlichia*, which causes human **ehrlichiosis**. **Human monocytic ehrlichiosis** (**HME**) is caused by *Ehrlichia chaffeensis* and was first described in 1987 from the southeastern and southcentral regions of the United States. The primary vector of HME is the lone star tick, *Amblyomma americanum*. In 1994, a second form of human ehrlichiosis was recognized, **human granulocytic ehrlichiosis** (**HGE**), caused by *Ehrlichia ewingii* and appears to be limited to Missouri, Oklahoma, and Tennessee. The vectors of HGE are *I. scapularis* and *Ixodes pacificus*. The symptoms of both types of ehrlichiosis are nonspecific such as fever, muscle pain, swollen lymph glands, and headache. They may vary from mild to severe. Antibiotics appear to be the treatment of choice.

While ticks are second only to mosquitoes as vectors for viruses, diseases ascribed to such arboviruses are confined mainly to Europe and Asia. Among such diseases are several types of encephalitis, with mortality rates varying from virtually 0% to as high as 25%–30%, depending upon the type of virus. Almost all are zoonotic, with small mammals and birds serving as sylvatic reservoirs and hard ticks serving as the usual vectors.

Mites

The larvae of harvest mites and allied forms (e.g., hard ticks) often remain on hosts for long periods of time, dropping off at random. They, too, infest entire, large areas frequented by their hosts and are not limited to nests, dens, and other abodes.

Mites are responsible for the transmission of *Rickettsia tsutsugamushi*, the causative agent for **scrub typhus**, also known as "mite disease" or tsutsugamushi fever (Fig. 18.25).

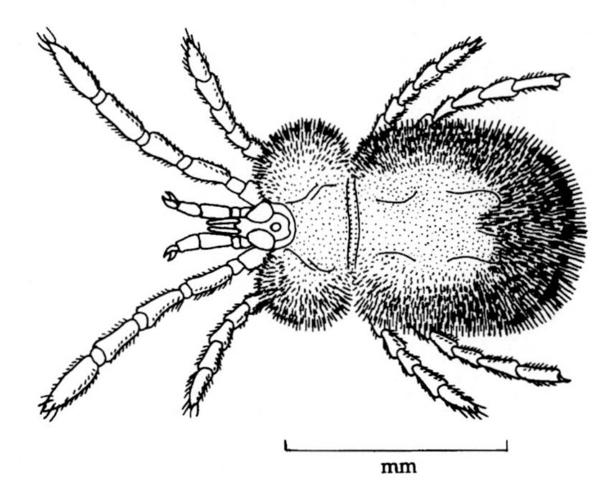

FIGURE 18.25 ***Trombicula*, a mite vector of scrub typhus.**

Scrub typhus occurs over large areas of Asia, including Japan and the Ukraine. Humans become infected from the bite of larval chigger mites of the genus *Leptotrombidium.* The larval mites that serve as vectors acquire the rickettsiae from natural reservoir hosts such as voles, rats, and other small mammals. The rickettsiae are then harbored in the mites' salivary glands awaiting transmission to human hosts. The infected mite carries the rickettsial population throughout its entire development from nymph to adult. Even though the adult mite is free-living and feeds on insect eggs and minute arthropods, it passes the rickettsiae on, via the egg, to the next generation.

The pathology of the disease varies, ranging from mild to grave. Symptoms usually appear after an incubation period of 4–10 days following the bite of an infected larval mite. An ulcer commonly forms at the site of the bite, and symptoms such as high fever and headache are common.

CLASSIFICATION OF THE ARTHROPODA

Phylum Arthropoda

Subphylum Uniramia

CLASS INSECTA

ORDER DIPTERA With functional forewings and reduced, knoblike hindwings (halteres), mouthparts variable (piercing-sucking or rasping/lapping) as is body form; complete metamorphosis.

Family Simulidae Bodies short and stout. Thorax much arched, giving humpbacked appearance; legs comparatively short; antennae 10- or 11-jointed, slightly longer than the head; ocelli absent; compound eyes on males large and contiguous, those on females widely separated; proboscis not elongate; palpi four-jointed (genus mentioned in text: *Simuliuin*).

Family Psychodidae Bodies mothlike; antennae slender, clothed with whorls of hair, long in males; wings with 9–11 long, parallel veins, with no cross-veins except at base (genera mentioned in text: *Phlebotomus, Lutzomyia*).

Family Culicidae Bodies slight; abdomen long and slender; wings narrow; antennae with 15 segments, plumose in males; proboscis long and slender; wings with fringe of scalelike setae on margins, compound eyes large, occupying a large portion of surface of head; ocelli lacking (genera mentioned in text: *Aedes*, *Anopheles*, *Culex*, *Mansonia*).

SUPERFAMILY BRACHYCERA

Family Tabanidae Bodies comparatively large, 7–30 mm long; wings well developed with veins evenly distributed; eyes large and widely separated (dichoptic) in females, contiguous in males (holoptic), antennae usually short and three-jointed, third joint with 4–8 annuli (genera mentioned in text: *Chrysops*, *Tabanus*).

SUBORDER CYCLORRAPHA

SUPERFAMILY SCHIZOPHORA (OR SCHIZOPHORA GROUP)

Family Muscidae Typical flies; hypopleural or pteropleural bristles present; basal abdominal bristles reduced, antennae plumose (genera mentioned in text: *Musca*, *Glossina*).

Family Calliphoridae Bodies, especially abdomens, are metallic green or blue, less frequently violet or copper colored; large flies; antennae are plumose; hypo- and pteropleural bristles present; posterior most posthumeral bristle present and more ventral than presutural bristle; second ventral abdominal sclerite lies with edges overlying or in contact with ventral edges of corresponding dorsal sclerites (genus mentioned in text: *Calliphora*).

ORDER HEMIPTERA Primarily ectoparasitic; bodies medium size; with piercing-sucking mouthparts forming a beak; antennae 6–10 segmented; eyes compound and large; with two pairs of wings; **corium** (thickened portion) present at base of outer wing; thinner extremities of outer wing overlap on dorsum when insect is at rest; wing venation reduced; abdomen lacks cercil; holometabolous metamorphosis.

Family Reduviidae Beak short, three-jointed, attached to tip of head; distal end of beak rests on prosternum in groove when not in use; ocelli present in winged species (with few exceptions), antennae four-jointed (genera mentioned in text: *Triatoma*, *Rhodnius*, *Panstrongylus*).

ORDER COLEOPTERA Few species endoparasitic in other insects; with two pairs of wings; forewings (elytra) veinless, hard, covering dorsal aspect of abdomen while in resting position, meeting along middorsal line; mouthparts of chewing type; antennae of 10–14 segments; compound eyes conspicuous; legs heavily sclerotized; holometabolous metamorphosis (genus mentioned in text: *Tenebrio*).

ORDER SIPHONAPTERA Adults small and wingless; ectoparasitic on birds and mammals; bodies laterally compressed; legs long, stout, and spinose; antennae short and clubbed, and fit in depressions alongside of head when not extended; mouthparts of piercing/sucking type; holometabolous metamorphosis.

Family Pulicidae Three thoracic tergites together longer than first abdominal tergite; head evenly rounded along margin; no vertical suture from dorsal margin of head to bases of antennae; abdominal tergites with one row of setae (genera mentioned in text: *Pulex*, *Xenopsylla*).

ORDER MALLOPHAGA Ectoparasitic on birds and mammals; bodies small to medium size, usually dorsoventrally flattened and wingless; with chewing mouthparts; antennae short, 3–5 segmented; with reduced compound eyes and no ocelli; thorax small; legs stout and short; no cerci on abdomen; hemimetabolous metamorphosis.

SUBORDER ISCHNOCERA Antennae 3–5 segmented, filiform, not concealed; maxillary palpi absent, mandibles vertically orientated; parasitic on birds and mammals.

Family Trichodectidae Antennae three-jointed; with one claw on tarsi; parasitic on mammals (genus mentioned in text: *Trichodectes*).

ORDER ANOPLURA Ectoparasites of mammals; bodies small to medium size, dorsoventrally flattened, and wingless; mouthparts modified as piercing-sucking organ that is retractile; antennae short, 3–5 segmented; some with legs terminating as hooked claws; thorax fused; abdomen of five to eight distinct segments; hemimetabolous metamorphosis.

Family Pediculidae Parasites of primates, including humans; bodies fairly robust, not covered with dense spines; abdomen armed with pleural plates (paratergites) and with tergal and sternal plates in most species; with well-developed eyes, comprised of pigment granules and a lens; with legs approximately equal in length or with first pair slightly smaller (genera mentioned in text: *Pediculus, Phthirus*).

Subphylum Chelicerata

ORDER ACARINA. Highly specialized Arachnida with body divided into proterosoma and hysterosoma which are distinguishable as boundary between second and third pairs of legs; segments of mouth and its appendages situated on capitulum (gnathostoma), more or less sharply set off from rest of body; typically with four pairs of legs; usually six podomeres of legs but may vary from two to seven; positions of respiratory and genital openings variable.

SUBORDER METASTIGMATA (IXODIDES). Large parasitic acarines known as ticks; mouth with recurved teeth modified for piercing; a tracheal spiracle located behind third or fourth pair of coxae.

Family Ixodidae (Hard Ticks) Body ovoid; scutum present in all stages; scutum of adult males extends to posterior margin of body; scutum of adult females, like that of larvae and nymphs, restricted to propodosomal zone; capitulum anterior and visible from dorsal view; festoons usually present; eyes, if present, situated dorsally on sides of scutum; segments of pedipalps fused, not movable; porose areas present on base of capitulum in females; stigmatal plates large, posterior to coxa IV; only females distended when engorged with blood; with marked sexual dimorphism (genera mentioned in text: *Dermacentor, Ixodes, Rhipicephalus*).

Family Argasidae (Soft Ticks) Integument leathery in nymphal and adult stages, and wrinkled, granulated, mammillated or having tubercles; scutum absent in all stages; capitulum either subterminal or protruding from anterior margin of body in nymphs and adults, subterminal or terminal in larvae; capitulum lies in distinctly or indistinctly marked depression (camerostome) in all stages; pedipalps freely articulate in all stages; porose area absent in both sexes; eyes usually absent (if present, on supracoxal folds); stigmata near coxa III lack stigmal plates; both sexes distended when engorged with blood; sexual dimorphism slight (genus mentioned in text: *Ornithodoros*).

Suggested Readings

Bennett, C. E. (1996). Ticks and lyme disease. *Advances in Parasitology, 36*, 344–405.
Busvine, J. R. (1979). Arthropods: vectors of disease. *Studies in biology*. London: Edward Arnold Publishers.
Hall, S. S. (2003). On the trail of the West Nile virus. *Smithsonian Magazine, 7*, 88–102.
Harwood, R. F., & James, M. T. (1979). *Entomology in human and animal health* (Seventh ed.). New York: Macmillan.
Jonas, G. (1978). Africa: taming of the tsetse. *RF Illustrated, 4*, 1–4.
McDaniel, B. (1979). *How to know the ticks and mites*. Dubuque, IA: William C. Brown.
Piesman, J. (1987). Emerging tick-borne diseases in temperate climates. *Parasitology Today, 3*, 197–199.

Spielman, A. (1988). Lyme disease and human babesiosis: evidence incriminating vector and reservoir hosts. In P. T. Englund, & A. Sher (Eds.), *The biology of parasitism* (pp. 147–165). New York: Alan R. Liss.

Zwiebel, L. J., & Takken, W. (2004). Olfactory regulation of mosquito–host interactions. *Insect Biochemistry and Molecular Biology*, *34*, 645–652.

Glossary

Acetabulate type of scolex that displays four suckers.
Acquired immunity a host's immune response to previous parasitic infection.
Adaptive immunity see *acquired immunity.*
Amastigote a form of hemoflagellate that develops intracellularly, characterized by subspherical shape and the presence of a very short flagellum.
Ambulatory buds redial structures that are believed to aid in their movement through tissue.
Amoebapore peptides released by *Entamoeba histolytica* that mediate lysis of cell membranes.
Amphids small depressions or pits located anteriorly on the body surface of most nematodes and believed to be chemoreceptors.
Anapolytic see *apolysis.*
Anthrax a bacterial disease of humans, cattle, and sheep, transmitted by tabanid flies.
Antibody a protein synthesized in response to an antigen or, in varying degrees, to molecules of similar structure; an antibody usually binds with a specific antigen.
Antifols drugs that block the synthesis of folic acid.
Antigen any substance, usually proteinaceous, capable, under appropriate conditions, of inducing the host to synthesize antibodies.
Antigen presenting cell (APC) specialized cell possessing Class II major histocompatibilty complex molecules that are involved in processing and presenting antigen fragments to helper T-cells.
Apical complex a combination of structures found in the apical region of sporozoites and merozoites of members of the phylum Apicomplexa.
Apical papilla sensitive anterior portion of a miracidium at which site glands empty to the exterior.
Apicoplast membranous organelle in sporozoites or merozoites of apicomplexans believed to be of photosynthetic origin.
Apolysis the release of gravid, or egg-filled, proglottids from tapeworm strobila.
Arbovirus member of a large group of viruses transmitted mainly by blood-sucking arthropods.
Autoinfection reinfection by a parasitic organism without its leaving the host.
Axoneme a microtubular element usually extending the length of a flagellum or cilium.
Axostyle a tube-shaped sheath of microtubules observed in many flagellates, that usually extends from a basal body to the posterior end.
B cell (B lymphocyte) a specialized lymphocyte that produces humoral or circulatory antibodies.
Balantidial dysentery symptoms resulting from invasion of *Balantidium coli* in mucosal epithelium of host.
Basal body (blepharoplast or kinetosome) an organelle, morphologically identical to a centriole, from which the flagellum or cilium originates.
Biological vector an arthropod used by a disease-producing organism for transmission and as a vehicle for reproduction and/or development.
Blackwater fever massive lysis of vertebrate erythrocytes that, at times, accompanies falciparum malaria.
Bladderworm see *cysticercus.*
Blepharoplast see *basal body.*
Bothriate scolex see *bothrium.*
Bothrium (plural, bothria) groove on the scolex of some tapeworms.
Bradyzoite stage in the life cycle of various coccidia such as *Toxoplasma.* Similar to a merozoite, it is found in pseudocysts in the tissues of nonfeline hosts.
Breakbone fever an arbovirus-caused disease. See *dengue.*
Brood (social) parasite organism that lays eggs in the nest of its host species.
Calabar (or fugative) swelling a transient, subcutaneous swelling caused by the nematode *Loa loa.*
Capitulum a small, anterior projection of acarines that bears the mouthparts.

CD4⁺ T cell helper T cell that promotes immune responses by secreting cytokines like IFN-γ and IL-2.
Cell-mediated reaction the effect produced by specialized cells, e.g., T lymphocytes, mobilized to arrest and, in most cases, eventually destroy a parasite.
Cellular reaction see *cell-mediated reaction*.
Cercaria a juvenile digenetic trematode produced by asexual reproduction in a sporocyst or redia.
Cercocystis a modified cysticercoid larva of *Hymenolepis nana* found in the intestinal villus of the definitive host.
Cercomer the posterior extension of procercoid and cysticercoid larvae, usually retaining the oncospheral hooks.
Chagas' disease a disease, also known as American trypanosomiasis, caused by *Trypanosoma cruzi*.
Chagoma local, edematous swelling due to early infection with *Trypanosoma cruzi*.
Chelicerae portion of the mouthparts of acarines.
Chemokine usually glycoprotein molecules of relatively low molecular weight released by or a part of a variety of cells which have effects on other cells (see *cytokines*).
Chigger a mite of the family Trombiculidae.
Chromatoidal bar a structure considered by many to be deposits of nucleic acids in members of the genus *Entamoeba*.
Circumsporozoite coat sporozoite surface coat in *Plasmodium*.
Cirrus the penis or ejaculatory duct of a flatworm.
Clonal selection theory adaptive responses derived from the selection of antigen specific lymphocytes from a large repertoire of preformed lymphocytes. Antigen specific lymphocytes respond to specific antigen stimulation through proliferation and differentiation into specific effector cells.
Coenurus a larval tapeworm of certain cyclophyllideans in which numerous scolices bud from an internal germinal epithelium.
Commensalism a symbiotic relationship in which neither the commensal nor the host is physiologically dependent upon the other.
Complement a group of about 20 serum proteins involved in a cascade of reactions which mediate immune functions. Classical complement pathway is initiated by the interaction of an antibody and a specific antigen.
Complementarity determining regions regions of both heavy and light chains in an antibody molecule those are highly variable in amino acid sequence when comparing one antibody to another.
Conoid a truncated cone of spirally arranged, fibrillar structures in the apical complex of certain members of the suborder Eimeriina, for example, *Toxoplasina gondii*.
Constant regions regions of antibodies that vary little in amino acid composition from one antibody molecule to another.
Copulatory spicules see *spicules*.
Coracidium (plural, coracidia) an oncosphere surrounded by a ciliated embryophore.
Carrion's disease (bartonellosis) a severe, often fatal, bacterial infection found among inhabitants of parts of South America and transmitted by sand flies.
Costa a striated rod, associated with basal bodies of many flagellates, that courses along the base of an undulating membrane.
Creeping eruption the irritation and rash caused by hookworm larvae in an unnatural host.
Critical density statistical computation indicating severity of malaria in a region.
Cutaneous larval migrans a condition caused by the migration of nematode larvae in the skin of an unnatural host. See *creeping eruption*.
Cyclodevelopmental biological transmission disease-producing organism undergoes cyclical changes in an arthropod but does not multiply in the arthropod.
Cyclopropagative biological transmission disease-producing organism reproduces and undergoes cyclical changes in the arthropod.
Cysticercoid the tapeworm larva, featuring a cercomer and a fully developed scolex enclosed within a multilayered cyst wall, that develops from the oncosphere of some cyclophyllideans.
Cysticercosis infection with cysticercus larvae.
Cysticercus a tapeworm larva, developing from the oncosphere of some cyclophyllideans, which is characterized by a fully developed scolex invaginated into a fluid-filled vesicle, or bladder, also called "bladder worm".
Cystogenous glands in some cercariae of digenetic trematodes, secretary cells that give rise to metacercarial cysts.
Cytokines any of several soluble chemical mediators, secreted by cells that react with other cells.
Dauer larva an arrested stage of a free-living nematode usually triggered by adverse environmental conditions.
Definitive host see *host, definitive*.

Delayed hypersensitivity increased reactivity to a specific antigen, mediated by cells rather than antibodies, usually requiring up to 24 h to reach maximum intensity. See *T cell.*

Dengue a disease caused by a mosquito-transmitted virus; also known as "break bone fever".

Didelphic refers to nematodes with two ovaries and two uteri.

Diffuse cutaneous leishmaniasis a diffuse infiltration of the host skin with parasites causing multiple nodules, which do not ulcerate.

Disseminated strongyloidiasis a severe life-threatening complication of *Strongyloides stercoralis* infection that can occur in immunocompromised individuals.

Domestic reservoir host reservoir host that lives close to humans, such as farm animals.

Dumdum fever see *Kala-azar*.

Ectoparasite a parasite that lives on the exterior surface of the host.

Ectopic site abnormal or unexpected site of infection.

Elephantiasis swollen body parts, usually extremities, resulting from long-time exposure to *Wuchereria bancrofti* infection.

Embryophore membranous structure associated with cestode egg shells.

Encephalitis infection of the brain usually by viruses or amoebas.

Endodyogeny endopolyogeny resulting in the formation of two daughter cells.

Endoparasite a parasite that lives inside the host.

Endopolyogeny formation of daughter cells while still enclosed in mother cell.

Epidemiology the study of the occurrence of a particular disease, including ecological factors, transmission, prevalence, and incidence.

Epimastigote a form of hemoflagellate, equipped with a short undulating membrane, in which the kinetoplast lies near, but anterior to, the nucleus.

Espundia a disease caused by *Leishmania braziliensis*, also known as mucocutaneous leishmaniasis, uta, pian bois, and chiclero ulcer.

Exflagellation rapid formation of microgametes of *Plasmodium* in the mosquito.

Exsheathing fluid secretion by nematode larva to aid in the molting process.

Fab region N-terminal region of an antibody molecule that contains the antigen binding site.

Facultative parasite an organism that, given the opportunity, can assume a parasitic existence.

Favism genetic deficiency of glucose-6-phosphate dehydrogenase in erythrocytes.

Fc region that portion of an antibody found on the heavy chain. It does not contain the antigen binding region.

Filariform larva third stage larva of many nematodes. It is usually the infective stage to the defintive host.

Flagellar pocket depression in cell surface from which a flagellum arises.

Flame cell specialized cell containing a ciliary tuft and situated at the end of a tubule-forming part of an osmoregulatory system.

Flame cell osmoregulatory system see *flame cell.*

Gametogony formation of gametes.

Gamogony see *gametogony.*

Gastrodermis the tissue lining the digestive tract, as found in digenetic trematodes.

Genital atrium a circumscribed area in the body wall of flatworms into which male and female genital ducts open; also known as common genital pore.

Genomics the systematic study of the complete DNA sequences (i.e., genome) of an organism.

Geographic medicine see *tropical medicine.*

Germ cell cycle asexual reproduction in digenea during which progeny arise through differentiation of germinal cells passed from one generation to the next.

Glycocalyx the carbohydrate-containing outer coat found on the free surface of most cells.

Glycosome a membrane-bound, microbody-like organelle, peculiar to the hemoflagellates, that contains enzymes essential for glycolysis.

Gonotyl the muscular sucker or specialized structure surrounding the genital pore of some digeneans.

Granuloma repaired area of an organ characterized by fibrous connective tissue.

Granulomatous amoebic encephalitis acute neurological disease usually caused by amoebas such as *Acanthamoeba* spp. and *Balamuthia* spp. in immunocompromised individuals.

Ground itch a localized irritation and rash caused by penetrating larvae and the accompanying bacteria of hookworms of humans.

Group accumulation of tachyzoites of *Toxoplasma gondii* in a host cell.

Gynaecophoric canal the ventral fold or groove in male schistosomes in which female worms are held *in copula*.
Halzoun the disease resulting from nasopharyngeal blockage due to the attachment of worms, such as pentastomids or young digenea, to buccal or pharyngeal membranes. Also caused by severe hypersensitivity to pentastomid.
Hemimetabolism incomplete metamorphosis in insects characterized by the following stages in the life cycle—egg, gradation of nymphal stages, and adult.
Hemocoel the major body cavity of arthropods, typically containing blood or hemolymph.
Hemoflagellate flagellate that resides in the host's blood and/or closely related tissues.
Hemorrhagic dengue see *dengue*.
Hemozoin granules, seen in erythrocytes infected with *Plasmodium malariae*, that may represent residues from incomplete hemoglobin digestion.
Heterogonic the term used to describe a life cycle in which free-living generations may alternate periodically with parasitic generations.
Hologonic refers to spermatogenesis zone extending the entire length of a nematode's testis.
Holometabolism complete metamorphosis in insects characterized by the following stages in the life cycle—egg, larva, pupa, and adult.
Holomyarian refers to a pattern of two rows of muscle cells in each quadrant of a nematode's body wall.
Homogonic the term used to describe a life style that is consistently either parasitic or free-living.
Host, definitive the host in which a parasite attains sexual maturity.
Host, intermediate the host in which a parasite undergoes developmental changes but does not yet reach sexual maturity.
Host, paratenic (or transfer) the host in which a parasite resides without further development; a transfer host.
Host, reservoir a host, usually nonhuman, in which a parasite lives and remains a source of infection but usually produces no symptoms.
Host specificity the extent to which a parasite can exist in more than one host species.
Human immunodeficiency virus (HIV) virus that is responsible for AIDS.
Human monocytic or granulocytic ehrlichiosis (HME, HGE) a disease caused by tick-transmitted bacteria belonging to the genus *Ehrlichia*.
Humoral reaction the effect produced when specialized molecules of the circulatory system interact with a parasite, usually immobilizing or destroying it.
Hydatid a larval tapeworm, of the cyclophyllidean genus *Echinococcus*, in which numerous scolices bud from secondary cysts.
Hydatid sand unattached protoscolices in a hydatid cyst.
Hydrogenosome a membrane-bound organelle in the cytoplasm of some flagellates (e.g., trichomonads) that is involved in carbohydrate metabolism, an endproduct of which may be molecular hydrogen.
Hypervariable region see *complementarity determining region*.
Hypnozoite the dormant stage of *Plasmodium vivax* and *Plasmodium ovale* in hepatocytes of the human host.
Hypobiosis a lag phase at some stage of development in the life cycle of a nematode.
Hypodermal cords longitudinal protrusions of the nematode hypodermis into the pseudocoel dividing it into quadrants.
Hypodermis the tissue that secretes the nematode cuticle.
Hypostome portion of the mouthparts of acarines.
Immunogen see *antigen*.
Immunoglobulin (Ig) a general term used to describe all antibody molecules which is based on structural aspects of their globular folding.
Infantile kala-azar highly fatal form of visceral leishmaniasis found along the Mediterranean specifically affecting infants.
Infection model predicts behavior of pathogen and the various hosts.
Infraciliature in a ciliophoran, numerous basal bodies interconnected by fibrils.
Innate immunity see *natural immunity*. A relatively rapid immune response that is not dependent on prior exposure to an antigen.
Interferon gamma (IFN-γ) a cytokine produced by $CD4^+$ T cells, $CD8^+$ T cells, and natural killer cells. One of its more important roles is to activate macrophages.
Interleukins any of a class of cytokines that act to stimulate, regulate, or modulate lymphocytes.
Intracystic bodies sporozoites of *Pneumocystis carinii*.
Invasive amoebiasis invasion of *Entamoeba histolytica* trophozoites to internal organs.
K-selection see *K-strategists*.

K-strategists (equilibrial species) organisms employing a survival strategy characterized by low reproductive capability, low mortality, long life span, and saturation of an environment.
Kala-azar a disease, also known as visceral leishmaniasis and Dumdum fever, caused by *Leishmania donovani*.
Kinetoplast a portion of the mitochondrion of the Kinetoplastidae containing KDNA. Usually associated with the kinetosome.
Kinetosome see *basal body*.
Laurer's canal a canal, originating on the surface of the oviduct near the seminal receptacle in some digenetic trematodes, that may represent a vestigial vagina.
Leishmaniasis a disease caused by members of the genus *Leishmania*.
Luminal amoebiasis infection with *Entamoeba histolytica* limited to the intestinal lumen.
Lysosomotropic agents a group of chemical compounds, such as chloroquine, that demonstrate an affinity for lysosomes.
Macrogamete "female" gamete.
Macrogametocyte the cell that gives rise to a macrogamete.
Macroparasite large parasite that does not develop within its defintive host, for example, cestodes, trematodes, and most nematodes.
Macrophage a large phagocytic cell of the mononuclear leukocyte series.
Major histocompatibility complex (MHC) a group of genes that encode surface glycoprotein molecules, which are required for antigen presentation to T lymphocytes.
Malpighian tubules excretory organs of terrestrial insects and most acarines.
Maurer's dots aggregates in cytoplasm of erythrocytes infected with *Plasmodium falciparum*.
Mechanical vector a passive carrier of a disease-producing agent.
Median bodies slightly curved structures of unknown function peculiar to members of the genus *Giardia*.
Megasyndrome enlargement of internal organs due to infection.
Mehlis' gland a group of unicellular glands that empty into the ootype region of flatworms.
Memory T cell antigen-specific cell that has a long life and is activated in response to secondary and repeated exposure to the antigen.
Merogony see *schizogony*.
Meromyarian refers to a pattern of two to five rows of muscle cells in each quadrant of a nematode's body wall.
Merozoite one of many cells resulting from schizogony.
Metabolic antigen antigenic substance secreted or excreted by the parasite.
Metacercaria the larval stage between cercaria and adult in the life cycle of many digenetic trematodes.
Metacestode see *cysticercus*.
Metacyclic the term used to describe a form of parasite infective to its vertebrate host, for example, metacyclic trypomastigote.
Metacystic trophozoite a small trophozoite of *Entamoeba* spp. that emerges from the cyst in the intestine of the host.
Metraterm the muscular, distal portion of the uterus of digenetic trematodes.
Microfilaria the juvenile, first-stage larva of filarial nematodes.
Microgamete "male" gamete.
Microgametocyte the cell that gives rise to microgametes.
Micronemes small, convoluted structures that lie parallel to rhoptries and appear to merge with them at the apices of sporozoites and merozoites.
Microparasite small parasite that multiplies within its host (e.g., a number of protozoans).
Microthrix (plural, *microthrices*) a specialized microvillus that projects from the outer, limiting membrane of the tapeworm tegument.
Miracidium the ciliated larva that emerges from the egg of digenetic trematodes.
Molecular mimicry production of or covering by host-like molecules, especially on the body surface of a parasite.
Monocytic locomotion inhibitory factor (MLIF) peptide released by *Entamoeba histolytica* that inhibits the motility of host monocytes and macrophages.
Monodelphic refers to a nematode with one ovary and one uterus.
Mutualism a symbiotic relationship in which each partner is physiologically dependent on the other.
Myoblast muscle cell region.
Nagana a disease of domestic ruminants, caused by *Trypanosoma brucei brucei*, *Trypanosoma congolense*, and *Trypanosoma vivax*.

Natural immunity a nonspecific host defense mechanism involved in the early phases of resistance to a pathogen. This response involves all aspects of the immune system not directly mediated by lymphocytes.
Natural killer cell large, granular non-B and non-T lymphocyte that plays an important role in innate immunity. They are capable of killing those cells infected with intracellular pathogens as well as some types of tumors.
Obligatory parasite parasite that is totally physiologically dependent upon its host.
Onchocercomas subcutaneous and visceral fibrous nodules containing adult *Onchocerca volvulus*.
Oncosphere the hexacanth embryo of a tapeworm.
Oocysta a rounded cyst containing sporoblasts.
Ookinete the motile, elongated zygote of many apicomplexans.
Ootype a specialized region of the flatworm oviduct that is surrounded by Mehlis' gland.
Open circulatory system the system, characteristic of arthropods and molluscs, in which blood flows slowly through large sinuses (hemocoel) back to a dorsal, tubular heart.
Operculum a lidlike structure at one end of the eggshell of many digenetic trematodes and some cestodes.
Opportunistic parasite parasitic infection that is normally asymptomatic but displays symptoms in immunocompromised hosts.
Opsonizing antibody antibody that binds a specific antigen and interacts with macrophages to induce phagocytosis.
Oriental sore a disease, also known as cutaneous leishmaniasis, caused by *Leishmania tropica*.
Parabasal body the Golgi complex of protozoans.
Parabasal filament a fibril running from the cisternae of the Golgi complex to one or more basal bodies.
Parasitism a symbiotic relationship in which only one of the organisms, the parasite, is physiologically dependent upon the other, the host.
Parasitoid usually dipterans that lay their eggs on or in living hosts and usually kill these hosts during development.
Parasitophorous vacuole within a cell, a vacuole containing a parasite (e.g., amastigote).
Paratenic host see *host, paratenic*.
Parenchyma in flatworms, mesodermal tissue filling all available body spaces.
Parthenogenesis development of an unfertilized egg to a new individual.
Pathogen associated molecular patterns (PAMPs) molecules associated with groups of pathogens that are recognized by cells of the innate immune system.
Pattern recognition receptors (PRRs) proteins expressed by cells of the innate immune system to identify pathogen-associated molecular patterns (PAMPs).
Pedipalps portion of the mouthparts of acarines.
Peristome the depressed area in some protozoans leading from the mouth.
Phasmids small, sensory pits, believed to be chemoreceptors, located posteriorly on the body surface of members of the nematode class Secernentea.
Phoresis a form of commensalism in which one organism (phoront) is mechanically carried by the other; the relationship is nonobligatory.
Phoront see *phoresis*.
Plasma cell an effector B cell that secretes into the circulation large numbers of antibodies of the same specificity as its cell surface receptors.
Plerocercoid the larval form that develops from the procercoid, as in *Diphyllobothrium latum*.
Polar rings electron-dense structures, circling the apical region, that constitute part of the apical complex of sporozoites and merozoites of apicomplexans, such as *Plasmodium* spp.
Polydelphic refers to nematodes with multiple ovaries and uteri.
Polyembryony the formation of multiple embryos from a single zygote with no intervening gamete stage.
Polymerase chain reaction (PCR) technique by which fragments of DNA can be made to replicate very rapidly.
Polymyarian refers to a pattern of multiple rows of muscle cells in each quadrant of a nematode's body wall.
Post kala-azar dermal leishmanoid disfiguring condition which may develop 1–2 years following inadequate treatment of kala-azar.
Premunition a form of acquired immunity dependent upon retention of the infective agent.
Primary amoebic meningoencephalitis acute, usually fatal, illness resulting from infection of the brain by amoebas like *Naegleria fowleri*.
Procercoid the larval form that develops from the coracidium.
Proglottid one segment in a tapeworm strobila complete with a full complement of reproductive organs.

Promastigote a hemoflagellate form bearing an anterior flagellum and a kinetoplast well anterior to the nucleus.
Propagative biological transmission disease-producing organism reproduces in an arthropod but undergoes no further development.
Protandrogony initial maturation of male organs which subsequently disappear, followed by maturation of female organs in hermaphroditic organisms. Also termed *protandry*.
Proteasome a large multifunctional protease complex which degrades intracellular proteins. It generates peptides that are presented by MHC class I molecules.
Proteomics the study of the entire set of proteins expressed by a genome, cell, tissue, or organism.
Protoscolex (plural, protoscolices) the immature scolex found in coenurus and hydatid larvae.
Pseudocyst a cluster of amastigotes of *Trypanosoma cruzi* in a muscle fiber.
r-selection see *r-strategists.*
r-strategists (opportunistic species) organisms employing a survival strategy characterized by high reproductive rates, high mortality, and short life span.
Ray bodies gametocytes of *Babesia* species.
Receptor cell a functionally specialized lymphocyte that produces a specific antibody.
Recrudescence a sudden increase in a previously persistent, low-level parasite population (e.g., *Plasmodium malariae*).
Recurrent flagellum flagellum that turns in opposite direction. Usually observed in cytostome of some flagellates.
Redia a larval form that arises asexually from within a sporocyst or a primary redia of a digenetic trematode.
Regulatory T cells (Treg) T cells that suppress activation of the immune system and shut down T cell-mediated immune reactions.
Renette a large, unicellular gland that empties to the exterior through a pore and serves as the basic component of the nematode excretory system.
Reservoir host see *host, reservoir.*
Resistance the ability of an organism to withstand infection.
Retroinfection reinfection by nematode larvae (e.g., *Enterobius vermicularis*) that hatch on skin and reenter the host's body.
Rhabditiform larva first stage larva of many nematodes.
Rhoptry part of the apical complex of the sporozoites and merozoites of apicomplexans, composed of electron-dense bodies extending posteriorly from the apex.
Romaña's sign early symptoms of Chagas' disease, consisting of unilateral, periorbital edema and conjunctivitis.
Rostellum the small, rounded projection, sometimes bearing hooks, on the apex of the scolex of some tapeworms.
Sand fly fever a viral infection transmitted by sand flies.
Schistosomule in a blood fluke, the juvenile stage that develops following cercarial penetration of the definitive host. Sometimes called a schistosomulum.
Schizogony (merogony) a form of asexual reproduction characterized by rapid organelle and nuclear divisions, followed by cytoplasmic divisions, resulting in many simultaneous daughter cells (e.g., merozoites).
Schizont (segmenter) a multinucleated cell undergoing schizogony, prior to cytoplasmic division.
Schüffner's dots fine granules distributed throughout erythrocytes infected with *Plasmodium vivax.*
Scolex the holdfast organ of tapeworms.
Scrub typhus rickettsial disease transmitted by chigger mites.
Secretory IgA molecule that is transported across intestinal cells and released to the intestinal lumen.
Sequential polyembryony the production of many embryos from the same zygote.
Shell-yolk glands clusters of cells each of which synthesizes globules essential to eggshell formation in many flatworms.
Signet ring stage early erythrocytic trophozoite of *Plasmodium* resembling a signet ring.
Sleeping sickness a disease caused by *Trypanosoma brucei rhodesiense* and *Trypanosoma brucei gambiense* in Africa; also, any one of many arboviruses transmitted by mosquitoes and causing encephalitis.
Social parasite see *brood parasite.*
Somatic antigen antigenic substance that is found on and is part of the surface of a parasite.
Sparganosis infection with plerocercoid larvae.
Spicules needlelike structures in nematodes, such as the copulatory spicules.
Sporoblasts in the oocyst of apicomplexans, cells that divide into sporozoites while still enclosed by the sporoblast membrane.

Sporocyst the larval stage of a digenetic trematode into which the miracidium metamorphoses, usually in a mollusc.
Sporogony multiple divisions of a zygote.
Sporozoite one of the daughter cells resulting from sporogony.
Stichocytes a series of unicellular glands surrounding the capillary like esophagus of many adenophorean nematodes, for example, *Trichuris trichiura.*
Strobila in tapeworms, a chain of segments formed by budding.
Strobilation the formation of a strobila by a tapeworm.
Subpellicular microtubular network spiral framework of microtubules just beneath the plasma membrane of many flagellates.
Swimmer's itch dermatitis resulting from the penetration of human skin by an animal schistosome cercaria.
Swollen belly syndrome (SBS) oftentimes fatal infection with *Strongyloides fuelleborni* in infants.
Sylvatic echinococcosis life cycle of *Echinococcus granulosus* confined to wild animals.
Sylvatic reservoir host reservoir host that lives in the wild.
Symbiont any organism that is involved in a symbiotic relationship.
Symbiosis a heterospecific interrelationship between two organisms.
Syncitium a multinucleated tissue with no cell boundaries.
Syngamy the union of gametes.
T cell (T lymphocyte) a specialized lymphocyte, processed through the thymus, that elicits cell-mediated reactions of various types.
Tachyzoite a form of merozoite in *Toxoplasma*, found in parasitophorous vacuoles of vertebrate hosts.
Tegument the syncytium that covers the surface of trematodes and cestodes.
Telegonic refers to spermatogenesis restricted to the blind end of a nematode's testis.
Threshold value the lowest level of conditions needed for an infection to persist in a population.
Toll-like receptors (TLRs) cell surface receptors that recognize certain molecular patterns on antigens that initiate innate immune responses.
Transmission coefficient predicts likelihood of new infections in a population.
Transovarial transmission the passage of a parasite from a mother to its progeny via the ovaries.
Tritrophic interactions functional integration of organisms on three trophic levels.
Trophozoite the motile, feeding stage of protozoans.
Tropical medicine area of human pathology attributable to diseases that occur most commonly in the tropics.
Trypomastigote a hemoflagellate form with an elongated undulating membrane and kinetoplast located posterior to the nucleus.
Tularemia a bacterial disease of humans transmitted by tabanid flies and for which rabbits often serve as sylvatic reservoir hosts.
Tumor necrosis factor macrophage or lymphocyte derived cytokine that serves several functions in the immune system.
Undulating membrane that portion of a plasma membrane or cytoplasm of a flagellate that is drawn away from the cell during the beating of a recurrent flagellum.
Vagina in tapeworms, a tubular organ that joins the oviduct and carries sperm from the genital atrium to the oviduct.
Variant antigenic types divergent surface antigens that arise as the result of the ability of trypomastigote populations circulating in the bloodstream to change the chemical composition of their glycocalyces.
Vector any agent, most commonly an arthropod, that actively transmits a disease-producing organism.
Vermicle the infective stage, analogous to sporozoite, of *Babesia* spp. from ticks.
Visceral larval migrans migration of second-stage ascarid larvae of nematodes in the internal organs of unnatural hosts.
Vitelline glands, Vitellaria see *shell-yolk glands*.
Vitelline reservoir common chamber into which the contents of the shell-yolk glands empty.
Vulva uterine opening in the body wall of female nematodes. In mammals, it is the external female genitalia.
Winterbottom's sign symptom of African sleeping sickness characterized by enlarged, sensitive cervical lymph nodes.
Xenodiagnosis diagnostic technique in which researchers identify a disease by infecting a laboratory animal and assessing its symptoms.
Yaws a fly-transmitted disease that is caused by the spirochete *Treponema*.

Yellow fever a viral disease transmitted by the mosquito *Aedes aegypti*.
Ziemann's dots pigment in erythrocytes infected with *Plasmodium malariae*.
Zika a hemorrhagic fever transmitted by several species of *Aedes*.
Zoitocyst tissue cyst containing bradyzoites in *Toxoplasma* species and related organisms.
Zoonosis any disease of animals that can be transmitted to humans.

Appendix A

Drugs for Parasitic Infections: Partial List of Generic and Brand Names

Albendazole—Albenza (GlaxoSmithKline)
Artemether—Artenam (Arenco, Belgium)
Artesunate—(Guilin No. 1 Factory, People's Republic of China)
Atovaquone—Mepron (GlaxoSmithKline)
Atovaquone/Proguanil—Malarone (GlaxoSmithKline)
Bacitracin—Many manufacturers
Bacitracin-zinc—(Apothekernes Laboratororium A.S., Oslo, Norway)
Benznidazole—Rochagan (Roche, Brazil)
Bithionol—Bitin (Tanabe, Japan)
Chloroquine HCl and Chloroquine phosphate—Aralen (Sanofi and others)
Crotamiton—Eurax (Westwood-Squibb)
Dapsone—(Jacobus)
Diethylcarbamazine citrate USP—(University of Iowa School of Pharmacy)
Diloxanide furoate—Furamide (Boots, United Kingdom)
Eflornithine (Difluoromethylornithine, DFMO)—Ornidyl (Aventis, France)
Furazolidone—Furoxane (Roberts)
Halofantrine—Halfan (GlaxoSmithKline)
Iodoquinol—Yodoxin (Glenwood) and others
Ivermectin—Stromectol (Merck)
Malathion—Ovide (Medicis)
Mebendazole—Vermox (McNeil)
Mefloquine—Lariam (Roche)
Meglumine antimoniate—Glucantime (Aventis, France)
Melarsoprol—Mel-B (Specia)
Metronidazole—Flagyl (Searle and others)
Miltefosine—(Zentaris)
Niclosamide—Yomesan (Bayer, Germany)
Nifurtimox—Lampit (Bayer, Germany)
Nitazoxanide—Cryptaz (Romark)
Ornidazole—Tiberal (Roche, France)
Oxamniquine—Vansil (Pfizer)
Paromomycin—Humatin (Monarch), Leshcutan (Teva Pharmaceutical Industries, Ltd., Israel; topical formulation not available in the United States)
Pentamidine isethionate—Pentam 300, NebuPent (Fujisawa)
Permethrin—Nix (GlaxoSmithKline); Elimite (Allergan)
Praziquantel—Biltricide (Bayer)
Primaquine phosphate USP—(Bayer)

Proguanil—Paludrine (Wyeth Ayerst, Canada; AstraZeneca, United Kingdom); in combination with Atovaquone as Malarone (GlaxoSmithKline)

Propamidine isethionate—Brolene (Aventis, Canada)

Pyrantel pamoate—Antiminth (Pfizer)

Pyrethrins and Piperonyl butoxide—Rid (Pfizer) and others

Pyrimethamine USP—Daraprim (GlaxoSmithKline)

Quinine dihydrochloride and sulfate—many companies

Sodium stibogluconate—Pentostam (GlaxoSmithKline, United Kingdom)

Spiramycin—Rovamycine (Aventis)

Suramin sodium—(Bayer, Germany)

Thiabendazole—Mintezol (Merck)

Tinidazole—Fasigyn (Pfizer)

Triclabendazole—Egaten (Novartis, Switzerland)

Trimetrexate—Neutrexin (US Bioscience)

Appendix B

Current Chemotherapeutic Regimen

Infection	Organism	Medication	Adult dosage	Pediatric dosage	Comments
Amoebiasis (Asymptomatic)	*Entamoeba histolytica*	Iodoquinol OR	650 mg PO TID × 20 days	30–40 mg/kg/day (max. 2 g) PO TID × 20 days	
		Paromomycin sulfate	25–35 mg/kg/day PO TID × 7 days	25–35 mg/kg/day PO TID × 7 days	
Amoebiasis (Symptomatic intestinal/ extraintestinal)	*Entamoeba histolytica*	Metronidazole OR	500–750 mg PO TID × 7–10 days	35–50 mg/kg/day PO TID × 7–10 days	
		Tinidazole followed by	2 g/day PO TID × 3 days	50–60 mg/kg/day (max 2 g) PO daily × 3–5 days	
		Iodoquinol OR	650 mg PO TID × 20 days	30–40 mg/kg/day (max 2 g) PO TID × 20 days	
		Paromomycin sulfate	25–35 mg/kg/day PO TID × 7 days	25–35 mg/kg/day PO TID × 7 days	
Amoebic meningoencephalitis	*Naegleria fowleri*	Amphotericin B OR	1.5 mg/kg/day (max 1.5 mg/ kg/day) BID IV × 3 days then 1 mg/kg/day daily IV × 11 days	See adult weight-based dosage	
		Amphotericin B OR	1.5 mg daily Intrathecal × 2 days then 1 mg/day QOD Intrathecal x 8 days	See adult weight-based dosage	
		Azithromycin OR	10 mg/kg/day (max 500 mg/ day) daily IV/ PO × 28 days	See adult weight-based dosage	
		Fluconazole OR	10 mg/kg/day (max 600 mg/ day) daily IV/ PO × 28 days	See adult weight-based dosage	
		Rifampin OR	10 mg/kg/day (max 600 mg/ day) daily IV/ PO × 28 days	See adult weight-based dosage	

(*Continued*)

Current Chemotherapeutic Regimen (*cont.*)

Infection	Organism	Medication	Adult dosage	Pediatric dosage	Comments
		Miltefosine OR	<45 kg: 50 mg BID PO × 28 days; >45 kg 50 mg TID PO × 28 days	See adult weight-based dosage	
		Dexamethasone	0.6 mg/kg/day QID IV × 4 days	See adult weight-based dosage	
Anisakiasis	*Anisakis* spp.	Albendazole (not FDA approved for this use)	400 mg PO BID × 6–21 days	Decreased dosage	
Ascariasis	*Ascaris lumbricoides*	Albendazole	400 mg PO single dose	400 mg PO single dose	
		Mebendazole	100 mg PO BID × 3 days or 500 mg PO single dose	100 mg PO BID × 3 days or 500 mg single dose	
		Ivermectin	150–200 mcg/kg PO single dose	150–200 mcg/kg PO single dose	
Babesiosis	*Babesia* spp.	Atovaquone Plus	750 mg PO BID × 7–10 days		
		Azithromycin	Day 1: 500–1000 mg/day PO then subsequent days: 250–1000 mg/day		
		Quinine Plus	650 mg PO TID × 7–10 days		
		Clindamycin	600 mg PO TID × 7–10 days or 300–600 mg IV QID		
Balantidiasis	*Balantidium coli*	Tetracycline	500 mg PO QID × 19 days	≥8 years old: 40 mg/kg/day (max 2 g) PO QID × 10 days	
		Metronidazole	500–750 mg PO TID × 5 days	35–50 mg/kg/day TID × 5 days	
		Iodoquinol (not FDA approved for this use)	650 mg PO TID × 20 days	30–40 mg/kg/day (max 2 g) PO TID × 20 days	
		Nitazoxanide (not FDA approved for this use)	500 mg PO BID × 3 days	4–11 years old: 200 mg PO BID × 3 days	
				1–3 years old: 100 mg PO BID × 3 days	
Chagas disease	*Trypanosoma cruzi*	Benznidazole	5–7 mg/kg/day PO BID × 60 days	<12 years old: 5–7.5 mg/kg/day PO BID × 60 days	
		Nifurtimox	8–10 mg/kg/day PO TID or QID × 90 days	≤10 years old: 15–20 mg/kg/day PO TID or QID × 90 days	

Infection	Organism	Medication	Adult dosage	Pediatric dosage	Comments
				11–16 years old: 12.5–15 mg/kg/day PO TID or QID × 90 days	
Cryptosporidiosis	*Cryptosporidium*	Nitazoxanide	300 mg PO BID × 3 days	1–3 years old: 100 mg PO BID × 3 days	
				4–11 years old: 200 mg PO BID × 3 days	
Cutaneous larval migrans	*Ancylostoma braziliense*	Albendazole	400 mg PO daily × 3–7 days	>2 years old: 400 mg PO daily × 3 days	
	Ancylostoma caninum	Ivermectin	200 mcg/kg PO single dose	>15 kg: 200 mcg/kg PO single dose	
Cyclosporiasis (formerly isosporiasis)	*Cyclospora cayetanensis*	Trimethoprim-Sulfamethoxazole (TMP-SMX)	TMP 160 mg/SMX 800 mg PO BID × 7–10 days	TMP 5 mg/kg/SMX 25 mg/kg PO BID × 7–10 days	
	Cyclospora infection	Ciprofloxacin	500 mg PO BID × 7 days		
		Iodoquinol	650 mg PO TID × 20 days	30–40 mg/kg/day (max 2 g) PO TID × 20 days	
Dientamoebiasis	*Dientamoeba fragilis*	Paromomycin	25–35 mg/kg/day PO TID × 7 days	25–35 mg/kg/day PO TID × 7 days	
		Metronidazole (not FDA approved for this use)	500–750 mg PO TID × 10 days	20–40 mg/kg/day PO TID × 10 days	
Dracunculiasis	*Dracunculus medinensis*	Near eradication no medications recommended			
Enterobiasis	*Enterobius vermicularis*	Pyrantel pamoate	11 mg/kg base (max 1 g) PO single dose; repeat in 2 weeks	11 mg/kg/base (max 1 g) PO single dose; repeat in 2 weeks	
		Mebendazole	100 mg PO single dose; repeat in 2 weeks	100 mg PO single dose; repeat in 2 weeks	
		Albendazole	400 mg PO single dose; repeat in2 weeks	400 mg PO single dose; repeat in 2 weeks	
Filariasis (lymphatic)	*Wuchereria bancrofti*	Diethylcarbamazine (DEC)—no longer FDA approved	6 mg/kg/day PO either single dose or 12-day course		
	Brugia malayi/				
	Loa 1c *Brugia timori*				

(*Continued*)

Current Chemotherapeutic Regimen (*cont.*)

Infection	Organism	Medication	Adult dosage	Pediatric dosage	Comments
Filariasis (Loiasis)	*Loa loa*	Albendazole (may be given before treatment with DEC)	8–10 mg/kg/day PO TID × 21 days	Same as adult dosage	
	Symptomatic loiasis		200 mg PO BID × 21 days	Same as adult dosage	
			200 mg PO daily × 4–8 weeks		
Filariasis	*Mansonella perstans*	Doxycycline (if *Wolbachia* are present)			
			250 mg PO TID × 5 days		
Giardiasis	*Giardia lamblia*	Metronidazole (not FDA approved for this)	2 g PO single dose	5 mg/kg PO TID × 5 days	
		Tinidazole	100 mg (max 300 mg/day) PO TID × 5 days	50 mg/kg (max 2 g) PO single dose	
		Quinacrine OR	100 mg PO QID × 7–10 days	2 mg/kg (max 300 mg/day) PO × 5 days or 50 mg/kg (max 2 g) single dose	
		Furazolidone OR	25–35 mg/kg/day PO TID × 7 days	25–35 mg/g/day PO TID × 7 days	
		Paromomycin		Same as adult dosage	
			400 mg PO single dose		
Hookworm infection	*Ancylostoma duodenale*	Albendazole OR	100 mg PO BID × 3 days or 500 mg PO single dose	Same as adult dosage	
	Necator americanus	Mebendazole OR	11 mg/kg (max 1 g) PO × 3 days	Same as adult dosage	
		Pyrantel pamoate		Same as adult dosage	
Leishmaniasis	*Leishmania*		≥12 years old and at least 30–44 kg: 50 mg PO BID × 28 days		
		Miltefosine	≥45 kg: 50 mg PO TID × 28 days	See adult dosing	
			20 mg/kg daily IV/IM × 28 days	See adult dosing	

Infection	Organism	Medication	Adult dosage	Pediatric dosage	Comments
		Pentavalent antimonial compounds (not FDA approved)	20 mg Sb/kg/ daily IV or IM × 20–28 days	see adult dosing	
		Sodium stibogluconate (provided by CDC)	3 mg/kg/day (days 1–5) and 3 mg/kg/day days 14 and 21	See adult dosing	
		Liposomal amphotericin B		See adult dosing	
Lice	*Pediculus humanus*		Topical treatment: single treatment—repeat in 9–10 days		
	Pediculus capitis	Pyrethrin with piperonyl butoxide OR	Topical treatment: single treatment—repeat in 9–10 days	>2 years old: same as adult	
	Pthirus pubis	Permethrin 1% OR	Topical treatment: single treatment—repeat in 7–9 days if live lice are present	>2 mos: same as adult	
		Malathion 0.5% OR	200 mcg/kg or 400 mcg/kg PO single dose; repeat in 9–10 days	>6 years old: same as adult	
		Ivermectin		>15 kg: same as adult	
Malaria—uncomplicated	Chloroquine-resistant *Plasmodium falciparum*		250 mg Atovaquone/100 mg proguanil 4 tabs PO daily × 3 days		
		Atovaquone-proguanil		62.5 mg Atovaquone/ 25 mg proguanil pediatric tablets:	
				5–8 kg: 2 ped tabs PO daily × 3 days	
				9–10 kg: 3 ped tabs PO daily × 3 days	
		OR		11–20 kg: 1 adult tab PO daily × 3 days	
				21–30 kg: 2 adult tabs PO daily × 3 days	
				31–40 kg: 3 adult tabs PO daily × 3 days	
				>40 kg: adult dosage	
			1 tab = 20 mg artemether/120 mg lumefantrine		

(Continued)

Current Chemotherapeutic Regimen (*cont.*)

Infection	Organism	Medication	Adult dosage	Pediatric dosage	Comments
		Artemether-lumefantrine	35 kg: 6 doses over 3 days:	5–<15 kg: 1 tab/dose	
			Day 1: 1st dose followed by 2nd dose 8 h later.	15–<25 kg: 2 tabs/dose	
			Day 2 and 3: 1 dose BID × 2 days	25–<35 kg: 3 tabs/dose	
				>35 kg: 4 tabs/dose	
			542 mg base (=650 mg salt) PO TID × 3 or 7 days		
		Quinine sulfate plus one of the following:	100 mg PO BID × 7 days	8.3 mg base/kg (=10 mg salt/kg) PO TID × 3 or 7 days	
		Doxycycline	250 mg PO QID × 7 days	2.2 mg/kg PO q 12 h × 7 days	
		Tetracycline	20 mg/base/kg/day TID × 7 days	2.5 mg/kg/day PO QID × 7 days	
		Clindamycin		20 mg/base/kg/day TID × 7 days	
			684 mg base (=750 mg salt) PO initial dose		
		Mefloquine	Followed by 456 mg base (=500 mg salt) PO single dose 6–12 h after initial dose.	13.7 mg base/kg (=15 mg salt/kg) PO initial dose	
			Total dose: 1250 mg salt	Followed by 9.1 mg base/kg (=10 mg salt/kg) PO 6–12 h after initial dose	
				Total dose: 25 mg salt/kg	
Malaria-uncomplicated	Chloroquine-sensitive *Plasmodium falciparum*		600 mg base (=1000 mg salt) PO immediately		
	Plasmodium malariae	Chloroquine phosphate	Followed by 300 mg base (=500 mg salt) PO 6, 24, 48 h afterward	10 mg base/kg PO immediately	
	Plasmodium knowlesi	OR	Total dose: 1500 mg base (=2500 salt)	Followed by 5 mg base/kg PO at 6, 24, 48 h after	

Infection	Organism	Medication	Adult dosage	Pediatric dosage	Comments
			620 mg base (=800 mg salt) PO immediately	Total dose: 25 mg base/kg	
		Hydroxychloroquine	Followed by 310 mg base (=400 mg salt) PO at 6, 24, 48 h afterward	10 mg base/kg PO immediately	
			Total dose: 1550 mg base (=2000 mg salt)	Followed by 5 mg base/kg PO at 6, 24, 48 h after	
				Total dose: 25 mg base/kg	
Malaria-uncomplicated	All areas:		600 mg base (=1000 mg salt) PO immediately		
	Plasmodium vivax	Chloroquine phosphate + Primaquine phosphate	Followed by 300 mg base (=500 mg salt) PO 6, 24, 48 h afterward	10 mg base/kg PO immediately	
	Plasmodium ovale		Total dose: 1500 mg base (=2500 salt)	Followed by 5 mg base/kg PO at 6, 24, 48 h after	
			Plus Primaquine phosphate: 30 mg base PO daily × 14 days	Total dose: 25 mg base/kg	
				Plus Primaquine phosphate: 0.5 mg base/kg PO daily × 14 days	
Malaria prophylaxis	All areas		Adult tab: 250 mg atovaquone/100 mg proguanil		
		Atovaquone-proguanil	1 PO daily begin 1–3 days before travel; continue until 7 days after leaving region.	Ped tab: 62.5 mg atovaquone/25 mg proguanil	
				5–8 kg: 1/2 ped tab daily	
				>8–10 kg: 3/4 ped tab daily	
				>10–20 kg: 1 ped tabs daily	
				>20–30kg: 2 ped tabs daily	

(*Continued*)

Current Chemotherapeutic Regimen (*cont.*)

Infection	Organism	Medication	Adult dosage	Pediatric dosage	Comments
				>30–40 kg: 3 ped tabs daily	
				40 kg: 1 adult tab daily	
	Chloroquine-sensitive areas		300 mg base (500 mg salt) PO weekly		
		Chloroquine phosphate	Begin 1–2 weeks before travel	>8 years old: 5 mg/kg base (8.3 mg/kg salt) weekly	
		OR	Take weekly on the same day, continue for 4 weeks after leaving area	Up to adult dosage of 300 mg base	
			310 mg base (400 mg salt) PO weekly		
		Hydroxychloroquine sulfate		5 mg/kg base (65 mg/kg salt) PO weekly	
				Up to Adult dosage of 310 mg base	
	Mefloquine-sensitive area		228 mg base (250 mg salt) PO weekly		
		Mefloquine	Begin 2 weeks before travel. Continue while in area	≤9 kg: 4.6 mg/kg base (5 mg/kg salt) PO weekly	
			Continue for 4 weeks after leaving area	>9–19 kg: 1/4 tab weekly	
				>19–30 kg: 1/2 tab weekly	
				>31–45 kg: 34/ tab weekly	
	Plasmodium vivax		30 mg base (52.6 mg salt) PO daily 1–2 days before travel. While in area	>45 kg: 1 tab weekly	
		Primaquine	Continue while in area	0.5 mg/kg base (o.6 mg/kg salt) up to adult dose PO daily 1–2 days before travel. While in area	

Infection	Organism	Medication	Adult dosage	Pediatric dosage	Comments
			After leaving area, continue for 14 days	Continue while in area and after leaving area, continue for 14 days.	1
Microsporidiosis	*Encephalitozoon bieneusi*	Albendazole	400 mg PO BID until immune system is restored for 6 months		
	Encephalitozoon intestinalis	Albendazole	400 mg PO BID until immune system is restored for 6 months		
	Trachipleistophora	Itraconazole + albendazole	400 mg daily + 400 mg BID		
	Anncaliia	Itraconazole + albendazole	400 mg daily + 400 mg BID	?	
Onchocerciasis	*Onchocerca volvulus*	Ivermectin OR	150 mcg/kg PO single dose every 6 months		
		doxycycline	200 mg PO daily × 6 weeks	Same as adult dosage	
				Same as adult dosage	
Schistosomiasis	*Schistosoma haematobium*	Praziquantel	40 mg/kg/day PO in 2 divided doses		
	Schistosoma mansoni	Praziquantel	40 mg/kg/day PO in 2 divided doses	>1 years old: same as adult dosage	
	Schistosoma japonicum	Praziquantel	60 mg/kg/day PO in 3 divided doses	>1 years old: same as adult dosage	
				>1 years old: same as adult dosage	
Strongyloidiasis	*Strongyloides stercoralis*	Ivermectin	200 mcg/kg PO × 1–2 days		
Disseminated strongyloidiasis/hyperinfection		Albendazole	400 mg PO BID × 7 days	Same as adult dosage	
		Ivermectin	200 mcg/kg PO until stool/sputum are negative for 2 weeks	Same as adult dosage	
Tapeworm (adult or intestinal stage)	*Diphyllobothrium latum*				
	Taenia saginata	Praziquantel OR	5–10 mgk/kg PO single dose	Same as adult dosage	
	Taenia solium	Niclosamide (not available in US)	2 g PO single dose	50 mg/kg PO single dose	

(*Continued*)

Current Chemotherapeutic Regimen (*cont.*)

Infection	Organism	Medication	Adult dosage	Pediatric dosage	Comments
	Dipylidium caninum				
Tapeworm (larval or tissue stage-cysticercosis)	*Taenia solium cysticercus*	Albendazole OR	15 mg/kg/day BID × 15 days		
	Cysticercus cellulosae	Praziquantel	50 mg/kg/ day × 15 days	Same as adult dosage	
Hydatid cyst	*Echinococcus granulosus*	Albendazole	400 mg PO BID × 1–6 months	Same as adult dosage	
				10–15 mg/kg/daily (max 800 mg) PO BID × 1–6 months	
Toxocariasis	*Toxocara* (see visceral larval migrans)	Albendazole OR	400 mg PO BID × 4 days		
		Mebendazole	100–200 mg PO BID × 5 days	Same as adult dosage	
				Same as adult dosage	
Toxoplasmosis	*Toxoplasma gondii*	Pyrimethamine Plus	100 mg PO loading dose then 25–50 mg PO daily+		
		Sulfadiazine	1 g PO QID+	Day 1: 2 mg/kg/day PO then 1 mg/kg PO daily	
				50 mg/kg PO BID	
Trematodes	*Clonorchis sinensis*	Praziquantel OR	75 mg/kg/day PO TID × 2 days		
		Albendazole	10 mg/kg/day PO × 7 days	Same as adult dosage	
				Same as adult dosage	
Fascioliasis	*Fasciola hepatica*	Triclabendazole (available through CDC in US)	10 mg/kg PO single dose		
			May repeat dose if severe case	Same as adult dosage	
	Fasciola buski	Praziquantel	10–20 mg/kg single dose OR	Same as adult dosage	
			25 mg/kg PO TID	Same as adult dosage	
				Same as adult dosage	
Opisthorchiasis	*Opisthorchis viverrini*	Praziquantel OR	75 mg/kg/day PO TID × 2 days		

Infection	Organism	Medication	Adult dosage	Pediatric dosage	Comments
		Albendazole (not FDA approved for this)	10 mg/kg/day PO × 7 days	Same as adult dosage	
				Same as adult dosage	
Paragonimus	*Paragonimus westermani*	Praziquantel OR	25 mg/kg PO TID × 2 days		
		Triclabendazole (available through CDC in US)	10 mg/kg PO × 1 or 2 doses	Same as adult dosage	
Trichinellosis	*Trichinella spiralis*	Albendazole OR	400 mg PO BID × 8–14 days		
		Mebendazole	200–400 mg PO TID × 3 days then 400–500 mg PO TID × 10 days	Same as adult dosage	
Trichomonas	*Trichomonas vaginalis*	Metronidazole OR	2 g PO single dose		
		tinidazole	2 g PO single dose	?	
				?	
Trichuriasis	*Trichuris trichiura*	Abendazole OR	400 mg PO daily × 3 days		
		Mebendazole OR	100 mg PO BID × 3 days	Same as adult dosage	
		Ivermectin	200 mcg/kg/day PO daily × 3 days	Same as adult dosage	
				Same as adult dosage	
Trypanosomiasis	*Trypanosoma cruzi*	See Chagas disease	See Chagas disease		
Hemolymphatic stage	*Trypanosoma brucei gambinese*	Pentamidine isethionate OR	4 mg/kg/day IM or IV × 7–10 days		
		Suramin	1 g IV on days 1, 3, 7, 14, 21	Same as adult dosage	
	Trypanosoma brucei rhodesiense	Suramin	100–200 mg IV single dose then 1 g IV on days 1, 3, 7, 14, 21	20 mg/kg IV on days 1, 3, 7, 14, 21	
				20 mg/kg IV on days 1, 3, 7, 14, 21	
Late stage with CYN involvement	*Trypanosoma brucei gambiense*	Melarsoprol OR	2–3.6 mg/kg/daily IV × 3 days		

(*Continued*)

Current Chemotherapeutic Regimen (*cont.*)

Infection	Organism	Medication	Adult dosage	Pediatric dosage	Comments
	Trypanosoma brucei rhodensiense		After 7 days, 3.6 mg/kg/day × 3 days	18–25 mg/kg total over 1 month	
			Give a 3rd series of 3.6 mg/kg/day after 10–21 days	Initial dose: 0.36 mg/kg IV increasing gradually to max 3.6 mg/kg at intervals of 1–5 days for total of 9–10 doses	
		Eflornithine	400 mg/kg/day QID × 14 days		
				Same as adult dosage	
Visceral larva migrans—toxocariasis	*Toxocara*	Albendazole OR	400 mg PO BID × 5 days		
		Mebendazole	100–200 mg PO BID × 5 days	Same as adult dosage	
				Same as adult dosage	

Index

D

E

F

O

T

Y

Z

Printed in the United States
By Bookmasters